Advances in Modern Agricultural Machinery

Advances in Modern Agricultural Machinery

Guest Editors

Mustafa Ucgul
Chung-Liang Chang

MDPI

Basel • Beijing • Wuhan • Barcelona • Belgrade • Novi Sad • Cluj • Manchester

Guest Editors

Mustafa Ucgul
Faculty of Science and Engineering
Southern Cross University
Lismore
Australia

Chung-Liang Chang
Department of Biomechatronics Engineering
National Pingtung University of Science and Technology
Pingtung County
Taiwan

Editorial Office
MDPI AG
Grosspeteranlage 5
4052 Basel, Switzerland

This is a reprint of the Special Issue, published open access by the journal *Agriculture* (ISSN 2077-0472), freely accessible at: https://www.mdpi.com/journal/agriculture/special_issues/F96M46U3XE.

For citation purposes, cite each article independently as indicated on the article page online and as indicated below:

Lastname, A.A.; Lastname, B.B. Article Title. *Journal Name* **Year**, *Volume Number*, Page Range.

ISBN 978-3-7258-3797-7 (Hbk)
ISBN 978-3-7258-3798-4 (PDF)
https://doi.org/10.3390/books978-3-7258-3798-4

Contents

About the Editors

Mustafa Ucgul

Dr. Mustafa Ucgul is a mechanical engineer with extensive expertise in computational mechanics and its applications in agricultural machinery design. With 18 years of research experience across Turkey and Australia, Dr Ucgul has specialized in discrete element modelling of soil-engaging tools, the design of agricultural machinery, and the thermal analysis of agricultural greenhouses and desalination systems. Since completing his PhD, he has contributed to multiple industry-linked projects.

Chung-Liang Chang

Dr. Chung-Liang Chang is an experienced academic with over 16 years of experience in teaching and research, specializing in robotics, guidance, navigation, and control systems. Currently, Dr. Chang serves in the Department of Biomechanical Engineering at National Pingtung University of Science and Technology, where his research primarily focuses on intelligent navigation, satellite positioning, mobile agricultural robot control, and advanced plant factory production systems. He has served as the convener of the Intelligent Agriculture Production Technology Alliance and has rich experience in industry–university cooperation. He is committed to fostering innovative learning environments and is actively engaged in mentoring students.

Preface

This Special Issue presents cutting-edge research focused on the design and application of modern agricultural machinery systems, addressing critical challenges in modern agricultural production. Given the rising global demand for food and increasing production costs, developing efficient machinery systems is essential to enhance productivity and sustainability.

The purpose of this Special Issue is to highlight innovative developments in the field, with contributions that cover a wide range of topics such as soil tillage, precision seeding, intelligent navigation, spraying technology, sensor integration, remote monitoring, and machinery automation. The research presented provides valuable insights into how these systems can be optimized to increase efficiency, reduce costs, and contribute to sustainable agricultural practices.

This Special Issue is intended for researchers, practitioners, and policymakers interested in agricultural technology and machinery systems. It brings together significant research contributions that will help shape the future of agricultural production, offering new perspectives on machinery design, implementation, and operational effectiveness.

We express our sincere gratitude to all the authors for their contributions, as well as to the peer reviewers and editorial team for their support and expertise. Their efforts have been integral in making this Special Issue a comprehensive resource for advancing agricultural machinery systems.

Mustafa Ucgul and Chung-Liang Chang
Guest Editors

agriculture

MDPI

Article

Design and Optimization of a Mixed-Flow Drying Chamber for Tiger Nuts Based on CFD-DEM Heat and Mass Transfer Model

Li Ding, Yufei Dou, Junying Li, Tan Yao, Aobo Ma, Yechao Yuan, Lele Wang and He Li *

College of Mechanical and Electrical Engineering, Henan Agricultural University, Zhengzhou 450002, China; dingli@henau.edu.cn (L.D.); douyufeii@stu.henau.edu.cn (Y.D.); lijunying@stu.henau.edu.cn (J.L.); yaotan@stu.henau.edu.cn (T.Y.); maaobo@stu.henau.edu.cn (A.M.); yuanyechao@stu.henau.edu.cn (Y.Y.); wanglele@henau.edu.cn (L.W.)

* Correspondence: lihenjx@henau.edu.cn

Abstract: In order to solve the problem of inconsistent moisture content in particles during the drying process of tiger nuts (Cyperus esculentus) due to uneven air flow and temperature distribution in the drying chamber, an open-hole corner box was designed based on the principle of negative pressure micro-perforated air supply. Using computational fluid dynamics (CFD) and discrete element method (DEM) simulation, coupled with the basic theory of interphase heat and mass transfer, a mathematical model for interphase heat and moisture coupling transfer was established. The effects of different aperture rates of corner boxes in the drying chamber, spatial location arrangement, and other related variables on the airfield distribution, temperature field distribution, tiger nut temperature, and moisture content changes were investigated. The results show that the average air velocity below the air inlet gradually increases as the opening ratio increases. When the opening rate is 0.33%, the wind field uniformity is better, and the inhomogeneity of the drying chamber wind field is improved. As the lateral distance increases, the consistency of the moisture content distribution increases and then decreases, and the flow rate of the tiger nuts gradually increases when the grain is discharged. The rate of decrease in water content decreases gradually with the increase in longitudinal distance. When the wind speed reaches 4 m/s, the drying chamber wind field is more uniform, and the water vapor diffusion efficiency at the outlet is basically the same. Therefore, the appropriate corner box has a horizontal distance of 320 mm and a longitudinal distance of 420 mm, providing a basis for the design of tiger nut drying equipment.

Keywords: CFD-DEM; heat transfer; mass transfer; mixed-flow grain dryers

Citation: Ding, L.; Dou, Y.; Li, J.; Yao, T.; Ma, A.; Yuan, Y.; Wang, L.; Li, H. Design and Optimization of a Mixed-Flow Drying Chamber for Tiger Nuts Based on CFD-DEM Heat and Mass Transfer Model. *Agriculture* **2024**, *14*, 541. https://doi.org/10.3390/agriculture14040541

Academic Editor: Lixia Hou

Received: 27 February 2024
Revised: 22 March 2024
Accepted: 22 March 2024
Published: 28 March 2024

1. Introduction

The tiger nut is a new type of oil crop. It integrates grain, oil, pasture, and feed, and is expected to become a high-quality resource to replace soybeans and solve the problem of food and oil security that has long plagued the world [1]. The benefits of tiger nut cultivation are much higher than those of major food crops, which are conducive to promoting farmers' income growth [2]. The drying of tiger nuts after harvest is one of the key links in the mechanization of tiger nut production. At present, the natural drying method is mainly used. However, it is no longer able to meet the requirements of the large-scale production and quality of tiger nuts, and there is an urgent need to develop special drying equipment for tiger nuts [3]. Mixed-flow drying towers have been widely used in grain drying, and the use of mixed-flow drying towers can effectively mitigate the impact of weather conditions on tiger nut drying operations. The drying chamber is the core part of the mixed-flow drying tower. However, tiger nuts are highly susceptible to uneven heating in the drying chamber of traditional mixed-flow drying towers [4–6]. Therefore, the structure and layout of the drying chamber's inner ventilation corner box are important factors that affect the airflow distribution.

Optimizing the drying chamber structure can improve the drying performance of the drying tower, reduce production costs, and cut down energy consumption. The researchers used mathematical modeling to conduct a large number of studies on the shape and spatial layout of the corner box. C. W. Cao et al. [7] investigated the effects of the duct size, shape, number of rows, and arrangement on the performance of a mixed-flow grain dryer. F. Weigler et al. [8] investigated the effects of dryer geometry and different corner box arrangements (horizontal and diagonal) on the wind and temperature fields using the CFD technique. Keppler, Istvan et al. [9] used the DEM technique to analyze the effects of the corner box shape and arrangement on particle flow uniformity. Qing Jiang et al. [10] used the CFD technique to optimize the drying tower with a variable cross-section corner box structure. The uniformity of the temperature field and airfield distribution in the drying chamber was improved. Che Gang et al. [11] designed a variable-diameter open-ended corner box using CFD technology in order to improve the air distribution relationship within the drying chamber and further improve the uniformity of rice drying. Anderson Rodrigo Visconcini et al. [12] studied the inlet and outlet airflow using CFD techniques. The material was set up as a laminar porous medium in an isothermal state. Luo, H et al. [13] used the DEM technique to analyze the flow state of paddy in the drying chamber and concluded that there is a velocity difference between adjacent paddy flow processes. The wind field and temperature field of the mixed-flow drying tower are more complicated. Merely analyzing the temperature field and wind field in the drying chamber does not directly indicate the drying effect, which is quite different from the actual situation. In order to better design the structure of the drying chamber, it is also necessary to obtain two important parameters, such as the moisture content and temperature of tiger nuts.

In order to optimize the design of the drying system, the use of CFD-DEM modeling is the most effective way to achieve this goal [14–20]. Bin Lan et al. [21] developed a CFD-DEM-IBM (immersed boundary method) method for the simulation of the particle drying process, which is useful for simulating the fluid flow, mass transfer, and heat transfer within a gas-fluidized bed. Rong Guo et al. [22] investigated the hydrodynamic and heat transfer characteristics of wet and dry particles in a fluidized bed using numerical modeling by introducing the effect of the liquid bridge force on the wet content of particles. H.Q. Che et al. [19] proposed a complete and reliable CFD-DEM model to study the coating process. Numerical methods focusing on particle–fluid thermal convection, coating fluid ejection, and evaporation of the ejected fluid on the particle surface were investigated. M. Sousani et al. [23] proposed an accelerated heat and mass transfer simulation using graphics processing unit (GPU) technology. The coupled model accurately captured the fluid-to-solid phase convective heat transfer process with a significant improvement in the simulation time. Khomwachirakul P et al. [24] used the CFD-DEM model to explore gas-particle motion behavior, average particle water content, average particle residence time, and particle residence time distribution. Aziz, H et al. [25] used CFD-DEM to simulate the drying process of pharmaceutical wet particles in a fluidized bed dryer. Water evaporation from the particle surface was considered. Obviously, CFD-DEM can explore the wind field and temperature field distribution in the drying chamber of the mixed-flow drying tower. More importantly, temperature and moisture content changes in tiger nut particles can be obtained.

This paper addresses the problem of inconsistent moisture content during the drying process of tiger nuts due to uneven airflow and temperature distribution. Based on the CFD-DEM coupled simulation and the basic theory of interphase heat and mass transfer, the mathematical model of interphase heat and moisture coupling transfer is established. The influences of different aperture rates of corner boxes and space position arrangements on the distribution of an airfield, temperature field, and the temperature and moisture content of tiger nuts in the drying chamber are investigated, providing a theoretical basis for the design of tiger nut drying equipment.

2. Materials and Methods

2.1. Mathematical Model

In order to facilitate the simulation and analysis of the wind field and temperature field distributions, as well as the temperature and humidity changes of tiger nuts in the drying room, using the CFD-DEM coupled simulation and the basic theory of interphase heat and mass transfer, a mathematical model of heat and humidity coupled transfer for tiger nuts was established.

During the actual drying process, the hot medium enters the drying chamber from the air inlet to carry out the drying operation. The exhaust gas leaves the drying section by the air outlet, and there are some uncertainties in this process. The theoretical control equations in the simulation refer to the ideal state. In order to simplify the model calculation, the following assumptions are made: it is assumed that the volume state of tiger nuts does not change during the drying process; the hot medium inside the drying section is considered a continuous, incompressible ideal gas; and the spontaneous heat generated by the respiration of the tiger nuts themselves is ignored.

2.1.1. Governing Equations for Solid Phase

The DEM is used to track the trajectory and rotation of each particle in the system through a time-stepped simulation. The motion of the mass is governed by Newton's second law of motion, and the equations of motion for translational and rotational motion are as follows [23]:

$$F = \begin{cases} m_p \ddot{u}_i \\ I\dot{\theta} \end{cases} \quad (1)$$

where F is the interparticle contact force; m_p is the mass of the particle; $\ddot{u}_i$ is the translational acceleration; I is the moment of inertia; and $\dot{\theta}$ is the corner acceleration. The equation used to calculate the force of different particle–particle interactions is as follows [23–26]:

$$F_n = \frac{4}{3} E^* \sqrt{R^*} \delta_n^{3/2} \quad (2)$$

$$F_n^d = -2\sqrt{5/6}\beta\sqrt{S_n m^*} \overrightarrow{U_n^{rel}} \quad (3)$$

$$F_t = -S_t \delta_t \quad (4)$$

$$F_n^d = -2\sqrt{5/6}\beta\sqrt{S_t m^*} \overrightarrow{U_t^{rel}} \quad (5)$$

where E^* is the effective Young's modulus; R^* is the particle equivalent radius; values of β are in the range of 0.01–0.03; m^* is the equivalent mass; $\overrightarrow{U_{n.t}^{rel}}$ is the normal/tangential component of the relative velocity; and $S_{n,t}$ is the normal/tangential stiffness.

2.1.2. Governing Equations for Gas Phase

The new dense discrete phase model (DDPM), which converts EDEM (engineering discrete element method, DEM Solution Ltd., Edin, UK) particle data into discrete phase model (DPM) injections, is used. Moreover, the volume fraction resistance, heat, and mass transfer are calculated by Fluent software (Fluent 2022, ANSYS, Inc., Canonsburg, PA, USA), thus improving the computational efficiency. Each phase satisfies the laws of the conservation of mass and conservation of momentum. The mass and momentum conservation equations for the fluid phase g are as follows [23]:

$$\frac{\partial}{\partial t}(\alpha_g fulllength \rho_g) + \nabla \cdot (\alpha_g fulllength \rho_g \vec{v}_g) = \sum_{q=1}^{nphases} (\dot{m}_{gp} - \dot{m}_{pg}) \quad (6)$$

where subscripts p and g denote the solid phase and fluid phase, respectively; α_g denotes the fluid volume fraction; $fulllength\rho_g$ denotes the density of the fluid phase; $\vec{V}_g$ denotes the velocity of the fluid phase; $\dot{m}_{gp}$ denotes the mass transferred from the fluid phase to the solid phase; and $\dot{m}_{pg}$ denotes the mass transferred from the solid phase to the fluid phase.

The fluid phase momentum conservation equation is as follows [23]:

$$\begin{aligned} &\frac{\partial}{\partial t}(\alpha_g fulllength\rho_g \vec{v}_g) + \nabla\cdot(\alpha_g fulllength\rho_g \vec{v}_g \vec{v}_g) \\ &= -\alpha_g \nabla_g + \nabla\cdot\left[\alpha_g \mu_g (\nabla_{\vec{V}_g} + \nabla_{\vec{V}_g} r)\right] + \alpha_g fulllength\rho_g \vec{g} + F_{vm,ift,user} \\ &+\sum_{pg}(\vec{K}_{pg}(\vec{v}_g - \vec{v}_p) + \dot{m}_{gp}\vec{v}_{gp} - \dot{m}_{pg}\vec{v}_{pg}) \\ &+\vec{K}_{DPM}(\vec{V}_{DPM} - \vec{v}_g) + S_{DPM,explicit} \end{aligned} \tag{7}$$

where μ_g is the shear viscosity of the particle phase g; $\vec{K}_{pg}$ is the interphase momentum exchange coefficient between the fluid and particle phases; $F_{vm,lift,user}$ is the lift force (due to the velocity gradients in the initial phase flow field); $\vec{v}_p$ and $\vec{v}_g$ are the particle and fluid velocities, respectively; $\vec{v}_{gp}$ is the interphase velocity and is dependent upon $\dot{m}_{gp}$; $S_{DPM,explicit}$ is the explicit component of the particle sink term; and $\vec{V}_{DPM}$ and $\vec{K}_{DPM}$ are the implicit terms of the particle averaged velocity of the considered discrete phase and interphase momentum exchange coefficient, respectively. In this study, the heat, mass, and momentum transfer between phases are considered, so the mass of the tiger nut particles is variable.

Modeling the drying process of particles containing water is a complex process. Scholars have provided a wealth of models for studying the mathematical modeling of drying processes. The following equation presents a simple kinetic model that considers the latent heat of the vaporization of water, as well as the different vaporization processes of surface-free water and internal bound water in solids. The mass transferred from the solid phase to the fluid phase can be expressed as follows [27]:

$$\dot{m}_{pg} = -\alpha\pi d^2(P_{sat} - P)\sqrt{\frac{M_w}{2\pi RT}} \tag{8}$$

$$P_{sat} = 133.322 \times 10^{0.6715} + 0.030(T - 273.15) - 0.0000798(T - 273.15)^2 \tag{9}$$

$$\alpha = \frac{Sh_p D_g}{d_p} \tag{10}$$

$$D_g = 5.05 \times 10^{-9} {T_g}^{1.5} \tag{11}$$

where α denotes the particle evaporation coefficient; M_w denotes the molecular weight of water (0.018 kg/mol for water); R denotes the universal gas constant (8.3145 J/mol K); T denotes the temperature of the particles, K; P denotes the gas pressure, pa; p_{sat} denotes the saturation pressure at temperature T, pa; d_p denotes the particle diameter, mm; D_g denotes the diffusion coefficient of water vapor in the gas phase; T_g the temperature of the gas phase, K.

The Ranz–Marshall model uses a similar approach to the Ranz–Marshall heat transfer coefficient model. The expression for the Sherwood number flowing through the spherical particles is the same as that for the Nusselt number in heat transfer, and the Prandtl number is replaced by the Schmidt number [27].

$$Sh_p = 2 + 0.6Re^{1/2}Sc_p^{1/3} \tag{12}$$

where the Schmidt number is defined as follows:

$$Sc_p = \frac{\mu_p}{fulllength\rho_p D_p} \tag{13}$$

where μ_P and $fulllength\rho_p$ are the dynamic viscosity and density of the fluid phase, respectively. The Ranz–Marshall model is based on the boundary layer theory and is used to stabilize the flow through spherical particles. It is usually applied under the following conditions.

$$0 \leq Sc_q < 250$$

$$0 \leq Re_q < 200$$

This work considers the heat transfer between 'particle–particle' and 'particle–fluid'. The energy control equation for a single particle is as follows [17,23–27]:

$$m_p C_p \frac{dT}{dt} = \sum (Q_{PF} + Q_{P_1P_2}) \tag{14}$$

$$Q_{PF} = h_{PF} A_p \Delta T_{PF} \tag{15}$$

$$Q_{P1P2} = h_c \Delta T_{P1P2} \tag{16}$$

$$h_c = \frac{4k_{P1}k_{P2}}{k_{P1} + k_{P2}} (\frac{3F_N r^*}{4E^*})^{1/3} \tag{17}$$

$$h_{PF} = k_F N_u / d_p \tag{18}$$

$$N_u = 2 + 0.6{R_e}^{1/2} Pr^{1/3} R_e < 200 \tag{19}$$

$$P_r = C_\mu / k \tag{20}$$

where h_{PF} is the convective fluid–particle heat transfer coefficient; A_P is the particle surface area, mm^2; ΔT_{PF} is the temperature difference between the fluid and the particles, K; $Q_{P_1P_2}$ is the inter-particle heat flux, W/m^2; h_c is the conductive heat transfer coefficient between two particles; $\Delta T_{P_1P_2}$ is their temperature difference, K; k_p is the thermal conductivity of the particles; F_N is the normal force, N; r^* is the geometric mean of the particles radii, mm. The bracketed term in the equation models the contact area between two particles. Finally, K_F is the gas thermal conductivity of the fluid; N_u is the Nusselt number; d_p is the particle diameter, mm; R_e is the Reynold's number based on the diameter of the individual phase and the relative velocity $\left|\vec{v}_p - \vec{v}_g\right|$; and P_r is the Prandtl number for the subsequent phase. The model presented does not consider conductive heat transfer from the particles to the geometry.

2.2. Particle Model

Yu tiger nut one seeds were used as test subjects. We randomly selected 1000 tiger nut seeds for the test, and the length, width, and height of each seed were measured with vernier calipers. The seeds were categorized into flat, spheroid, and spherical, according to their shapes, as shown in Figure 1. Each seed was measured five times, and the average value was taken as the triaxial dimensions of the length, width, and height of the seeds. The triaxial dimensions of the seeds were also statistically analyzed to determine the distribution of the mean diameter of the seeds, as shown in Table 1.

The sphericity of tiger nut seeds is as follows:

$$S_p = \frac{(L \cdot D \cdot H)^{1/3}}{L} \times 100\% \quad (21)$$

where S_p is sphericity, %; L is seed length, mm; D is seed width, mm; and H is seed height, mm.

Figure 1. Schematic diagram of tiger nut seed's triaxial size. In the figure, L_0 is the length of tiger nut seeds; w_0 is the width of tiger nut seeds; and t_0 is the tiger nut seed thickness.

Table 1. Three-axis size of tiger nut seeds.

Shape and Size		L_0	W_0	t_0
Flat	Average value (mm)	14.18	9.12	6.76
	Standard deviation (mm)	0.91	0.82	0.61
Spheroid	Average value (mm)	12.24	10.18	9.56
	Standard deviation (mm)	1.1	0.88	0.92
Spherical	Average value (mm)	9.31	8.14	7.94
	Standard deviation (mm)	0.78	0.85	0.79

From Equation (21), the tiger nut seed sphericity is 84%. To make the simulation more realistic, the three common tiger nut seed shapes and sizes described above are used, as shown in the top row in Figure 2. In the DEM simulation, seed particles are filled by a combination of multi-spherical surfaces to generate a particle model in the EDEM 2022 software.

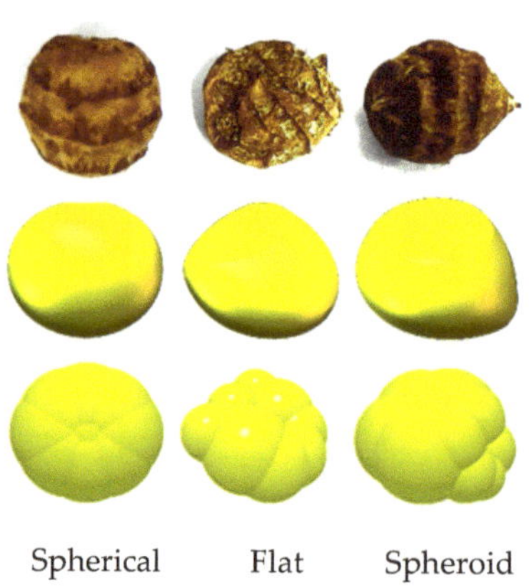

Figure 2. Tiger nut seed particle physical map and discrete element model.

2.3. *Geometric Model*

2.3.1. Corner Box

As shown in Figure 3, the tiger nut mixed-flow drying tower consists of a slow suction section, air inlet duct, drying chamber, natural gas burner, hot air blower, discharge screw conveyor, feed inlet, bucket elevator, equalizer, feed screw conveyor, corner box opening, corner box, air inlet, air outlet, and so on. When the mixed-flow drying tower works, tiger nuts enter the drying tower from the feed opening and are lifted to the top of the drying tower by the bucket elevator and feed screw conveyor. Under the grain equalizer, the tiger nuts fall into the inside of the tower. When the tiger nuts completely fill the drying tower, the hot air enters the drying room from the inlet corner box to exchange moisture and heat with the tiger nuts, and the water vapor is discharged from the drying room through the outlet corner box. When the temperature of hot air reaches the set value, the discharge screw conveyor and six impellers start to run, and the tiger nuts fall to the bottom of the tower and return to the bucket elevator through the discharge screw conveyor. Tiger nuts continuously move from top to bottom through the slow suction section, the drying chamber of the operation process, until the moisture content of the tiger nuts meets the drying requirements.

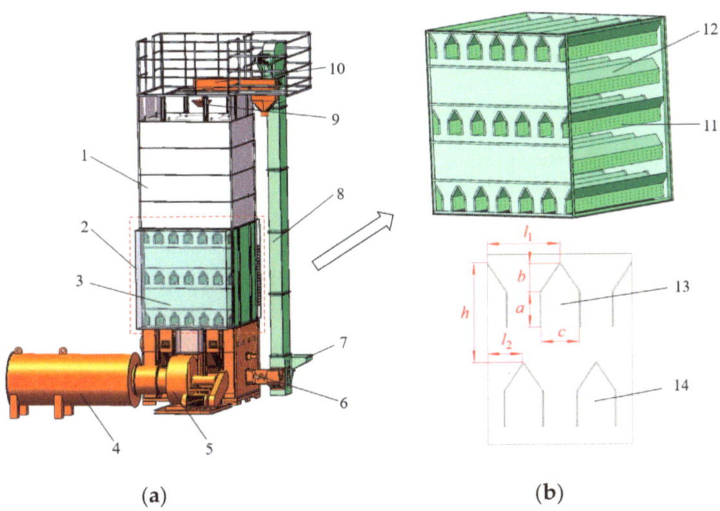

Figure 3. (**a**) Overall structure of tiger nut mixed-flow drying tower; (**b**) corner box arrangement and structure diagram. In the figure, 1. Slow supply section; 2. inlet air duct; 3. drying chamber; 4. natural gas burner; 5. hot-air blower; 6. discharge screw conveyor; 7. inlet; 8. bucket elevator; 9. grain equalizer; 10. inlet screw conveyor; 11. corner box openings; 12. corner box; 13. air inlet; 14. air outlet.

The design of the corner tube is based on the Danish CIMBRIA and Swedish SVEGMA manuals, which cover machine type and drying technology [28]. The design of corner boxes adheres to the principle that relates wind pressure and medium flow rate. The corner box has a length of 2000 mm, a width c of 160 mm, a height a of 150 mm, and a height b of 120 mm, as shown in Figure 3b, drawing on the design concept of an air distribution board in a fluidized bed dryer. The open-hole corner box is designed with the principle of negative pressure microporous air supply. The formula for calculating the opening rate of the corner tube in the vertical box dryer is as follows:

$$\alpha = \left(\frac{\zeta \rho_g {\mu_1}^2}{2g(\Delta P_D)_{SC}} \right)^{1/2} \tag{22}$$

Among them:

$$R_{SC} = 0.01 + 0.2\left(1 - e^{(-0.5S/H)}\right)$$

$$\Delta P_B = L_{\mathrm{mf}}(1-\varepsilon_{\mathrm{mf}})(fulllength\rho_s - fulllength\rho_g)$$

$$\varepsilon_{\mathrm{mf}} = \left(\frac{1}{14\varphi_s}\right)^{1/3}$$

where α is the opening rate, %; ξ is denotes the side resistance coefficient of the corner box, 1.5~2.5; $fulllength\rho_g$ is the hot air density, measured in kg/m^3, determined to be 0.972 kg/m^3; μ_1 is no-load airflow velocity, m/s; $(\Delta P_D)_{SC}$ is the corner box side critical pressure drop, Pa; R_{SC} is the parameter related to the bed diameter, D, and bed height, L. It can be approximated as the parameter related to the size of the drying chamber structure of the vertical dryer. S is the bottom circumference of the drying section, mm; H is the drying section height, mm; ΔP_B is the theoretical bed pressure drop, Pa; L_{mf} is the static height of the tiger nut in the drying chamber, and is approximately the thickness of the valley, mm; $fulllength\rho_s$ is the particle density of the tiger nut, 1.455×10^3 kg/m^3; $\varepsilon_{\mathrm{mf}}$ is the porosity of minimum fluidization velocity; $fulllength\varphi_s$ is the sphericity of solid particles, with tiger nut particles being irregular in shape, and its sphericity is 0.84.

According to the principle of a low aperture rate, the along-travel resistance of the open hole corner box is balanced with the static pressure difference across the micro-aperture to regulate the uniformity of the lateral wind field. The appropriate openings for the side panels of the corner box are calculated to be 1.9% to 4.7%. The three openings are determined to be 1.9%, 3.3%, and 4.7%, respectively, based on the percentage size of the orifice plate. The micropore diameter is 5 mm, and the size is smaller than the diameter of the tiger nut to prevent blockage during drying. Considering the size range of the side panels of the corner box, the number of micro-perforations corresponding to different opening rates is designed to be 290, 504, and 718, respectively. The schematic structure of the open-hole corner box is shown in Figure 4.

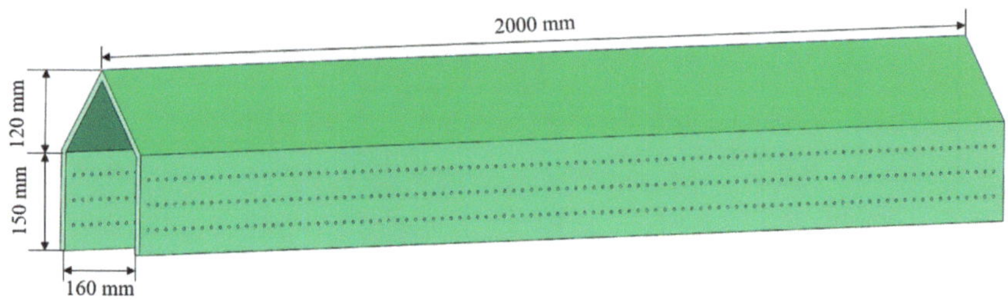

Figure 4. Schematic diagram of the corner box structure.

2.3.2. Corner Box Layout Analysis

In order to address the problems of clogging and uneven heating of tiger nuts in the drying chamber of the drying tower, the horizontal and vertical spacing of the corner boxes in the drying chamber are designed by optimizing the design. This improves the movement, heat condition, and airfield distribution of tiger nut particles inside the drying chamber of the drying tower. From the literature [29], it is known that the horizontal spacing of the corner box in the drying tower chamber ranges from 200 to 500 mm, and the vertical spacing ranges from 290 to 550 mm. Considering the size of the drying tower chamber and the sizes of the tiger nuts on the corner box spacing design, spacings of 240 mm, 320 mm, and 480 mm are selected for the lateral simulation tests of the drying tower chamber, and 320 mm, 420 mm, and 520 mm are selected as the three vertical levels. In order to reduce the number of coupled simulation calculations, the length of the drying chamber is determined to be 960 mm, and the height is 800 mm when studying the impact of horizontal spacing. When studying the effect of vertical spacing, the length of the drying chamber is determined to be 900 mm, and the height values are determined to be 640, 840, and 1040 mm, respectively, as shown in Figure 5.

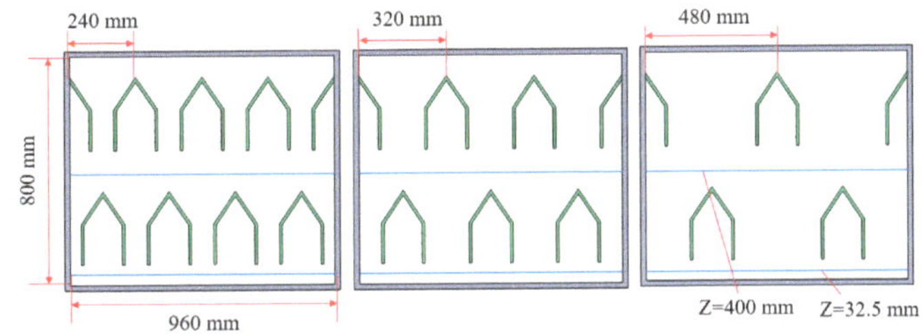

(a)

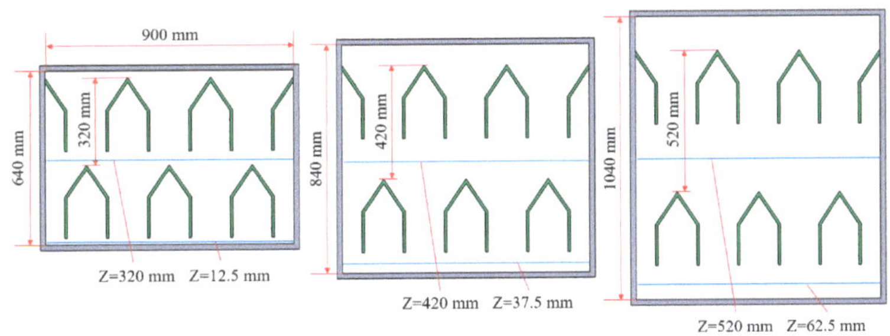

(b)

Figure 5. (**a**) Schematic diagram of the distribution of boxed corners in the three lateral positions; (**b**) schematic diagram of the distribution of boxed corners in the three longitudinal positions.

2.4. Simulation Conditions

CFD-DEM is based on ANSYS Fluent 2022 and EDEM 2022 software. First, Fluent meshing in ANSYS is utilized to perform meshing, and mesh encryption is performed on the corner box opening portion. The corner box inlet is set as the velocity inlet boundary condition and the corner box outlet is set as the pressure outlet boundary condition, as shown in Figure 6. The standard k-ε turbulence model in ANSYS, the SST k-omega equation, is used as the turbulence model. The Ranz and Marshall heat transfer model and multicomponent model [15–19] are used because heat and mass transfer are considered in the work. In the component set, two compositions are identified: the evaporable component of water and a non-evaporable component of tiger nuts; the composition ratio is 45% and 55%, respectively. They represent the initial water content of the tiger nut and the other parameters of the tiger nut, as referred to in the literature [30]. The inlet air velocity and temperature are set according to preliminary tests and by consulting references [28,31,32]. All CFD input parameters and boundary conditions are shown in Table 2.

In EDEM, the material of the drying tower is set as a steel plate, and the detailed parameters of DEM are shown in Table 3. During the working process, the seeds may slide, roll, collide, extrude, and perform other motions in the device. Therefore, intrinsic (e.g., Poisson's ratio, density, etc.) and operating (e.g., the coefficient of rolling friction, coefficient of sliding friction, etc.) parameters need to be set for both the seed and the drying device. Heat and mass transfer between the seeds also need to be taken into account. So, opening the heat conduction model and temperature update model, as well as setting parameters, such as thermal conductivity and specific heat capacity in EDEM 2022 software, are necessary. The water content of the tiger nut particles is set in the particle parameters, which are added to the particle factory by Fluent after coupling. The DEM parameters used for the tiger nut particles are referenced in [33,34] in Table 3. Before the coupling simulation, it is necessary to pre-fill the drying chamber with tiger nuts in order to make the tiger nuts completely stationary at the time of coupling; the particle generation time is 3 s, and the total simulation time is 10 s.

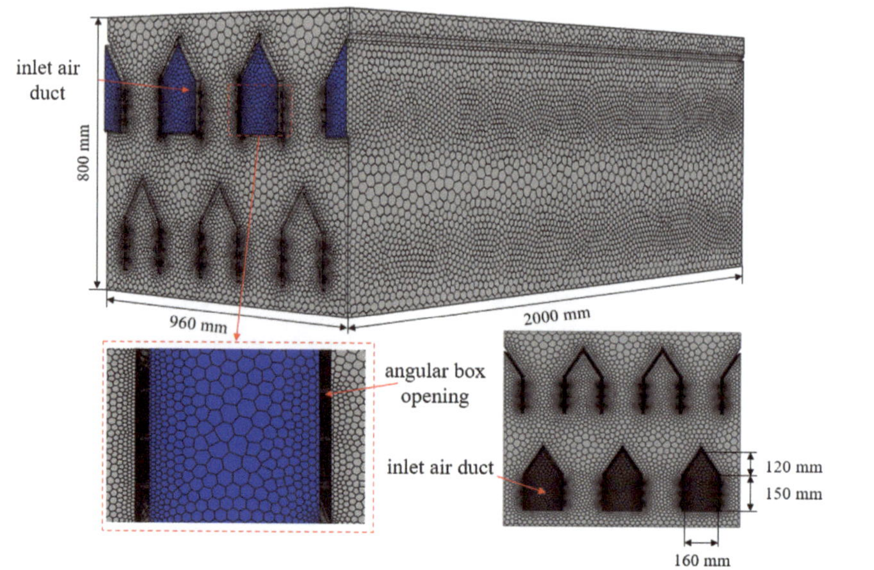

Figure 6. Mesh model of drying cell with a lateral distance of 320 mm in DEM and CFD.

Table 2. CFD input parameters.

	CFD Input Parameters	
Air	Type of fluid	Air
	Density (kg/m^3)	1.225
	Specific heat capacity (j/Kg·K)	4506.43
	Viscosity (Pa·s)	1.789×10^{-5}
	Wind speed (m/s)	2, 4, 6
	Wind temperature(K)	338
	Thermal conductivity (W/m·K)	0.0242
DPM injection	Initial temperature (K)	288
	Drag law	Gidaspow
	Water content of particles (%)	45
Walls	Adiabatic and no slip	
	Turbulence model	Standard k-ε
	CFD cell type	Hybrid tetrahedral hexahedral
	CFD time step (sec)	0.05

Table 3. EDEM Input parameters.

EDEM Input Parameters	
Particle density (kg/m^3)	1.186×10^3
Particle diameter (mm)	16
Particle shear modulus (Pa)	2.77×10^7
Particle Poisson's ratio	0.18
Geometry density (kg/m^3)	7850
Geometry shear modulus (Pa)	1×10^{10}
Geometry Poisson's ratio	0.3
Particle–particle coefficient of restitution	0.48
Particle–particle static friction coefficient	0.34
Particle–particle rolling friction coefficient	0.1
Particle–geometry coefficient of restitution	0.619
Particle–geometry static friction coefficient	0.254
Particle–geometry rolling coefficient	0.072

Table 3. *Cont.*

EDEM Input Parameters	
EDEM time step (s)	1×10^{-4}
Particle gravity (m/s^2)	9.81
Thermal conductivity (W/(m·k))	0.106
Specific heat capacity (J/(kg·k))	15,920

2.5. Experiment

2.5.1. Materials

The raw material for the experiment was fresh tiger nuts harvested in October 2023 from Shangqiu City, Henan Province, and the variety was Yu tiger nut I. The tiger nuts were ellipsoidal, with an average mass of 1.13 g/pc, an average geometric diameter of 10.99 mm, a density of 1.17 g cm^{-3}, a wet basis moisture content of 45% (103 ± 2 °C constant weight method), and an angle of repose of 32.91°. The tiger nuts were cleaned using a cylindrical primary cleaning sieve to remove dust and cilia from the surface and obtain uniform and full tiger nuts. The moisture content was determined according to the direct drying method outlined in GB5009.3-2016, regarding the determination of moisture in food. The tiger nuts were dried in an oven at 105 °C until a constant mass was achieved. The drying equipment used for the tiger nut samples was a JK-KB1700 thin layer drying test bed (National Engineering Laboratory of Grain Storage and Transportation, Changchun, China), with a wind temperature range between 20 °C and 100 °C (±1 °C), a humidity range between 20 and 80% RH (±4% RH), and a wind speed range between 1.0 and 3.2 m/s. There are two methods to express the moisture content of a material, namely the dry basis method and the wet basis method. The dry basis expression is calculated on the basis of the solid dry matter in the material, and the wet basis expression is calculated on the basis of the material's mass. The wet basis moisture content was used in this test, and the wet basis moisture content can be expressed by Equation (23):

$$M_w = \frac{m_w}{m_s + m_w} \tag{23}$$

where M_w is the wet basis moisture content, %; m_w is the mass of moisture contained in tiger nuts, *g*; and m_s is the mass of dry matter contained in tiger nuts, *g*.

2.5.2. Experiment Equipment

In order to verify the accuracy of the simulation, drying experiments were conducted using the experimental platform shown in Figure 7. The experimental equipment consisted of four parts, i.e., a mixed-flow dryer, an airflow velocity measurement system, a temperature acquisition system, and an information processing system.

In order to verify the simulation results of the temperature field and wind field inside the drying chamber, the K-type thermocouple sensor and the hot-wire air velocity sensor were fixed on the bottom surface of the inlet corner box (Z = 400 m), and the bottom surface of the outlet corner box (Z = 32.5 mm), respectively. As shown in Figure 5, the blue part shows the locations of 98 sampling points, including 49 monitoring points below the air inlet and 49 monitoring points below the air outlet.

The wind speed acquisition system includes the FY-CJ2 data collector (FY-CJ2, Wuhan Fuyuan Flying Fortress Electronic Technology Co., Ltd., Wuhan, China), a hot-wire wind speed sensor (Wuhan Fuyuan Flying Fortress Electronic Technology Co., Ltd., China), an industrial-grade USB-to-RS485 modular protocol converter, and an information processing system (ASUS ASUS Flying Fortress 6 ZX80G, ASUS, Taiwan, China). The temperature acquisition system consists of an LK1048U Multi-Circuit Temperature Inspector (LK1048U, Changzhou Blu-ray Electronics Co., Ltd., Changzhou, China), a K-Type Thermocouple Sensor (K-Type, Shanghai Song guide Heating Sensors Co., Ltd., Shanghai, China), and an industrial-grade USB-to-RS485 Module Protocol Converter and Information Processing System (ASUS ASUS Flying Fortress 6 ZX80G).

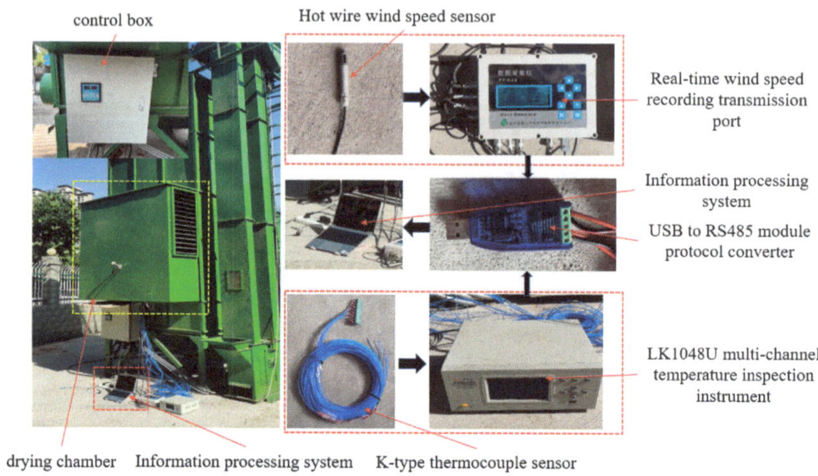

Figure 7. Tiger nut drying validation testbed.

2.5.3. Experiment Method

The harvested tiger nuts are decontaminated and mixed thoroughly. Before the test, the moisture content of tiger nuts was measured three times using the JK-KB1700 thin-layer drying test bench, and the average value was taken as the initial moisture content of tiger nuts for the test and simulation. In the test, the horizontal spacing of the corner boxes in the drying chamber was 320 mm, the vertical spacing was 420 mm, and the opening rate of the corner boxes was 4.7%. The tiger nut drying test was carried out at the three-area experimental center of Henan Agricultural University, with an ambient temperature of about 15 °C, an ambient humidity of 15–25%, and cloudy weather. The drying validation test platform used the developed mixed-flow drying tower for tiger nuts, model 5HH-7.5. The hot air temperature and air velocity were monitored by temperature and humidity sensors, and air velocity sensors were mounted below the air intake chamber and the air intake angle-shaped tube. The heat source equipment was first turned on to bring the wind speed and heat medium temperature up to the required level, after which, grain loading was carried out. The timing started at the end of grain loading, the drying duration of each test was 1 h, and the test was conducted 8 times. At the end of the test, the moisture content of tiger nuts was measured three times, and the average value was taken as the moisture content of tiger nuts after drying.

3. Results and Discussion

3.1. *Simulation vs. Experimental Results*

In order to determine the accuracy of the fluid model, the average temperatures of 49 monitoring points below the air inlet and 49 monitoring points below the air outlet were calculated, as shown in Figure 8, for the average temperature of the monitoring points over time, similar to the hot air temperature variation curve in reference [13]. In order to compare the differences between the simulation and experimental results, an analysis of variance (ANOVA) was performed on the temperatures at the monitoring points below the air inlet and outlet. ANOVA is used as a test of significance for the difference between the means of two or more samples. It is used to determine the magnitude of the influence of controllable factors on the results of the study by analyzing the magnitude of the contribution of different sources of variation to the total variation of the study [35]. The results are shown in Tables 4 and 5. The results show that the simulated mean values of the monitoring points inside the drying chamber are all greater than the experimental mean values. This is because the simulation of the drying chamber box was simplified and set as an adiabatic wall, and the interference of external environmental changes was ignored, so the error occurred. However, the trend in the simulation results is consistent with the experimental results. The ANOVA results show that there is no statistically significant difference between the experimental and simulated results ($p > 0.01$), as shown in Tables 4 and 5.

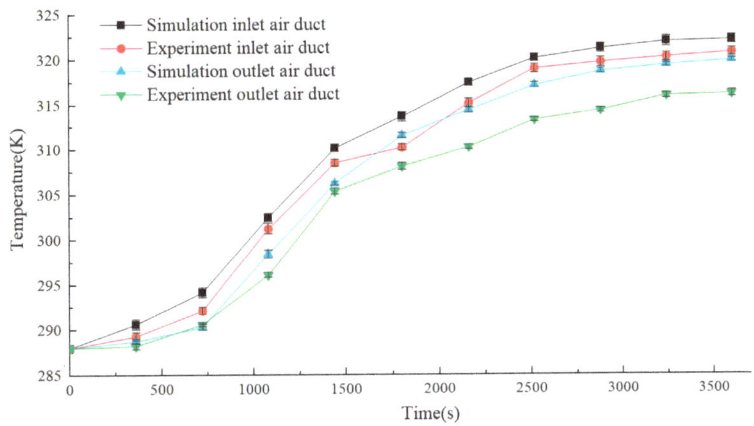

Figure 8. Variation in average temperature over time at simulation and experimental monitoring points.

Table 4. Analysis of variance for inlet air temperature.

Origin of Variance	Sum of Squares	df	Mean Squares	F Value	*p* Value
Different groups	15.59608	1	15.59608	0.11269	0.740983
Interior groups	2491.165	18	138.3981		
Total	2506.762	19			

Extremely significant at $p < 0.001$.

Table 5. Analysis of variance for outlet air temperature.

Origin of Variance	Sum of Squares	df	Mean Squares	F Value	*p* Value
Different groups	34.11533	1	34.11533	0.267978	0.610995
Interior groups	2291.513	18	127.3063		
Total	2325.628	19			

Extremely significant at $p < 0.001$.

In order to compare the simulation and test wind field distribution more intuitively, the measured wind speed data array is interpolated and smoothed using Origin 2022 software. The distribution of the wind velocity field under the inlet angle box and outlet angle box for the simulation and test is obtained, as shown in Figure 9. From Figure 9, the simulation results are slightly higher than the test results. The wind field distribution is similar to that of reference [11]. This is because the drying chamber box is idealized in the simulation and is a closed environment except for the air inlet and outlet, while it is difficult to achieve such a closed environment in the test. Moreover, errors are generated due to the measurement accuracy of the sensors and the influence of the external environment on the test data. In order to compare the difference between the simulated and experimental wind velocity field distributions, a repeatable two-factor ANOVA was performed on the wind velocity field, as shown in Tables 6 and 7. The ANOVA results show that there is no significant difference ($p > 0.01$) between the experimental and simulated results.

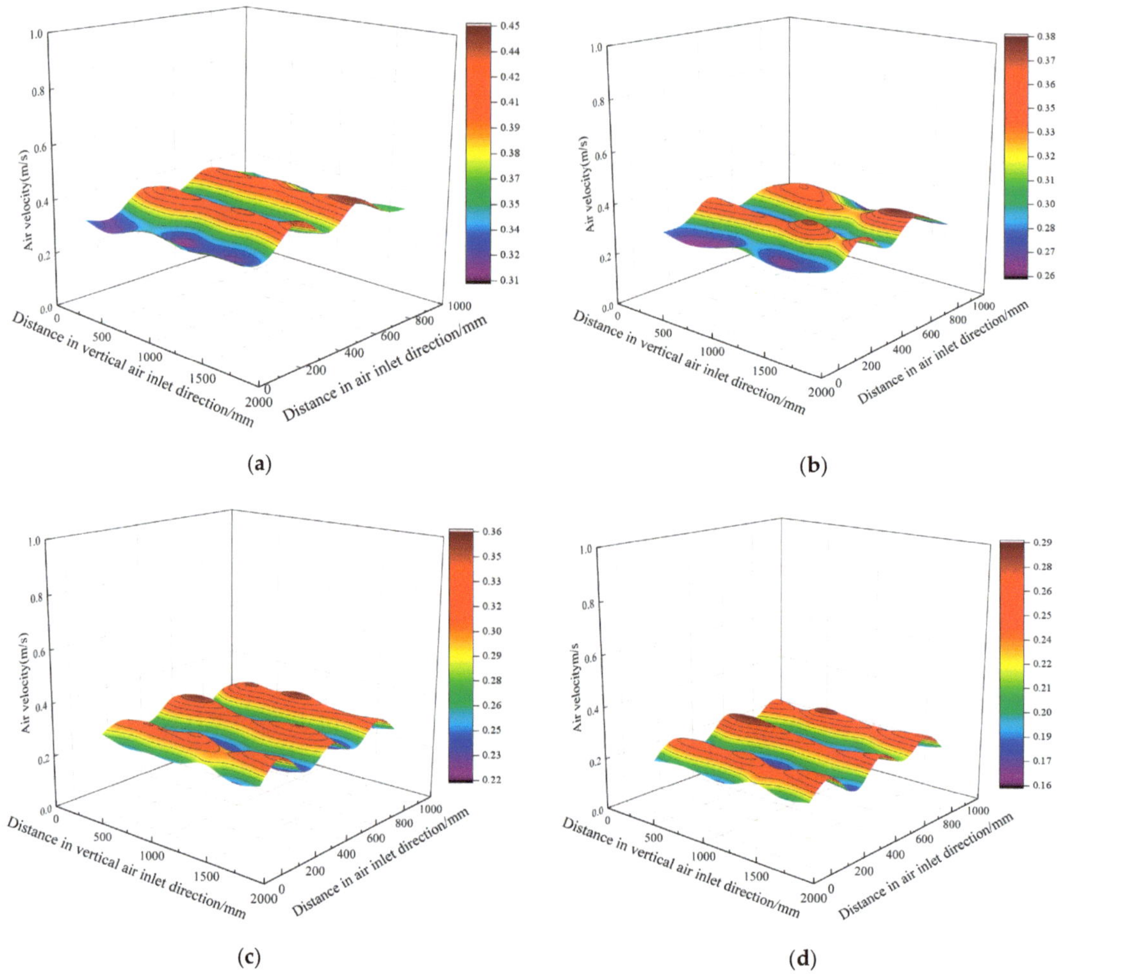

Figure 9. Wind speed distribution in drying room; (**a**) simulated wind speed distribution at air inlet; (**b**) wind speed distribution at air inlet test; (**c**) simulated wind speed distribution at air outlet; (**d**) wind speed distribution at air outlet test.

Table 6. Analysis of variance for inlet air velocity.

Origin of Variance	Sum of Squares	df	Mean Squares	F Value	p-Value
Sample	0.002778	6	0.000463	0.222059	0.967812
Columns	0.117978	6	0.019663	9.432044	7.03×10^{-7}
Interaction	0.009594	36	0.000266	0.127835	1
Interior-groups	0.10215	49	0.002085		
Total	0.232499	97			

Extremely significant at $p < 0.001$.

To compare the variations in the water content of tiger nuts, 300 tiger nut seeds were used in each experiment, with their temperature measured and averaged. In the simulation experiment, 300 seeds were randomly selected to calculate the average value. The comparison between the experimental results and simulation results is shown in Figure 10. It can be seen that the water content of tiger nuts in each group decreased by about 3.0%; the simulation results are consistent with the experimental results and in accordance with the general law of drying [36]. The results show that the above CFD-DEM model can simulate the hot air drying of tiger nuts in the drying chamber of the drying tower.

Table 7. Analysis of variance for outlet air velocity.

Origin of Variance	Sum of Squares	df	Mean Squares	F Value	*p*-Value
Sample	0.159396	6	0.026566	14.18718	2.75×10^{-9}
Columns	0.001663	6	0.000277	0.148012	0.988613
Interaction	0.010264	36	0.000285	0.152264	1
Interior-groups	0.091754	49	0.001873		
Total	0.263077	97			

Extremely significant at $p < 0.001$.

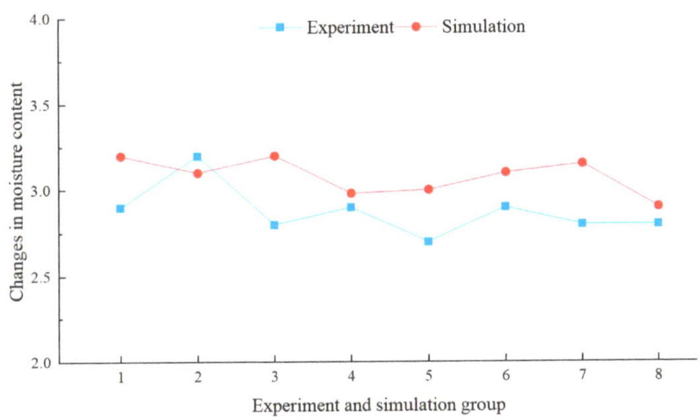

Figure 10. Comparison between experimental and simulation results.

3.2. *Effect of Opening Rate*

In order to further study the influence of the open corner box on the uniformity of the flow field in the drying chamber, the transverse distance of the fixed corner box is 320 mm, the longitudinal distance is 420 mm, and the wind speed at the inlet is 4 m/s. Moreover, 20 mm below the inlet corner box is selected as the measurement point, and the wind speed distribution trend is obtained, as shown in Figure 11. The average wind speed below the unopened corner box is 0.68 m/s, and the standard deviation is 0.58 m/s. The wind speed is higher near the air inlet and outlet and lower in the middle position, and the overall distribution of wind speed is not uniform. This is due to the higher wind pressure at the air inlet and outlet. In the open-hole corner tube flow, part of the gas can be released through the microporous sides of the side plate of the corner box, resulting in a significantly lower internal wind speed compared to the unopened-hole corner tube; this tends to stabilize the effect of wind distribution. The average wind speed under the corner box with 0.19% pore opening rate is 0.52 m/s, and the standard deviation is 0.43 m/s. This is because the pressure at the air inlet and outlet with a small pore opening rate is still larger than the pressure in the middle, the wind speeds at both ends are higher than that in the middle, and the wind speeds at both ends are lower. The average wind speed below the corner box with 4.7% openings is 0.41 m/s, and the standard deviation is 0.35 m/s. This is because more hot air flows into the side holes of the corner box, resulting in a significant reduction in wind speed and an increase in wind speed inhomogeneity. The advantage of the corner box with an opening rate of 0.33% is obvious, and the difference in wind speed along the inlet direction is small, with the average wind speed below the corner box being 0.47 m/s and the standard deviation being 0.12 m/s, which makes the wind field uniformity better. Therefore, it is preferable to adopt the corner box with an opening rate of 0.33% as the main ventilation structure in the drying room.

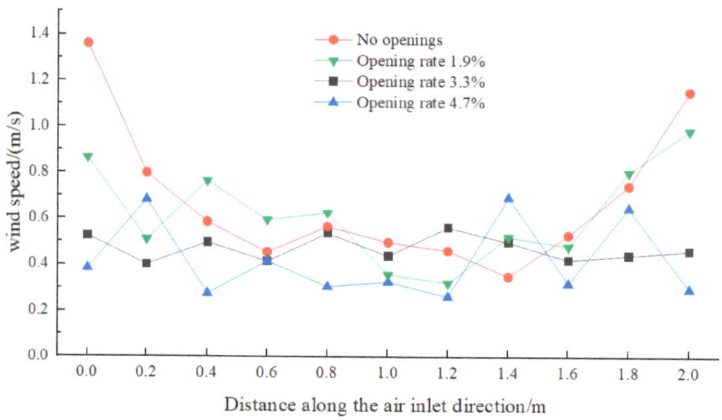

Figure 11. Distribution of wind speed for corner boxes with different aperture ratios.

3.3. *Effects of Spatial Location*

The grain layer's resistance to hot air will be different for different longitudinal and transverse spacings of the corner boxes in the drying chamber. The resistance of the grain layer is also related to the type of seed. A suitable layer resistance will reduce the energy consumed by the drying system. Therefore, there is a reasonable range of corner box layouts that depend on the grain being dried in the drying tower.

3.3.1. Effects of Lateral Distance

In Figure 12, the temperature and wind field distributions at different corner box horizontal spacings are displayed. The temperature and wind speed distributions in the figure are similar to those in references [11,31]. It can be seen that as the horizontal distance increases, the temperature and wind speed below the exit gradually decrease, and the temperature gradient and wind speed gradient change. When the horizontal spacing of the corner box is equal to 240 mm, the temperature field and wind field are not uniformly distributed. The wind speed is higher, and the temperature is higher near the exit direction. This is due to the high number of corner-mounted boxes, whose walls form a wall effect with the tiger nut seeds. The void ratio of tiger nut seeds on and near the walls of the corner boxes is always greater than that inside the drying bed. Because the resistance is relatively small, the fluid flow rate near the wall must be greater than inside the bed. The number of seeds is relatively small, and tiger nut seeds to the hot air lateral resistance is reduced. As a result, excessive wind speeds and temperatures are formed in some places. When the corner box spacing increases to 320 mm, the temperature and wind speed distributions at the two cross-sections are more uniform; the region basically realizes uniform wind distribution. In the air inlet corner of the box below, the temperature distribution is more uniform, meeting the requirements of the drying tower air temperature. When the spacing between the corner box is 480 mm, the temperature at the outlet and the temperature near the inlet has a certain gap, and the gradient of wind speed is larger in the two cross-sections. This is because the spacing between the two corner boxes is too large, and the tiger nut seeds have a greater lateral resistance to the hot air, resulting in a lower wind speed and temperature in the middle of the two corner boxes.

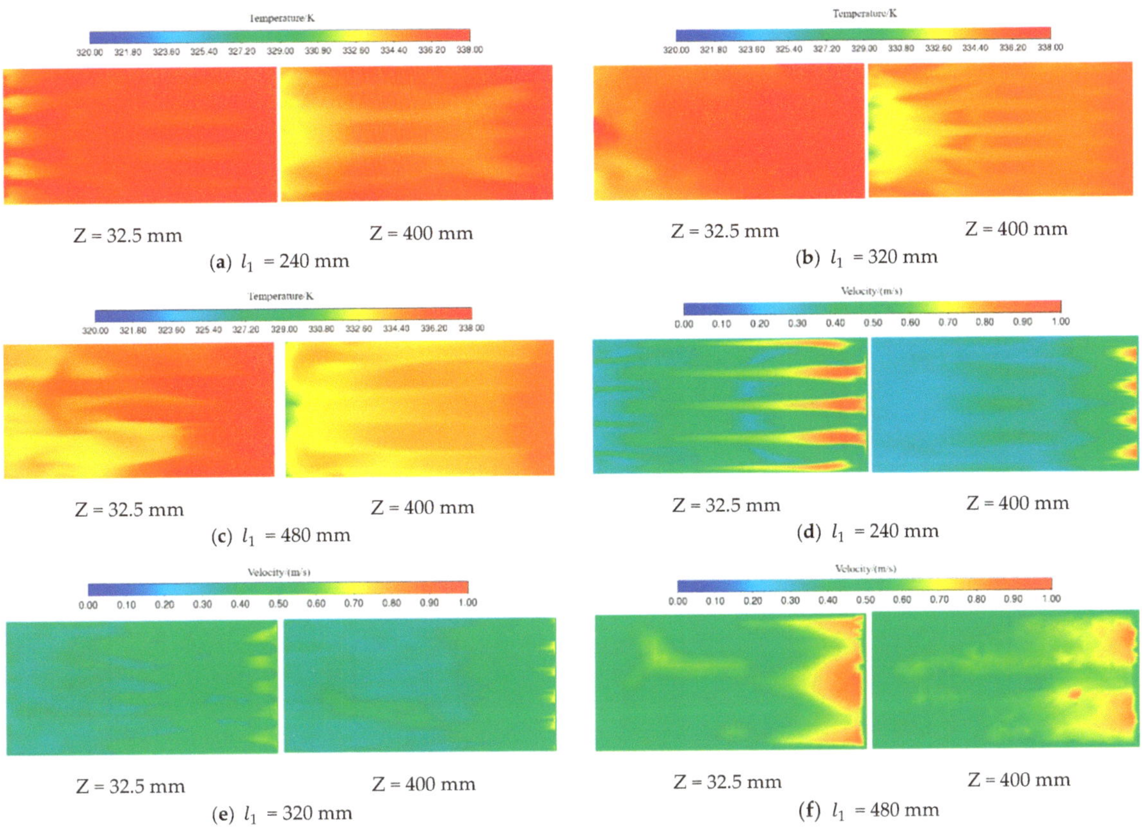

Figure 12. Temperature field and wind field distribution at different lateral distances.

In order to further analyze the effect of the horizontal spacing of the corner box on the heat condition of the particles, as shown in Figure 5, 100 tiger nut seeds are selected at cross-sections Z = 32.5 and Z = 400 mm, respectively, to determine the variation in temperature and moisture content over time, and the results are shown in Figure 13. The trends in temperature and moisture content are similar to those in reference [13,36]. From Figure 13, it can be seen that with the increase in time, the tiger nut temperature first increases rapidly and then tends to stabilize, and the water content gradually decreases. This is because, first of all, the wet particles are preheated and heated, and at the same time vaporize a small amount of water, so the warming is faster in the early stage, and the water content decreases slowly. Then, all the heat transferred from the hot airflow to the pellet is used to vaporize the water, and the surface temperature of the pellet remains basically unchanged while the water is vaporized at a certain rate. When the horizontal spacing of the cassettes is equal to 240 mm, the final average temperatures of the tiger nuts are 298.20 K and 296.51 K, with high extreme temperatures and coefficients of variation. The final average moisture content values of tiger nuts are 38.1% and 41.5%. This is because—at this time—the corner box hot air volume is larger, the corner box is directly below the higher temperature field, and the wind speed is larger, so this part of the surface of the tiger nut air flow rate is larger, and the drying rate is accelerated. When the horizontal spacing of the box is equal to 320 mm, the final average temperature values of the tiger nut are

297.5 K and 295 K; moreover, the final average moisture content values of the tiger nut are 39.7% and 41.1%. Overall, the temperature and moisture content distribution of tiger nuts in each region are basically the same. When the horizontal spacing of the boxes is equal to 480 mm, the final average temperature values of tiger nuts are 296 K and 294.21 K, and the final average moisture content values are 40.3% and 43.1%. At this time, the extreme difference and coefficient of the variation in temperature and water content are high. This is because the number of tiger nuts is large, and the distribution distance of the corner box is far away, so it is not evenly heated.

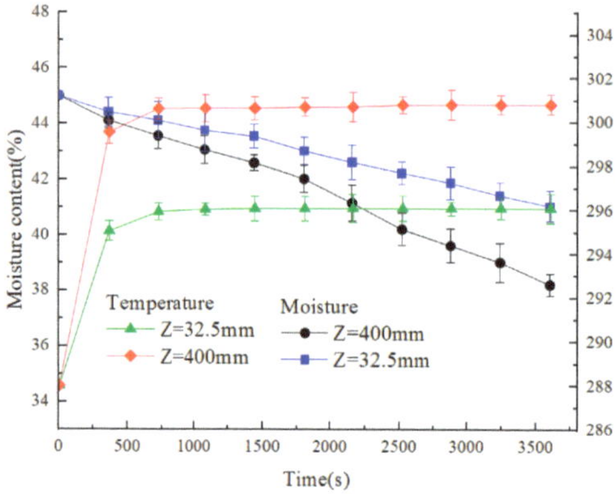

(**a**) l_1 = 240 mm

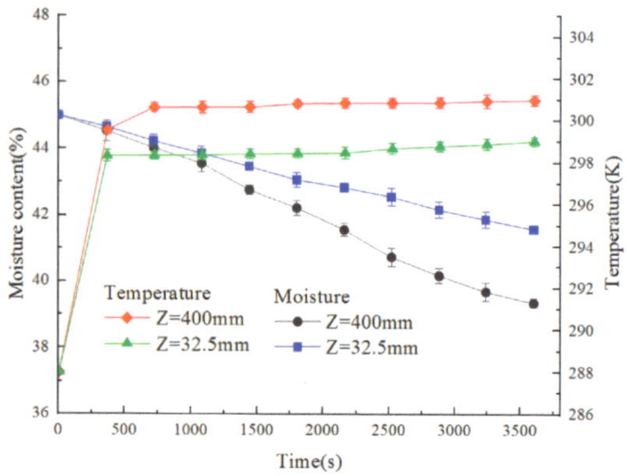

(**b**) l_1 = 320 mm

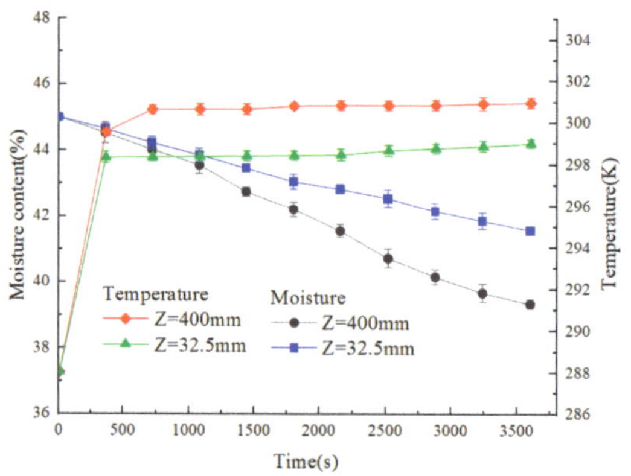

(**c**) l_1 = 480 mm

Figure 13. Changes in temperature and moisture content of tiger nut seeds at different lateral distances.

In order to investigate the effect of the lateral spacing of the corner-mounted box on the flow velocity of tiger nuts during grain discharge, the flow characteristics of tiger nuts are analyzed, as shown in Figure 14. As in Figure 14a, the flow velocities of tiger nut particles in contact with the corner box or the wall are comparatively lower than the flow velocities in other regions. The tiger nut flow velocities in the region between neighboring corner tubes are higher than those between the corner box and the inner wall, and there is a significant velocity difference between the particles. These effects were confirmed by flow experiments by K. L. Iroba and Jochen Mellmann et al. [37,38]. In order to assess the impact of different lateral distances of the corner box on the falling velocity of tiger nut particles, seed velocity monitors were set up at A, and B to monitor the vertical downward velocity of the seeds. As can be seen in Figure 14b–d, the seed falling velocity gradually accelerates with the increase in the lateral distance. When the lateral distance is equal to 240 mm, the average flow velocities at A and B are, respectively, 0.199 m/s and 0.271 m/s. The flow velocity of the tiger nuts on the side close to the inner wall of the drying chamber is significantly lower than that in the middle area of the two corner boxes. When the lateral distance is equal to 320 mm, the average flow rate values at A and B are 0.281 m/s and 0.267 m/s, respectively, and the flow rates of tiger nuts in the two areas are basically the same. When the lateral distance is equal to 480 mm, the average flow velocities at A and B are 0.314 m/s and 0.318 m/s. The difference in flow rates between the two regions is not significant, but there are large fluctuations in flow rates. By analyzing this, it can be seen that the shape of the tiger nut is irregular and the friction of the wall during the flow hinders the movement of the grains. If the residence time of the grain is less than the drying time, the grain will be under-dried. If the residence time of the grain is higher than the drying time, the grain will be over-dried. Increasing the transverse distance can, to a certain extent, prevent the thermal damage caused by localized overheating within the grains during the drying process, which can correspondingly improve the drying quality of tiger nuts. Therefore, it is best to choose a transverse distance of about 320 mm for corner boxes.

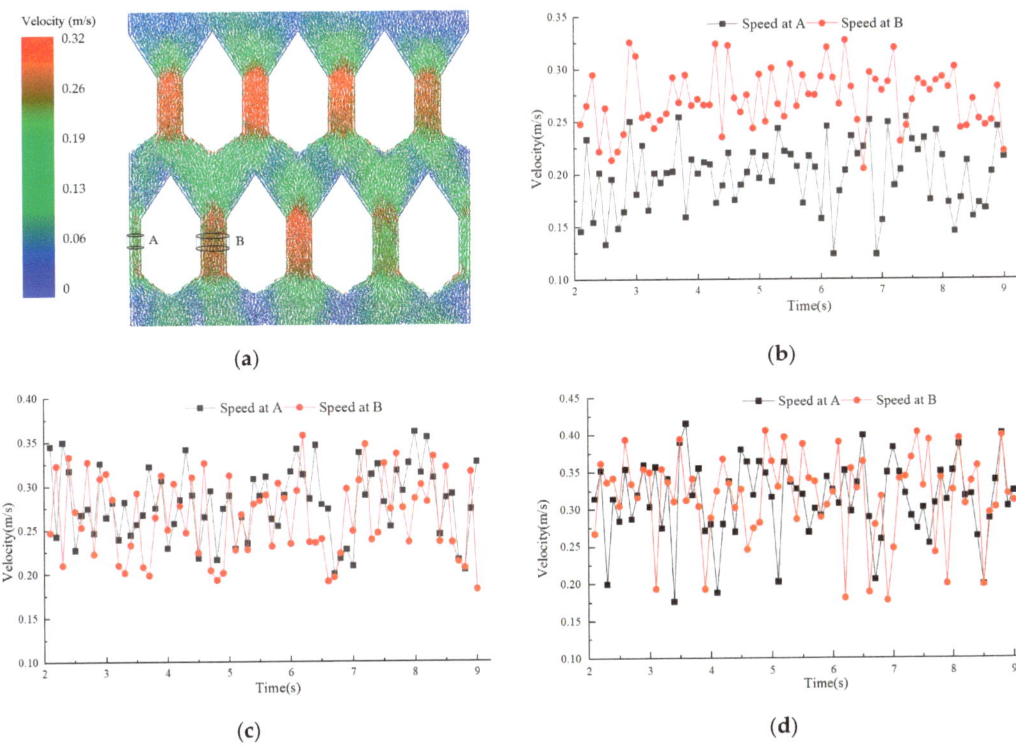

Figure 14. Grain velocity analysis. (**a**) schematic diagram of grain velocity distribution in drying chamber; (**b**) The change of seed velocity when the lateral distance is 240 mm; (**c**) The change of seed velocity when the lateral distance is 320 mm; (**d**) The change of seed velocity when the lateral distance is 480 mm.

3.3.2. Effect of Longitudinal Distance

As shown in Figure 15, the temperature field and wind field distribution under different corner box vertical spacings are shown. When the vertical spacing of the corner box is equal to 320 mm, there is a significant difference in temperature in the same plane.

In the horizontal direction, there is a laminar distribution, with the temperature on the exit side significantly higher than the import direction. If the vertical spacing of the corner box is equal to 520 mm, a drying dead zone is formed near the bottom of the air inlet, where the hot air temperature and wind speed are significantly lower than in other areas. When the vertical distance of the corner box is equal to 420 mm, the temperature difference between the two sections is small, and the hot air temperature is basically uniformly distributed. It can be seen that with the increase in the longitudinal distance, the gap between the average wind speeds of the two planes gradually becomes larger. This is because, due to the role of gravity, the closer to the bottom, the greater the static pressure between the tiger nut grain layer and the tiger nut particles. At this time, the denser accumulation of tiger nut particles results in a lower porosity of the tiger nut grain layer and a greater resistance to airflow; an increase in airflow resistance leads to a decrease in wind speed.

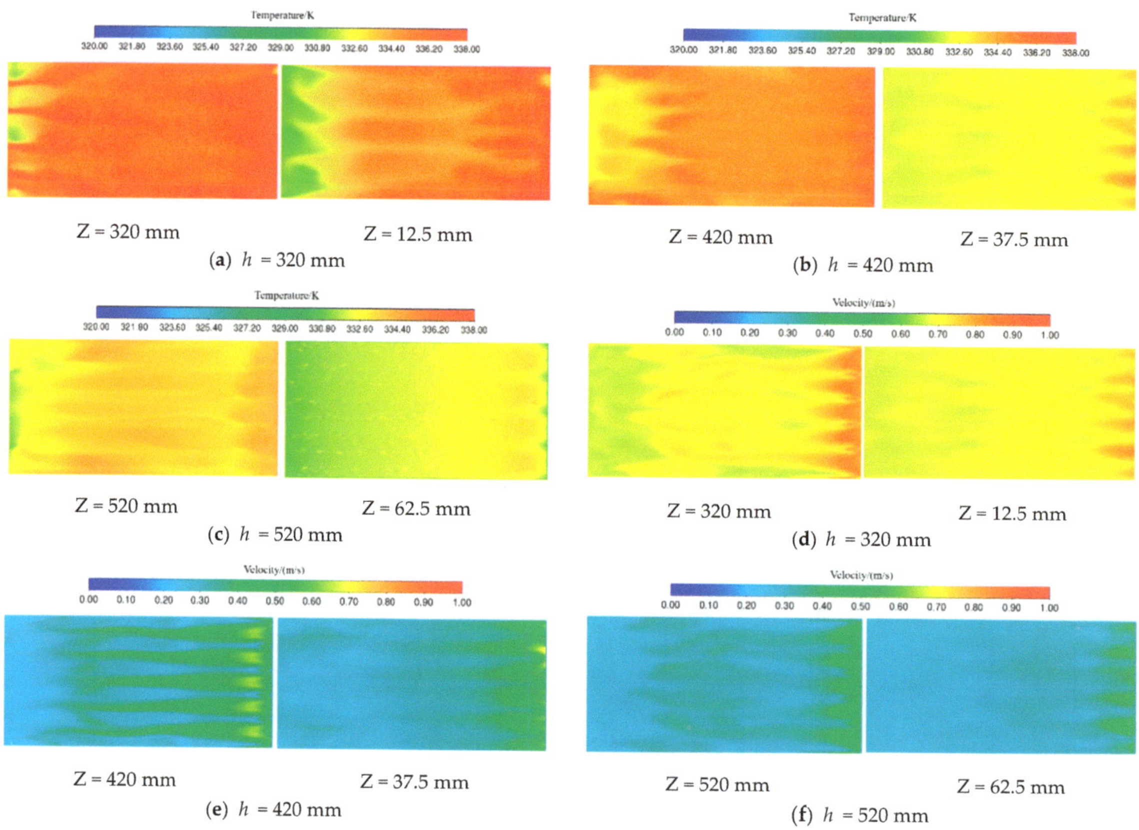

Figure 15. Temperature and wind field distribution at different longitudinal distances.

In order to further analyze the impact of the horizontal spacing of corner boxes on the heat conditions of particles, 100 tiger nut seeds are selected at cross-section Z = 320, Z = 12.5 mm; Z = 420, Z = 37.5 mm; Z = 520, Z = 62.5 mm, respectively, to find out the changes in temperature and moisture content over time; the results are shown in Figure 16. From the figure, it can be seen that with the increase in time, the temperature of the tiger nut first increases rapidly and then tends to stabilize, and the water content gradually decreases. With the increase in longitudinal distance, the gap between the temperature and moisture content of tiger nuts at two sections of the same vertical distance gradually increases. When the vertical distance between the corner boxes is equal to 320 mm, the final average temperature values of the tiger nuts are 305.14 K and 301.52 K, the extreme temperature and coefficient of variation are higher, and the average water content values are 36.4% and 38.7%. As the vertical spacing of corner boxes is equal to 420 mm, the final average temperatures of tiger nuts are 301.6 K and 296.4 K, and the average water content values are 39.6% and 41.3%. Overall, the temperature and moisture content distribution of tiger nuts in each region are basically the same. The average final temperatures of tiger nuts are 295.8 K and 292.9 K, and the average moisture content values are 39.5% and 42.3% when corner boxes are vertically spaced, equaling 520 mm. This is due to the large number of tiger nuts, and the distribution distance of the corner-mounted boxes is wide, making them unevenly heated. Therefore, it is better to choose a longitudinal spacing of corner boxes around 420 mm. This is because the increase in the thickness of the grain layer leads to an increase in airflow resistance and a decrease in the porosity between the tiger nuts. As a result, the airflow space inside the drying chamber decreases, leading to a decrease in air velocity. The water vapor evaporated from the surface of the tiger nut is taken away at a reduced rate, and the time required to reach equilibrium within the tiger nut grows, slowing down the drying rate.

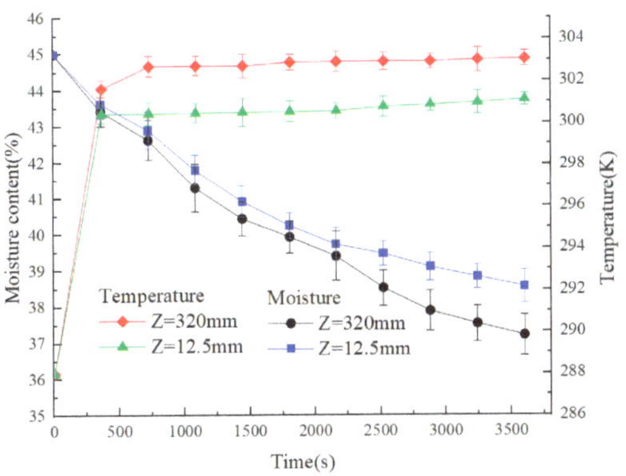

(**a**) h = 320 mm

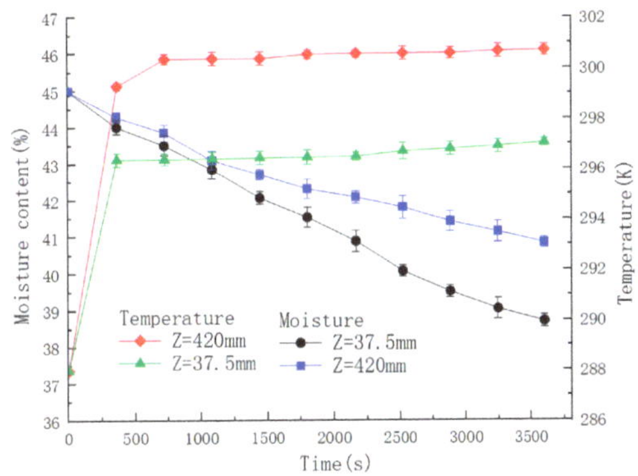

(**b**) h = 420 mm

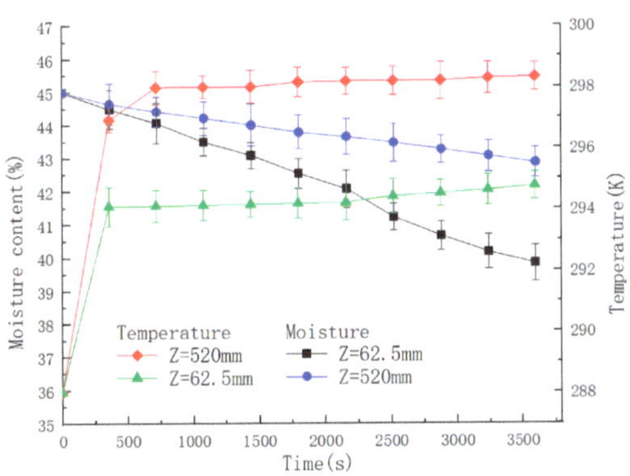

(**c**) h = 520 mm

Figure 16. Changes in the temperature and moisture content of tiger nut seeds with different longitudinal distances.

3.4. Effect of Wind Speed on Drying Effect

Wind speed is an important operating parameter and indicator used to measure energy consumption. The wind speeds selected in this paper are 2, 4, and 6 m/s. Different tiger nut particle thicknesses have different ventilation resistance values and drying rates.

In order to select the airspeed that is applicable to the designed dryer, in addition to analyzing the drying characteristics, it is also necessary to study the ventilation resistance so as to determine the optimal drying airspeed for a certain tiger nut particle thickness. According to the previous simulation test results, a horizontal spacing of 320 mm and a longitudinal spacing of 420 mm were selected for the layout of the drying chamber's corner box space structure.

Figure 17 shows the wind field distribution at different wind speeds. From the figure, it can be seen that with the inlet wind speed increase, the wind speed inside the drying chamber gradually increases, but the uniformity of the wind field first increases and then decreases. Reference [31] reaches similar conclusions in the experimental part. When the wind speed is 2 m/s, the wind field distribution map in the middle position of the flow field line distribution is messy, resulting in the formation of a large vortex area, with large differences in wind speed at different positions. When the wind speed is 4 m/s, the wind speed distribution is uniform, and the wind speed difference between each part is small. When the wind speed is 6 m/s, the wind speed in the drying room near the air outlet side is greater than the air inlet side. This is mainly due to the airflow in the corner of the box moving in a straight line without any obstacles caused by the air volume being gathered in the second half of the duct.

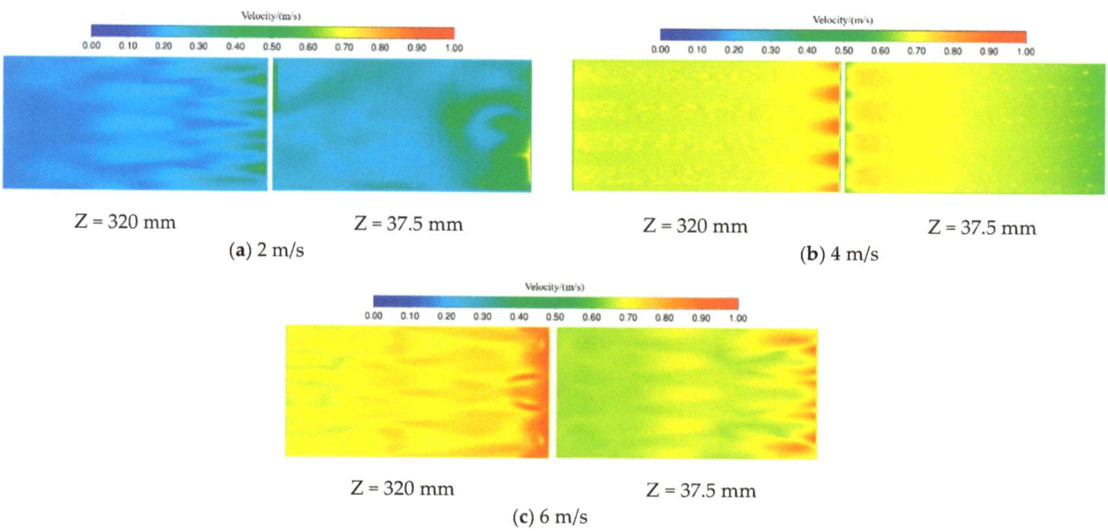

Figure 17. Wind field distribution under different wind speeds.

To study the impact of wind speed on the drying rate, 100 tiger nut seeds that were dried for 1 h were selected at cross-sections Z = 37.5 mm and Z = 320 mm, respectively, and the average values of temperature and moisture content were found, as shown in Figure 18a. From the figure, it can be seen that with the inlet wind speed increase, the final temperature and drying rate of the tiger nut gradually increase, and the moisture content gradually decreases. However, from 2 m/s to 4 m/s, when the temperature and moisture content of tiger nuts have greater changes, the drying rate obviously accelerates; 4 m/s to 6 m/s is when the temperature and moisture content of tiger nuts have little effect. When the wind speed is 2 m/s and 6 m/s, the temperature difference and moisture content difference between two sections of tiger nuts are larger, with a larger standard deviation, indicating that uniformity is poor. At 4 m/s, the uniformity is good.

To further demonstrate the impact of wind speed on the drying effect, the diffusion efficiency of water vapor at the air outlet is extracted, as shown in Figure 18b. It can be seen that the diffusion efficiency of water vapor increases gradually with the wind speed increase at the inlet. From 2 m/s to 4 m/s, the diffusion efficiency increases significantly. From 4 m/s to 6 m/s, only the diffusion efficiency in the middle two outlets increases, and there is no significant change in the two side outlets. When the wind speeds are 4 m/s and 6 m/s, there is a significant difference in the diffusion efficiency of water vapor in the four air outlets. When the wind speed is 2 m/s, the diffusion efficiency of water vapor in the four air outlets is basically the same. Therefore, increasing the wind speed can effectively improve the drying rate of tiger nut particles and reduce the drying time of tiger nut particles. This is mainly because the greater the wind speed, the more air flows over the surface of the tiger nut in a unit of time, and the thinner the flow boundary layer, reducing the moisture content in the air. Therefore, by reducing the resistance to water vapor diffusion from the tiger nut surface to the air, moisture evaporation accelerated, and the time required for drying is reduced. However, when the fan wind speed is too high, it will consume a lot of energy, resulting in waste, and when the outlet wind speed is too high, it may exceed the suspension velocity of the particles, and the particles may overflow from the position of the air outlet, leading to losses in drying. Therefore, the best wind speed is about 4 m/s.

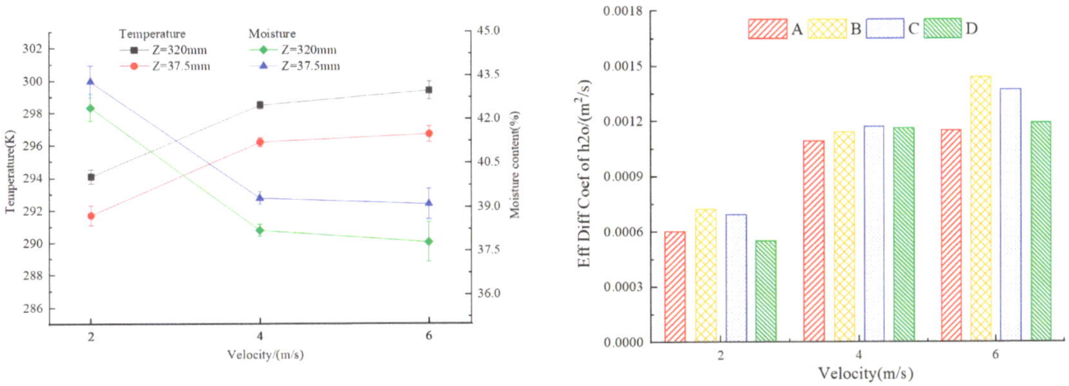

(**a**) (**b**)

Figure 18. Temperature and moisture content changes of tiger nut seeds (**a**) and the water vapor diffusion rate at the outlet under different wind speeds (**b**).

4. Conclusions

(1) Using the principle of negative pressure micro-perforated air supply, the open-hole corner box is designed. The tiger nut model is established using EDEM, and the tiger nut drying process is simulated using CFD-DEM coupled simulation to verify the accuracy of the simulation model. Analysis of variance (ANOVA) is used to analyze the temperature and wind speed data of the simulation and test; it is concluded that there is no statistically significant difference between the test and simulation results. Therefore, the tiger nut drying process can be simulated better by using this model.

(2) The wind fields of three corner boxes with different aperture ratios are compared and analyzed. The results show that the average wind speed below the air inlet increases gradually with the opening ratio increase. When the opening ratio is 0.33%, the wind field uniformity is better.

(3) The influence of corner box distribution on changes in the water content of tiger nuts is analyzed. The consistency of water content distribution increases and then decreases with increasing lateral distance, and the flow rate of tiger nuts during grain discharge gradually increases. The rate of water content decrease gradually decreases with the longitudinal distance increase. When the lateral distance is 320 mm and the longitudinal distance is 420 mm, the water content of the grain is basically consistent.

(4) The effects of different inlet wind speeds on the drying effect are analyzed. With the increase in wind speed, the temperature of the seeds warms up faster, the water content decreases faster, and the diffusion rate of water vapor increases faster. When the wind speed is 4 m/s, the temperature and water content of tiger nuts basically change in the same way, and the water content has good consistency. Therefore, this can provide a theoretical basis for the design of tiger nut drying equipment.

Author Contributions: Conceptualization, L.D. and Y.D.; methodology, L.D. and Y.D.; validation, Y.D., T.Y., J.L. and A.M.; formal analysis, Y.D., Y.Y. and J.L.; resources, L.D.; data curation, Y.D.; writing—original draft preparation, Y.D.; writing—review and editing, L.D., L.W. and H.L.; supervision, L.D. and H.L.; project administration, L.D. and H.L.; funding acquisition, L.D. and H.L. All authors have read and agreed to the published version of the manuscript.

Funding: Major Science and Technology Special Project of Henan Province: Technical Innovation and Integrated Application of Wheat and Corn Disaster Response under Abnormal Weather by Integrating Multi-source Data, National Modern Agricultural Industrial Technology System Project (CARS-04), Key Technologies R & D Program of Henan Province (222102110032), Henan Science and Technology R&D Program Joint Fund Project (232103810019).

Institutional Review Board Statement: Not applicable.

Data Availability Statement: The data used to support the results of this study are available from the corresponding authors upon request.

Acknowledgments: The authors would like to express their sincere gratitude to their colleagues and laboratories and to the reviewers who provided helpful suggestions for this manuscript.

Conflicts of Interest: The authors declare no conflicts of interest.

References

1. Ezeh, O.; Gordon, M.; Niranjan, K. Enhancing the recovery of tiger nut (*Cyperus esculentus*) oil by mechanical pressing: Moisture content, particle size, high pressure and enzymatic pre-treatment effects. *Food Chem.* **2016**, *194*, 354–361. [CrossRef]
2. Qu, Z.; Han, M.; Lv, Y.; Zhou, Z.; Lv, Z.; Wang, W.; He, X. Design and Test of a Crawler-Type Tiger-Nut Combine Harvester. *Agriculture* **2023**, *13*, 277. [CrossRef]
3. Yu, Y.; Lu, X.; Zhang, T.; Zhao, C.; Guan, S.; Pu, Y.; Gao, F. Tiger Nut (*Cyperus esculentus* L.): Nutrition, Processing, Function and Applications. *Foods* **2022**, *11*, 601. [CrossRef] [PubMed]
4. Jokiniemi, H.T.; Ahokas, J.M. Drying process optimisation in a mixed-flow batch grain dryer. *Biosyst. Eng.* **2014**, *121*, 209–220. [CrossRef]
5. Mellmann, J.; Iroba, K.; Metzger, T.; Tsotsas, E.; Mészáros, C.; Farkas, I. Moisture content and residence time distributions in mixed-flow grain dryers. *Biosyst. Eng.* **2011**, *109*, 297–307. [CrossRef]
6. Scaar, H.; Franke, G.; Weigler, F.; Delele, M.; Tsotsas, E.; Mellmann, J. Experimental and numerical study of the airflow distribution in mixed-flow grain dryers. *Dry. Technol.* **2016**, *34*, 595–607. [CrossRef]
7. Cao, C.; Yang, D.; Liu, Q. Research on modeling and simulation of mixed flow grain dryer. *Dry. Technol.* **2007**, *25*, 681–687. [CrossRef]
8. Weigler, F.; Scaar, H.; Mellmann, J. Investigation of particle and air flows in a mixed-flow dryer. *Dry. Technol.* **2012**, *30*, 1730–1741. [CrossRef]
9. Keppler, I.; Kocsis, L.; Oldal, I.; Farkas, I.; Csatar, A. Grain velocity distribution in a mixed flow dryer. *Adv. Powder Technol.* **2012**, *23*, 824–832. [CrossRef]
10. Jiang, Q.; Zhang, Y.; Yan, S.; Xu, L. Optimal design of an corner box for a mixed flow grain dryer. *Appl. Eng. Agric.* **2021**, *37*, 555–562. [CrossRef]
11. Gang, C.; Ruili, G.; Lin, W. Optimized design and experimental study on rice negative pressure mixed flow drying chamber. *Trans. CSAE* **2021**, *37*, 87–96. [CrossRef]
12. Visconcini, A.R.; Andrade, C.M.G.; de Souza Costa, A.M. Fluid flow simulation of industrial fixed bed mixed-flow grain dryer using k-ω SST turbulence model. *Int. J. Agric. Biol. Eng.* **2021**, *14*, 226–230. [CrossRef]
13. Luo, H.; Li, C.; Zhang, Y. Design and experimental study of 5HP-25 type grain dryer. *Trans. CSAE* **2021**, *37*, 279–289. [CrossRef]
14. Ma, H.; Zhou, L.; Liu, Z.; Chen, M.; Xia, X.; Zhao, Y. A review of recent development for the CFD-DEM investigations of non-spherical particles. *Powder Technol.* **2022**, *412*, 117972. [CrossRef]
15. Handayani, S.U.; Wahyudi, H.; Agustina, S.; Yulianto, M.E.; Ariyanto, H.D. CFD-DEM Study of heat and mass transfer of ellipsoidal particles in fluidized bed dryers. *Powder Technol.* **2023**, *425*, 118535. [CrossRef]

16. Wang, P.; Wang, H.; Zhang, R.; Hu, R.; Hao, B.; Huang, J. Numerical Simulation of an Online Cotton Lint Sampling Device Using Coupled CFD–DEM Analysis. *Agriculture* **2024**, *14*, 127. [CrossRef]
17. Luo, X.; Yu, J.; Wang, B.; Wang, J. Heat transfer and hydrodynamics in stirred tanks with liquid-solid flow studied by CFD–DEM method. *Processes* **2021**, *9*, 849. [CrossRef]
18. Yuan, Y.; Wang, J.; Zhang, X.; Zhao, S. Effect of Rotary Speed on Soil and Straw Throwing Process by Stubble-Crushing Blade for Strip Tillage Using DEM-CFD. *Agriculture* **2023**, *13*, 877. [CrossRef]
19. Che, H.; Wang, H.; Xu, L.; Ge, R. Investigation of gas-solid heat and mass transfer in a Wurster coater using a scaled CFD-DEM model. *Powder Technol.* **2022**, *406*, 117598. [CrossRef]
20. Xiao, Y.; Ma, Z.; Wu, M.; Luo, H. Numerical Study of Pneumatic Conveying of Rapeseed through a Pipe Bend by DEM-CFD. *Agriculture* **2022**, *12*, 1845. [CrossRef]
21. Lan, B.; Zhao, P.; Xu, J.; Zhao, B.; Zhai, M.; Wang, J. CFD-DEM-IBM simulation of particle drying processes in gas-fluidized beds. *Chem. Eng. Sci.* **2022**, *255*, 117653. [CrossRef]
22. Guo, R.; Bai, J.; Wu, F.; Wang, J.; Ma, X.; Hui, Z. CFD–DEM simulation of wet granular-fluid flows and heat transfer in an integral multi-jet spout-fluidized bed. *Powder Technol.* **2022**, *403*, 117384. [CrossRef]
23. Sousani, M.; Hobbs, A.M.; Anderson, A.; Wood, R. Accelerated heat transfer simulations using coupled DEM and CFD. *Powder Technol.* **2019**, *357*, 367–376. [CrossRef]
24. Khomwachirakul, P.; Devahastin, S.; Swasdisevi, T.; Soponronnarit, S. Simulation of flow and drying characteristics of high-moisture particles in an impinging stream dryer via CFD-DEM. *Dry. Technol.* **2016**, *34*, 403–419. [CrossRef]
25. Aziz, H.; Ahsan, S.; De Simone, G.; Gao, Y.; Chaudhuri, B. Computational modeling of drying of pharmaceutical wet granules in a fluidized bed dryer using coupled CFD-DEM approach. *Aaps Pharmscitech* **2022**, *23*, 59. [CrossRef] [PubMed]
26. Thornton, C. *Granular Dynamics, Contact Mechanics and Particle System Simulations*; A DEM Study; Particle Technology Series; Springer International Publishing: Cham, Switzerland, 2015; p. 24. [CrossRef]
27. Shi, D. Advanced Simulation of Particle Processing: The Roles of Cohesion, Mass and Heat Transfer in Gas-Solid Flows. Ph.D. Thesis, University of Pittsburgh, Pittsburgh, PA, USA, 2008.
28. Che, G.; Chen, W.; Wu, C.; Li, H.; Jin, Z.; Wan, L. Design and experiment of large-scale 5HFS-10 type automatic control negative pressure grain dryer. *Trans. CSAE* **2017**, *33*, 267–275. [CrossRef]
29. Pan, Y. *Modern Drying Technology*; Chemical Industry Press: Beijing, China, 2007.
30. Usman, D.; Adanu, E.; Jahun, B.; Ibrahim, K. Effect of moisture content variation on thermo-physical properties of brown variety Tigernut (*Cyperus esculentus*). *Arid. Zone J. Eng. Technol. Environ.* **2019**, *15*, 714–724.
31. Chen, Z.; Che, G.; Wan, L.; Wang, H.; Qu, T.; Zhang, Q. Numerical simulation and experiment of four-way ventilation mixed flow drying section for rice. *Trans. CSAE* **2022**, *38*, 237–247. [CrossRef]
32. Li, H.; Niu, X.; Chai, J.; Guo, C.; Sun, Y.; Li, J.; Li, C. Optimization of hot air drying process for tiger nut and analysis of fatty acid composition of tiger nut oil. *Int. J. Agric. Biol. Eng.* **2021**, *14*, 228–236. [CrossRef]
33. Chen, Y.; Gao, X.; Jin, X.; Ma, X.; Hu, B.; Zhang, X. Calibration and Analysis of Seeding Parameters of *Cyperus esculentus* Seeds Based on Discrete Element Simulation. *Trans. Chin. Soc. Agric. Mach.* **2023**, *54*, 58–69.
34. He, X.; Lv, Y.; Qu, Z.; Wang, W.; Zhou, Z.; He, H. Parameters Optimization and Test of Caterpillar Self-Propelled Tiger Nut Harvester Hoisting Device. *Agriculture* **2022**, *12*, 1060. [CrossRef]
35. Chen, P.; Han, Y.; Jia, F.; Zhao, D.; Meng, X.; Li, A.; Chu, Y.; Zhao, H. Investigation of the mechanism of aerodynamic separation of rice husks from brown rice following paddy hulling by coupled CFD-DEM. *Biosyst. Eng.* **2022**, *218*, 200–215. [CrossRef]
36. Lin, J.; Bao, M.; Li, H.; Yang, J. Characteristic simulation and working parameter optimization of the aggregate dryer. *Dry. Technol.* **2023**, *41*, 590–604. [CrossRef]
37. Iroba, K.; Weigler, F.; Mellmann, J.; Metzger, T.; Tsotsas, E. Residence time distribution in mixed-flow grain dryers. *Dry. Technol.* **2011**, *29*, 1252–1266. [CrossRef]
38. Mellmann, J.; Weigler, F.; Scaar, H. Research on procedural optimization and development of agricultural drying processes. *Dry. Technol.* **2019**, *37*, 569–578. [CrossRef]

agriculture

MDPI

Review

Lateral Overturning and Backward Rollover of Agricultural Tractors: A Review

Moon-Kyeong Jang [1,2], Seung-Jun Kim [1,2], Beom-Soo Shin [1,2,*] and Ju-Seok Nam [1,2,*]

1 Department of Biosystems Engineering, Kangwon National University, 1 Kangwondaehak-gil, Chuncheon 24341, Republic of Korea; moon2842@kangwon.ac.kr (M.-K.J.); gracenav4@naver.com (S.-J.K.)
2 Interdisciplinary Program in Smart Agriculture, Kangwon National University, 1 Kangwondaehak-gil, Chuncheon 24341, Republic of Korea
* Correspondence: bshin@kangwon.ac.kr (B.-S.S.); njsg1218@kangwon.ac.kr (J.-S.N.)

Abstract: Tractor accidents caused by lateral overturning and backward rollover during agricultural activities and general driving are common. In this study, various research cases were analyzed to identify the factors influencing the lateral overturning and backward rollover of tractors and to examine their static and dynamic stability. Studies on the analysis of the major causes of these incidents and evaluation of tractor safety were compiled. Test methods, including actual tests and simulations, were categorized, and the characteristics of lateral overturning and backward rollover safety of tractors in different studies were examined. Additionally, safety improvement measures were proposed by identifying and summarizing the causes of accidents involving agricultural machinery. Tractor safety was evaluated primarily by conducting actual tractor and simulation tests. These tests were classified into field tests, tests on scale models, spreadsheet programs, and 3D simulation programs. The primary causes of lateral overturning and backward rollover were unstable center of gravity, extremely high driving speed, and ground conditions. Given the considerable number of studies dedicated to evaluating tractor safety, various technologies aimed at preventing lateral overturning and backward rollover incidents are expected to be applied to tractors in the future. The production and testing of safe agricultural machinery are expected to contribute to a reduction in accident rates.

Keywords: backward rollover; lateral overturning; review; safety; tractor

Citation: Jang, M.-K.; Kim, S.-J.; Shin, B.-S.; Nam, J.-S. Lateral Overturning and Backward Rollover of Agricultural Tractors: A Review. *Agriculture* **2024**, *14*, 334. https://doi.org/10.3390/agriculture14030334

Academic Editors: Mustafa Ucgul and Chung-Liang Chang

Received: 26 January 2024
Revised: 17 February 2024
Accepted: 17 February 2024
Published: 20 February 2024

1. Introduction

The proportion of adults aged ≥65 years is increasing in rural areas of South Korea, reaching 42.3% in 2020; the female population is also increasing [1]. Central European countries, including Poland, Hungary, and the Czech Republic, face a need for countermeasures owing to the rapid increase in the elderly population in agriculture, which plays a significant role in the economy [2]; in 2010, over 50% of all agricultural workers in Europe were aged ≥55, with only 6% aged ≤35 years, indicating a trend toward population aging [3]. To address this issue, the penetration rate of agricultural machinery has been increasing [4,5]. In South Korea, the penetration rate of agricultural tractors increased by 39.65%, with the number of units rising from 65,909 in 2020 to 92,041 in 2021. The global agricultural machinery market, valued at approximately USD 132.5 billion in 2019, is expected to reach USD 160.3 billion in 2024, reflecting an average annual growth rate of over 3.9% [6]. However, owing to the widespread use of tractors, the number of accidents is increasing [7]. Consequently, the accident rate of tractors has been investigated in several countries, including South Korea, Spain, and Turkey [8–13]. Reports indicate that more than half of tractor-related deaths are caused by rollover accidents [14]. In South Korea, tractor accidents constituted 12.3% of agricultural machinery accidents in 2020, with farmer injuries caused by lateral overturning and backward rollover accidents in riding-type agricultural machinery accounting for the largest share (34.1%) [6]. Tractors are a major cause of agricultural accidents in South Korea and in many advanced countries [15]. In the United States and EU, agricultural activities are associated with approximately three times more accidents than other activities, and 80% of these accidents involve agricultural machinery, with tractors being the leading cause [16]. In Spain, an analysis of approximately 200 agricultural accidents between 2004 and 2013 revealed that 69% of machinery-related accidents were caused by tractors and that 30% of those were rollover accidents [17]. Similarly, in Turkey, an analysis of 85 deaths resulting from tractor overturning and rollover from 2000 to 2007 revealed that 53 deaths (61.6%) occurred on fields, farmland, and ridges [8]. Results indicate that the causes of lateral overturning and backward rollover of tractors include an excessive tip angle (beyond the static safety limit), unstable and rough ground, loading conditions, and extremely high driving speed [14,18–20].

In this study, previous studies on the analysis of lateral overturning and backward rollover of tractors were summarized to enhance safety and identify factors causing tractor accidents. The characteristics of each study were examined and summarized to determine the latest trends in tractor safety evaluation, thereby providing insights into reducing agricultural tractor accidents.

2. Study and Limitations of Lateral Overturning/Backward Rollover Safety Using Rollover Protective Structure (ROPS)

To reduce the number of tractor accidents, the installation of a ROPS, a mechanical structure designed to mitigate impact on the driver during accidents, has been proposed (Figure 1). Myers and Hendricks [21] analyzed the mortality rate resulting from tractor accidents in the United States and advocated for the installation of a ROPS in agricultural tractors to decrease mortality rates. The ROPS originated in Sweden and New Zealand in the 1950s. During 1959–1978, numerous countries, including Norway, Finland, New Zealand, and the United States, introduced regulations mandating the installation of a ROPS, leading to a significant reduction in tractor rollover risks. Moreover, the number of deaths decreased depending on the type and utilization method of the ROPS [22]. Several studies were conducted to enhance ROPS safety. Latorre-Biel et al. [23] developed an energy dispersion disc system capable of dispersing energy when the deformation of the ROPS begins upon impact. They found that the disc could reduce the stress applied to the ROPS and prevent its failure by rotating when deformation begins. Sun et al. [24] utilized the Bullet physics engine to evaluate the relationship between vertical/lateral linear velocity and roll angular velocity based on root mean square error during accidents on embankment slopes and uniform slopes. The estimated error was 0.7, and the collision simulation results of Bullet were consistent with those of the Organization for Economic Cooperation and Development (OECD) simulation for critical ROPS height. Chen et al. [25] demonstrated that the ROPS can be designed by performing dynamic simulations. They also introduced the lateral stiffness coefficient (LSC) as an additional safety criterion for evaluating ROPS safety, highlighting that safety decreases when the LSC is excessively high or low. They noted that driver safety cannot be guaranteed in the event of an accident, even when the ROPS meets static test standards. Ayers et al. [26] emphasized the development of a foldable ROPS for certain tractors without an installed ROPS and highlighted that the ROPS is not properly used in certain tractors. Hunter and Owen [27] mentioned that although the installation of a ROPS cannot perfectly prevent driver injury caused by tractor rollover, it is a crucial safety measure.

Therefore, research is required to evaluate the safety of tractors and decrease the accident rate. Studies on safety are mainly conducted by performing authorized tractor tests and theoretical analysis. Recently, however, safety has been analyzed by employing various methods, such as simulations and scale models (Figure 2).

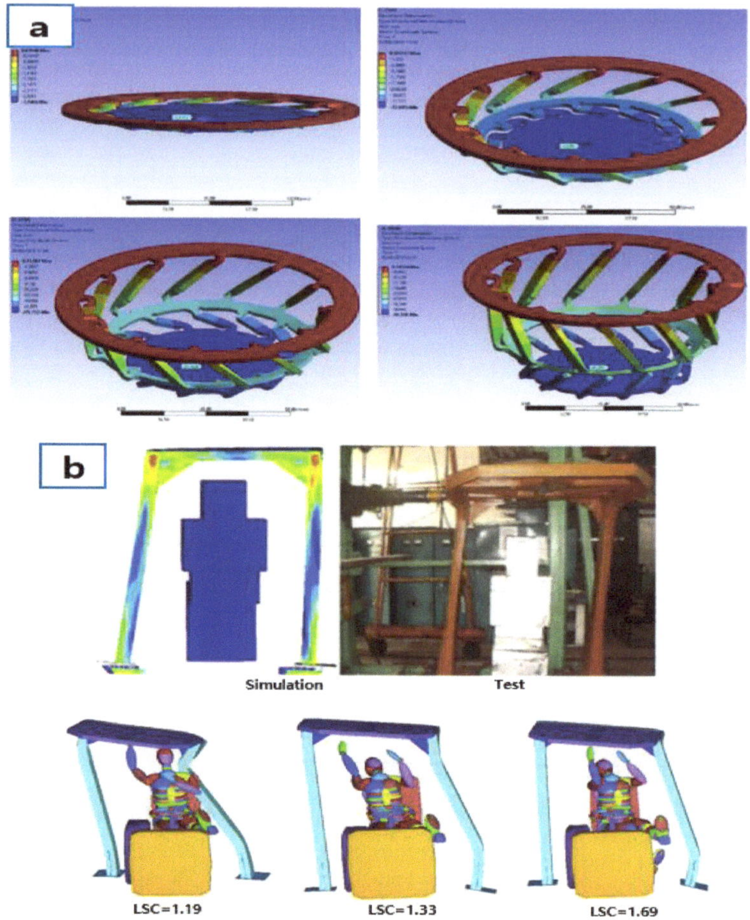

Figure 1. Various studies on ROPS safety: (**a**) sequence of deformation of an energy absorption disc as simulated by using the finite element method (Reprinted with permission from Ref. [23]. 2019, Latorre-Biel, J.I.); (**b**) ROPS deformation and motion state of the dummy when a collision occurs (Reprinted with permission from Ref. [25]. 2012. Chen, C.)

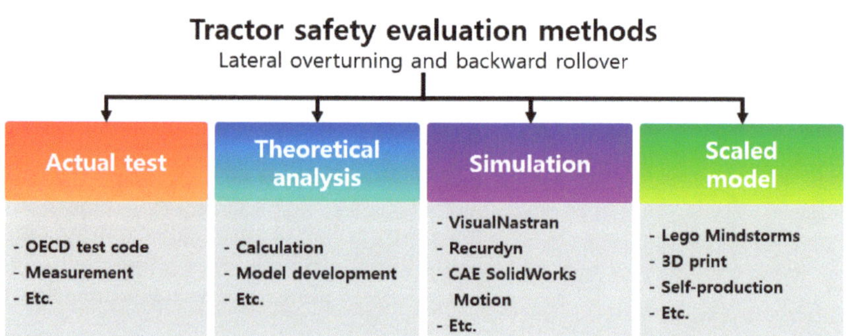

Figure 2. Tractor safety evaluation methods.

3. Mathematical Models for Lateral Overturning/Backward Rollover

Theoretical analyses have been performed to assess the lateral overturning and backward rollover of tractors and to develop new models based on established theories such as tractor engineering and chassis dynamics. For theoretical analyses, the variables evaluated by conducting OECD standardized tests are predominantly utilized.

Guzzomi [28] emphasized the need for a new model that allows for better in-depth analyses than existent primary overturn models. Consequently, a new model was devised to analyze situations where both the front and rear wheels on the upper side of a slope are lifted from the ground, leading to lateral overturning. To predict the overturning angle and tire contact force of a tractor equipped with a front axle pivot, a model was presented based on two rigid bodies—the front body consisted of the front axle and front wheels, and the rear body consisted of the remaining chassis and rear wheels. The model analysis results demonstrated that secondary overturning can be prevented by applying brakes to all tires of the tractor. Baker and Guzzomi [29] categorized the center of gravity of a tractor into two parts (front and rear) and analyzed the primary overturning process of the tractor under the influence of the mass of the front part. They noted that safety is influenced by the position of the tractor's rear center of gravity. When both centers of gravity were considered, an increase in the mass of the front part decreased safety as the rear center of gravity moved to a relatively unstable position. Previati et al. [30] presented three mathematical models for tractor safety analysis: one assuming tires are rigid bodies, another considering the vertical and horizontal stiffness of tires, and a third incorporating suspension on the front axle. They highlighted the significant effect of tire stiffness on vehicle accidents, noting that safety can be overestimated by up to 15% when this factor is disregarded. They stated that the installation of suspension on the front axle significantly improves safety by increasing the static sidelong falling angle by up to 20% and can considerably increase the safety of the tractor when transporting asymmetric implements. The safety of the tractor was notably low when one of the wheels was lifted from the ground. Li et al. [31] analyzed the behavior of the tractor based on its velocity, ground slope, and maximum static friction coefficient. They developed safety indices for lateral overturning and slip, accounting for sensitivity through bounce displacement and acceleration, as well as pitch direction angle and acceleration. Lateral overturning safety was observed to decrease as the tractor's velocity and ground slope increased, with the maximum static friction coefficient significantly impacting tractor slip. Choi et al. [32] constructed a mathematical model for analyzing the static lateral overturning safety of tractors equipped with asymmetric harvesters and compared it with simulation and actual tests. They found that the developed model exhibited smaller errors than did existent mathematical models and underscored the significant impact of the movement of the coordinates of the tractor's center of gravity on tractor driving safety. Although experiments were conducted to evaluate the safety of tractors through mathematical models, there were many studies that actually evaluated the safety of tractor manufacturing and structure. In addition, there is a disadvantage that it is not as accurate as tests using actual tractors and simulation tests. The theoretical analysis for tractor safety evaluation is summarized in Table 1.

Table 1. Theoretical analysis for evaluating the safety of tractors.

Target	Variables	Reference
Lateral overturning	Contact force of the tire	[28]
Lateral overturning	Center of gravity of front and rear bodies Potential energy of fixed-chassis tractors	[29]
Lateral overturning and backward rollover	Tire stiffness (vertical and lateral) Center of gravity Ground slope	[30]
Lateral overturning	Driving speed Ground slope Maximum static friction coefficient	[31]
Lateral overturning	Center of gravity	[32]

4. Examining Lateral Overturning/Backward Rollover Factors by Conducting Tests

4.1. Real Test

The OECD test code was developed in 1959 to evaluate the performance and safety of tractors, and agricultural tractors must pass static and dynamic safety tests based on the OECD guidelines. Tests are generally conducted to evaluate the engine output, fuel consumption, and ROPS. The center of gravity and overturning angle are the most important factors in tractor safety evaluation [33,34]. The center of gravity and overturning angle tests, which examine tractor safety under static conditions, require less time than safety evaluations under dynamic conditions, and position changes need not be considered [35]. Tests based on OECD guidelines can increase the export of agricultural tractors through the sale of verified products and enhance safety through improvements in product technology [36].

Chisholm [37] repeated the rollover test 30 times from a 2 m height using the tractor's mass, track width, cab length, and velocity, as well as the contact of the tire, as variables. Cameras and sensors were used to analyze tractor behavior during the rollover process. The analysis results revealed that the tractor's vibration and tire friction force significantly affect behavior and that the friction coefficient and camber angle of the tire located on the lower side of the slope influence tractor accidents. Fabbri and Molari [35] stated that the method of measuring the center of gravity height of a tractor under dynamic conditions is complex and inaccurate, and they proposed a method for measuring it under static conditions. They measured variables such as the mass and reaction force of the tractor and the ground slope for the tractor by using a mass system composed of four cubes; they derived the center of gravity height by applying an equilibrium equation and compared it with the result obtained under dynamic conditions. They could easily measure the variables under static conditions and stated that the center of gravity height can be derived relatively easily even though they could not maintain the error of $\leq$3 mm stipulated in OECD guidelines. Gravalos et al. [38] investigated safety tendencies for the rear track width of the tractor and additional weight applied to the rear wheel. They statically placed a tractor on a specially prepared test bench and measured the reaction force of the wheels under a load. They stated that safety increases as the rear track width of the tractor working along contour lines increases and that the use of weight on the rear wheel located on the upper side of the slope facilitates moderately safe operation of the tractor by significantly reducing the load moving to the wheel on the lower side of the slope. Bietresato and Mazzetto [39] developed a tiltable platform capable of generating various ground slopes to enable static and dynamic safety tests for agricultural machinery. They determined that the center of gravity, which changes during tests, can be identified, unlike in existent safety evaluations, and that relatively accurate safety evaluations would be possible because the centrifugal force is considered. Kang et al. [40] measured the static sidelong falling angle while increasing the track width of a three-wheel riding-type tractor to derive its static sidelong falling angle. When the left and right overturning angles of the tractor were compared, the safety for overturning to the right side was found to be low because the center of gravity (the oil pump and fuel cell) was located on the right side. They also found that safety for backward rollover decreased as the front wheel lift angle increased. Testing using a real tractor is relatively accurate, but the cost of testing and the probability of a safety accident occurring are high. Therefore, it is necessary to be careful when conducting tests using a real tractor to evaluate the safety of lateral overturning and backward rollover. Real test for tractor safety evaluation is summarized in Table 2.

Table 2. Real test for evaluating the safety of tractors.

Target	Variables	Reference
Lateral overturning	Mass of the tractor Track width Length of cab Contact of the tire Tractor's velocity	[37]
Mass center of the tractor	Mass of the tractor Reaction force of the tractor Ground slope	[35]
Lateral overturning	Tractor's rear track width Additional weight on the rear wheels	[38]
Mass center of the tractor	Center of gravity changes during the test	[39]
Lateral overturning	Tractor's rear track width	[40]

4.2. Scaled Model Test

The safety of tractors is mainly evaluated by conducting authorized tests, which are time-consuming and costly [41]. Safety evaluation through simulations is also time-consuming, costly, and not very reliable, owing to the input of inaccurate properties [42]. Tests based on scale models have been used in various applications, including airplanes, tanks, and agricultural machinery, because test conditions for the external environment can be easily implemented [43]. The model used to evaluate the safety of tractors is shown in Figure 3.

Spencer [44] analyzed conditions in which six different tractors on a slope lose safety when they are equipped with implements and used for towing. Mathematical models were developed considering the reaction force and momentum of the tractors equipped with implements, and they were verified by conducting experiments on actual tractors and scale models. The safety of four-wheel drive tractors was found to exceed that of two-wheel drive tractors, and safety was observed to significantly decrease when tractors moving along contour lines turned toward a downhill slope. Koc et al. [45] presented a method for monitoring the safety of a tractor located on a side slope using a smartphone application. They created tractor scale models using the LEGO Mindstorms kit and conducted tests where rollover was caused while driving. They used the dimensions, acceleration, and angular velocity data of tractors to predict rollover situations. They calculated a safety index using the data and developed and verified a system that can send emergency messages by calculating the index change. Li et al. [46] developed a wheel loader dynamic model with seven degrees of freedom using the Lagrange method. They constructed a 1/16-scale model to test the developed model and verified the safety for turning on flatland and slopes and rollover over obstacles by introducing the lateral transfer ratio (LTR) index. An LTR value of zero indicates that the vehicle is safe because both sides are in contact with the ground, whereas an LTR value of 1 means that it is unsafe because the left or right tire is lifted from the ground. During the turning of the wheel loader, the LTR increased when the slope of the ground increased and when the wheel loader passed over obstacles, which resulted from the centrifugal force, roll direction angle, and vertical acceleration. In addition, Li et al. [47] evaluated lateral overturning safety by deriving the roll angle, LTR, and primary overturning index. They designed an experiment using the Taguchi method for the tire type, slope and roughness of the ground, forward ballast weight, and track width. Tests were conducted by preparing 1/16-scale models. The evaluation results revealed that predicting lateral overturning using the roll angle is difficult and that the primary overturning index can more effectively predict lateral overturning compared with the LTR. Li et al. [48] constructed the scale model of a 2WD tractor and performed an experiment to analyze the effects of the tire stiffness coefficient, ballast weight, and rear axle track width on the safety of the tractor. They created the ground with roughness grades E and F and a ground slope of 10° using a 3D printer and measured the reaction force of the tractor tire using sensors. They then evaluated safety by deriving the front-axle-based safety index and rear-axle-based safety index. The evaluation results showed that safety increases as the tire stiffness coefficient, front ballast weight, and rear wheel track width increase. Jang et al. [49] derived the static sidelong falling angle by employing a tractor scale model constructed using a 3D printer and compared it with that obtained from the authorized performance test. A 1/20-scale model was constructed, and the center of gravity of the actual tractor was implemented by attaching additional structures. They prepared a small platform capable of measuring the static sidelong falling angle and measured the angle at the moment the front and rear wheels located on the upper side of the slope were separated from the ground. A comparison with the static sidelong falling angle of the actual tractor showed that deriving the angle by using the scale model with an error of 2.18% is possible. Tests using a scaled model can be performed relatively easily but have the disadvantage of making it difficult to implement actual phenomena. Additionally, it is relatively difficult to imitate the center of gravity or shape of an actual tractor. Scaled model test for tractor safety evaluation is summarized in Table 3.

Table 3. Scaled model for evaluating the safety of tractors.

Equipment	Measurement Item	Reference
LEGO Mindstorms	Roll angle of the tractor Lateral dynamic stability index	[45]
Self-production	Lateral transfer ratio	[46]
Self-production	Roll angle of the tractor Lateral-load transfer ratio Phase I overturn index	[47]
Self-production	Force-based index	[48]
3D print	Static sidelong falling angle	[49]

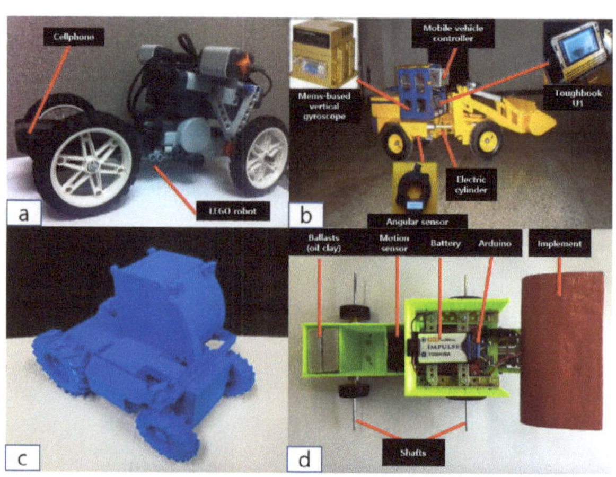

Figure 3. Methods of scaled models: (**a**) LEGO Mindstorms (Reprinted with permission from Ref. [45]. 2012, Koc, A.B.); (**b**) self-production (Reprinted with permission from Ref. [46]. 2013, Li, X.); (**c**) 3D print [49]; (**d**) self-production (Reprinted with permission from Ref. [47]. 2016, Li, Z.).

5. Examining Lateral Overturning/Backward Rollover Factors by Conducting Tests

Because safety tests using actual tractors are not only time-consuming and costly but also pose the risk of accidents, simulations have attracted attention as an alternative [50]. Simulation technology for tractor safety evaluation was introduced in the 1960s, and it accelerated research on lateral overturning and backward rollover safety for agricultural tractors [51]. Simulations facilitated studies on implementing various situations, unlike actual tractor experiments, and research has been actively conducted to derive and analyze various factors and identify indicators. In the case of simulation, tests can be performed more quickly and safely compared with tests using actual tractors. Compared with a scaled model test, it has the advantage of being easier to implement for external factors. However, care must be taken when simulating the model because incorrect results may be derived due to incorrect input of properties.

5.1. 2D Simulation

Ahmadi [50] analyzed the effects of a tractor's velocity, ground slope, and friction coefficient between the wheel and ground on the lateral overturning safety of the tractor when position disturbance of the tractor occurs. In addition, equations for lateral overturning and slip were established for the Mitsubishi tractor, and indicators that determine the stability index were developed. Safety decreased as the tractor's velocity and ground slope increased, and it was more affected by the slip of the tractor than by lateral overturning. Therefore, the slip phenomenon needs to be considered for the Mitsubishi tractor. In addition, Ahmadi [52] developed a mathematical model for identifying the longitudinal safety of a tractor equipped with a plow when it performs agricultural work on a slope. To lift the plow for plowing and transport, the effects of the force and torque transmitted to the tractor from the plow on the longitudinal safety index of the tractor were determined. No significant difference in longitudinal safety was observed when the plow penetrated soil; however, longitudinal safety decreased significantly, causing the risk of backward rollover when the plow was lifted for transport. Demšar et al. [53] developed a mathematical model for analyzing lateral overturning safety based on the center of gravity, track width, wheelbase, and front axle attachment height of a tractor located on a slope. They found that lateral overturning and backward rollover safety increased as the center of gravity height decreased. They also found that backward rollover safety improved as the wheelbase increased and that lateral overturning safety increased as the track width increased. They, however, stated that the turning radius increased and steering became relatively difficult if the track width increased. Li et al. [54] developed a mathematical model for analyzing lateral overturning when a tractor passes over an obstacle on a slope. The moment the front and rear wheels of the tractor pass over an obstacle was divided into four sections: (1) from the moment the front tire comes into contact with the obstacle to the moment it passes over the obstacle, (2) from the moment the front wheel passes the obstacle to the moment the rear wheel comes into contact with the obstacle, (3) from the moment the rear wheel comes into contact with the obstacle to the moment it passes over the obstacle, and (4) the moment after the rear wheel has passed over the obstacle. The entire process was analyzed by using a mathematical model. The center of gravity moved to an unstable position when the rear wheel passed over the obstacle compared with when the front wheel passed over it, and the direction of the tractor significantly changed when the front wheel passed over the obstacle. Additionally, an increase in ground slope had a considerable impact on the direction of the tractor. Shim et al. [55] developed an integrated implement suitable for the Korean agricultural environment and analyzed its static and dynamic safety when it was attached to a tractor. Its static and dynamic backward rollover safeties were analyzed by deriving the front wheel reaction force when the implement was lifted at static conditions and in a towing operation situation for tractors under 48 kW and those under 92 kW. They reported that backward rollover can be caused by the attachment of the implement, owing to the load movement, and suggested that installing a bucket loader that serves as a weight at the front of a tractor can improve safety. Arote et al. [56] evaluated the lateral overturning safety of a tractor equipped with a bucket loader when the tractor moves on a slope along contour lines based on the lift height of the loader. They introduced a tractor safety index (TSI) and categorized grades as follows: very poor for a TSI of less than 0, poor for a TSI between 0 and 2, good for a TSI between 2 and 4, and excellent for a TSI of more than 4 They found that lateral overturning safety increases as the loader is closer to the ground and the track width of the tractor increases. 2D simulation for tractor safety evaluation is summarized in Table 4.

Table 4. 2D simulation study for evaluating the safety of tractors.

Toolkit	Target	Variables	Reference
Microsoft Excel	Lateral overturning	Driving speed Ground slope Friction coefficient of wheel and ground	[50]
Microsoft Excel	Backward rollover	Plow depth Center of gravity	[52]
Self-production	Lateral overturning and backward rollover	Center of gravity Track width Wheelbase	[53]
Microsoft Excel	Lateral overturning	Driving speed Ground slope	[54]
Microsoft Excel	Backward rollover	Center of gravity	[55]
Microsoft Excel	Lateral overturning	Center of gravity	[56]

5.2. 3D Simulation

Park et al. (2002) [57] evaluated the lateral overturning safety of tractors moving on a slope along contour lines based on the ground slope, obstacle height, and timber load conditions by using VisualNastran, a dynamic analysis software program. Timber loading decreased safety, and the risk of lateral overturning increased as the obstacle height and ground slope increased. Huang et al. [58] used dynamics simulation software (Recurdyn) and analyzed the safety of a tractor by performing tire impact analysis when the tractor passed over a cylindrical obstacle and moved uphill (20° and 44°) and downhill (20° and 44°). They found that the maximum impact of the front wheel increased as the uphill slope increased and decreased as the downhill slope increased. When the tractor moved over the cylindrical obstacle, the impact of the front wheel decreased as the driving speed of the tractor decreased. Chowdhury et al. [59] investigated the lateral overturning safety of a tractor equipped with a radish collector. They derived the center of gravity of the tractor by performing mathematical calculations and obtained the overturning angle based on the load conditions and folded position of the radish collector. The simulation results showed that the average left and right overturning angles decreased by approximately 15° when a load was applied on the radish collector compared with when no load was applied, and no significant difference in the overturning angle was observed depending on the folded position of the radish collector. Hwang et al. [60] determined the critical driving speeds of a tractor that cause lateral overturning and backward rollover based on the internal steering angle and front wheel lift angle and compared them with the results obtained through theoretical equations for lateral overturning and backward rollover. The error between the simulation results and the theoretical equations was found to be less than 5%, and the critical driving speeds of the tractor for lateral overturning and backward rollover decreased as the internal steering angle and front wheel lift angle increased. Jang et al. [61] established the lateral overturning and backward rollover tendencies of tractors based on ground slope and obstacle geometry and height. The simulation results showed that the critical speed decreased as the ground slope and obstacle height increased, thereby lowering safety against lateral overturning and backward rollover. Backward rollover occurred at a low-ground slope, and lateral overturning occurred at a high-ground slope. Jang et al. [62] determined relative safety against lateral overturning and backward rollover when implements were attached to the same tractor and analyzed safety based on the change in the center of gravity depending on the attachment of implements. The analysis results showed that the types of lateral overturning and backward rollover changed based on a certain ground slope and that safety decreased as the ground slope and obstacle height increased. Additionally, the safety of the tractor with implements was lower than that of the tractor with no implements. Lysych [63] analyzed the static sidelong falling angle of a tractor equipped with front and rear implements by employing multi-body dynamics simulations. When the rear implement was attached, the static sidelong falling angle decreased as the implement weight increased. When both the front and rear implements were attached, the static sidelong falling angle increased. In a follow-up study, when a tractor equipped with both front and rear implements moved on the ground with single linear, single sequential, group linear, and group sequential obstacles, the lift height of the wheel center and the center of gravity displacement and linear velocity of the tractor were derived. The single linear obstacle did not affect the safety of the tractor. Safety can be completely compromised and an accident may occur owing to vibrations when the tractor passes over group sequential obstacles [41]. 3D simulation for tractor safety evaluation is summarized in Table 5.

Table 5. 3D Simulation study for evaluating the safety of tractors.

Toolkit	Measurement Items	Variables	Reference
VisualNastran	Critical driving speed	Ground slope Height of the obstacle Load conditions Driving speed	[57]
Recurdyn	Impact of tire	Ground slope Presence of obstacle Driving speed	[58]
Recurdyn	Static sidelong falling angle	Load conditions Folding conditions of collector's conveyor Overturning side	[59]
Recurdyn	Critical driving speed Critical angular velocity	Inner steering angle Floating angle of front wheel	[60]
Recurdyn	Critical driving speed Rotational angle of the center of gravity Reaction force of tire Angular velocity of the tractor Vertical displacement of the tractor	Ground slope Height of the obstacle Shape of the obstacle	[61]
Recurdyn	Critical driving speed Rotational angle of the center of gravity Reaction force of tire Angular velocity of the tractor Vertical displacement of the tractor	Ground slope Height of the obstacle Shape of the obstacle Attachment of the implement	[62]
CAE SolidWorks Motion	Contact force of the tire	Ground slope Type of the implement	[63]
CAE SolidWorks Motion	Lift height of the tires Displacement of the center of gravity Linear speed of the tractor	Type of the obstacle	[41]

6. Discussions on the Safety of Lateral Overturning/Backward Rollover

Studies have presented various methods for reducing the number of accidents caused by the lateral overturning and backward rollover of tractors. Lowering the center of gravity for the tractor chassis and preventing the center of gravity from being excessively biased while driving is crucial. In addition, applying brake and suspension systems to all tires improves safety. An increase in tractor track width can prevent lateral overturning and backward rollover, but an excessive increase is unfavorable for driving and steering. When an implement is attached to a tractor, the excessive lift of the implement may cause backward rollover. Excessive steering for turning may cause lateral overturning, and the substantial lift of the front wheel from the ground may cause backward rollover. When a tractor moves on the ground with a high slope, turning and a reduction in velocity are required. The tractor's velocity needs to be lowered as ground roughness increases. Safety decreases as the obstacle height on the ground increases, and the tractor's lateral overturning and backward rollover risk increases when obstacles are sequentially located. Solutions that can improve the safety of tractors are shown in Table 6.

Table 6. Measures to reduce tractor accidents.

Target	Solutions	References
Tractor	Four-wheel drive tractors are safer than two-wheel drive tractors The tractor's center of gravity must be lowered Increasing the track width is better Attaching brakes to all wheels of the tractor is better Front axle suspension increases safety A bucket loader should be attached to the front when lifting an implement attached to the rear The front mass of the tractor should not be excessively increased	[28–30,32,44,50,53,55,56]
Operation conditions	When attaching the plow, depth should not be excessive The implement should be lifted cautiously As the inner steering angle increases, the safety of the tractor decreases As the front wheel lift angle increases, the safety of the tractor decreases Using weight on the rear wheel located at the top of a slope increases safety	[46,52,60]
Ground conditions	When the ground slope increases, the driving speed must be lowered The higher the obstacle, the lower the safety Maximum coefficient of static friction is the main cause of sideslip The greater the roughness of the ground, the lower the safety The presence of continuous obstacles reduces safety	[31,37,46,50,54,57,58,61–63]

Recently, the number of studies for developing new models through theoretical analyses and comparing them with simulation results has been increasing. This trend may stem from the considerable time and cost involved in evaluating driving safety by using actual tractors or conducting dynamic experiments, coupled with the high risk of accidents owing to unexpected outcomes. Theoretical analyses have been geared toward creating new models based on established theories such as tractor engineering and chassis dynamics to enhance existing theories. Typically, theoretical findings are validated by performing tests and simulations on actual models and scale models. Simulation studies have been focused on the analysis of the impact of the tractor's center of gravity, driving speed, and ground conditions on safety by incorporating the machine's actual specifications. Experiments using scale models offer relative ease of preparation and execution compared with other methods; these involve various sensors, such as reaction force, angular velocity, and proximity sensors. In most studies, verifications through two or more experiments have been integrated, although some were based on a single experiment. Factors such as an unstable center of gravity, tractor geometry and velocity, and ground slope and roughness contribute to the lateral overturning and backward rollover of tractors. A comprehensive consideration of these factors is deemed necessary.

Although various causes of tractor accidents have been identified through studies, it is still difficult to perfectly embody actual situations. Therefore, there is a need to develop a tractor accident model that can more accurately embody phenomena that occur in actual fields. In fact, several studies were believed to be able to derive more diverse figures, but they were not achieved. It is believed that new results can be derived by combining multiple studies. It is essential to develop equipment and models that can effectively experiment with various studies, and it is also necessary to develop a platform that can comprehensively derive various results. Additionally, the information derived from research is not reflected in or applied to actual tractors. Therefore, an active attitude is needed to apply the derived contents to actual tractors.

7. Conclusions

In this study, we analyzed research trends on the lateral overturning and backward rollover of tractors, which frequently cause accidents in agricultural fields, and analyzed the main causes of accidents. The evaluation of tractor safety involves diverse approaches, including tests on actual tractors, safety assessment through theoretical analyses, simulations, and experiments using scale models. Tests on actual tractors encompass static conditions involving measurements of the center of gravity and tests on static sidelong falling angles. Factors that impede the safety of a tractor include external factors (ground slope, obstacle, load conditions, etc.), tractor operating conditions (driving speed, plow depth, turning radius, etc.), and the tractor's center of gravity (wheelbase, track width, attachment of implement). etc.). Therefore, in order to increase the safety of tractors, it is necessary to consider safety when manufacturing tractors, and it is necessary to set operating conditions by considering external factors. There is a need to apply findings to actual tractors based on various research results in the future. The insights from this study are expected to elucidate the latest research trends in tractor safety evaluation and provide valuable information to agricultural workers.

Author Contributions: Conceptualization, M.-K.J.; investigation, M.-K.J. and S.-J.K.; data curation, M.-K.J. and S.-J.K.; writing, original draft preparation, M.-K.J.; writing, review, and editing, B.-S.S. and J.-S.N.; visualization, M.-K.J.; supervision, B.-S.S. and J.-S.N.; project administration, B.-S.S. and J.-S.N.; funding acquisition, J.-S.N. All authors have read and agreed to the published version of the manuscript.

Funding: This work was supported by the Korea Institute of Planning and Evaluation for Technology in Food, Agriculture and Forestry (IPET) through the Machinery Mechanization Technology Development Program for Field Farming Program, funded by the Ministry of Agriculture, Food and Rural Affairs (MAFRA) (RS-2023-00235957). Also, this study was carried out with the support of 'R&D Program for Forest Science Technology (2023475A00-2325-BB01)' provided by Korea Forest Service (Korea Forestry Promotion Institute).

Institutional Review Board Statement: Not applicable.

Data Availability Statement: The data are not publicly available due to privacy.

Conflicts of Interest: The authors declare no conflicts of interest.

References

1. Kim, J.T.; Han, H.W.; Oh, J.S.; Chung, W.J.; Cho, S.J.; Park, Y.J. Structural Design of Garlic Plants Footplate Considering Physical Characteristics of Elderly Women. *J. Biosyst. Eng.* **2020**, *45*, 16–23. [CrossRef]
2. Kocsis, J.; Klara, M. A General Overview of Agriculture and Profitability in Agricultural Enterprises in Central Europe. In *Managing Agricultural Enterprises*; Palgrave Macmillan: Cham, Switzerland, 2018; pp. 243–265. [CrossRef]
3. *European Commission Agriculture in the EU Statistical and Economic Information Report 2010*; Agriculture in the European Union: Maastricht, The Netherlands, 2011; ISBN 9789279193026.
4. Kim, B.; Lee, J.; Kim, S. Improvement of the Distribution System for Used Agricultural Machinery. *J. Biosyst. Eng.* **2020**, *45*, 318–324. [CrossRef]
5. Kim, S.J.; Gim, D.H.; Jang, M.K.; Hwang, S.J.; Kim, J.H.; Yang, Y.J.; Nam, J.S. Development of Regression Model for Predicting the Maximum Static Friction Force of Tractors with a Front-End Loader. *J. Biosyst. Eng.* **2023**, *48*, 329–338. [CrossRef]
6. *KAMICO Agricultural Machinery Yearbook in Republic of Korea*; KSAM: Jeonju, Republic of Korea, 2022; pp. 1–16.
7. Kwak, H.Y.; Son, B.C. A Review of Studies on Injury and Safety of the Agricultural Machine. *J. Rehabil. Welf. Eng. Assist. Technol.* **2017**, *11*, 223–229.
8. Dogan, K.H.; Demirci, S.; Sunam, G.S.; Deniz, I.; Gunaydin, G. Evaluation of Farm Tractor-Related Fatalities. *Am. J. Forensic Med. Pathol.* **2010**, *31*, 64–68. [CrossRef]
9. Karbeyaz, K.; Şimşek, Ü.; Yilmaz, A. Deaths Related to Tractor Accidents in Eskişehir, Turkey: A 25-Year Analysis. *J. Forensic Sci.* **2019**, *64*, 1731–1734. [CrossRef]
10. Suutarinen, J. Tractor Accidents and Their Prevention. *Int. J. Ind. Ergon.* **1992**, *10*, 321–329. [CrossRef]
11. Jones, C.B.; Day, L.; Staines, C. Trends in Tractor Related Fatalities among Adults Working on Farms in Victoria, Australia, 1985-2010. *Accid. Anal. Prev.* **2013**, *50*, 110–114. [CrossRef] [PubMed]
12. Facchinetti, D.; Santoro, S.; Galli, L.E.; Pessina, D. Agricultural Tractor Roll-over Related Fatalities in Italy: Results from a 12 Years Analysis. *Sustainability* **2021**, *13*, 4536. [CrossRef]
13. Etherton, J.R.; Myers, J.R.; Jensen, R.C.; Russell, J.C.; Braddee, R.W. Agricultural Machine-Related Deaths. *Am. J. Public Health* **1991**, *81*, 766–768. [CrossRef] [PubMed]
14. Abubakar, M.S.; Ahmad, D.; Akande, F.B. A Review of Farm Tractor Overturning Accidents and Safety. *Pertanika J. Sci. Technol.* **2010**, *18*, 377–385.

15. Rondelli, V.; Casazza, C.; Martelli, R. Tractor Rollover Fatalities, Analyzing Accident Scenario. *J. Saf. Res.* **2018**, *67*, 99–106 [CrossRef] [PubMed]
16. Fargnoli, M.; Lombardi, M.; Haber, N.; Puri, D. The Impact of Human Error in the Use of Agricultural Tractors: A Case Study Research in Vineyard Cultivation in Italy. *Agriculture* **2018**, *8*, 82. [CrossRef]
17. Jarén, C.; Ibarrola, A.; Mangado, T.; Adin, A.; Arnal, P.; López-Maestresalas, A.; Ríos, A.; Arazuri, S. Fatal Tractor Accidents in the Agricultural Sector in Spain during the Past Decade. *Agronomy* **2022**, *12*, 1694. [CrossRef]
18. Ali, M.; Lee, Y.S.; Chowdhury, M.; Khan, N.A.; Swe, K.M.; Rasool, K.; Kabir, M.S.N.; Lee, D.H.; Chung, S.O. Analysis of Driving Stability and Vibration of a 20-KW Self-Propelled 1-Row Chinese Cabbage Harvester. *J. Biosyst. Eng.* **2021**, *46*, 48–59. [CrossRef]
19. Vigoroso, L.; Caffaro, F.; Cavallo, E. Warning against Critical Slopes in Agriculture: Comprehension of Targeted Safety Signs in a Group of Machinery Operators in Italy. *Int. J. Environ. Res. Public Health* **2019**, *16*, 611. [CrossRef]
20. Hunter, A.G.M.; Owen, G.M. Tractor Overturning Accidents on Slopes. *J. Occup. Accid.* **1983**, *5*, 195–210. [CrossRef]
21. Myers, J.R.; Hendricks, K.J. Agricultural Tractor Overturn Deaths: Assessment of Trends and Risk Factors. *Am. J. Ind. Med.* **2010**, *53*, 662–672. [CrossRef]
22. Springfeldt, B. Rollover of Tractors—International Experiences. *Saf. Sci.* **1996**, *24*, 95–110. [CrossRef]
23. Latorre-Biel, J.I.; Ballesteros, T.; Arana, I.; Alfaro, J.R. Development of an Inexpensive Rollover Energy Dissipation Device to Improve Safety Provided by ROPS. *Biosyst. Eng.* **2019**, *185*, 88–102. [CrossRef]
24. Sun, C.; Nakashima, H.; Shimizu, H.; Miyasaka, J.; Ohdoi, K. Physics Engine Application to Overturning Dynamics Analysis on Banks and Uniform Slopes for an Agricultural Tractor with a Rollover Protective Structure. *Biosyst. Eng.* **2019**, *185*, 150–160. [CrossRef]
25. Chen, C.; Wang, G.; Zhang, Y.; Zhang, Y.; Si, J. Effect of Lateral Stiffness Coefficient of Loader ROPS on Human Injury in a Lateral Rollover Incident. *Biosyst. Eng.* **2012**, *113*, 207–219. [CrossRef]
26. Ayers, P.; Khorsandi, F.; Wang, X.; Araujo, G. ROPS Designs to Protect Operators during Agricultural Tractor Rollovers. *J. Terramechanics* **2018**, *75*, 49–55. [CrossRef]
27. Owen, G.M.; Hunter, A.G.M. A Survey of Tractor Overturning Accidents in the United Kingdom. *J. Occup. Accid.* **1983**, *5*, 185–193. [CrossRef]
28. Guzzomi, A.L. A Revised Kineto-Static Model for Phase I Tractor Rollover. *Biosyst. Eng.* **2012**, *113*, 65–75. [CrossRef]
29. Baker, V.; Guzzomi, A.L. A Model and Comparison of 4-Wheel-Drive Fixed-Chassis Tractor Rollover during Phase I. *Biosyst. Eng.* **2013**, *116*, 179–189. [CrossRef]
30. Previati, G.; Gobbi, M.; Mastinu, G. Mathematical Models for Farm Tractor Rollover Prediction. *Int. J. Veh. Des.* **2014**, *64*, 280–303. [CrossRef]
31. Li, Z.; Mitsuoka, M.; Inoue, E.; Okayasu, T.; Hirai, Y. Development of Stability Indicators for Dynamic Phase I Overturn of Conventional Farm Tractors with Front Axle Pivot. *Biosyst. Eng.* **2015**, *134*, 55–67. [CrossRef]
32. Choi, K.H.; Kim, S.M.; Hong, S. Analysis of Static Stability by Modified Mathematical Model for Asymmetric Tractor-Harvester System: Changes in Lateral Overturning Angle by Movement of Center of Gravity Coordinates. *Biosyst. Eng.* **2017**, *42*, 127–135.
33. Liberatori, S. *Standardisation, Testing and Certification Together for a New Combined Success*; Club of Bologna: Rome, Italy, 2013; pp. 1–7.
34. Bietresato, M.; Carabin, G.; Vidoni, R.; Gasparetto, A.; Mazzetto, F. Evaluation of the Stability of an Articulated Farm Tractor Using Mounted Implements on Hillsides. In Proceedings of the International Conference Rural Health & Ragusa SHWA, Lodi, Italy, 8–11 September 2015; pp. 267–275.
35. Fabbri, A.; Molari, G. Static Measurement of the Centre of Gravity Height on Narrow-Track Agricultural Tractors. *Biosyst. Eng.* **2004**, *87*, 299–304. [CrossRef]
36. Solovyev, R.; Cheranev, S.; Gerasimov, M.; Kolomeichenko, A.; Zotov, P.; Sovin, K. Need for Supporting Laboratory for Testing Agricultural and Forestry Tractors. *IOP Conf. Ser. Earth Environ. Sci.* **2022**, *954*, 012076. [CrossRef]
37. Chisholm, C.J. Experimental Validation of a Tractor Overturning Simulation. *J. Agric. Eng. Res.* **1979**, *24*, 395–415. [CrossRef]
38. Gravalos, I.; Gialamas, T.; Loutridis, S.; Moshou, D.; Kateris, D.; Xyradakis, P.; Tsiropoulos, Z. An Experimental Study on the Impact of the Rear Track Width on the Stability of Agricultural Tractors Using a Test Bench. *J. Terramechanics* **2011**, *48*, 319–323. [CrossRef]
39. Bietresato, M.; Mazzetto, F. Increasing the Safety of Agricultural Machinery Operating on Sloping Grounds by Performing Static and Dynamic Tests of Stability on a New-Concept Facility. *Int. J. Saf. Secur. Eng.* **2018**, *8*, 77–89. [CrossRef]
40. Kang, N.R.; Choi, I.S.; Lee, W.J.; Woo, J.K.; Kim, Y.K.; Choi, Y.; Hyun, C.S.; Yoo, S.N. Sideways Overturning and Overturning Angle Test for a Three-Wheel Riding-Type Cultivator. *J. Biosyst. Eng.* **2019**, *44*, 12–17. [CrossRef]
41. Lysych, M.N. Study Driving Dynamics of the Machine-Tractor Unit on a Virtual Stand with Obstacles. *J. Phys. Conf. Ser.* **2020**, *1515*, 042079. [CrossRef]
42. Carson, J.S. Introduction to Modeling and Simulation. In Proceedings of the Winter Simulation Conference, Orlando, FL, USA, 4 December 2005; pp. 16–23. [CrossRef]
43. Calle, M.A.G.; Salmi, M.; Mazzariol, L.M.; Kujala, P. Miniature Reproduction of Raking Tests on Marine Structure: Similarity Technique and Experiment. *Eng. Struct.* **2020**, *212*, 110527. [CrossRef]
44. Spencer, H.B. Stability and Control of Two-Wheel Drive Tractors and Machinery on Sloping Ground. *J. Agric. Eng. Res.* **1978**, *23*, 169–188. [CrossRef]

45. Koc, A.B.; Liu, B.; Langley, G.C. Teaching Tractor Roll-over Stability Using Lego Mindstorms. In Proceedings of the 2012 ASABE Annual International Meeting, Dallas, TX, USA, 29 July–1 August 2012; pp. 2220–2229. [CrossRef]
46. Li, X.; Wang, G.; Yao, Z.; Qu, J. Dynamic Model and Validation of an Articulated Steering Wheel Loader on Slopes and over Obstacles. *Veh. Syst. Dyn.* **2013**, *51*, 1305–1323. [CrossRef]
47. Li, Z.; Mitsuoka, M.; Inoue, E.; Okayasu, T.; Hirai, Y.; Zhu, Z. Parameter Sensitivity for Tractor Lateral Stability against Phase I Overturn on Random Road Surfaces. *Biosyst. Eng.* **2016**, *150*, 10–23. [CrossRef]
48. Li, Z.; Mitsuoka, M.; Inoue, E.; Okayasu, T.; Hirai, Y.; Zhu, Z.; Song, Z.; Mao, E. Identification of Configurational Effectiveness on Improvement of Tractor Stability on Uneven Roads. *J. Fac. Agric. Kyushu Univ.* **2018**, *63*, 113–117. [CrossRef]
49. Jang, M.K.; Hwang, S.J.; Shin, C.S.; Nam, J.S. A Novel Approach to Determine Static Falling down Sidelong Angle of Tractor Using a 3D Printed Miniature Model. *Appl. Sci.* **2022**, *12*, 43. [CrossRef]
50. Ahmadi, I. Dynamics of Tractor Lateral Overturn on Slopes under the Influence of Position Disturbances (Model Development). *J. Terramechanics* **2011**, *48*, 339–346. [CrossRef]
51. Kim, K.U.; Rehkugler, G.E. Review of Tractor Dynamics and Stability. *Trans. Am. Soc. Agric. Eng.* **1987**, *30*, 615–623. [CrossRef]
52. Ahmadi, I. Effect of Transition of Mounted Chisel Plow from Operation to Transport on Longitudinal Stability of Tractor (Part I-Modeling). *Agric. Conspec. Sci.* **2013**, *78*, 343–351.
53. Demšar, I.; Bernik, R.; Duhovnik, J. A Mathematical Model and Numerical Simulation of the Static Stability of a Tractor. *Agric. Conspec. Sci.* **2012**, *77*, 143–150.
54. Li, Z.; Mitsuoka, M.; Inoue, E.; Okayasu, T.; Hirai, Y. Dynamic Analysis of Agricultural Wheel Tractor Driving on Uneven Surface under the Influences of Speed and Slope Angle. *J. Fac. Agric. Kyushu Univ.* **2014**, *59*, 339–343. [CrossRef]
55. Shim, S.; Nam, Y.J.; Koo, Y. Conceptual Design and Rearward Stability Analysis of an Integrated Tractor Implement for Flat Ridge Preparation. *J. Agric. Life Sci.* **2018**, *52*, 111–122. [CrossRef]
56. Arote, P.; Sudarsanam, S.; Muthuraman, M.; Kharade, P. The Dynamic Stability Index Calculator for Agricultural Tractors Equipped with Front End Loader. *AIP Conf. Proc.* **2019**, *2148*, 030031. [CrossRef]
57. Park, H.; Kim, K.; Kim, J.; Song, T.; Park, M.; Cho, K. Sideways Overturning Analysis of Forwarder Using a Multibody Dynamics Analysis Program. *KSAM* **2002**, *27*, 185–194.
58. Huang, W.; Xu, F.; Ge, J.; Zhang, C. Simulated Analysis of a Wheeled Tractor on Soft Soil Based on RecurDyn. In *IFIP Advances in Information and Communication Technology*; Springer: Berlin/Heidelberg, Germany, 2012; pp. 332–342. [CrossRef]
59. Chowdhury, M.; Islam, M.N.; Iqbal, M.Z.; Islam, S.; Lee, D.H.; Kim, D.G.; Jun, H.J.; Chung, S.O. Analysis of Overturning and Vibration during Field Operation of a Tractor-Mounted 4-Row Radish Collector toward Ensuring User Safety. *Machines* **2020**, *8*, 77. [CrossRef]
60. Hwang, S.J.; Jang, M.K.; Nam, J.S. Application of Lateral Overturning and Backward Rollover Analysis in a Multi-Purpose Agricultural Machine Developed in South Korea. *Agronomy* **2021**, *11*, 297. [CrossRef]
61. Jang, M.K.; Hwang, S.J.; Kim, J.H.; Nam, J.S. Overturning and Rollover Characteristics of a Tractor through Dynamic Simulations: Effect of Slope Angle and Obstacles on a Hard Surface. *Biosyst. Eng.* **2022**, *219*, 11–24. [CrossRef]
62. Jang, M.K.; Hwang, S.J.; Nam, J.S. Simulation Study for Overturning and Rollover Characteristics of a Tractor with an Implement on a Hard Surface. *Agronomy* **2022**, *12*, 3093. [CrossRef]
63. Lysych, M.N. A Study of the Static Lateral Stability of a Tillage Machine-Tractor Unit on a Virtual Stand. *J. Phys. Conf. Ser.* **2020**, *1515*, 042033. [CrossRef]

agriculture

MDPI

Article

Field Sprayer with Application Rate Control Using Fast Response Proportional Valves under Variable Speed Conditions

Hayrettin Karadol [1,*], Ali Aybek [1], Mustafa Ucgul [2], Hamza Kuzu [1] and Mahit Gunes [3]

1 Biosystems Engineering Department, Faculty of Agriculture, Kahramanmaras Sutcu Imam University, Kahramanmaras 46040, Turkey; aaybek@ksu.edu.tr (A.A.); hamzakuzu@ksu.edu.tr (H.K.)
2 Faculty of Science and Engineering, Southern Cross University, East Lismore, NSW 2084, Australia; mustafa.ucgul@scu.edu.au
3 Department of Electrical and Electronics Engineering, Kahramanmaras Sutcu Imam University, Kahramanmaras 46100, Turkey; mgunes@ksu.edu.tr
* Correspondence: hayrettinkaradol@gmail.com or hayrettinkaradol@ksu.edu.tr

Abstract: In modern agriculture, which is characterised by dynamic field environments, challenges are faced in maintaining consistent application rates due to varying tractor speeds, field conditions, and certain calibration errors. Conventional control systems, which rely on slower valves, have difficulty adapting to these dynamic field conditions. By contrast, the integration of fast-acting proportional valves improves the precision and flexibility of flow rate adjustment during spraying applications. This research focused on evaluating the accuracy of spraying applications under different tractor speed conditions through field experiments and data analysis. This study involves a field sprayer with boom wings divided into right and left sections, where the flow rate of the liquid to each section is controlled by proportional valves with a 3 s full opening and closing time, dependent on speed information. Using a closed-loop control system consisting of a flow meter, proportional valve, and PLC, the valves are controlled by the PLC's internal PID blocks. Observations reveal that as the tractor speed increases to a certain level, the system effectively adjusts the application rate close to the target value and maintains control against the changing ground speed during all field tests. The study included five different application tests, with target application rates of 100, 150, 200, 250, and 300 L ha^{-1}, with each repeated three times, resulting in a total of 15 field tests at different ground speeds. During these tests, the data were meticulously recorded every second, covering the tractor speed, flow rate, and pressure values for both right and left boom sections, along with regulator pressure, proportional valve opening rates, and application rates. The durations for each application rate were documented alongside instances within specified periods where error boundaries of $\pm$10% were exceeded. During the total test duration of 9734 s, the actual application rate value exceeded error boundaries during only 209 s. Within the application durations, the speed variation intervals ranged from 5.10 to 10.23 km h^{-1}, 4.64 to 9.91 km h^{-1}, 3.68 to 7.89 km h^{-1}, 4.80 to 8.21 km h^{-1}, and from 4.90 to 8.69 km h^{-1}. The absolute percentage mean application errors were recorded as 2.81%, 2.68%, 2.28%, 2.14%, and 2.51% for respective application rates. Furthermore, statistically significant correlations ($p < 0.01$) were identified among the variables (speed, valve opening rate, flow rate, pressure) in both the right and left boom sections across all application rates.

Keywords: field sprayer; flow rate control; PID; pesticide

Citation: Karadol, H.; Aybek, A.; Ucgul, M.; Kuzu, H.; Gunes, M. Field Sprayer with Application Rate Control Using Fast Response Proportional Valves under Variable Speed Conditions. *Agriculture* **2024**, *14*, 361. https://doi.org/10.3390/agriculture14030361

Academic Editor: Shuo Yan

Received: 14 January 2024
Revised: 20 February 2024
Accepted: 22 February 2024
Published: 23 February 2024

1. Introduction

The continuous advancement of pesticide application technologies has yielded a diverse range of pesticides that are crucial to meeting today's consumer demands and ensuring ample food supplies. However, these pesticides, while instrumental in crop protection and growth, pose certain environmental and health risks to living organisms, including the human population [1–3]. Maintaining a uniform application rate per unit area is critical to the efficient use of pesticides with field sprayers, as their uneven distribution compromises treatment efficacy and contributes to residues of non-target pesticides [4]. Application errors can result in pesticide residues found in everyday consumer products, such as cooked meals, water, wine, fruit juices, and animal feeds [5]. Several factors related to equipment and application methods can affect the coverage and effectiveness of pesticide application. In particular, parameters such as ground speed and the type of nozzle/droplet size are crucial in influencing spray coverage, quality, efficacy, and potential droplet drift. Higher ground speeds have been shown to lead to a heightened generation of smaller droplets, consequently raising the risk of spray drift. Several studies have documented greater pesticide drift and reduced coverage associated with increases in the travel speed of the tractor [6–10]. In addition, resistance to pests can develop over time, which impairs the effectiveness of pest control. Successful pesticide application hinges on the utilisation of calibrated sprayers operated by trained individuals [11]. However, the inadequacies resulting from improperly calibrated equipment result in the waste of thousands of gallons of pesticides and billions of dollars. [12–14]. A study conducted in North Dakota in the United States revealed significant variations in pesticide application rates, with 60% of applicators delivering over or under their intended rate by more than 10% and with several instances exceeding 30% [15]. The study primarily attributed these issues to factors such as worn nozzle tips, imprecisely calibrated equipment, or the inability to sustain necessary flow rates during field application [16,17].

Efficient sprayer rate control is deemed essential for the accurate application of pesticides. Traditional pressure-based field sprayers that lack control systems require operators to continually manage both tractor speed and system pressure in order to maintain a consistent application rate. Inherent variations in the forward speed of tractors during application, coupled with operational errors, make maintaining a target rate significantly challenging. However, achieving the desired target application rate is crucial for optimal pesticide coverage and efficacy [18], as well as economic viability.

The application rate (L ha^{-1}) is determined by the flow rate (L min^{-1}) and the tractor speed (km h^{-1}). Therefore, the precision of the feedback signals from flow rate and speed sensors and the accuracy of the valves that control the flow are critical to minimising application errors. To measure the flow rate in the pressure line of a sprayer system, there are several types of liquid flowmeters, such as turbine, vortex, electromagnetic, and ultrasonic flowmeters. When different flow measurement systems are examined, ultrasonic flowmeters are widely used due to their high sensitivity [19–21].

Over time, spray rate controllers have been integrated into agricultural sprayers in order to effectively manage application rates in cases where the ground speed changes during field operations [22–24]. Some such control methods are now widely available for use in agricultural spraying control systems and related fields as follows: PID (Proportional Integral Derivative) control [25,26], fuzzy control [27–29], neural network control [30], and intelligent control systems [31,32].

These systems primarily employ proportional valves to control liquid flow in the pressure line. However, the slow opening and closing times of these valves, extending up to 15 s, indicate unacceptable motion dynamics [33]. As the sprayer equipment accelerates or decelerates, the limitations of the rate controller and system become more apparent, potentially leading to under or over-application [34]. The current study aims to effectively control the application rate in field sprayers through the utilisation of fast-response proportional valves in order to reduce errors due to speed changes.

2. Materials and Methods

The field tests were conducted on a flat area of 1.5 hectares on the campus of Kahramanmaras Sutcu Imam University, Turkey (37′35″18° N–36′48″47° E). The sprayer system used in this study consisted of a piston-diaphragm pump (71 L min^{-1}, max 50 bar), a 400 L tank capacity, and a 10 m boom width. The boom is divided into two sections (right and left), with each section controlled independently. A total of 20 flat-fan spray nozzles

(Agroer Co. Ltd., Istanbul, Turkey) were mounted on the boom at spacing intervals of 0.5 m The colour codes and flow rates of the nozzles were as follows: yellow 0.65–1.03 L min^{-1}, blue 0.97–1.53 L min^{-1}, and red 1.29–2.09 L min^{-1} (see Figure 1a). An appropriate nozzle type was selected for each different application test based on the information specified in the product catalogue. The hardware of the flow rate control system consisted of a PLC (Programmable Logic Controller) (S7 1200, 6ES7214-1AG40-0XB0, 14 digital inputs, 10 digital outputs, and 2 analogue input) with an analogue module (SM 1234, 4 analogue inputs, 2 analogue outputs), flow meters (Christian Bürkert GmbH & Co. KG., Karlsruhe, Germany, 8081, 0.5–50 L min^{-1}), proportional valves (Christian Bürkert GmbH & Co. KG., Karlsruhe, Germany 3280, with a fully opening time of 3 s and a 4–20 mA input signal), and pressure sensors (Dwyer Instruments, LLC., USA, Wyoming, a 0–10 bar for right and left boom sections, and 0–100 bar for regulator output). Proportional valves and flow meters were mounted between the regulator output point and both boom sections (see Figure 1b). Power for the PLC, sensors, modem, and valves was supplied from the tractor's battery. Due to the different input voltage levels of the equipment used, a 12–220 V inverter device and a 220–24 V converter device were used.

(**a**)

(**b**)

Figure 1. Sprayer system with application rate controller. (**a**) Boom sections and pressure sensors; (**b**) proportional valve and flow meter connection.

The speed of the tractor was measured using a proximity sensor attached to the front wheel (see Figure 2). To record the speed, 32 pieces of metal were attached evenly to the wheel at regular intervals. To acquire speed information, the pulse values detected by the proximity sensor, determined according to the distance covered by the wheel in one revolution, were fed into the PLC High-Speed Counter (HSC) input. As a result of this operation, speed in kilometres per hour was obtained by utilising mathematical functions within the PLC program.

Figure 2. Tractor speed measurement.

The output signals of the flow meter (4–20 mA), the output signals of the pressure sensor of the boom section (4–20 mA), and the output signals of the pressure sensor of the regulator (0–10 V) were successively applied to the analogue input terminals of the PLC module (AI0, AI1, AI2, AI3, and AI4). PID (Proportional Integral Derivative) controller blocks, which are commonly used in PLC programming, were used to control the proportional valves.

The target application rate was calculated depending on the tractor's speed, the flow rate, and the boom width for each boom section, and the target application rate was then applied to the PID input terminals. The control signal (4–20 mA) generated by the PID controller using these parameters was then applied to the proportional valves through the analogue output terminals (AQ0 and AQ1). The overall structure of the system used is illustrated as shown in Figure 3.

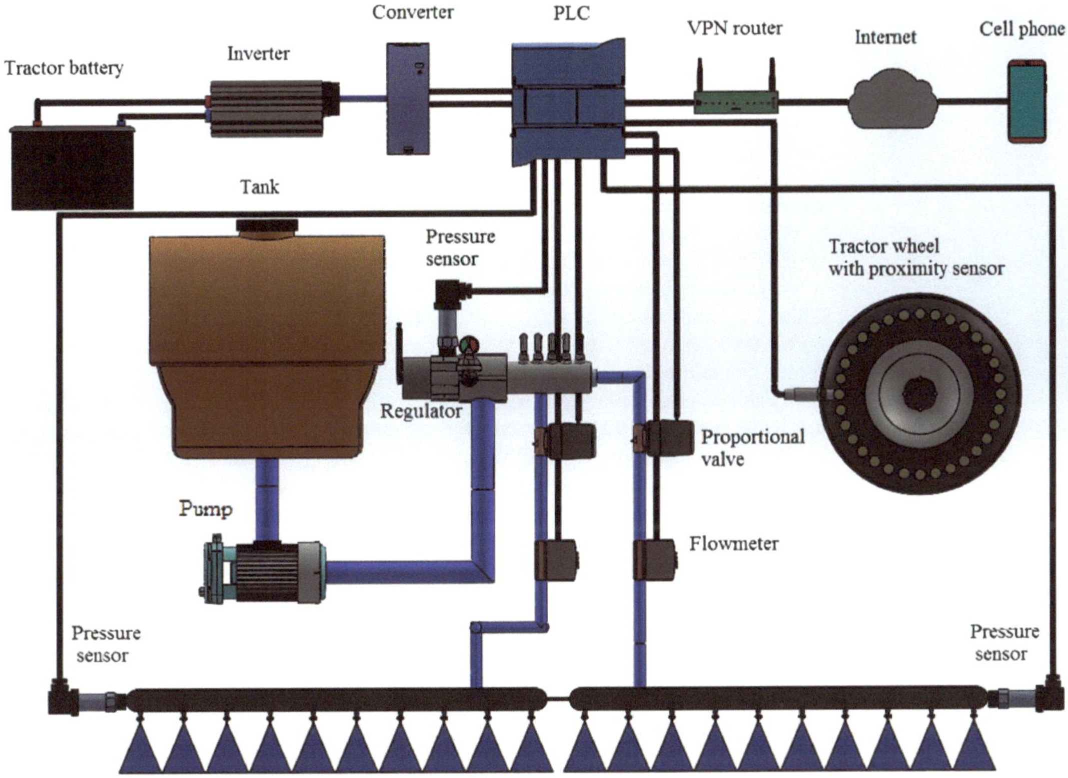

Figure 3. General structure of the system.

Five different target application rate tests (100, 150, 200, 250, and 300 L ha^{-1}) were conducted, each with three repetitions, resulting in a total of 15 different field tests conducted at varying tractor speeds of between 5 and 10 km h^{-1}. The sequential process followed for each test was as follows: (1) sprayer tank filled with water, (2) target application rate determined, (3) sending the pulse signal to the corresponding input of the controller for data recording, (4) starting the PTO (power take-off), (5) performing the application in the field at different ground speeds for about 10 min, and (6) transferring the data from the PLC memory to the computer in a 'csv' file format. When calculating the application error, the deviation between the actual application rate and the target application rate was evaluated, as shown in Equation (1) for the absolute application rate error as a percentage.

$$\text{Absolute Application Rate Error } (\%) = \frac{(|\text{Actual Application Rate} - \text{Target Application Rate}|)}{\text{Target Application Rate}} \times 100 \quad (1)$$

The recorded data encompassed tractor speed, the flow rate of the right boom section (FRoRS), the flow rate of the left boom section (FRoLS), the valve opening rate of the right boom section (VORoRS), valve opening rate of the left boom section (VORoLS), the application rate of the right boom section (ARoRS), the application rate of the left boom section (ARoLS), regulator pressure (RP), right boom section pressure (RSP), and left boom section pressure (LSP). Data logging occurred every second through the data logger block within the PLC program, which comprised the following five instructions: (1) 'DataLogCreate' to generate a data log file, (2) 'DataLogOpen' to access an existing data log file, (3) 'DataLogWrite' to record a data record, (4) 'DataLogClose' to conclude an open data log, and (5) 'DataLogNewFile' to create a new data log with similar properties as an existing file but with a new name [35].

The data analysis revealed that sudden changes in real-time sensor data have a disruptive impact on the output values of the PID blocks responsible for regulating the proportional valve opening rate. To mitigate this, the data obtained from the sensors in real-time were smoothed using the "Moving Average" block. This block calculated the (Moving) Average based on the set window width, where the window width indicates the maximum number of values last read (set to 100). As soon as the maximum number of values was reached, "WindowSizeReached" was set, and each newly read value replaced the oldest value (FIFO principle; First Input First Output) [36].

Web server software (TIA Portal V.15) was developed to monitor real-time data over the internet, which is transmitted via a mobile 4G VPN router (Robustel Co. Ltd., Guangzhou, China). An index.htm file was created to display the sensor data directly via the PLC web server (PWS). This file displayed the outputs of the sensor data assigned to the variables in real-time via the server interface.

Prior to undertaking the tests, a fine-tuning test was conducted by setting the target application rate to 100 L ha^{-1} in order to determine the PID controller coefficients based on flow rate variations at different ground speeds (5–10 km h^{-1}). PID fine-tuning was employed so as to achieve the desired system response, to minimise errors, and also to improve stability within the control system. The P component responds to the current error, which is the difference between the desired setpoint and the actual process variable, and the I component considers the cumulative sum of past errors over time, which helps eliminate any steady-state error and brings the system to the desired setpoint. The D component predicts the future trend of the error based on its current rate of change, helping to dampen oscillations and prevent overshooting. This test involved continuously varying the tractor's speed within certain limits. After approximately 20 min of fine-tuning, the obtained PID coefficients presented in Table 1 were achieved.

Table 1. Determined PID coefficients during fine-tuning process.

Coefficient	Right Section	Left Section
Proportional gain	10.20958	17.89999
Integral action time (s)	1.967492	2.348299
Derivative action time (s)	0.528921	0.590065

3. Results

Figure 4 depicts the application rate of the right boom section (ARoRS), the application rate of the left boom section (ARoLS), the target application rate (TAR), and the error boundary (EB) value for 100 L ha^{-1} target application rate tests. In this test period, the interval speed variations ranged from 5.10 to 10.23 km h^{-1}. Of the 1726 s duration of this test period, the actual application rate exceeded the error boundaries for 49 s, representing an absolute mean application error value of 2.81%. Notably, the initial speed change, which was greater than subsequent changes in travel speed during the application, delayed the controller in approaching the set point for the target application rate. The proportional valve's complete closure during the initial movement resulted in no liquid passing through the flow meter, even when the tractor was in motion. In order to address this, an alert signal was sent to the flow meter output for a 3 s duration when the travel speed reached 1 km h^{-1}. In this scenario, the valve opened slightly, allowing liquid to enter the flow meter.

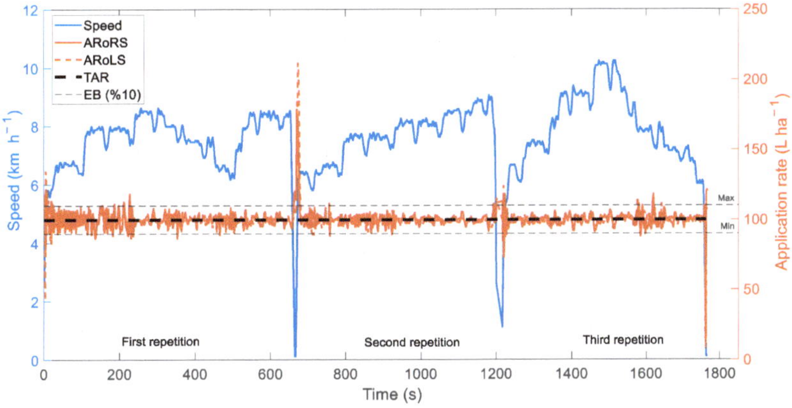

Figure 4. Speed vs. actual application rate (100 L ha^{-1}).

Figure 5 shows the graph of the actual application rate corresponding to the different ground speed values for the target application rate of 150 L ha^{-1}. In this test period, the interval speed variations ranged from 4.64 to 9.91 km h^{-1}. Of the 1810 s duration of this test period, the actual application rate exceeded the error boundaries for 46 s, representing an absolute mean application error value of 2.68%.

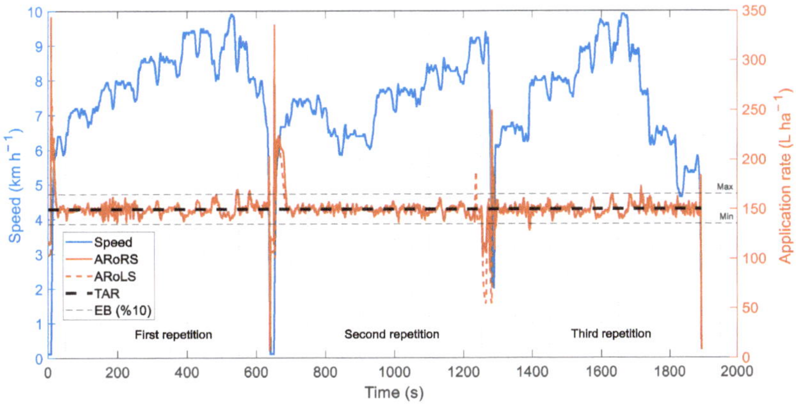

Figure 5. Speed vs. actual application rate (150 L ha^{-1}).

Figure 6 shows the graph of the actual application rate corresponding to the varying speed values for the target application rate of 200 L ha^{-1}. In this test period, the interval speed variations ranged from 3.68 to 7.89 km h^{-1}. Of the 1973 s duration of this test period, the actual application rate exceeded the error boundaries for 21 s, representing how the absolute mean application error value was found to be 2.28%. During this test period, the actual application rate consistently remained within acceptable levels due to smoother speed changes.

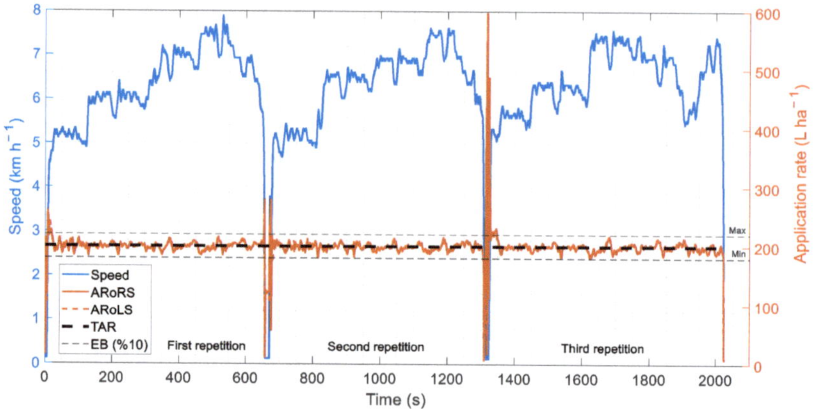

Figure 6. Speed vs. actual application rate (200 L ha^{-1}).

Figure 7 shows the graph of the actual application rate corresponding to the varying speed values for the target application rate of 250 L h^{-1}. In this test period, the interval speed variations ranged from 4.80 to 8.21 km h^{-1}. Of the 1848 s duration of this test period, the actual application rate exceeded the error boundaries for 40 s, representing how the absolute mean application error value was found to be 2.14%. During the second and third repetitions of this test period, it was observed that where there was a sudden change in speed (approx. 1 km h^{-1}), the application rate value exceeded the error limits for a few seconds.

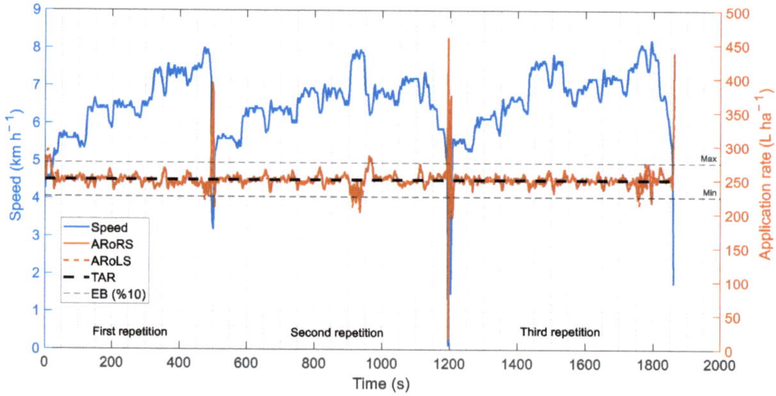

Figure 7. Speed vs. actual application rate (250 L ha^{-1}).

Figure 8 shows the graph of the actual application rate corresponding to the varying speed values for the target application rate of 300 L h^{-1}. In this test period, the interval speed variations ranged from 4.90 to 8.69 km h^{-1}. Of the 2168 s duration of this test period, the actual application rate exceeded the error boundaries for 53 s, representing how the absolute mean application error value was found to be 2.51%.

Figure 9 presents the regulator pressure, right boom section pressure (RSP), and left boom section pressure (LSP) for the 200 L ha^{-1} target application rate. Throughout the application, the regulator pressure (RP) maintained a steady level of approximately 6–7 bars, while the pressures associated with the boom sections varied in the range of 1.5–3.5 bars, contingent on speed changes and, consequently, flow rate. Notably, the precise regulation of the liquid pressurised by the pump prior to its application to the proportional valves significantly contributed to enhancing the stability of the system.

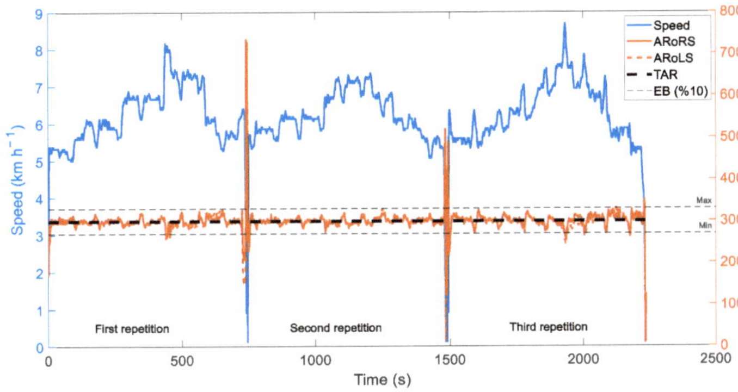

Figure 8. Speed vs. actual application rate (300 L ha^{-1}).

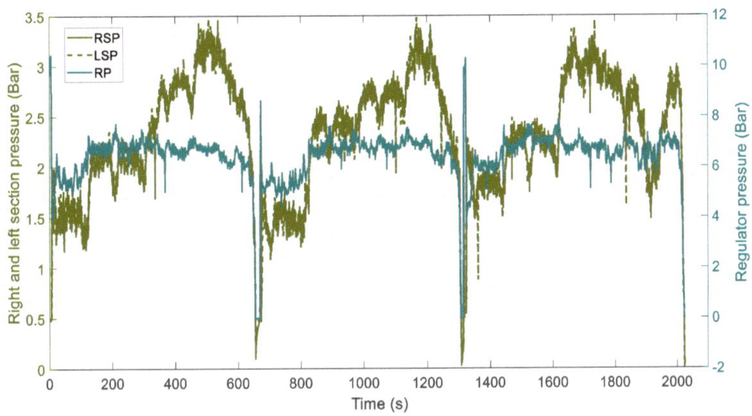

Figure 9. Boom sections and regulator pressure (200 L ha^{-1}).

Figure 10 shows the relationship between the opening rate of the proportional valves (VORoRS and VORoLS) and the speed and flow rates in the right and left boom sections (FRoRS and FRoLS). The left *y*-axis represents the travel speed of the tractor and the flow rate of the boom sections, while the right *y*-axis depicts the valve opening rate of the boom sections. The results indicate that the proportional valves adeptly regulated the flow rate, responding effectively to changes in speed.

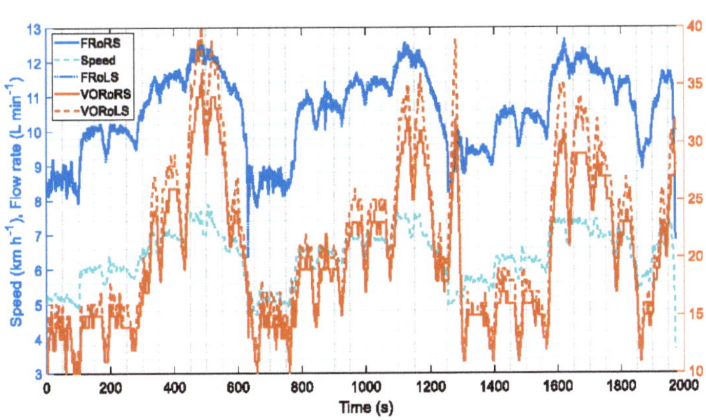

Figure 10. Speed and flow rate vs. proportional valve opening rate (200 L ha^{-1}).

The correlation analysis performed for different target application rates (100, 150, 200, 250, and 300 L ha^{-1}) provides valuable insights into the relationships between the variables in both the right and left boom sections. Speed exhibits robust positive correlations, VOR, flow rate, and pressure, indicating a simultaneous increase in these parameters as the application rate increase. This correlation is particularly pronounced at higher application rates. The positive correlations between speed, and flow rate emphasise the interdependence between these two factors. This relationship indicates that the actual application value is maintained over the entire test period. VOR shows positive correlations with flow rate and pressure, and the correlation becomes more stable at higher application rates. The correlation between speed and VOR, as well as the correlation between speed and flow rate, indicate a strong interaction of factors in the system. The system regulates the liquid delivered to the nozzles to ensure uniform application, potentially smoothing out fluctuations caused by changes in speed and reflectance. Similarly, pressure shows positive correlations with speed, VOR and flow rate with these correlations as the application rates increase. These results indicate the synchronised behaviour of the variables, especially in scenarios with an increased application rate (see Tables 2–6).

According to the analysis of variance performed on the basis of the randomised complete block design, the analysis of variance performed for the factor boom sections and application rate indicates a significant effect on the speed variable. This means that changes in the application rate have a significant influence on the speed variable. As far as the VOR variable is concerned, the analysis of variance, which examines the interaction of the factor boom sections and application rate with VOR, shows a significant change in the VOR variable. This indicates that VOR can vary considerably in both the boom sections and the application rates. For the variable flow rate, the analysis of variance carried out of the factor boom sections and application rate shows a significant effect on the variable flow rate. This means that changes in the application rate have a noticeable impact on the flow rate. As with the pressure variable, the variance analysis carried out with the boom sections, and the factors for the application rate show a significant change in the pressure variable. This emphasises the significant impact that changes in application rate have on pressure (see Table 7).

The results revealed from the variance analysis offer a more detailed explanation of the relationships identified in the correlation analysis. For instance, when examining the interaction of speed with actual application rate and boom section factors, it was observed that these factors have a significant impact on speed. Similarly, the effects of VOR, flow rate, and pressure are associated with both the boom section factors and the actual application rate. The lack of significance for the boom sections and the significant influence of application rate (F-statistic: 2224.353, p-value: 0.000) indicate that changes in the actual application rate have a pronounced effect on VOR, while boom sections are not shown to significantly impact this relationship. The variance analysis indicates that the actual application rate significantly affects the flow rate (F-statistic: 16,613.497, p-value: 0.000), which emphasises the impact of changes in the actual application rate on the variability of the flow rate. The boom sections, however, were not found to have any significant influence on this relationship.

Table 2. Correlation analysis of values of variables for the target application rate (100 L ha^{-1}).

Variables	Right Boom Section					Left Boom Section				
	Speed	**VOR**	**Flow Rate**	**Pressure**	**Mean ± SE**	**Speed**	**VOR**	**Flow Rate**	**Pressure**	**Mean ± SE**
Speed (km h^{-1})	1	0.696 **	0.959 **	0.787 **	7.55 ± 0.03	1	0.348 **	0.946 **	0.670 **	7.55 ± 0.03
VOR (%)		1	0.658 **	0.819 **	5.75 ± 0.05		1	0.356 **	0.638 **	6.56 ± 0.07
Flow rate (L min^{-1})			1	0.770 **	6.29 ± 0.03			1	0.677 **	6.31 ± 0.03
Pressure (Bar)				1	1.49 ± 0.01				1	1.36 ± 0.01

** $p < 0.01$.

Table 3. Correlation analysis of values of variables for target application rate (150 L ha^{-1}).

Variables	Right Boom Section					Left Boom Section				
	Speed	**VOR**	**Flow Rate**	**Pressure**	**Mean ± SE**	**Speed**	**VOR**	**Flow Rate**	**Pressure**	**Mean ± SE**
Speed (km h^{-1})	1	0.471 **	0.951 **	0.859 **	7.55 ± 0.03	1	0.472 **	0.962 **	0.867 **	7.55 ± 0.03
VOR (%)		1	0.578 **	0.720 **	12.59 ± 0.16		1	0.544 **	0.703 **	13.94 ± 0.17
Flow rate (L min^{-1})			1	0.898 **	9.47 ± 0.04			1	0.882 **	9.47 ± 0.04
Pressure (Bar)				1	1.94 ± 0.01				1	1.93 ± 0.01

** $p < 0.01$.

Table 4. Correlation analysis of values of variables for target application rate (200 L ha^{-1}).

Variables	Right Boom Section					Left Boom Section				
	Speed	**VOR**	**Flow Rate**	**Pressure**	**Mean ± SE**	**Speed**	**VOR**	**Flow Rate**	**Pressure**	**Mean ± SE**
Speed (km h^{-1})	1	0.765 **	0.972 **	0.894 **	6.25 ± 0.02	1	0.768 **	0.977 **	0.897 **	6.25 ± 0.02
VOR (%)		1	0.762 **	0.853 **	19.47 ± 0.15		1	0.777 **	0.852 **	20.51 ± 0.16
Flow rate (L min^{-1})			1	0.883 **	10.37 ± 0.04			1	0.890 **	10.37 ± 0.04
Pressure (Bar)				1	2.32 ± 0.01				1	2.32 ± 0.01

** $p < 0.01$.

Table 5. Correlation analysis of values of variables for target application rate (250 L ha^{-1}).

Variables	Right Boom Section					Left Boom Section				
	Speed	VOR	Flow Rate	Pressure	Mean ± SE	Speed	VOR	Flow Rate	Pressure	Mean ± SE
Speed (km h^{-1})	1	0.677 **	0.696 **	0.658 **	6.33 ± 0.02	1	0.504 **	0.738 **	0.663 **	6.33 ± 0.02
VOR (%)		1	0.629 **	0.641 **	24.16 ± 0.24		1	0.569 **	0.642 **	30.96 ± 0.47
Flow rate (L min^{-1})			1	0.923 **	14.17 ± 0.04			1	0.929 **	14.11 ± 0.05
Pressure (Bar)				1	1.84 ± 0.01				1	1.82 ± 0.01

** $p < 0.01$.

Table 6. Correlation analysis of values of variables for target application rate (300 L ha^{-1}).

Variables	Right Boom Section					Left Boom Section				
	Speed	VOR	Flow Rate	Pressure	Mean ± SE	Speed	VOR	Flow Rate	Pressure	Mean ± SE
Speed (km h^{-1})	1	0.677 **	0.916 **	0.898 **	6.20 ± 0.02	1	0.592 **	0.891 **	0.888 **	6.20 ± 0.02
VOR (%)		1	0.688 **	0.721 **	25.58 ± 0.28		1	0.524 **	0.588 **	37.22 ± 0.57
Flow rate (L min^{-1})			1	0.918 **	15.47 ± 0.04			1	0.905 **	15.39 ± 0.04
Pressure (Bar)				1	2.19 ± 0.01				1	2.16 ± 0.01

** $p < 0.01$.

Table 7. Analysis of variance for the variables.

Variations	VK	SD	KO	F	p
Speed	Boom sections	1	0.000	0.000	1.000
	Actual application rate	4	1849.475	1466.551	0.000 **
	Actual application rate × Boom sections	4	0.000	0.000	1.000
	Error	19,288	1.261		
VOR	Boom sections	1	89,485.513	501.684	0.000 **
	Actual application rate	4	396,758.542	2224.353	0.000 **
	Actual application rate × Boom sections	4	23,481.152	131.643	0.000 **
	Error	19,288	178.370		
Flow rate	Boom sections	1	3.112	1.013	0.314
	Actual application rate	4	51,068.190	16,613.497	0.000 **
	Actual application rate × Boom sections	4	1.630	0.530	0.713
	Error	19,288	3.074		
Pressure	Boom sections	1	7.771	28.302	0.000 **
	Actual application rate	4	433.479	1578.791	0.000 **
	Actual application rate × Boom sections	4	2.434	8.866	0.000 **
	Error	19,288	0.275		

** $p < 0.01$.

4. Discussion

When applying pesticides under field conditions, the error rate (the deviation between the actual application rate and the target application rate) should not exceed 10% [37,38]. In modern application rate control systems used for commercial purposes, it has been observed that the proportional valve, which is responsible for regulating the flow rate, has a limited ability to respond to control signals. In this study, if errors in the system approaching the set point at the beginning of each test are not considered (from the moment the error value is less than 10 percent), the application times for all tests and the duration of exceeding the error limit (±10%) within these application times are shown as follows: 1726 × 49, 1810 × 46, 1973 × 21, 1848 × 40, and 2168 × 53 s, respectively. The range of speed changes within these application times was 5.10–10.23, 4.64–9.91, 3.68–7.89, 4.80–8.21, and 4.90–8.69 km h^{-1}, respectively. Despite the approximately 2-fold change in speed in all tests, the absolute mean percentage application errors were 2.81, 2.68, 2.28, 2.14, and 2.51, respectively. In another study conducted by Atcıoglu [39] using a commercially available application rate control system, the percentage rate of change in application rates compared to the percentage rate changes for the same actual application rate values were as follows: 17–24.5, 4.9–13.9, 9.2–13.9, 3.5–6.6, and 5.8–7.9. It can be said that the proportional valves, which are responsible for flow control and respond effectively to the control signal, maintain the actual application rate around the set point as the rate changes.

5. Conclusions

Pesticide application errors can be exacerbated when proportional valves, responsible for flow control in sprayer systems, fail to respond rapidly to control signals. In the current study, a closed-loop control system consisting of flow meters and proportional valves was implemented in order to mitigate such errors in pesticide applications using field sprayers. The boom was subdivided into independent right and left sections, with each being autonomously controlled. It was observed that application discrepancies between these two sections were statistically insignificant, which was expected. These findings highlight the potential of fast-response proportional valves to enhance the performance and adaptability of agricultural sprayer systems. Consequently, as the tractor speed increases, the valve opening rate rises, converging the application rate toward the set point. Once the travel speed of the tractor reaches a certain level, the system effectively controls the actual application rate. Instances of a sudden increase or decrease in the tractor's speed temporarily push the actual application rate value beyond its acceptable limits; however, it has also been shown to swiftly return within its acceptable boundaries. Despite the intentional changes to travel speed applied in all tests conducted in the current study, considering that speed variations during pesticide applications are typically smoother, this system was shown to adeptly maintain application errors within acceptable limits.

Author Contributions: Conceptualisation, H.K. (Hayrettin Karadol) and A.A.; methodology, H.K. (Hayrettin Karadol), A.A. and H.K. (Hamza Kuzu); software, H.K. (Hayrettin Karadol) and M.G.; investigation, H.K. (Hayrettin Karadol), A.A. and H.K. (Hamza Kuzu); writing—original draft preparation, H.K. (Hayrettin Karadol), A.A. and M.U.; writing—review and editing, A.A., M.U. and H.K. (Hamza Kuzu); visualisation, H.K. (Hayrettin Karadol); supervision, H.K. (Hayrettin Karadol), A.A. and M.U.; project administration, H.K. (Hayrettin Karadol), A.A., H.K. (Hamza Kuzu) and M.G. All authors have read and agreed to the published version of the manuscript.

Funding: This project numbered 1210683, was funded by The Scientific and Technological Research Council of Turkey.

Institutional Review Board Statement: Not applicable.

Data Availability Statement: The data presented in this study are available upon request from the authors.

Conflicts of Interest: The authors declare no conflicts of interest.

References

1. Hayes, T.B.; Case, P.; Chui, S.; Chung, D.; Haeffele, C.; Haston, K.; Lee, M.; Mai, V.P.; Marjuoa, Y.; Parker, J.; et al. Pesticide mixtures, endocrine disruption, and amphibian declines: Are we underestimating the impact? *Environ. Health Perspect.* **2006**, *114*, 40–50. [CrossRef]
2. Alewu, B.; Nosiri, C. Pesticides and human health. In *Pesticides in the Modern World—Effects of Pesticides Exposure*; Stoytcheva, M., Ed.; InTech: Nappanee, IN, USA, 2011; pp. 231–250.
3. Khan, B.A.; Nadeem, M.A.; Nawaz, H.; Amin, M.M.; Abbasi, G.H.; Nadeem, M.; Ayub, M.A. Pesticides: Impacts on agriculture productivity, environment, and management strategies. In *Emerging Contaminants and Plants: Interactions, Adaptations and Remediation Technologies*; Aftab, T., Ed.; Springer International Publishing: Cham, Switzerland, 2023; pp. 109–134.
4. Chen, Y.; Zheng, J. Control system for precision pesticide application based on variable rate technology. *Trans. Chin. Soc. Agric. Eng.* **2005**, *21*, 69–72.
5. Chourasiya, S.; Khillare, P.S.; Jyethi, D.S. Health risk assessment of organochlorine pesticide exposure through dietary intake of vegetables grown in the periurban sites of Delhi, India. *Environ. Sci. Pollut. Res. Int.* **2015**, *22*, 5793–5806. [CrossRef]
6. Miller, P.; Smith, R. The effects of forward speed on the drift from boom sprayers. In Proceedings of the Brighton Crop Protection Conference, Weeds, Brighton, UK, 17–20 November 1997.
7. Taylor, W.; Anderson, P.; Cooper, S. The use of air assistance in a field crop sprayer to reduce drift and modify drop trajectories. In Proceedings of the Brighton Crop Protection Conference, Weeds, Brighton, UK, 20–23 November 1989.
8. Hofman, V.; Solseng, E. Reducing Spray Drift. In *North Dakota State University Extension Service*; Agricultural and Biosystems Engineering: Fargo, ND, USA, 2001.
9. Nuyttens, D.; De Schampheleire, M.; Baetens, K.; Sonck, B. The influence of operator-controlled variables on spray drift from field crop sprayers. *Trans. ASABE* **2007**, *50*, 1129–1140. [CrossRef]
10. Virk, S.; Prostko, E.; Kemerait, R.; Abney, M.; Rains, G.; Powell, C.; Carlson, D.; Jacobs, J.; Tyson, W. On-Farm Evaluation of Nozzle Types for Peanut Pest Management Using Commercial Sprayers. *Peanut Sci.* **2021**, *48*, 87–96. [CrossRef]
11. Doruchowski, G.; Hołownicki, R.; Godyń, A.; Świechowski, W. Sprayer calibration training—Concept and performance. *Jul. Kuhn Arch.* **2012**, *4*, 228–233.
12. Ozkan, H.E. Sprayer performance evaluation with microcomputers. *Appl. Eng. Agric.* **1987**, *3*, 36–41. [CrossRef]
13. Sharda, A.; Fulton, J.P.; McDonald, T.P.; Zech, W.C.; Darr, M.J.; Brodbeck, C.J. Real-time pressure and flow dynamics due to boom section and individual nozzle control on agricultural sprayers. *Trans. ASABE* **2010**, *53*, 1363–1371. [CrossRef]
14. Butts, T.R.; Butts, L.E.; Luck, J.D.; Fritz, B.K.; Hoffmann, W.C.; Kruger, G.R. Droplet size and nozzle tip pressure from a pulse-width modulation sprayer. *Biosyst. Eng.* **2019**, *178*, 52–69. [CrossRef]
15. Hofman, V.; Solseng, E. *Spray Equipment and Calibration*; North Dakota State University: Fargo, ND, USA, 2004.
16. Gerling, J.F. *A Chemical Industry's View of Application Needs*; ASAE Paper No. 85-1094; ASABE: St. Joseph, MI, USA, 1985.
17. Grisso, R.D.; Dickey, E.C.; Schulze, L.D. The Cost of Misapplication of Herbicides. *Appl. Eng. Agric.* **1989**, *5*, 344–347. [CrossRef]
18. Sapkota, M.; Virk, S.; Rains, G. Spray Deposition and Quality Assessment at Varying Ground Speeds for an Agricultural Sprayer with and without a Rate Controller. *AgriEngineering* **2023**, *5*, 506–519. [CrossRef]
19. Tsukada, K.; Tsuzuki, N.; Kikura, H. A Study of Air-coupled Ultrasonic Flowmeter Using Beam Focusing. *Energy Procedia* **2015**, *71*, 352–359. [CrossRef]

20. Mandard, E.; Kouame, D.; Battault, R.; Remenieras, J.; Patat, F. Methodology for developing a high-precision ultrasound flow meter and fluid velocity profile reconstruction. *IEEE Trans. Ultrason. Ferroelectr. Freq. Control* **2008**, *55*, 161–172. [CrossRef]
21. Tsukada, K.; Kikura, H. Flowrate Measurement on Metal Pipes by Air-coupled Ultrasound. *Int. J. Comput. Methods Exp. Meas.* **2016**, *4*, 583–593. [CrossRef]
22. Ayers, P.D.; Rogowski, S.M.; Kimble, B.L. An investigation of factors affecting sprayer control system performance. *Appl. Eng. Agric.* **1990**, *6*, 701–706. [CrossRef]
23. Luck, J.D.; Sharda, A.; Pitla, S.K.; Fulton, J.P.; Shearer, S.A. A case study concerning the effects of controller response and turning movements on application rate uniformity with a self-propelled sprayer. *Trans. ASABE* **2011**, *54*, 423–431. [CrossRef]
24. Liu, H.; Zhu, H.; Shen, Y.; Chen, Y.; Ozkan, H.E. Development of digital flow control system for multi-channel variable-rate sprayers. *Trans. ASABE* **2014**, *57*, 273–281.
25. Guzmán, J.L.; Rodríguez, F.; Sánchez, F.; Berenguel, M. Robust pressure control in a mobile robot for spraying tasks. *Trans. ASABE* **2008**, *51*, 715–727. [CrossRef]
26. Karadöl, H. Flow and Pressure Control of a Field Sprayer Using PID Controller. *J. Anim. Plant Sci.* **2023**, *33*, 562–569.
27. Liu, Z.Z.; Hong, T.S.; Li, Z.; Song, S.; Yue, X.; Fan, Z. Simulation of flow control valve based on fuzzy control. *Trans. Chin. Soc. Agric. Eng.* **2009**, *25*, 83–86.
28. Shi, Y.; Zhang, C.M.; Liang, A.B.; Yuan, H. Fuzzy control of the spraying medicine control system. In Proceedings of the International Conference on Computer and Computing Technologies in Agriculture, Wuyishan, China, 18–20 August 2007; Springer: Boston, MA, USA, 2007; pp. 1087–1094.
29. Song, L.P.; Dong, Z.M.; Liang, L.J.; Xing, S. Variable universe adaptive fuzzy PID control of spray flow valve. *Trans. Chin. Soc. Agric. Eng.* **2010**, *26*, 114–118.
30. Wang, X.D.; Xu, Y.L.; Meng, X.T.; He, R.; Zhai, Y.T. Simulation and experiment of precision variable spraying system based on BAS PID control. *J. Chin. Agric. Mech.* **2020**, *41*, 62–66. [CrossRef]
31. Felizardo, K.R.; Mercaldi, H.V.; Oliveira, V.A.; Cruvinel, P.E. Modeling and predictive control of a variable-rate spraying system. In Proceedings of the 2013 8th EUROSIM Congress on Modelling and Simulation, Cardiff, UK, 10–13 September 2013; pp. 202–207.
32. Hussain, N.; Farooque, A.A.; Schumann, A.W.; McKenzie-Gopsill, A.; Esau, T.; Abbas, F.; Acharya, B.; Zaman, Q. Design and Development of a Smart Variable Rate Sprayer Using Deep Learning. *Remote Sens.* **2020**, *12*, 4091. [CrossRef]
33. Anthonis, J.; De Baerdemaeker, J.; Ramon, H. Application Techniques for Crop Protection. In *Precision Agriculture, in CIGR Handbook of Agricultural Engineering Volume VI Information Technology*; Munack, A., Ed.; CIGR—The International Commission of Agricultural Engineering; ASABE: St. Joseph, MI, USA, 2006; Chapter 5; pp. 289–294.
34. Luck, J. Technologies for Improving Sprayer Field Performance and Efficiencies. Available online: https://ocj.com/2021/01/technologies-for-improving-sprayer-field-performance-and-efficiencies/ (accessed on 10 September 2023).
35. Siemens. Process Data Acquisition and Monitoring. Available online: https://cache.industry.siemens.com/dl/files/156/64396156/att_992804/v1/64396156_S7-1200_DataLogging_DOC_v3d0d1_en.pdf (accessed on 20 April 2023).
36. Siemens. LGF Moving Average. Available online: https://support.industry.siemens.com/cs/document/109479728/library-of-general-functions-(lgf)-for-simatic-step-7-(tia-portal)-and-simatic-s7-1200-s7-1500?dti=0&lc=en-CL (accessed on 2 May 2023).
37. Reitz, S.; Palyi, B.; Ganzelmeier, H.; Aszlo, A. Performance of Electronic Controls for Field Sprayers. *J. Agric. Eng. Res.* **1997**, *68*, 399–407. [CrossRef]
38. BSI EN ISO. Agricultural and Forestry Machinery. Environmental Requirements for Sprayers. Horizontal Boom Sprayers. Available online: https://shop.bsigroup.com/ProductDetail?pid=000000000030376269 (accessed on 12 October 2020).
39. Atcıoglu, T. Adaptation an Electronic Sprayer Controller to a Domestic Field Sprayer and Determination of Its Performance. Master's Thesis, Institute of Natural and Applied Sciences, University of Cukurova, Adana, Türkiye, 2006. (In Turkish).

agriculture

MDPI

Article

Robust Guidance and Selective Spraying Based on Deep Learning for an Advanced Four-Wheeled Farming Robot

Chung-Liang Chang *, Hung-Wen Chen and Jing-Yun Ke

Department of Biomechatronics Engineering, National Pingtung University of Science and Technology, Neipu 91201, Taiwan; gueest6622@gmail.com (H.-W.C.); jingyunko@gmail.com (J.-Y.K.)
* Correspondence: chungliang@mail.npust.edu.tw; Tel.: +886-8-7703202 (ext. 7586)

Abstract: Complex farmland backgrounds and varying light intensities make the detection of guidance paths more difficult, even with computer vision technology. In this study, a robust line extraction approach for use in vision-guided farming robot navigation is proposed. The crops, drip irrigation belts, and ridges are extracted through a deep learning method to form multiple navigation feature points, which are then fitted into a regression line using the least squares method. Furthermore, deep learning-driven methods are used to detect weeds and unhealthy crops. Programmed proportional–integral–derivative (PID) speed control and fuzzy logic-based steering control are embedded in a low-cost hardware system and assist a highly maneuverable farming robot in maintaining forward movement at a constant speed and performing selective spraying operations efficiently. The experimental results show that under different weather conditions, the farming robot can maintain a deviation angle of 1 degree at a speed of 12.5 cm/s and perform selective spraying operations efficiently. The effective weed coverage (EWC) and ineffective weed coverage (IWC) reached 83% and 8%, respectively, and the pesticide reduction reached 53%. Detailed analysis and evaluation of the proposed scheme are also illustrated in this paper.

Keywords: agricultural robot; deep learning; selective spraying; autonomous navigation; agricultural practices

Citation: Chang, C.-L.; Chen, H.-W.; Ke, J.-Y. Robust Guidance and Selective Spraying Based on Deep Learning for an Advanced Four-Wheeled Farming Robot. *Agriculture* **2024**, *14*, 57. https://doi.org/10.3390/agriculture14010057

Academic Editor: Jiyu Li

Received: 5 December 2023
Revised: 24 December 2023
Accepted: 26 December 2023
Published: 28 December 2023

1. Introduction

In the rapid development of smart technologies, their integration into agriculture has been critical to combating labor scarcity and an aging workforce. Contemporary practices in crop management, including plant monitoring, watering, pesticide spraying, and fertilizing, are still often performed manually using machines and tools. This requires farmers to focus on field operations for a long time, which results in long working hours and high labor costs. The proven efficacy of automation in performing monotonous tasks has seen its adoption across various sectors, with its applicability in agriculture being equally comprehensive, encompassing activities like weeding and harvesting [1]. Traditional small-scale agricultural robots were designed to navigate based on sensor fusion methods, which are suitable for structured environments [2]. Relying on the improvement of precision agriculture technology, the real-time kinemetric (RTK) global navigation satellite system (GNSS) (RTK-GNSS) and machine vision have played an important role in automatic guidance technology. RTK-GNSS has been pivotal in providing precise metrics for positioning, velocity, and timing, which assists users in planning the robot's movement path [3,4]. In particular, after the user can define multiple specific positions to form a path in a known field environment, the robot can then autonomously move to that point to perform field operations and reduce errors through heading control [5,6].

With the improvement of computer computing performance, machine vision technology has been used to identify, track, and measure targets and perform image processing [7]. Its technology has low development costs, is easy to maintain, and has wide applicability [8–14]. Morphology-based methods have been used to extract guide lines from rice field images, enabling autonomous weeding robots to operate without damaging crops [9,10]. This approach initially involved grayscale or CIE-Lab color space images, followed by Otsu thresholding and thinning processes to extract the edges of plant objects. The traditional crop row detection method often fails due to the influence of excessive ambient light or crop occlusion and other issues. Therefore, the method of using soil distribution information to find guidance points has been proven to be able to correctly find crop lines [11]. Post-identification Hough transform operations are utilized for edge detection, and the median lines between them serve as navigation lines, guiding unmanned vehicles through field operations autonomously [12]. The method of obtaining path fitting from grayscale images of specific areas of interest through image segmentation, navigation point extraction, and predicted point Hough transforms has been proven to be effective to improve the computational efficiency of the traditional Hough transform [13]. Meanwhile, this method can also solve the problem of insufficient accuracy caused by using the least squares method.

Based on the above description, the Hough transform and least squares method were the most commonly used path fitting methods in crop row identification. Among them, Hough processing easily extracts feature edge lines and then obtains the crop lines. The least squares rule is a statistical method of regression analysis which can fit a navigation path with acceptable accuracy. Although both methods can detect row guide lines, different environmental conditions, such as variations in color, brightness, saturation, contrast, reflection, shadow, occlusion, and noise in the same scene, can lead to the failure of guidance line extraction [15,16]. Secondly, differences in wind intensity in the field will also cause plant movement, which will blur the plant image and cause inaccurate crop center point detection [17–19].

With the advancements in high-speed computing technology, employing deep learning for navigation line extraction has gained traction. A U-Net deep learning model has been used to detect crops without interruption in the field, given favorable weather conditions, suitable lighting, few weeds, and neatly arranged crops. Finally, the Hough transform operation was used to identify the guidance lines [20]. The crop row extraction method based on Tiny-YOLOv4 can quickly detect multiple objects in an image and extract crop feature points within the frame through binarization operations and mean filtering operations. Finally, regression analysis with the least squares method was used to fit a guidance line [21]. An object detection method combining YOLO-R and density-based spatial clustering of applications with noise (DBSCAN) can quickly identify the number of crop rows and the crops in each row. Crop row lines can be found through the least squares method, and under different rice growth stages, the crop row recognition rate reaches at least 89.87% [22].

Deep learning methods have also been frequently employed in robotics for the identification of weeds and crops. Various research endeavors highlight the effectiveness of these advanced techniques in precision agriculture. Among them, the YOLO-based method has been commonly used to detect weeds in the field [23–28]. Ruigrok et al. [23] used the trained YOLO v3 model to detect weeds and spray them. The results showed that 96% of the weeds were controlled, but about 3% of the crops were sprayed by mistake. Twelve types of weeds in rice fields were detected by the trained YOLO v4 model, with an accuracy of 97% and an average detection time of 377 ms [24]. A weeding robot with deep learning developed by Chang et al. [25] could remove weeds at a speed of 15 cm per second, with an efficiency rating of 88.6%. The trained YOLO v5 model was utilized for weed (*Solanum rostratum* Dunal) detection, and the accuracy and recall rate of the model were 95% and 90%, respectively [26]. The YOLO-sesame model was used to identify weeds and crops in sesame fields, and its results showed a mean accuracy (mAP) of 96.1% at a frame rate of 36.8 frames per second (FPS) [27]. Utilizing the YOLO v3 model for weed detection, as detailed by Ruigrok et al., they trained it on image data from 20 different fields and tested it in 5 different arable fields [28]. The results indicated that increasing the variance in training data while keeping the sample size constant could reduce the generalization error during detection. Five deep learning models were used to detect weeds in soybean fields, with a custom five-layer CNN architecture showing a high detection accuracy of 97.7% and the lowest latency and memory usage [29].

With the four-wheel steering mechanism and flexible steering control method, the robot can move on any terrain on the site with high maneuverability and avoid slipping. Common steering control methods are based on Ackerman steering principle technology [30,31] combined with proportional–integral–derivative (PID), fuzzy logic, and sliding mode control [5,6,12,32].

The purpose of efficient guidance and control systems for agricultural robots is to accurately perform tasks such as spraying and weeding. However, a large part of existing research is still limited to the field of offline simulation or laboratory experiments, or they were only used to demonstrate crop row guidance performance, with little empirical evidence to support their applicability in real-world agricultural task operations.

In the real field, the surface appearance of field soil is constantly changing in farmland. During fallow periods, farmland may exhibit only furrows or what are referred to as drip irrigation belts interspersed between ridges. In contrast, the planting season may present a mix of crops and ridges without the consistent presence of irrigation belts or, in some instances, exclusively crops, contingent upon individual agricultural practices. Many studies often focus on feature extraction of single objects in the field. Once the features of objects in the field are unclear or do not exist, the method used often loses the guidance line, especially in low-light environments. Compounding these challenges is the reliance on a singular type of object for training datasets, thereby critically hampering the universality and adaptability of the detection models. Aside from that, open field images are often used for crop line detection. In practice, these images are often exposed, causing the detection model to be unable to identify crop row lines. It is uncertain whether these methods can be used for detection during robot motion or achieve the same detection performance. Furthermore, field testing and validation of these integrated approaches for steering control and task execution remains challenging.

In this study, the proposed scheme was used to automatically detect potential guidance lines on field ridges with deep learning and least squares regression, using a PID controller and fuzzy logic controller (FLC) to maintain the travel speed and heading angle. By adopting the one-stage object detection framework, the robot operation system was tailored for various object recognition tasks such as crop identification, drip irrigation belt detection, ridge recognition, weed detection, and identification of crops with nutrient deficiencies. It was also specifically designed to analyze and compare the object detection performance of the trained models at different FPS and obtain the real-time processing performance of the detection model in a field under different weather conditions. In terms of field operations, the smart sprayers were designed to spray nutrient-deficient crops as well as weeds.

The organization of this paper is as follows. Section 2 introduces the methodology, including the motion model of a farming robot, guidance line generation, methods for controlling the speed and heading of the robot, and spraying operation. Section 3 describes the configuration of each module within the robot.Section 4 discusses the experimental results, including tests of autonomous guidance of the robot, identification of weeds and unhealthy crops, and tests of selective spraying system performance. Finally, Section 5 provides the conclusions, summarizing the main findings of this study.

2. Autonomous Navigation and Selective Spraying Scheme

2.1. Motion Model

Given the constant and relatively slow travel speed of the robot, along with its rigid tires, its motion state at any given moment can be described using a bicycle model [33] as shown in Figure 1a. The global X-Y coordinate plane is a fixed horizontal plane upon which the robot moves and is used to describe its motion. It was assumed that O, O_f, and O_r represent the center of gravity of the robot, the center of the rear wheel, and the center of the front wheel, respectively. The distance between the center of the front wheel and the robot's center of gravity is denoted as L_f, while L_r is the distance between the center of the rear wheel and the center of gravity of the robot. The slip angle was represented by α, the heading angle by θ, and the speed at the center of gravity by v, with its component velocities being $\dot{x}$ and $\dot{y}$. This motion model is based on front-wheel steering, assuming the direction of the rear wheels is parallel to the robot body. The kinematic model of the robot is represented by $(\dot{x}, \dot{y}, \dot{\theta})$:

$$\dot{x} = v\cos(\theta + \alpha) \tag{1}$$

$$\dot{y} = v\sin(\theta + \alpha) \tag{2}$$

$$\dot{\theta} = \frac{v \cos \alpha \tan \delta}{L_f + L_r} \tag{3}$$

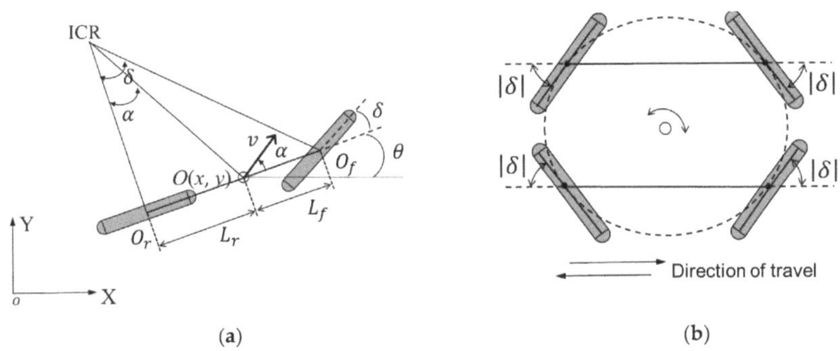

Figure 1. The motion mode of the robot. (**a**) A bicycle model for straight-line movement of the robot. (**b**) The steering angle of the four wheels in spin-on-the-spot mode. $|\cdot|$ stands for an absolute value operation.

The velocity v is given by $v = \frac{v_f \cos \delta + v_r}{2 \cos\alpha}$, where v_f and v_r represent the velocities of the front and rear wheel, respectively. The slip angle α is calculated as $\alpha = \tan^{-1}\left(\frac{L_r \tan \delta}{L_f + L_r}\right)$. When the robot moves in a straight line, the steering angle δ of both of the front wheels is within the range from $-\delta_{max}$ to $+\delta_{max}$, and the steering angle of both the rear wheels is fixed at zero. Moreover, during turning maneuvers, all wheels assumed the same steering angle $|\delta|$, putting the robot in a state of on-the-spot rotation (Figure 1b).

2.2. Guidance Line Generation

First, the top in-view image of the field is captured by a digital camera. The target objects in the image consist of ridges, crops, or a drip irrigation belt (Figure 2a). In order to form a guidance line, the original image is divided into two sub-images through a masking operation, which covers the long strip area in the center of the image with a white color (see Figure 2b). Subsequently, deep learning techniques are utilized to detect the ridges, crops, or drip irrigation belt within the image, as shown in Figure 2c. Meanwhile, the center point of each object is also extracted.

The tool used for labeling the ridges, crops, and drip irrigation belt in the image was LabelImg, which executes the LabelImg script through Anaconda. This tool marks the positions of objects in each image and generates an XML file containing information about the objects and their positions, providing training data for the dataset. In this study, YOLO v4 was used as the object detector [34], and its architecture is based on YOLO v3, which was proposed by Joseph Redmon [35]. YOLO v4 is proficient at identifying small objects at high speeds while maintaining a certain level of recognition accuracy. The architecture of YOLO v4 includes three core parts: the backbone, the neck, and the detection head. As shown in

Figure 3, CSPDarknet53 serves as the backbone network for the object detector. Its structure is based on DenseNet, which functions to connect layers in a convolutional neural network and adds a cross-stage partial network (CSPNet) [36]. Splitting and merging techniques are used to obtain a more efficient flow of gradient information and improve the accuracy of gradient calculations. The deep features of the image are then introduced into the neck layer, which separates the smallest-scale features from the backbone and pools multiple sizes to increase the receptive field.

Figure 2. An xample of object detection in the image of the ridge. (**a**) The original image. (**b**) The original image divided into two sub-images through masking processing. (**c**) An example of object detection results. Red dot = center point of object; green box = detected objects (ridges, crops, or drip irrigation belt).

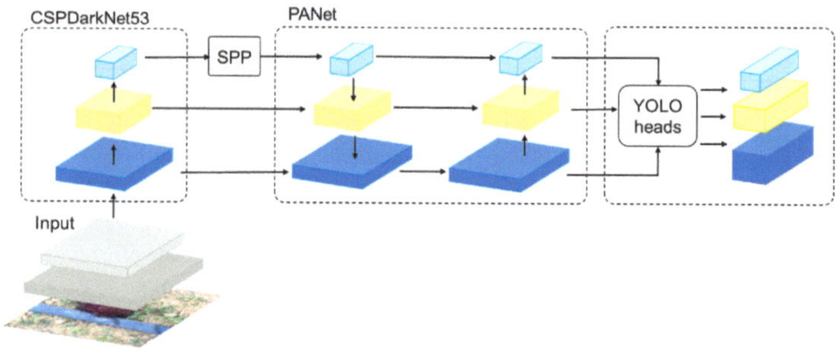

Figure 3. The framework of YOLO v4.

The Path Aggregation Network (PANet) uses feature maps formulated through spatial pyramid pooling (SPP) [37] and CSP-Darknet53 at each level to perform multiple scaling operations sequentially. It transfers spatial information from the lower layers to the top ones with minimal loss to achieve more precise localization. YOLO v4, similar to YOLO v3, employs one-stage object detectors as detection heads. These YOLO heads are used for fusion and interaction with feature maps of different scales to detect objects.

In YOLO v4, the "bag of specials" (BoS) [34] and "bag of freebies" (BoF) [38] tools are deployed to improve the network performance. The use of BoS tools increases the inference time but can significantly enhance the performance of the network. In contrast, BoF contains several data augmentation techniques that improve the model accuracy without increasing the inference time. The complete intersection over union (CIOU) loss, drop block regularization, cutMix, mosaic augmentation techniques, etc. are packaged in BoF. The BoS features include Mish activation, SPP, a spatial attention module (SAM), DIOU-NMS [39], and PANet blocks.

The loss function is an important indicator for evaluating the quality of the detection model in object detection [40]. In YOLO v4, the total loss, P_{Loss}, comprising object classification loss (L_{OC}), confidence loss (L_{OF}), and regression loss (L_{OCI}), is defined by Equation (4):

$$P_{Loss} = \epsilon_1 L_{OC} + \epsilon_2 L_{OF} + \epsilon_3 L_{OCI} \quad (4)$$

where ϵ_1, ϵ_2, and ϵ_3 represent the balancing coefficients, which are usually all set to one. L_{OC} and L_{OF} are measured using the cross-entropy operation, similar to YOLO v3 [35]. L_{OCI} is predicted based on the CIOU algorithm, which calculates the positional loss between the predicted bounding box (φ') and ground truth (φ), as illustrated in Figure 4, while $\overline{\varphi}$ denotes the minimum outer bounding box encompassing both the predicted bounding box and the ground truth:

$$\mathrm{CIOU} = \mathrm{IOU}(\varphi, \varphi') - \frac{u^2}{c^2} - \varepsilon\rho \quad (5)$$

where u represents the distance between the center point of φ' and φ while c is the diagonal distance between $\overline{\varphi}$ and $\varepsilon = \rho/(1 - \mathrm{IOU}) + \rho$. The intersection over union (IOU) is given by $\mathrm{IOU} = |\varphi \cap \varphi'|/|\varphi \cup \varphi'|$. The symbols "$\cap$" and "$\cup$" depict the intersection and union operation, respectively. The adjustment factor ρ is shown in Equation (6):

$$\rho = \frac{\left(\tan^{-1}(W_g/H_g) - \tan^{-1}(W_p/H_p)\right)^2}{0.25\pi^2} \quad (6)$$

where W_g and H_g denote the width and height of φ, respectively, while W_p and H_p denote the width and height of φ', respectively. In Equation (4), the regression loss component L_{OCI} can be measured as $1 - \mathrm{CIOU}$. Furthermore, the accuracy of the bounding box is presented through the IOU of the predicted box and the actual box. The confidence level of the bounding box is measured by Equation (7):

$$C_f = P_r(\mathrm{Obj}) \times \mathrm{IOU},\ P_r(\mathrm{Obj}) \in \{0, 1\} \quad (7)$$

where $P_r(\mathrm{Obj})$ denotes the probability of an "object" being present within a bounding box. When that bounding box contains the target object, $P_r(\mathrm{Obj}) = 1$; otherwise, if the bounding box does not contain the target object, then $P_r(\mathrm{Obj}) = 0$.

Figure 4. Illustration of the prediction bounding box and the ground truth. The red dotted box (φ') represents the prediction bounding box, the green dotted box (φ) indicates the ground truth, and $\overline{\varphi}$ denotes the minimum outer bounding box encompassing both φ' and φ.

During the movement of the robot, challenges such as uneven terrain and variable external lighting may cause the center point of the object to be distorted or disappear. These factors can disrupt the accuracy of object bounding box determination in deep learning applications, leading to deviations in center point detection. To address this, the object center points are extracted from multiple images, and a least square regression analysis is performed. This analysis aims to determine the regression line that fits the distribution of these points while minimizing the sum of the squared vertical distances between the line and the points [41]. Given k data pairs, represented as $\{(x_i, y_i), i = 1, \ldots, k\}$, the relation between x_i and y_i is discerned, factoring in an error term ε_i to account for any uncertainties or deviations:

$$y_i = ax_i + b + \varepsilon_i \tag{8}$$

Assuming that $\hat{a}$ and $\hat{b}$ represent the approximations of parameters a and b, these are utilized in a subsequent problem, aiming to achieve the most suitable fit for the data points:

$$\min_{a,b} \sum\nolimits_{i=1}^{k} \hat{\varepsilon}_i^2 \tag{9}$$

If b is assumed to be zero, then the ordinary least squares method is used for estimating $\hat{a}$ to minimize Equation (9):

$$\hat{a} = \frac{\sum_{i=1}^{k} x_i y_i}{\sum_{i=1}^{k} x_i^2} \tag{10}$$

Once the slope $\hat{a}$ is obtained, this indicates that a regression line has been formed (see the example in Figure 5).

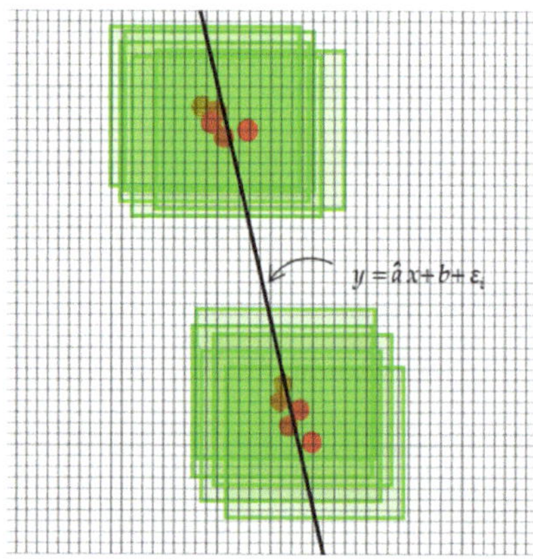

Figure 5. An example of generating a regression line using the least squares method with multiple data points in pixels, with each represented by a red dot. The green frame and red dot represent the continuously detected object frame and its center point respectively.

It should be noted that during the object recognition process using the YOLO v4 model, detected objects of the same type are assigned corresponding identification numbers, and the center point position of each object is continuously recorded.

Common performance indicators for evaluating YOLO v4 models include the precision (PR), recall (RCA), and F1-score, as outlined in Equations (11)–(13), respectively. Among these, truth positive (TP) represents samples where the model correctly predicts a positive outcome, false positive (FP) represents samples where the model incorrectly predicts a positive outcome, true negative (TN) represents samples where the model correctly predicts a negative outcome, and false negative (FN) represents samples where the model incorrectly predicts a negative outcome. It is worth noting that for each image, if the IoU exceeds a predetermined threshold, then it is classified as TP; otherwise, it is classified as FP:

$$\mathrm{PR} = \frac{\mathrm{TP}}{\mathrm{TP} + \mathrm{FP}} \tag{11}$$

$$\mathrm{RCA} = \frac{\mathrm{TP}}{\mathrm{TP} + \mathrm{FN}} \tag{12}$$

$$\text{F1-score} = \frac{2(\mathrm{PR} \times \mathrm{RCA})}{\mathrm{PR} + \mathrm{RCA}} \tag{13}$$

2.3. Guidance and Control

2.3.1. Heading Control Using FLC

As described in the previous section, linear regression fitting is performed on the centers of multiple objects, measured in pixels, to obtain a regression line (see Figure 6). This process can generate up to three regression lines. Additionally, the average heading angle can be calculated using Equation (14):

$$\theta = \sum_{n=1}^{N} \tan^{-1} \widehat{a}_n / N \quad (14)$$

where N represents the number of object categories.

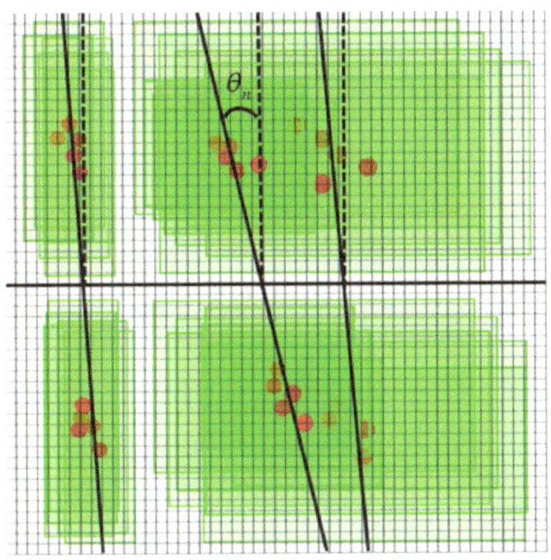

Figure 6. Illustration of multiple regression line and heading angle. The dotted line represents the vertical line in the image. The black line signifies the regression line.

In practical operation, these regression lines, constructed from pixel coordinates, are used to estimate their slope. When the guidance line is parallel to any vertical line on the image, this indicates that the robot is moving straight forward without deviating from the desired heading. However, the presence of positive or negative slopes, as well as changes in the slope, typically signify a deviation in the heading to the right or left, respectively.

Fuzzy logic is a well-known technique that involves expert opinion in decision making and is particularly suitable for finding effective solutions when information is insufficient. In this study, an FLC is utilized to adjust the heading angle of the robot. Its components include fuzzification, fuzzy decision making, defuzzification, and a knowledge base [42].

First, the role of fuzzification is to map input crisp values, denoted as "v", to fuzzy sets. This involves defining a linguistic term $\underset{\sim}{A}$ to represent a fuzzy set, viewed as a membership function. The most common membership functions are triangular and trapezoidal. As shown in Figure 7a, the triangular function and its mathematical representation consist of three parameters. The lower boundaries on the left and right are represented by α and β, respectively, while γ denotes the peak of the triangle. When the input crisp value "v" falls between α and γ, its degree of membership $\mu_{\underset{\sim}{A}}(v)$ is nonzero, while it is one when "v" equals γ and zero when "v" is less than α or greater than β. This implies that the closer "v" is to γ, the higher its degree of membership. Similarly, Figure 7b depicts the trapezoidal membership function and its mathematical representation, consisting of four parameters: α', γ_1, γ_2, and β'.

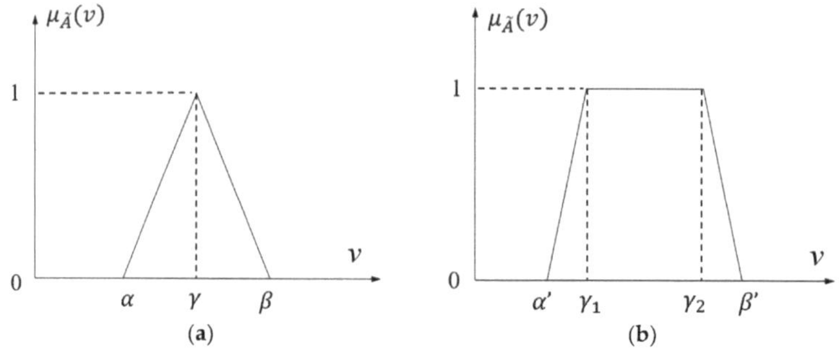

Figure 7. Illustration of membership function and their mathematical expressions. (**a**) Triangular membership function. (**b**) Ladder membership function.

In the FLC, the heading angle (θ) and the rate of change of the heading angle ($\dot{\theta}$) are the two input variables. The heading angle corresponds to three fuzzy linguistic terms: left offset (LO), middle (M), and right offset (RO). Additionally, the rate of change of the heading angle is represented by three fuzzy linguistic terms: negative (N), zero (Z), and positive (P). The output variable is the steering angle (δ), defined by the fuzzy linguistic terms left (L), mid (M), and right (R). Table 1 compiles the input and output variable values, corresponding to the fuzzy logic statements, and the parameter values of their membership functions for the FLC.

Table 1. Parameters of membership function of input and output variables in FLC.

Input Variable						**Output Variable**		
Heading angle (θ)			Rate of change of the heading angle ($\dot{\theta}$)			Steering angle (δ)		
Crisp interval		Linguistic labels	Crisp interval		Linguistic labels	Crisp interval		Linguistic labels
Triangular $[\alpha, \gamma, \beta]$	Ladderr $[\alpha', \gamma_1, \gamma_2, \beta']$		Triangularr $[\alpha, \gamma, \beta]$	Ladderr $[\alpha', \gamma_1, \gamma_2, \beta']$		Triangularr $[\alpha, \gamma, \beta]$	Ladderr $[\alpha', \gamma_1, \gamma_2, \beta']$	
–	[−100, −100, −20, 0]	LO	–	[−100, −100, −25, 0]	N	–	[−17, −17, −7, 0]	L
[−10, 0, 10]	–	M	[−20, 0, 20]	–	Z	[−5, 0, 5]	–	M
–	[0, 20, 100, 100]	RO	–	[0, 25, 100, 100]	P	–	[0, 7, 17, 17]	R

Second, the knowledge base consists of rules that primarily use [IF—THEN] statements to describe the relationships between the input and output variables. When multiple input variables are involved in the FLC, [IF—AND—THEN] statements are used. In this study, nine rules have been defined based on two input variables (θ and $\dot{\theta}$) and one output variable (δ):

Rule 1: IF (θ is LO) AND ($\dot{\theta}$ is N) THEN (δ is L);
Rule 2: IF (θ is LO) AND ($\dot{\theta}$ is Z) THEN (δ is M);
Rule 3: IF (θ is LO) AND ($\dot{\theta}$ is P) THEN (δ is R);
Rule 4: IF (θ is M) AND ($\dot{\theta}$ is N) THEN (δ is L);
Rule 5: IF (θ is M) AND ($\dot{\theta}$ is Z) THEN (δ is M);
Rule 6: IF (θ is M) AND ($\dot{\theta}$ is P) THEN (δ is R);
Rule 7: IF (θ is RO) AND ($\dot{\theta}$ is N) THEN (δ is L);
Rule 8: IF (θ is RO) AND ($\dot{\theta}$ is Z) THEN (δ is M);
Rule 9: IF (θ is RO) AND ($\dot{\theta}$ is P) THEN (δ is R).

For example, Rule 1 states that when θ belongs to the "left offset" category, and $\dot{\theta}$ indicates a "negative rate of change", the wheels should turn left.

Subsequently, the fuzzy decision and defuzzification are established. For decision making, the Mamdani model, also known as the "max-min composition method", is employed. The principle of this model involves selecting the minimum membership degree corresponding to the fuzzy set of the antecedent condition in the activated rules and assigning it to the corresponding fuzzy set in the consequent condition based on the input values ("minimum" operation). The output fuzzy sets of all the activated rules are then combined using a union operation (i.e., the "maximum" operation). After the comprehensive inference, the final result comprises a series of fuzzy sets with varying degrees. For defuzzification, the centroid method is used to convert these fuzzy values into crisp values.

2.3.2. Speed Control Using a PID Controller

The PID controller is commonly used to regulate the speed of a robot. It maintains the stable motion of the robot by adjusting three parameters: the proportional gain (K_p), integral gain (K_i) and differential gain (K_d). When the "P" term increases, the output in response to an error also increases, and vice versa. However, using only the "P" term, the system may exhibit a steady state error. To eliminate this offset, the "I" term was introduced, which works by integrating the error over time to accelerate the system's response in reaching the target state. As time progresses, the "I" term accumulates, meaning that even with smaller errors, its contribution grows due to the passage of time until the steady state error is eliminated. On the other hand, the "D" term adjusts based on the rate of change of the error relative to time. In this study, four sets of PID controllers were used to control the rotation speed of the four motors of a four-wheeled robot, with each dedicated to one motor. This approach ensures precise adjustments, responding to the specific needs and conditions of each wheel and thereby enhancing the robot's overall stability and performance in varying operational contexts. The output of the PID controller, denoted as $u(t)$, is calculated as follows:

$$u(t) = \underbrace{K_p e(t)}_{P} + \underbrace{K_i \int_0^t e(t)dt}_{I} + \underbrace{K_d \frac{d}{dt}e(t)}_{D} \quad (15)$$

where $e(t)$ represents the error between the desired velocity and estimated velocity at time t. The PID parameters are defined in Table 2. Initially, a trial-and-error method is employed to determine the parameter K_p that brings the system to a marginally stable state. This process yields a proportional gain parameter, denoted as K_{pc}. K_{pc}, in combination with a proportional coefficient, which is used to set the parameter K_i. Additionally, the system's time period, denoted as T_c, is measured and employed to obtain the parameter K_d.

Table 2. Definition of parameters of PID controller for speed control of a robot.

Parameters	K_p	K_i	K_d
Value	0.3 K_{pc}	0.2 K_{pc}	0.06 $K_{pc}T_c$

2.4. Selective Spraying

The design concept of selective spraying is depicted in Figure 8. A farm robot carries a camera and three nozzles for dispensing herbicide. The camera is positioned approximately 40 cm above the ground to ensure a comprehensive top-down field of view (depicted as the bold black box in Figure 8). The nozzles are positioned approximately 20 cm above the ground. The spray areas are divided into three sub-areas, represented by the symbols "❶", "❷", and "❸", with each corresponding to a set on the ridge. As mentioned in Section 2.2, the YOLO v4 model was used to detect drip irrigation belts, crops, and field ridges from images. A separate YOLO v4 model is also trained to specifically detect weeds and unhealthy crops, identifying the presence of tiny weeds on the ridge. When weeds or unhealthy crops are detected, the center point position of the object in pixels is estimated. As the robot travels, these tracked center points within the image cross trigger lines on the screen of the camera, as shown by the dotted line in Figure 9, activating the corresponding nozzles to deliver chemicals or nutrient solutions to the targeted objects (weeds or unhealthy crops). The middle nozzle can be used to deliver nutrient solutions or pesticides. It is important to note that the nozzle type and pressure are adjusted to cover their corresponding areas. The distance between the camera and nozzles, denoted as s_d, is crucial. When the sprayer is active, the center point of the object now crossing the spray line can be ignored to reduce the possibility of spray failure. During spraying, the distance traveled by the robot can be regarded as $s_W = vt_s$, where t_s represents the spraying time. Light blue indicates the sprayed areas, and w_R, w_M, and w_L denote the width of the right-hand, middle, and left-hand ridge in cm, respectively.

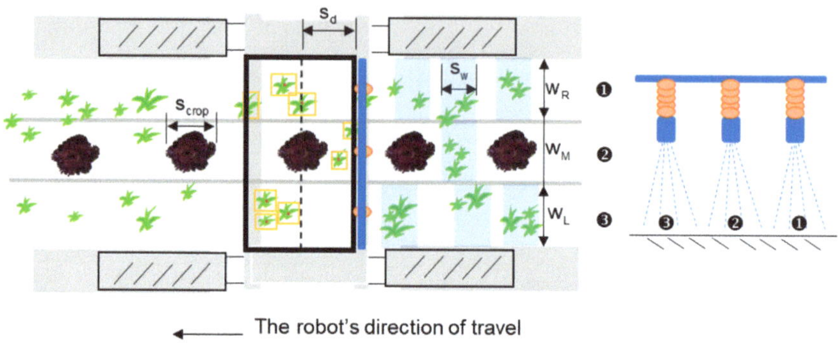

Figure 8. Design concept of selective spraying based on deep learning for an agricultural robot (depicted in gray color) in the field. Symbols "❶", "❷", and "❸" represent the right, middle, and left spray areas, respectively, each corresponding to one of the three nozzles. Purple color indicates lettuce; green color signifies weeds; and light blue color denotes areas that have been sprayed.

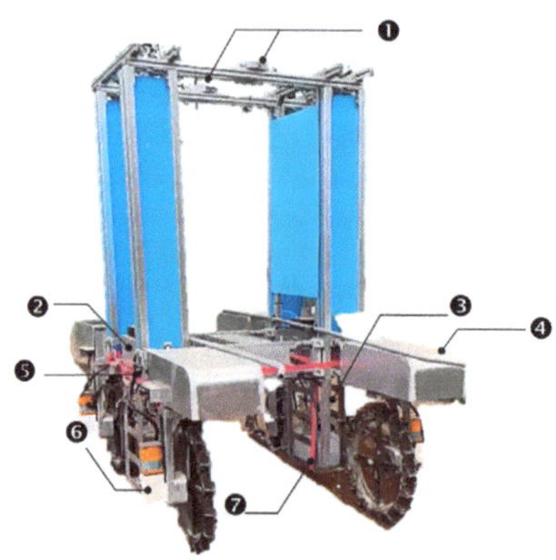

Figure 9. The experimental platform of a farming robot. Key components are labeled as follows: ❶ = two GNSS antennas; ❷ = control box; ❸ = installation space of spraying module; ❹ = linear electric actuator; ❺ = battery; ❻ = DC blushless motor; and ❼ = water tank.

In practice, the nozzles deliver chemicals to the weeds as soon as the center point of a detected weed object crosses the spray line. Due to varying weed sizes and dispersion of weeds, the object detector, even with lower recognition ability, still ensures that most of the weeds are covered with herbicides. The relationship between t_s, the width of the object $s_{'\text{object}'}$, the delay time for starting the sprayer t_{delay}, and the speed of the robot v is as follows:

$$s_d - \lambda s_{'\text{object}'} < (t_s + t_{delay})v < s_d + (1-\lambda)s_{'\text{object}'} \quad (16)$$

where $0 \leq \lambda < 1$ is the regulation factor and object $\in$ (crop, weed). Two evaluation metrics, the effective weed coverage (EWC) and ineffective weed coverage (IWC), are used to evaluate the coverage area of effective spraying and the area of ineffective spraying, given a known weed detection rate. It is assumed that there are three nozzles with fixed heights and arrangements:

$$\text{EWC }(\%) = \frac{N_{T_R} s_w w_R + N_{T_M} s_w w_M + N_{T_L} s_w w_L}{L(w_R + w_M + w_L) - s_{crop} w_M} \tag{17}$$

$$\text{IWC }(\%) = \frac{N_{F_R} s_w w_R + N_{F_M} s_w w_M + N_{F_L} s_w w_L}{L(w_R + w_M + w_L) - s_{crop} w_M} \tag{18}$$

where L represents the total length of the selective spraying experimental site and N_{T_R}, N_{T_M}, and N_{T_L} represent the number of times the right, middle, and left nozzles correctly deliver the pesticide to the weeds, respectively. Conversely, N_{F_R}, N_{F_M}, and N_{F_L} represent the number of times the right, middle, and left nozzles incorrectly deliver the pesticide, respectively. These values are estimated through spray control systems and experiments. It is particularly noteworthy that the process of spraying unhealthy crops is distinct from that of spraying weeds. The spraying rate sprayC (%) $= N_U/N_C$ is used to evaluate the spraying efficiency, where N_U and N_C represent the number of sprayed unhealthy crops and the actual number of unhealthy crops, respectively. Finally, the amount of pesticide consumed by selective spraying C_{sel} and the amount of traditional spraying C_{full} are compared to determine the pesticide reduction ratio C (%) $=(C_{full} - C_{sel})/C_{full}$.

3. Description of the Farming Robot

A four-wheel-drive (4WD) and four-wheel-steering (4WS) farming robot was utilized to evaluate the autonomous navigation and selective spraying with deep learning approach. The mechanism of the robot and software and hardware system configuration are explained below.

3.1. Mechatronics System

The experimental platform, shown in Figure 9, features a chassis composed of multiple modular mechanical components [43]. The height of the robot is available in two types: 80 cm and 200 cm. Its width is adjustable through a platinum connecting element. The shock absorber, forming a double A-arm shock absorber module, is 21 cm long with a 4 cm compression stroke. The wheels, made of hard rubber, have a diameter of 65 cm and a width of 8 cm. Each wheel is powered by a DC brushless motor (model: 9B200P-DM, TROY Enterprise Co., Ltd., Wugu, New Taipei City, Taiwan) coupled to a reduction gear set (model: 9VD360H, TROY Enterprise Co., Ltd., Wugu, New Taipei City, Taiwan) with a 360:1 reduction ratio and controlled by a motor drive (model: AGV-BLD-1S-200W, TROY Enterprise Co., Ltd., Wugu, New Taipei City, Taiwan). Additionally, four steering drives (model: CSBL1400, CSIM Inc., Xinzhuang, New Taipei City, Taiwan) are connected to four servo motors (model: CS60-150C8AE, CSIM Inc., Xinzhuang, New Taipei City, Taiwan) built into linear electric actuators (model: LOE-40-100-C-L5, LIND Inc., Taiping, Taichung County, Taiwan). The robot's embedded board and peripheral electronic components are housed inside a control box. Two sets of RTK-GNSS modules (model: C099-F9P, u-Blox Inc., Thalwil, canton of Zürich, Switzerland) with two antennas (model: VEXXIS Antennas GNSS-502, NovAtel Inc., Issy-les-Moulineaux, France) are installed on the front and rear brackets at the top of the robot. An embedded board (model: Jetson Xavier NX, NVIDIA Inc., Sunnyvale, CA, USA), serving as the main controller of the robot's operating system, executes deep learning algorithms for selective spraying and enables autonomous operation based on programmed instructions. A camera (model: BRIO 4K Ultra HD, Logitech Inc., Lausanne, Switzerland) is mounted under the central frame to capture images of field ridges, crops, weeds, or drip irrigation belts. The spray module, housed in a waterproof box attached to the side bracket, is connected by hoses to nozzles at the rear of the central bracket. The nozzles, directed toward the ground, cover the left, center, and right areas of the camera's field of view. Data transfer connectivity utilizes Universal Serial Bus (USB), General-Purpose Input/Output (GPIO), RS485, and RS-232 protocols, providing an interface between the robot's operating system and electronic components such as the camera, GNSS receivers, drivers, spraying module, and other peripherals (Figure 10). The detailed specifications of the electronic components are presented in Table 3.

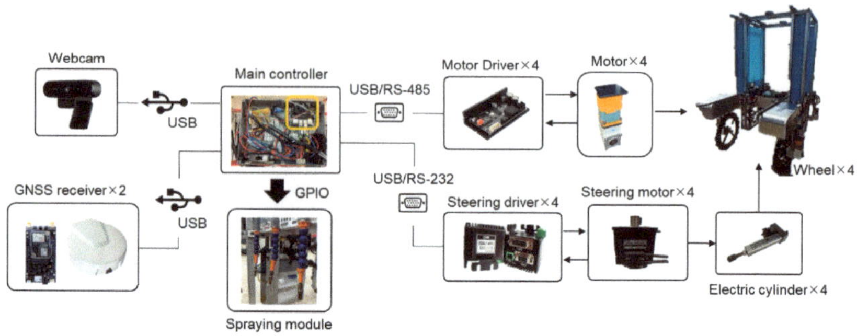

Figure 10. Hardware architecture of robotic control system.

Table 3. Component specification for an agricultural robot.

Description	Value or Other Details
Mechanism body	
Length	2.4 m
Width	1.06 m
Height	2 m
Wheel diameter	0.65 m
Maximum weight	440 kg
Drive components	
Drive method	4 WS/4 WD
Maximum speed	1 Km/h
Motors (input voltage, gear ratio; torque)	DC 24 V, 1/360, 0.8 N-m
Linear electric actuator (input voltage, gear ratio; torque)	DC 24 V, 1/5, 0.64 N-m
Battery (voltage, capacity)	DC 24 V, 30 Ah
Sprayer	
Electromagnetic valve (volt, pressure)	DC 12 V, 0~10 Kg/cm^2
Pump (input voltage, power, pressure, volumetric flow rate)	DC 24 V, 70 W, 1.0 MPA, 4.5 L/min
Copper nozzle (diameter)	1 mm/0.1 mm
Water container	20 L
Electronics	
GNSS receiver (voltage, accuracy, band)	3.3/5 V, 0.01 m–2.5 m, L1/L2C
Antennas (input voltage, signal type, noise figure, type, connector)	3.3 V–18 V, GPS/GLONASS/Galileo/BeiDou, 2.5 dB, active, TNC
Camera (maximum resolution, mega pixel, focus type, FoV, connection type)	4 K/30 FPS, 13, autofocus, 90 degree, USB

3.2. Steering Mechanism

The linear electric actuator, characterized by its high output torque, is ideally suited for assembly in agricultural robots, particularly for steering control. It comprises a servo motor and a screw mechanism, which convert the rotational motion of the motor shaft into the linear motion of the piston rod. This steering mechanism is used to adjust the steering angle δ (as shown in Figure 11), which is defined by Equation (19):

$$\delta = \cos^{-1}\left(\frac{r^2 + f^2 - d'^2}{2rf}\right) - \cos^{-1}\left(\frac{r^2 + f^2 - d^2}{2rf}\right) \quad (19)$$

where r denotes the length from the center point of the link slider to the endpoint of the piston rod and f represents the distance between the center point of the link slider and the base end of the electric actuator, while d and d' signify the original length and the extended length from the front end to the base end of the electric actuator, respectively. In this study, $r = 8.07$ cm, $f = 46.8$ cm, and $d = 39.6$ cm.

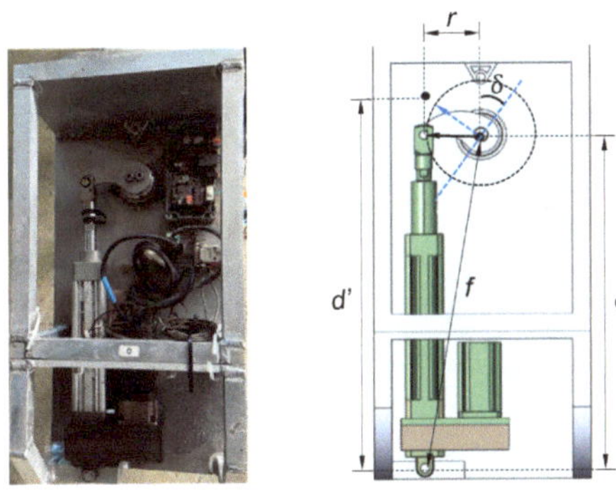

(**a**) (**b**)

Figure 11. Steering mechanism of robot. (**a**) Appearance of steering mechanism. (**b**) Relationship between the steering and linear electronic actuators.

3.3. Spraying Module

The circuit of the spray module is shown in Figure 12a. The main controller operates the spray program and sends a start or stop command to the relay module through the GPIO interface. This process controls the activation or deactivation of the solenoid valve, thereby regulating the timing of the spraying. A webcam is utilized to capture images on the ridge. The control box, shown in the upper part of Figure 12b, includes relays, DC-DC converters, transformers, and peripheral circuit components. The internal components of the spray box, depicted in the lower part of Figure 12b, comprise pumps, solenoid valves, connectors, and plastic water pipes. Five sets of spray connectors are installed on both the left and right sides of the box. Both the external appearance and the internal configuration of the spray nozzle used in the agricultural robot are shown in Figure 13.

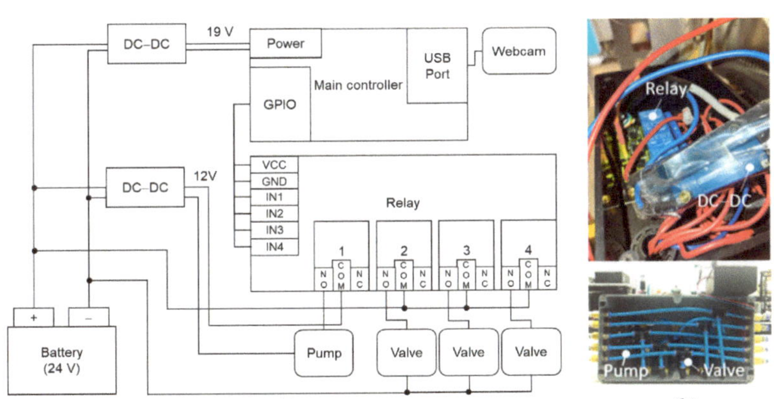

(**a**) (**b**)

Figure 12. Spray module. (**a**) A diagram illustrating the spraying circuit. (**b**) The internal configuration of the control box showing peripheral components (**top**) and the arrangement of water pipe connections inside the spray box (**bottom**).

Figure 13. The external appearance and the internal configuration of the spray nozzle.

4. Experiments and Results

4.1. Environmental Conditions and Parameters

The experimental site was an empty field located in front of the Department of Biomechanical Engineering building (longitude: 120.6059°, latitude: 22.6467°). The experiments were conducted from summer to autumn. The field spans approximately 10 m in length, with individual ridge widths measuring 80 cm. Due to the limited farming area, we only allowed the farming robot to move between two ridges (as shown in Figure 14). The experimental site is surrounded by green trees. The left and right wheels of the robot straddled the sides of a strip-shaped farmland, with both ends of the farmland serving as turning points (marked by star-shaped dots). When the position of the robot fell into the set range to be turned, the robot stopped and performed a 90 degree on-the-spot rotation in place to keep the head of the robot facing the forward direction. The motion behavior was repeated until the robot returned to the origin point (see "ST/END" in Figure 14).

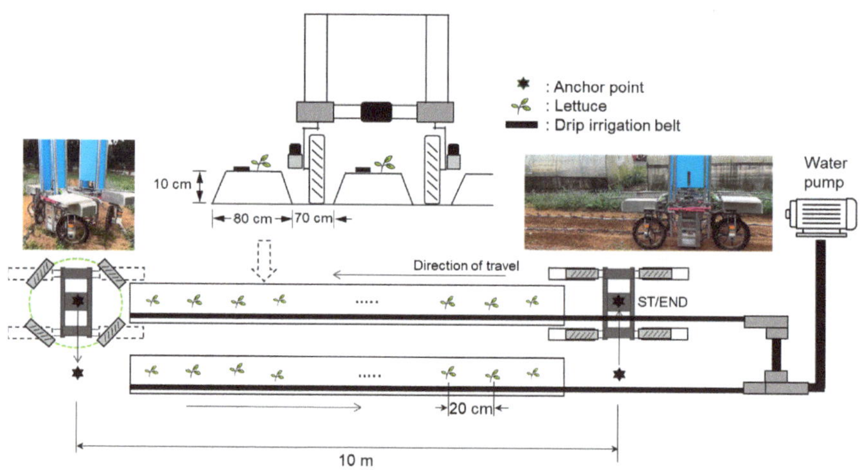

Figure 14. Schematic of movement behavior of the robot in the field. Star-shaped dots represent turning points, as well as start and end (ST/END) points.

The GNSS receiver, enhanced with RTK capabilities, produces navigation data in a format established by the National Marine Electronics Association (NMEA), offering highly accurate longitudinal and latitudinal details [44]. Two GNSS receivers were employed to record the robot position, separated by 0.65 m. The latitude and longitude of the location obtained from the two receivers were each converted into two TWD97-based positions [45]. Generally speaking, a typical video frame rate is 15–30 FPS [46]. Considering hardware limitations and expected objects, in this study, FPS values between 1 and 13 were evaluated with an image size of 416 × 416.

The experimental periods were divided into morning (9:00–11:00 a.m.), noon (12:00–2:00 p.m.), and afternoon (3:00–5:00 p.m.). The weather conditions during these periods varied and may have been sunny (9800~35,000 lux), partly sunny (3000~9800 lux), or cloudy (0~3000 lux). The experimental period spanned 3 months. Since lettuce crops are harvested approximately 20–30 days after sowing, a total of three harvests were performed during the trial, and the cultivated land was reseeded after each harvest. During the planting process, the amount of watering for each crop was adjusted, resulting in differences in growth conditions.

4.2. Preliminary Test

Red leaf lettuce (model: HV-067, Taiwan Known-You Seed Co., Ltd., Dashu Kaohsiung city, Taiwan) was selected as the target crop. Two YOLO v4 models were trained and employed for guidance line detection (DetModel #1) and the detection of weeds and unhealthy crops (DetModel #2). The users captured image samples randomly at the experimental site every day using cameras. These images encompassed weeds, both healthy and abnormal crops, drip irrigation belts, and ridges.

A total of 5800 images were collected in the experimental farm area of Pingtung University of Science and Technology during the autumn and winter of 2023. A multi-channel data logger (model: WatchDog 1650, Spectrum Technologies, Inc., Aurora, IL, USA) was used to record the light intensity. These images were taken through the camera on the robot platform at different times and under different weather conditions. Each image showed drip irrigation belts, field ridges, crops, unhealthy crops, or weeds. These images were then processed through image argumentation to obtain a total of 9500 images, which were used to build DetModel #1 and DetModel #2. The images were divided into a training set, test set, and validation set according to the ratio of 7:2:1. The images in the training set were manually annotated by using the open-source image annotation tool LabelImg to mark objects within the images. The abnormal growth of crops was characterized by symptoms such as wrinkled leaves, as depicted in Figure 15a,b. The type of weed at the experimental site is also shown in Figure 15c.

(a) (b) (c)

Figure 15. Leaf appearance for unhealthy crops and weeds. (**a**) Wrinkled leaves (sample 1). (**b**) Wrinkled leaves (sample 2). (**c**) The type of weed.

The hyperparameters for both YOLO v4 models (DetModel #1 and #2) were configured as follows: the batch size set to 64, subdivisions set to 32, image size of 416 × 416 (width × height), decay rate of 0.0005, momentum of 0.949, learning rate set to 0.001, and maximum number of batches set to 10,000.

The training iterations for DetModel #1 and DetModel #2 were stopped after reaching 10,000 and 6000 iterations, with P_{Loss} values of 1.0719 and 14.8856, respectively. The mean average precision (mAP) values of DetModel #1 and DetModel #2 were 99.0% and 92.7%, respectively (Figure 16).

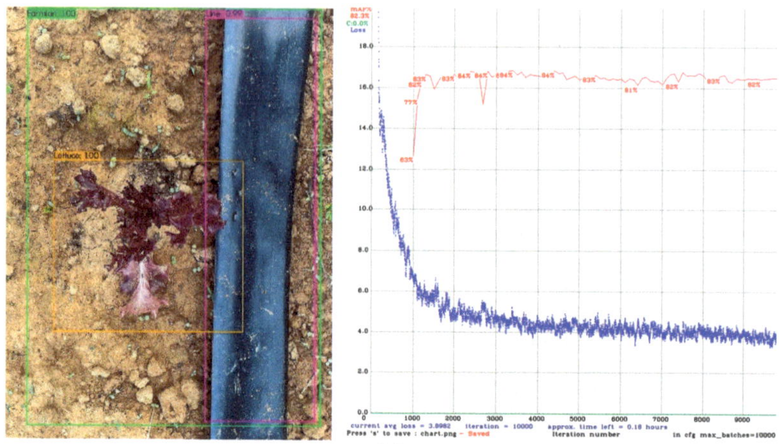

(a)

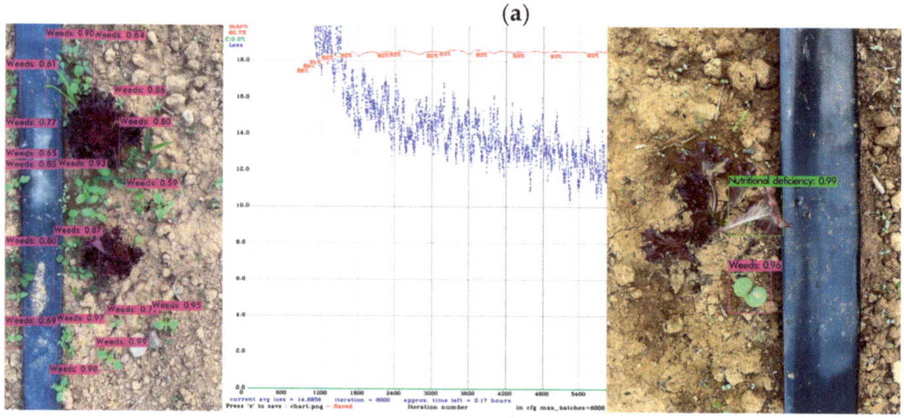

(b)

Figure 16. Identification of objects in the ridge using trained YOLO v4 models. (**a**) Detection results for the ridge, drip irrigation belt, and crops using DetModel #1. (**b**) Detection results for weeds and unhealthy crops using DetModel #2, featuring weed detection results (purple bounding box on the left side), the loss convergence curve (in the middle), unhealthy crops (green box on the right side), and weeds (purple bounding box).

The performance comparison results of the two detection models, DetModel #1 and DetModel #2, under different weather conditions are shown in Table 4. The results indicate that the recognition rates of DetModel #1 for detecting drip irrigation belts, crops, and ridges ranged from about 96 to 99%, 93 to 98%, and 93 to 97% respectively. Under the sunny conditions and during time periods from 3:00 to 5:00 p.m., DetModel #1 achieved the best recognition rate of 99% for identifying drip irrigation belts. On cloudy days and between 3:00 and 5:00 p.m., the average precision for ridges dropped to 93%. Overall, using DetModel #1, the average accuracy was about 98%. Secondly, the accuracy of DetModel #2 for detecting unhealthy crops and weeds ranged from 84 to 92% and 86 to 92%, respectively. Among these, the highest accuracy rate for weed detection was 93% on sunny days between 9:00 and 11:00 a.m. In contrast, on cloudy days during the same time period, the detection rate fell to 84%. Using DetModel #2, the average accuracy for weeds and unhealthy crops was about 88%.

Table 4. Comparison of the performance of detection models in identifying different types of objects under different weather conditions.

Name of Model	Type of Object	Evaluation Matric (%)	Sunny			Partly Sunny			Cloudy		
			9:00~ 11:00 a.m.	12:00~ 2:00 p.m.	3:00~ 5:00 p.m.	9:00~ 11:00 a.m.	12:00~ 2:00 p.m.	3:00~ 5:00 p.m.	9:00~ 11:00 a.m.	12:00~ 2:00 p.m.	3:00~ 5:00 p.m.
DetModel #1	Drip irrigation belt	PR	98.1 ± 0.1	94.3 ± 0.7	98.7 ± 0.9	97.3 ± 0.9	94.3 ± 0.9	97.2 ± 0.9	97.2 ± 0.3	95.1 ± 0.4	96.4 ± 0.7
		Recall	96.2 ± 0.5	94.3 ± 0.8	97.4 ± 0.9	95.1 ± 0.8	94.1 ± 1.1	96.6 ± 0.7	96.1 ± 0.5	93.7 ± 0.6	94.7 ± 0.6
		F1 score	97.1 ± 0.1	94.3 ± 0.7	98.0 ± 0.9	96.2 ± 0.8	94.2 ± 0.9	96.9 ± 0.8	96.6 ± 0.4	94.4 ± 0.5	95.5 ± 0.6
	Crop	PR	97.8 ± 0.2	93.2 ± 0.5	97.6 ± 0.1	97.8 ± 0.9	94.3 ± 1.1	97.1 ± 1.0	97.8 ± 0.7	95.0 ± 0.8	97.4 ± 0.3
		Recall	96.1 ± 0.1	93.1 ± 0.7	96.9 ± 0.4	96.8 ± 0.6	94.6 ± 0.9	96.7 ± 0.8	96.1 ± 0.6	94.1 ± 0.6	96.3 ± 0.3
		F1 score	96.9 ± 0.1	93.1 ± 0.6	97.2 ± 0.3	97.3 ± 0.7	94.4 ± 1.0	96.9 ± 0.9	96.9 ± 0.6	94.5 ± 0.7	96.8 ± 0.3
	Ridge	PR	95.3 ± 0.2	93.4 ± 0.8	97.1 ± 0.9	95.2 ± 1.1	94.3 ± 1.2	96.9 ± 1.1	94.8 ± 0.3	94.2 ± 0.9	93.2 ± 1.2
		Recall	96.1 ± 0.4	95.1 ± 0.7	96.1 ± 0.6	94.7 ± 0.7	94.4 ± 0.1	96.8 ± 0.8	94.1 ± 0.2	94.2 ± 0.7	93.3 ± 0.2
		F1 score	95.7 ± 0.3	94.2 ± 0.7	96.6 ± 0.7	94.9 ± 0.9	94.3 ± 0.2	97.2 ± 0.9	94.4 ± 0.2	94.2 ± 0.7	93.2 ± 0.3
DetModel #2	Unhealthy crop	PR	90.3 ± 2.0	88.3 ± 2.2	91.2 ± 1.8	90.1 ± 0.5	87.8 ± 1.2	90.8 ± 1.2	90.2 ± 1.1	88.4 ± 1.8	85.1 ± 1.4
		Recall	81.3 ± 0.1	82.4 ± 0.1	82.4 ± 0.1	83.1 ± 0.4	83.2 ± 0.9	82.2 ± 0.9	81.1 ± 1.0	81.3 ± 1.2	80.3 ± 1.1
		F1 score	85.6 ± 0.2	85.2 ± 0.2	86.4 ± 0.2	86.5 ± 0.4	85.4 ± 0.9	86.3 ± 0.9	85.4 ± 1.1	84.7 ± 1.3	82.6 ± 1.2
	Weed	PR	92.1 ± 1.1	88.3 ± 1.5	90.1 ± 1.8	89.7 ± 1.4	90.8 ± 1.2	90.2 ± 1.9	90.2 ± 1.1	88.7 ± 1.2	86.6 ± 1.5
		Recall	84.3 ± 0.1	81.2 ± 1.1	82.4 ± 1.6	83.4 ± 1.3	84.1 ± 1.1	82.4 ± 1.6	83.3 ± 0.9	82.4 ± 1.0	80.3 ± 1.1
		F1 score	88.0 ± 0.2	84.6 ± 1.1	86.1 ± 1.6	86.4 ± 1.3	87.3 ± 1.1	86.1 ± 1.7	86.6 ± 0.9	85.4 ± 1.1	83.3 ± 1.1

4.3. Experimental Results

Two scenarios, comprising the autonomous guidance and selective spraying experiments, were conducted to evaluate the robustness of the proposed scheme.

4.3.1. Scenario 1

A total of 60 crops were planted in two rows of farmland. The speed of the robot was 12.5 cm/s. During the autonomous navigation of the farming robot, data such as the velocity of each motor, heading angle, and the output value of the FLC were recorded. The experimental time was from 9:00 to 11:00 a.m., and the weather conditions were sunny. Three types of guidance lines were measured to estimate the deviation angle ($N = 3$), and the angles were averaged to determine the real-time heading angle of the robot. PID control and a fuzzy logic-based steering control program embedded in the guidance system were executed to continuously maintain the speed of the robot and correct its heading angle. The movement trajectory of the robot was obtained by two GNSS receivers as shown in Figure 17a. The change in velocity of each wheel of the robot is shown in Figure 17b. The control parameters of the PID controller for the four motor drivers, K_p, K_i, and K_d, were all set to 0.5, 0.1, and 0.6, respectively. Specifically, when the robot moved forward, the required motor speed for the four wheels was maintained at approximately 700 rpm (regardless of the reduction ratio). A speed overshoot occurred briefly when the motor speed value was switched. Additionally, this phenomenon also occurred when the velocity of the four motors was maintained at about 1000 rpm during in-place rotation for the robot.

Secondly, the results of using different guidance lines for estimating the heading angle and correcting it through the FLC were compared, as shown in Figure 18a. The speed of the robot was 12.5 cm/s. The changes in the heading angles, measured from the regression lines generated by fitting different types of objects, were observed. It was found that when the crop line was used as the reference guidance line, it showed the largest change in heading angle (Figure 18a). Conversely, using the irrigation line as a guidance line resulted in minimal variation in the heading angle. Figure 18b shows the steering angle obtained by using the FLC when the crop line was the guidance line. The output value of the FLC ranged between ±2 degrees, with a few output values reaching ±6 degrees. These larger steering angle peaks appeared within the unit time range of 50~60, 120~130, or 225~235. These values reflect the heading angle and its changes during the corresponding time intervals when the robot traveled along the crop line (Figure 18a). The reason for these instantaneous changes in angles is that the planting position of the crops was deviated due to human factors, or crops that were not growing well (elongated) caused the center point of the labeled object to deviate too far from the expected position.

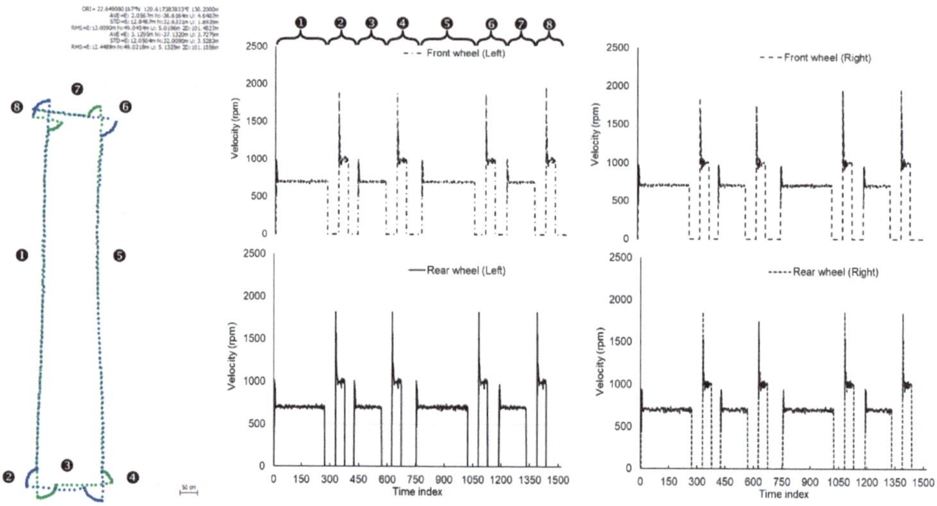

Figure 17. The velocity control of the farming robot in the farmland using PID control. (**a**) The movement trajectory of the robot (depicted by blue and green dotted lines) obtained using two GNSS receivers. (**b**) A comparison of the speed variation range and motion behavior of each wheel in relation to the positioning trajectory shown in (**a**). Numbers 1 through 8 correspond to the respective rotational speed changes of the four motors when the robot is in motion.

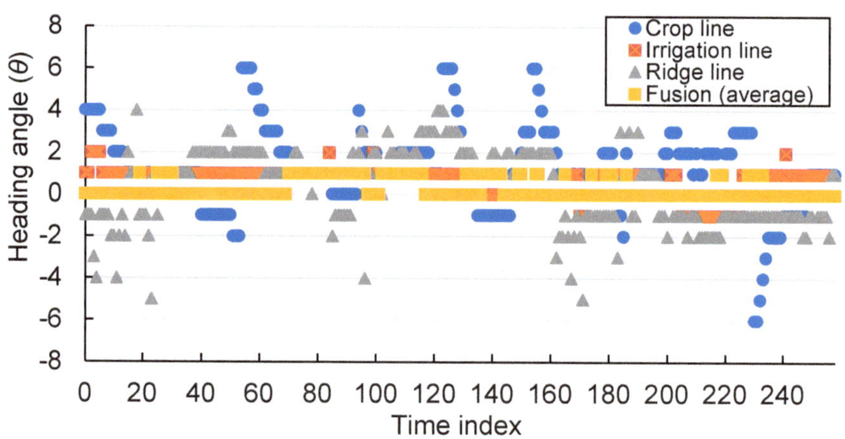

(**a**)

Figure 18. *Cont.*

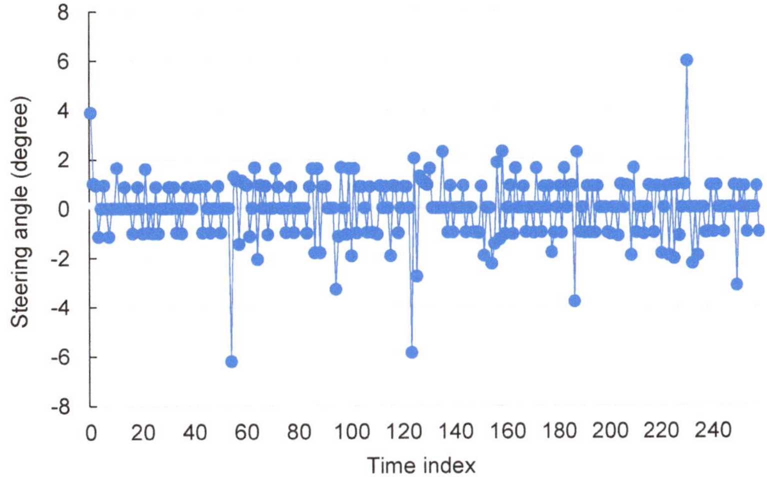

(**b**)

Figure 18. Changes in heading angle using different types of guidance lines. (**a**) The average angular variation in the heading angle was approximately 1 degree, indicated by the orange square. (**b**) The steering angle obtained using the FLC when the crop line served as a navigation line.

On the other hand, the impact of different speeds of the robot on the fitting results of the regression under varying weather conditions was observed. As shown in Figure 19, it is evident that the use of DetModel #1 enabled object detection and achieved a mean average precision (mAP) of 97% when the speed of the robot was 12.5 cm/s. However, as the speed of the robot increased, the mAP gradually decreased. Specifically, at a speed of 19 cm/s, the mAP dropped to below 75%, and when the speed of the robot reached 35 cm/s, the mAP decreased to below 50%. At a travel speed of 12.5 cm/s, there was no significant difference in the mAP under different weather conditions.

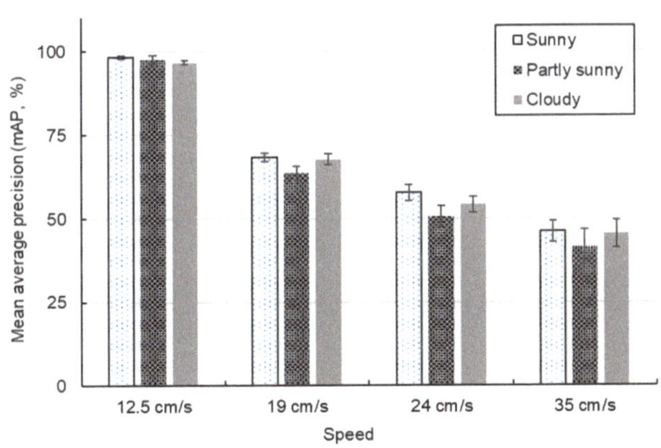

Figure 19. Comparison of the robot's object detection performance at different robot speeds.

A snapshot of the results from using DetModel #1 to continuously detect objects of different categories (FPS = 7) and using the least squares method to generate guidance lines under different weather conditions is shown in Figure 20. Although Figure 20a,b displays an uneven brightness distribution, three lines were still generated: the irrigation line (red), the crop line (orange), and the field border line (blue). The same results were also observed in low-brightness environments, as demonstrated in Figure 20c.

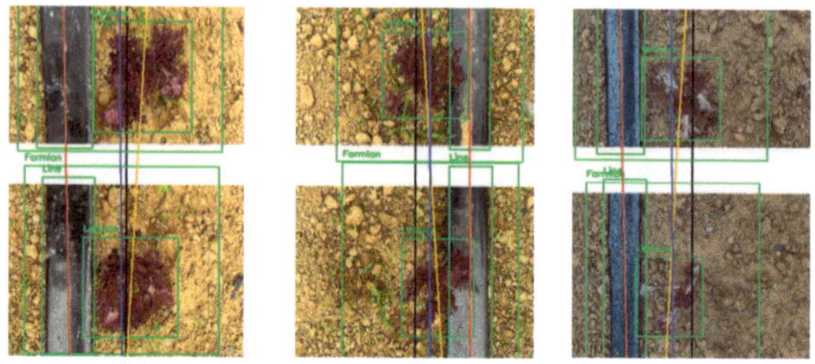

(a) (b) (c)

Figure 20. Snapshots showcasing guidance line generation and object detection results under various weather conditions (FPS = 7; speed = 12.5 cm/s). (**a**) Sunny weather. (**b**) Partly sunny conditions. (**c**) Cloudy conditions. Black line represents the vertical line in the center of the image; orange line indicates crop line; red line denotes irrigation line; blue line signifies ridge line; and green frame highlights the detected object frame.

A guidance line was generated after fitting the measured data through least squares regression. The effect of the number of FPS on the reliability of regression line generation was evaluated. Videos recorded by remotely controlling the robot while it traveled at different speeds in the field were used to evaluate the impact of different frame rates on changes in the heading angle. The heading angles obtained from the three fitted regression lines were averaged and compared with the FPS values. When the speed of the robot was maintained at 12.5 cm/s, and the FPS was greater than seven, the variation range in heading angle was about two degrees (Figure 21). Similar results were also presented when the speed of the robot was equal to 19 cm/s and the FPS was equal to 11 and 13. When the speed of the robot was maintained at 24 cm/s, and the FPS was greater than 5, its heading angle changed by about 3.5–5.5 degrees. When the robot speed was 35 cm/s, the FPS had to be increased to above nine to obtain the heading angle. However, in this case, the range of variation in the heading angle was the largest compared with other scenarios.

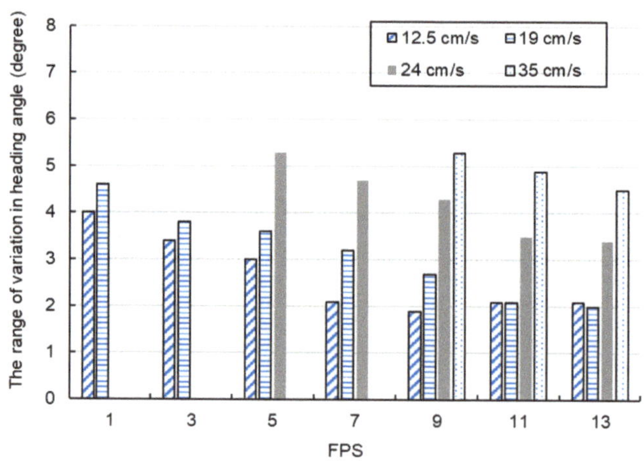

Figure 21. Variation range in heading angle versus FPS. The heading angle could not be obtained when the speed of the robot was 19 cm/s, 24 cm/s, or 35 cm/s.

4.3.2. Scenario 2

In the selective spraying experiment, water-soluble pigments were used as the spraying solution. After each spraying experiment, weeds were removed manually. About a week later, once the weeds had regrown, the spraying experiment was carried out again. Equations (16) and (17) were used to estimate the EWC and IWC, respectively. The realism of the sprayed area was determined by visually inspecting whether water dripped onto the weeds or crops. This procedure was executed in three replicates, where the total amount of solution applied in each test was quantified through a water storage bucket on the side of the robot.

According to the weed detection results presented in Table 4, the weather conditions during this experiment were sunny, and the time period was from 9:00 to 11:00 a.m. There were two strips of cultivated land. Among them, $w_R = w_L = 12$ cm, $w_M = 15$ cm, $s_d = 35$ cm, and $s_{crop} = 12$ cm. When the speed of the robot was 12.5 cm/s, 19 cm/s, 24 cm/s and 35 cm/s, the delay time for starting the sprayer was set to 2.5 s, 1.5 s, 1 s, and 0.5 s, respectively. The spray time of the sprayer was set to 0.5 s ($t_s = 0.05$) for each operation. When the spraying program was executed in the robot operating system, it recorded the number of executions of the left, middle, and right nozzles, estimated the spraying area and volume, and compared them with the volume obtained using traditional spraying methods to calculate the pesticide reduction. It is important to note that during the weeding spraying experiments, the robot was remotely controlled by a human at a constant speed, and only the central nozzle sprayed the weeds. Unhealthy crops were not sprayed during these experiments. A windless environment was ensured when performing the spray tests.

The impact of different FPS values on the detection performance of DetModel #2 for weeds and unhealthy crops was also evaluated, with the speed of the robot set to 12.5 cm/s. Figure 22 shows that when the FPS was greater than seven, the average accuracy of detecting weeds and unhealthy crops ranged from 89% to 92% and from 88% to 90%, respectively. However, when the FPS was less than seven, the accuracy of detecting weeds and unhealthy crops dropped significantly.

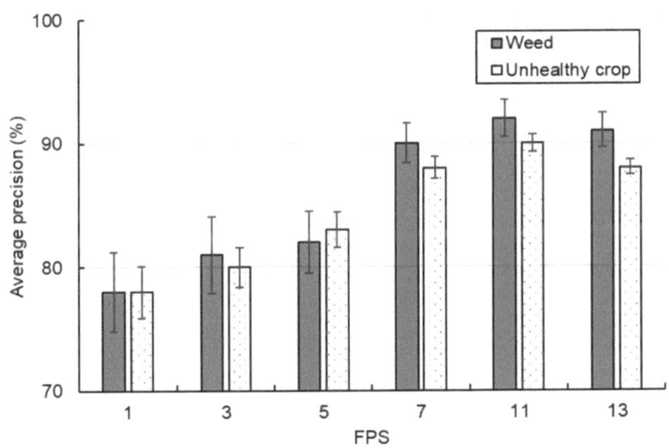

Figure 22. Comparison of the detection accuracy for weeds and unhealthy crops at different FPS values.

The results for the EWC, IWC, and pesticide reduction rate at different robot speeds with a detection time of 143 ms per image for DetModel #2 are illustrated in Table 5. Pesticide reduction refers to the decrease in pesticide usage achieved by selective spraying compared with traditional uniform spraying. It can be seen from this table that as the speed of the robot increased, the EWC gradually decreased, and the IWC gradually increased. However, limited by the detection performance of DetModel #2, when the robot speed reached 35 cm/s, the EWC was at its lowest, and the IWC was at its highest. In this scenario, although pesticide usage was significantly reduced (about 63%), most weeds were not sprayed.

Table 5. Comparison of effective and ineffective weeding coverage and pesticide reduction rates at different robot speeds (t_s= 0.5 s).

Speed	EWC (%)	IWC (%)	Pesticide Reduction (%)
12.5 cm/s	82.9 ± 1.9	8.4 ± 0.9	53.2 ± 2.9
19 cm/s	73.0 ± 2.8	15.2 ± 1.9	36.1 ± 3.6
24 cm/s	73.3 ± 5.3	19.4 ± 4.2	30.7 ± 3.6
35 cm/s	23.7 ± 5.3	27.3 ± 5.2	63.4 ± 5.8

The performance of selective spraying for unhealthy crops was evaluated. Each spray time of the sprayer was also set to 0.5 s (t_s= 0.5 s). The delay time for starting the sprayer was the same as that in the weed spraying experiment. This spraying experiment was conducted at different time periods and under different weather conditions, with the experimental procedures for each time period being repeated three times. The comparison results of the spray rate of the robot at different speeds are demonstrated in Figure 23. When the speed of the robot was 12.5 cm/s, 19 cm/s, 24 cm/s, and 35 cm/s, the spraying rates (SprayC) were between about 85 and 92%, 83% and 88%, 70 and 88%, and 40 and 65%, respectively. A snapshot of an unhealthy crop after spraying is shown in Figure 24. These images were all captured by cameras after selective spraying tests under different weather conditions. The bounding boxes in these images were labeled offline with DetModel #2 and confirm the spray behavior on unhealthy crops or weeds.

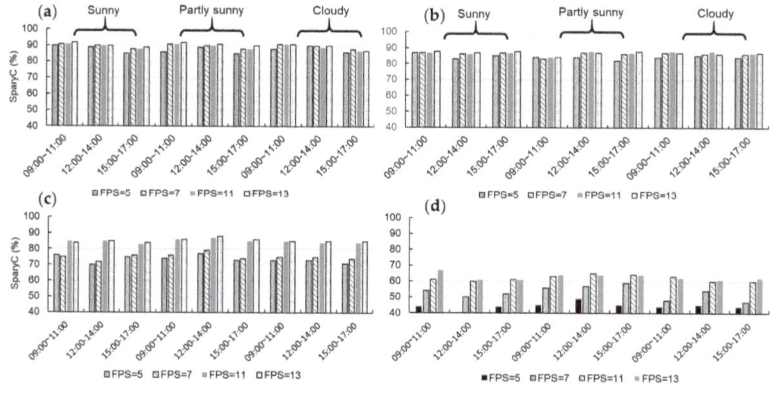

Figure 23. Comparison of spray rates (SprayC) for the robot at different speeds and FPS values: (**a**) 12.5 cm/s; (**b**) 19 cm/s; (**c**) 24 cm/s; and (**d**) 35 cm/s.

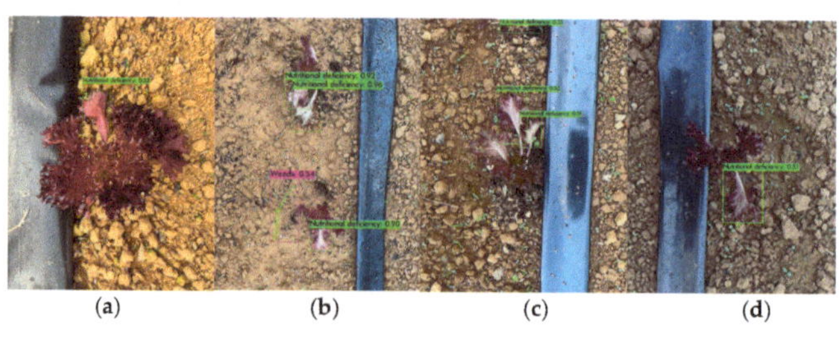

(a) (b) (c) (d)

Figure 24. Spraying results under different weather conditions (speed of 12.5 cm/s; FPS = 7): (**a**) sunny (zoomed in); (**b**) partly sunny; (**c**) cloudy (Example 1); and (**d**) coludy (Example 2). The bounding boxes were marked by DetModel #2 after spraying.

4.4. Discussion

Based on the comprehensive analysis of our experimental results, we have summarized the following points:

1. The fitting result of the guidance line is related to the speed of the robot, the detection performance of DetModel #1, and the FPS value. As shown in Table 4, the accuracy of using DetModel #1 to detect ridges, crops, or drip irrigation belts ranged from 92 to 99%, with an average accuracy (AP) of about 95%. The accuracy of identifying drip irrigation belts was the highest, reaching up to 99%, followed by crops at 98% and field ridges at 97%. Since the training set samples came from images captured in a real environment from different angles and under various weather conditions, such realistic datasets can enhance the object recognition ability of the model [47,48]. The accuracy of navigation line extraction depends on the performance of the trained model in detecting targets. When the same type of object of the image is successfully detected multiple times, a line can be fitted using the least squares method. Soil images under different climate conditions were collected, and even overexposed images were used as training samples for modeling. The experimental results show that the trained model can effectively improve its generalization performance, especially under different climate conditions. In practical operation, as long as the center points of at least two objects can be detected, the navigation line can be extracted. When the robot traveled at a speed of 12.5 cm/s, and the FPS is set to 7, its mAP could reach 98%. As the robot speed increased, the mAP gradually decreased. When comparing the relationship between the mAP and heading angle variation in Figures 19 and 21, it can be observed that a lower mAP would lead to an increased range of heading angle variation, producing more dubious heading angles. Increasing the FPS can help reduce false detection for objects caused by instantaneous strong ambient light and maintain a certain mAP even as the speed of robot increases. However, it also increases the computing load of the system, leading to the risk of overheating of the hardware system. Under conditions that meet various climatic requirements while maintaining the average object detection performance, the average detection time of DetModel #1 was 143 ms/per image.
2. Under different weather conditions, DetModel #2 was used to identify unhealthy crops and weeds, achieving average PRs of about 84% and 93%, respectively. In the afternoon on a cloudy day, the PR was slightly reduced to 84~89%, demonstrating that the deep learning model exhibited better adaptability to images with weaker light intensities. This adaptability improves upon the limitations of traditional machine learning techniques [49]. However, the limited dynamic illumination range of RGB cameras can easily lead to image color distortion when this limit is exceeded or not met [50], making object recognition difficult, especially with overexposed or insufficiently bright images. Rapid movement can result in specific frames failing to capture the target. Objects often have features significantly different from those encountered during the training process, making them difficult detect [25]. Changes in ambient lighting alter the tones and shadows in the image frame, affecting an object's color, and pixel edge blur or shadows can also significantly impact detection accuracy [51,52]. Adding more feature information to the training of the deep learning model can improve its detection performance [53]. In this study, more field images, including soil types in low-light environments and even overexposed images, were used as samples. The experimental results show that the trained deep learning model had better generalization performance.
3. Limited by the soil hardness and site flatness, in this study, the K_p value of the PID controller was set to 0.5, enabling the four DC motors to reach the required speed quickly. Although there is a short-term small oscillation when the motor starts, its impact on the traveling speed of the robot is minimal. On the other hand, the FLC can smoothly adjust the heading angle and improve the two-wheel speed difference control method [54]. The experimental results show that when the robot moved in a

guidance line at a speed of 12.5 cm/s, the variation in the heading angle stayed within one degree. Although no drip irrigation belt was present on the field, the average variation range of the heading angle remained within ±3.5 degrees (Figure 18a). Assuming that the planting position of the crops in the image was not on the vertical line in the center of the field ridge, it would also affect the estimated heading angle when the robot moved. For instance, in Figure 18a, at time indices of about 55, 130, 125, and 232, the heading angle estimated based on the crop line changed instantly by more than four degrees. Fortunately, this result has to be averaged with the heading angles obtained from the other regression lines, preventing ridge damage due to overcorrection of the wheels and avoiding misjudgment of the heading angle. When the robot rotated, a higher motor speed output value (1000 rpm) was used, ensuring greater torque was instantly available and allowing the four drives to power the motors. The advantage of the 4 WD/4 WS system is that it enables the robot to achieve a turning radius of zero. Compared with common car drive systems, this steering mode reduces the space required for turning and improves land use efficiency. Moreover, relying on RTK-GNSS receivers for centimeter-level autonomous guidance of a farming robot is a common practice [55]. This centimeter-level positioning error is acceptable when a fixed ambiguity solution is obtained [56]. However, there are risks in autonomous guiding operations on narrow farming areas. If the positioning signal is interrupted, or the positioning solution is a floating solution, then the robot may damage the field during movement.

4. During spraying test operation, setting the delay time of the sprayer according to the speed of the robot can avoid ineffective spraying. In this study, when the robot speed was 12.5 cm/s, with a sprayer time of about 0.5 s and a delay time of 2.5 s, the EWC and IWC reached 83% and 8% respectively, and the pesticide reduction reached 53%. Similar results were also presented in [57], with a 34.5% reduction in pesticide usage compared with traditional uniform spraying methods. As the speed of the robot increased, the EWC decreased, and the IWC increased. At a speed of 35 cm/s, the average precision was too low, which indirectly resulted in a reduction in the number of pesticide applications. On the other hand, the unhealthy crops were about 7–12 cm in diameter. Even with the appropriate delay time set, there were still some unhealthy crops or weeds that could not be sprayed correctly.

5. Conclusions

This study offers insights into the challenges and potential solutions for agricultural robots in real-world applications. Agricultural robots must rely on high-precision positioning systems to complete autonomous operations in the field. In practice, 4WS/4WD agricultural robots are utilized to navigate strip fields, turning on the spot to transition to other farming areas. Firstly, the experimental results confirmed that the deep learning model can detect drip irrigation belts, field ridges, crops, unhealthy crops, and weed objects with an accuracy rate ranging from 83 to 99%. This enables the implementation of various operations in climatically diverse environments, such as inter-row line tracking and selective spraying. In terms of robot guidance line tracking, this study presents an innovative method that estimates the robot's heading angle using multiple regression lines. This method has been proven to be more reliable than conventional crop row extraction techniques, significantly reducing the risk of oversteering and potential damage to crops and field ridges during the robot's movement. The deviation angle was within one degree at a speed of 12.5 cm/s, assisted by multiple guidance lines and using a PID controller and FLC. Minimizing human factors, such as misaligned crop planting and uneven crop spacing, will enhance the accuracy of the heading angle.

Secondly, an excessively fast robot movement speed causes the object detection rate to decrease. The reason involves limited computing resources leading to insufficient FPS values. By allowing the robot to operate at a slower speed, we ensure the maintenance of a certain level of object detection performance for selective spraying operations. At a frame processing rate of 7 FPS and a robot speed of 12.5 cm/s, the mAP for detecting weeds and unhealthy crops ranged from 93 to 98%. The accuracy of spraying unhealthy crops could reach up to 92%. Considering the sprayer's fixed delay time, the autonomous robot achieved an effective weed coverage rate of 83% and a pesticide saving rate of 53% while operating at a speed of 12.5 cm/s. The applicability of these advancements to low-cost hardware expands their impact across various agricultural settings, particularly benefiting small-scale and resource-limited farming. Future research will concentrate on integrating adaptive FPS and minimizing spray start latency in autonomous agricultural robots, enabling them to perform a variety of tasks in a decentralized fashion.

Author Contributions: Conceptualization, C.-L.C.; methodology, C.-L.C.; software, C.-L.C., J.-Y.K. and H.-W.C.; verification, J.-Y.K. and H.-W.C.; data management, J.-Y.K. and H.-W.C.; writing—manuscript preparation, C.-L.C. and H.-W.C.; writing—review and editing, C.-L.C.; visualization, C.-L.C.; supervision, C.-L.C.; project management, C.-L.C.; fund acquisition, C.-L.C. All authors have read and agreed to the published version of the manuscript.

Funding: This research was supported by the National Science and Technology Council (grant number NSTC 112-2221-E-020-013).

Institutional Review Board Statement: Not applicable.

Informed Consent Statement: Not applicable.

Data Availability Statement: The data that support the findings of this study are available from the corresponding author, C.-L.C., upon reasonable request.

Acknowledgments: Many thanks are due to the editors and reviewers for their valuable comments to refine this paper.

Conflicts of Interest: The authors declare no conflicts of interest.

References

1. Spykman, O.; Gabriel, A.; Ptacek, M.; Gandorfer, M. Farmers' perspectives on field crop robots—Evidence from Bavaria, Germany. *Comput. Electron. Agric.* **2021**, *186*, 106176. [CrossRef]
2. Wu, J.; Jin, Z.; Liu, A.; Yu, L.; Yang, F. A survey of learning-based control of robotic visual servoing systems. *J. Franklin Inst.* **2022**, *359*, 556–577. [CrossRef]
3. Kato, Y.; Morioka, K. Autonomous robot navigation system without grid maps based on double deep Q-Network and RTK-GNSS localization in outdoor environments. In Proceedings of the 2019 IEEE/SICE International Symposium on System Integration (SII), Paris, France, 14–16 January 2019; pp. 346–351.
4. Galati, R.; Mantriota, G.; Reina, G. RoboNav: An affordable yet highly accurate navigation system for autonomous agricultural robots. *Robotics* **2022**, *11*, 99. [CrossRef]
5. Chien, J.C.; Chang, C.L.; Yu, C.C. Automated guided robot with backstepping sliding mode control and its path planning in strip farming. *Int. J. iRobotics* **2022**, *5*, 16–23.
6. Zhang, L.; Zhang, R.; Li, L.; Ding, C.; Zhang, D.; Chen, L. Research on virtual Ackerman steering model based navigation system for tracked vehicles. *Comput. Electron. Agric.* **2022**, *192*, 106615. [CrossRef]
7. Tian, H.; Wang, T.; Liu, Y.; Qiao, X.; Li, Y. Computer vision technology in agricultural automation—A review. *Inf. Process. Agric.* **2020**, *7*, 1–19. [CrossRef]
8. Leemans, V.; Destain, M.F. Application of the Hough transform for seed row localisation using machine vision. *Biosyst. Eng.* **2006**, *94*, 325–336. [CrossRef]
9. Choi, K.H.; Han, S.K.; Han, S.H.; Park, K.-H.; Kim, K.-S.; Kim, S. Morphology-based guidance line extraction for an autonomous weeding robot in paddy fields. *Comput. Electron. Agric.* **2015**, *113*, 266–274. [CrossRef]
10. Zhou, X.; Zhang, X.; Zhao, R.; Chen, Y.; Liu, X. Navigation line extraction method for broad-leaved plants in the multi-period environments of the high-ridge cultivation mode. *Agriculture* **2023**, *13*, 1496. [CrossRef]
11. Suriyakoon, S.; Ruangpayoongsak, N. Leading point-based interrow robot guidance in corn fields. In Proceedings of the 2017 2nd International Conference on Control and Robotics Engineering (ICCRE), Bangkok, Thailand, 1–3 April 2017; pp. 8–12.
12. Bonadiesa, S.; Gadsden, S.A. An overview of autonomous crop row navigation strategies for unmanned ground vehicles. *Eng. Agric. Environ. Food* **2019**, *12*, 24–31. [CrossRef]
13. Chen, J.; Qiang, H.; Wu, J.; Xu, G.; Wang, Z. Navigation path extraction for greenhouse cucumber-picking robots using the prediction-point Hough transform. *Comput. Electron. Agric.* **2021**, *180*, 105911. [CrossRef]
14. Ma, Z.; Tao, Z.; Du, X.; Yu, Y.; Wu, C. Automatic detection of crop root rows in paddy fields based on straight-line clustering algorithm and supervised learning method. *Biosyst. Eng.* **2021**, *211*, 63–76. [CrossRef]

15. Shi, J.; Bai, Y.; Diao, Z.; Zhou, J.; Yao, X.; Zhang, B. Row detection-based navigation and guidance for agricultural robots and autonomous vehicles in row-crop fields: Methods and applications. *Agronomy* **2023**, *13*, 1780. [CrossRef]
16. Zhang, S.; Wang, Y.; Zhu, Z.; Li, Z.; Du, Y.; Mao, E. Tractor path tracking control based on binocular vision. *Inf. Process. Agric.* **2018**, *5*, 422–432. [CrossRef]
17. Mavridou, E.; Vrochidou, E.; Papakostas, G.A.; Pachidis, T.; Kaburlasos, V.G. Machine vision systems in precision agriculture for crop farming. *J. Imaging* **2019**, *5*, 89. [CrossRef]
18. Gu, Y.; Li, Z.; Zhang, Z.; Li, J.; Chen, L. Path tracking control of field information-collecting robot based on improved convolutional neural network algorithm. *Sensors* **2020**, *20*, 797. [CrossRef]
19. Pajares, G.; García-Santillán, I.; Campos, Y.; Montalvo, M.; Guerrero, J.M.; Emmi, L.; Romeo, J.; Guijarro, M.; González-de-Santos, P. Machine-vision systems selection for agricultural vehicles: A guide. *J. Imaging* **2016**, *2*, 34. [CrossRef]
20. de Silva, R.; Cielniak, G.; Gao, J. Towards agricultural autonomy: Crop row detection under varying field conditions using deep learning. *arXiv* **2021**, arXiv:2109.08247.
21. Hu, Y.; Huang, H. Extraction method for centerlines of crop row based on improved lightweight Yolov4. In Proceedings of the 2021 6th International Symposium on Computer and Information Processing Technology (ISCIPT), Changsha, China, 11–13 June 2021; pp. 127–132.
22. Ruan, Z.; Chang, P.; Cui, S.; Luo, J.; Gao, R.; Su, Z. A precise crop row detection algorithm in complex farmland for unmanned agricultural machines. *Biosyst. Eng.* **2023**, *232*, 1–12. [CrossRef]
23. Ruigrok, T.; van Henten, E.; Booij, J.; van Boheemen, K.; Kootstra, G. Application-specific evaluation of a weed-detection algorithm for plant-specific spraying. *Sensors* **2020**, *20*, 7262. [CrossRef]
24. Hu, D.; Ma, C.; Tian, Z.; Shen, G.; Li, L. Rice Weed detection method on YOLOv4 convolutional neural network. In Proceedings of the 2021 International Conference on Artificial Intelligence, Big Data and Algorithms (CAIBDA), Xi'an, China, 28–30 May 2021; pp. 41–45.
25. Chang, C.L.; Xie, B.X.; Chung, S.C. Mechanical control with a deep learning method for precise weeding on a farm. *Agriculture* **2021**, *11*, 1049. [CrossRef]
26. Wang, Q.; Cheng, M.; Huang, S.; Cai, Z.; Zhang, J.; Yuan, H. A deep learning approach incorporating YOLO v5 and attention mechanisms for field real-time detection of the invasive weed *Solanum rostratum* Dunal seedlings. *Comput. Electron. Agric.* **2022**, *199*, 107194. [CrossRef]
27. Chen, J.; Wang, H.; Zhang, H.; Luo, T.; Wei, D.; Long, T.; Wang, Z. Weed detection in sesame fields using a YOLO model with an enhanced attention mechanism and feature fusion. *Comput. Electron. Agric.* **2022**, *202*, 107412. [CrossRef]
28. Ruigrok, T.; van Henten, E.J.; Kootstra, G. Improved generalization of a plant-detection model for precision weed control. *Comput. Electron. Agric.* **2023**, *204*, 107554. [CrossRef]
29. Razfar, N.; True, J.; Bassiouny, R.; Venkatesh, V.; Kashef, R. Weed detection in soybean crops using custom lightweight deep learning models. *J. Agric. Food Res.* **2022**, *8*, 100308. [CrossRef]
30. Qiu, Q.; Fan, Z.; Meng, Z.; Zhang, Q.; Cong, Y.; Li, B.; Wang, N.; Zhao, C. Extended Ackerman steering principle for the co-ordinated movement control of a four wheel drive agricultural mobile robot. *Comput. Electron. Agric.* **2018**, *152*, 40–50. [CrossRef]
31. Bak, T.; Jakobsen, H. Agricultural robotic platform with four wheel steering for weed detection. *Biosyst. Eng.* **2004**, *87*, 125–136. [CrossRef]
32. Tu, X.; Gai, J.; Tang, L. Robust navigation control of a 4WD/4WS agricultural robotic vehicle. *Comput. Electron. Agric.* **2019**, *164*, 104892. [CrossRef]
33. Wang, D.; Qi, F. Trajectory planning for a four-wheel-steering vehicle. In Proceedings of the 2001 ICRA. IEEE International Conference on Robotics and Automation, Seoul, Republic of Korea, 21–26 May 2001; Volume 4, pp. 3320–3325.
34. Bochkovskiy, A.; Wang, C.-Y.; Liao, H.M. YOLOv4: Optimal speed and accuracy of object detection. *arXiv* **2020**, arXiv:2004.10934.
35. Redmon, J.; Farhadi, A. YOLOv3: An incremental improvement. *arXiv* **2018**, arXiv:1804.02767.
36. Wang, C.Y.; Liao, H.Y.M.; Wu, Y.H.; Chen, P.Y.; Hsieh, J.W.; Yeh, I.H. CSPNet: A new backbone that can enhance learning capability of CNN. In Proceedings of the 2020 IEEE/CVF Conference on Computer Vision and Pattern Recognition Work-shops (CVPRW), Seattle, WA, USA, 14–19 June 2020; pp. 1571–1580.
37. He, K.; Zhang, X.; Ren, S.; Sun, J. Spatial pyramid pooling in deep convolutional networks for visual recognition. *IEEE Trans. Pattern Anal. Mach. Intell.* **2015**, *37*, 1904–1916. [CrossRef] [PubMed]
38. Zhang, Z.; He, T.; Zhang, H.; Zhang, Z.; Xie, J.; Li, M. Bag of freebies for training object detection neural networks. *arXiv* **2019**, arXiv:1902.04103.
39. Zheng, Z.; Wang, P.; Liu, W.; Li, J.; Ye, R.; Ren, D. Distance-IoU loss: Faster and better learning for bounding box regression. *AAAI Tech. Track Vis.* **2020**, *34*, 12993–13000. [CrossRef]
40. Roy, A.M.; Bhaduri, J. Real-time growth stage detection model for high degree of occultation using DenseNet-fused YOLOv4. *Comput. Electron. Agric.* **2022**, *193*, 106694. [CrossRef]
41. Chang, C.L.; Chen, H.W. Straight-line generation approach using deep learning for mobile robot guidance in lettuce fields. In Proceedings of the 2023 9th International Conference on Applied System Innovation (ICASI), Chiba, Japan, 21–25 April 2023.
42. Lee, C.C. Fuzzy logic in control system: Fuzzy logic controller. *IEEE Trans. Syst. Man Cybern. Syst.* **1990**, *20*, 404–418. [CrossRef]
43. Yu, C.C.; Tsen, Y.W.; Chang, C.L. Modeled Carrier. TW Patent No. I706715, 11 October 2020.

44. Bennett, P. *The NMEA FAQ (Fragen und Antworten zu NMEA)*, Ver. 6.1; Sepember 1997. Available online: https://www.geocities.ws/lrfernandes/gps_project/Appendix_E_NMEA_FAQ.pdf (accessed on 30 January 2023).
45. Shih, P.T.-Y. TWD97 and WGS84, datum or map projection? *J. Cadastr. Surv.* **2020**, *39*, 1–12.
46. Lee, J.; Hwang, K.I. YOLO with adaptive frame control for real-time object detection applications. *Multimed. Tools Appl.* **2022**, *81*, 36375–36396. [CrossRef]
47. Hasan, R.I.; Yusuf, S.M.; Alzubaidi, L. Review of the state of the art of deep learning for plant diseases: A broad analysis and discussion. *Plants* **2020**, *9*, 1302. [CrossRef]
48. Arsenovic, M.; Karanovic, M.; Sladojevic, S.; Anderla, A.; Stefanovic, D. Solving current limitations of deep learning based approaches for plant disease detection. *Symmetry* **2019**, *11*, 939. [CrossRef]
49. Zhang, Y.; Chen, H.; He, Y.; Ye, M.; Cai, X.; Zhang, D. Road segmentation for all-day outdoor robot navigation. *Neurocomputing* **2018**, *314*, 316–325. [CrossRef]
50. Liu, J.; Wang, X. Plant diseases and pests detection based on deep learning: A review. *Plant Methods* **2021**, *17*, 22–35. [CrossRef] [PubMed]
51. Jiao, L.; Zhang, R.; Liu, F.; Yang, S.; Hou, B.; Li, L.; Tang, X. New generation deep learning for video object detection: A survey. *IEEE Trans. Neural Netw. Learn. Syst.* **2021**, *33*, 3195–3215. [CrossRef] [PubMed]
52. Li, D.; Wang, R.; Xie, C.; Liu, L.; Zhang, J.; Li, R.; Wang, F.; Zhou, M.; Liu, W. A recognition method for rice plant diseases and pests video detection based on deep convolutional neural network. *Sensors* **2020**, *20*, 578. [CrossRef]
53. Altalak, M.; Ammad uddin, M.; Alajmi, A.; Rizg, A. Smart agriculture applications using deep learning technologies: A survey. *Appl. Sci.* **2022**, *12*, 5919. [CrossRef]
54. Chang, C.L.; Chen, H.W.; Chen, Y.H.; Yu, C.C. Drip-tape-following approach based on machine vision for a two-wheeled robot trailer in strip farming. *Agriculture* **2022**, *12*, 428. [CrossRef]
55. del Rey, J.C.; Vega, J.A.; Pérez-Ruiz, M.; Emmi, L. Comparison of positional accuracy between RTK and RTX GNSS based on the autonomous agricultural vehicles under field conditions. *Appl. Eng. Agric.* **2014**, *30*, 361–366.
56. Han, J.H.; Park, C.H.; Park, Y.J.; Kwon, J.H. Preliminary results of the development of a single-frequency GNSS RTK-based autonomous driving system for a speed sprayer. *J. Sens.* **2019**, *2019*, 4687819. [CrossRef]
57. Gonzalez-de-Soto, M.; Emmi, L.; Perez-Ruiz, M.; Aguera, J.; Gonzalez-de-Santos, P. Autonomous systems for precise spraying—Evaluation of a robotised patch sprayer. *Biosyst. Eng.* **2016**, *146*, 165–182. [CrossRef]

agriculture

MDPI

Article

Stress Simulation on Four-Bar Link-Type Transplanting Device of Semiautomatic Vegetable Transplanter

Sri Markumningsih [1,†], Seok-Joon Hwang [2,3,†], Jeong-Hun Kim [2,3], Moon-Kyeong Jang [2,3] and Ju-Seok Nam [2,3,*]

1 Department of Agricultural and Biosystems Engineering, Faculty of Agricultural Technology, Universitas Gadjah Mada, Yogyakarta 55281, Indonesia; sri_markumningsih@ugm.ac.id
2 Department of Biosystems Engineering, Kangwon National University, Chuncheon 24341, Republic of Korea; human51@kangwon.ac.kr (S.-J.H.); jhkim1995@kangwon.ac.kr (J.-H.K.); moon2842@kangwon.ac.kr (M.-K.J.)
3 Interdisciplinary Program in Smart Agriculture, Kangwon National University, Chuncheon 24341, Republic of Korea
* Correspondence: njsg1218@kangwon.ac.kr; Tel.: +82-33-250-6497
† These authors contributed equally to this work.

Abstract: The aim of this study is to analyze the stress exerted on a four-bar link-type transplanting device using two distinct methods: stress measurement performed during a field test and stress simulation. A field test is conducted to measure stress using a strain gauge positioned at 15 specific points on the transplanting device. Subsequently, the measured strain data are converted into calculated stress data. In another method, stress is simulated using specialized multibody dynamic simulation software. The simulation results are compared with the stress measured during field tests to verify the simulation model. Based on the results, the maximum stress derived from the simulation correlates with the measured results, although notable discrepancies are shown, particularly at strain gauge positions 11 and 13. The maximum stress derived from the simulation is used to calculate the static safety factor of the transplanting device. The peak stress derived from the simulation aligns with the measured results, although significant discrepancies are observed at positions corresponding to strain gauges 4 and 10. The maximum stress (150.82 MPa) is observed on the link of the transplanting device, and the static safety factor determined via the simulation is 1.39.

Keywords: four-bar link-type; static safety factor; stress simulation; transplanting device; vegetable transplanter

Citation: Markumningsih, S.; Hwang, S.-J.; Kim, J.-H.; Jang, M.-K.; Nam, J.-S. Stress Simulation on Four-Bar Link-Type Transplanting Device of Semiautomatic Vegetable Transplanter. *Agriculture* **2024**, *14*, 42. https://doi.org/10.3390/agriculture14010042

Academic Editors: Chung-Liang Chang and Mustafa Ucgul

Received: 24 October 2023
Revised: 21 December 2023
Accepted: 22 December 2023
Published: 26 December 2023

1. Introduction

In modern agriculture, the development of efficient and reliable mechanized equipment is pivotal in enhancing productivity and reducing labor-intensive processes [1–3]. The vegetable transplanter is a prime example of this innovation which facilitates the transplantation of seedlings with high precision and speed. The classification of vegetable transplanters into semiautomatic and fully automatic types depends on how the seedlings are extracted and positioned within the seedling cylinder. In a fully automatic vegetable transplanter, the seedlings are supplied and loaded into the seedling cylinder automatically. On the other hand, in a semiautomatic vegetable transplanter, the operator manually provides and positions the seedlings within the seedling cylinder, requiring human involvement in the seedling handling process [4,5]. The transplanting device is the main part of a vegetable transplanter, tasked with planting seedlings into the soil. Various types of vegetable transplanters have been created, distinguished by their transplanting device design, including wheel, rotary, four-bar link, and cam types [6,7]. Among these, the four-bar link mechanism is notably prevalent and widely employed. This is attributed to the simpler structure of the four-bar link-type transplanting device relative to other types, the low manufacturing cost, and greater user-friendliness. Therefore, four-bar link-type transplanters are widely used on farms.

When a vegetable transplanter is operating, the transplanting device receives large and repeated stresses. This stress may result in material breakdown, diminished efficiency, and the emergence of safety risks [8]. However, to ensure the seamless functionality and safe operation of vegetable transplanters, a comprehensive load and safety analysis is imperative. Such analyses are equally essential for understanding the behavior of transplanting devices under different operating conditions and for identifying potential design improvements [9,10]. Sri et al. [11] conducted a study on four-bar link vegetable transplanters, analyzing loads and safety. Strain gauges were placed on the device to measure strains, which were converted to stress data for static safety factor and fatigue life estimation. The experiments covered four engine speeds and ten planting distances, revealing that increased speed and distance heightened stresses. The results showed a reduced static safety factor and fatigue life with higher engine speed and extended planting distance. Despite consistently exceeding 1.0, the static safety factor decreased. The shortest fatigue life (49,153.3 h) occurred at link A (S_14), with 750 rpm engine rotational speed and 0.35 m planting distance.

The first step in assessing the load and safety of a machine component is to evaluate stress levels. The prevalent approach for gauging the stress on machine components involves the utilization of sensors, such as strain gauges or load cells, which are validated through field trials [12]. The selection of sensors is contingent upon the intricacy of the apparatus and the precision necessary to precisely capture the stress distribution. Nevertheless, sensors can only appraise stress at specific installation sites. A multitude of sensors must be deployed to ascertain stress distribution across multiple points. Consequently, stress assessments via field experiments can be time-consuming and incur substantial expenses [13,14].

An alternative approach for ascertaining the stress affecting a machine component is simulation. The initial stage of stress simulation entails the construction of a precise virtual prototype model. Using this model, simulations can be conducted to acquire stress-related data across all sections of the machine component, as opposed to solely specific locations, as demonstrated in stress measurements performed experimentally [15,16]. Diverse commercially available dynamic simulation software programs are promising for predicting stress dispersion within intricate structures and highlighting regions with high stress concentrations. Simulation results can provide insights into the behavior of an apparatus under varying loads and operational circumstances, thereby facilitating the recognition of potential design enhancements to improve performance and safety [17,18].

Numerous studies have explored stress simulations in various agricultural machinery. Plouffe et al. [19] examined the impact of different components and adjustments on a moldboard plow's performance in clay soil. They combined modeling using the finite element method (FEM) with experimental observations. Makange et al. [20] conducted a FEM analysis on a nine-tine cultivator to identify potential weaknesses in the shovel element under varying loads and speeds in medium-black soil. Their findings revealed maximum and minimum principal stresses of 5.1726 and 0.20944 megapascals (MPa), respectively, with a total deformation of 0.076953 mm. Importantly, the highest stress remained below the yield point, indicating no tine failure due to deformation. Similarly, Yurdem et al. [14] investigated stresses in a three-bottom moldboard plough using strain gauges on different parts of the moldboard frame. They compared these measurements with outcomes from finite element simulations. The study concluded that reducing the moldboard frame thickness did not lead to excessive stresses, and the observed strains closely matched the simulated data. Kesner et al. [15] developed a computational model for a tillage machine to analyze these loads. The results showed good agreement between experimental stress measurements and simulation data from the model. Therefore, the methods used in this study can be applied in designing and improving tillage machinery. In a separate study, Islam et al. [21] analyzed the stress resistance of the gear mechanism in the picking device of an automatic pepper transplanter. They aimed to determine optimal materials and dimensions and predict fatigue life based on damage assessment. To the best of our knowledge, there has not been a specific virtual stress simulation model tailored for four-bar link-type transplanting devices, despite extensive research in this area.

The aim of this study is to investigate the stress distribution and analyze the safety of four-bar link-type transplanting devices of semiautomatic vegetable transplanters through simulation. The specific objectives are (1) to develop a three-dimensional (3D) model of the transplanting device using a commercial dynamic simulation program; (2) to perform an analysis through simulation; and (3) to validate the precision of the created virtual model by contrasting simulation data with experimental data. The results of this study offer valuable insights to the manufacturers and designers of transplanting devices for improving the safety, performance, and reliability of these devices. In addition, the results of this study can be used as basic data to establish the design process or design guidelines for four-bar link-type planting devices.

2. Materials and Methods

2.1. Vegetable Transplanter Used in Current Study [17]

A four-bar link-type vegetable transplanter was used in this study. The shapes and specifications of the vegetable transplanter are shown in Figure 1 and Table 1, respectively. It comprises an engine to supply power, a transmission to transfer power to both the wheel and transplanting device, seedling cylinders to place seedlings, and transplanting devices to plant seedlings supplied from the seedling cylinder into the soil.

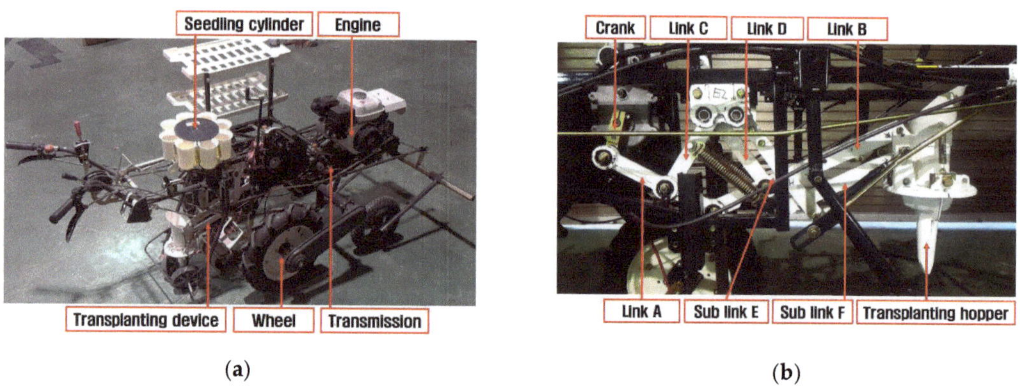

Figure 1. Shape of four-bar link-type vegetable transplanter used in this study: (**a**) Overall view; (**b**) Four-bar link-type transplanting device.

Table 1. Specifications of the four-bar link-type vegetable transplanter used in this study.

Item	Specification
model/manufacturer/nation	KTP-30/TONGYANGMOOLSAN, Seoul, South Korea
length/width/height (mm)	2125/1180/1510
weight (kg)	199
engine: rated power (kw)/rated sped (rpm)	3.4/1800
planting distance (mm)	300–500
maximum working speed (m/s)	0.4
working efficiency (h/10a)	1.5–2.0
length/width/height (mm)	2125/1180/1510
weight (kg)	199

The transplanter with the four-bar link-type transplanting device operates as follows. A user determines the transplanting speed and supplies the seedlings to the seedling cylinders manually. The transplanter operates in the forward direction and plants the seedlings into the ground one by one due to the motion of the transplanting device. That motion makes the transplanting hopper of the four-bar link-type transplanting device move up and down in a certain trajectory. When the hopper is at the top, it is located just below one of the seedling cylinders. Then, the seedling is dropped into the transplanting hopper by the opening of the seedling cylinder. When the transplanting hopper is at the bottom and it is located at a certain depth in the ground, the transplanting hopper is opened and the seedlings are planted into the ground. The transplanting work is completed after covering the planted seedlings with soil. During the transplanting process, the transplanter places a load on the transplanting device. The range of transplanting frequency is between 0.5 and 1 Hz, depending on the planting distance. This value was obtained based on the results of measurements from previous research [17]. In addition, the transplanter establishes contact with the soil, and a relatively higher load is exerted on the transplanting device than on other components. The main components of the transplanting device are the links, the crank, and the transplanting hopper. Due to the kinematic design, the movement of the link affects the operation cycle and trajectory of the transplanting hopper. The crank supplies the power transmitted from the engine and transmission to the transplanting device. The material properties of the four-bar link-type transplanting device are listed in Table 2.

Table 2. Mechanical properties of the four-bar link-type transplanting device (steel alloy 1020).

Item	Specification
density, ρ (kg/m^3)	7.85×10^3
modulus of elasticity, E (GPa)	207
Poisson's ratio, ν	0.3
yield strength, S_y (MPa)	210
yield strength in shear, S_{sy} (MPa)	105
ultimate strength, S_{ut} (MPa)	380
fatigue strength of 10^6 cycles, S_n (MPa)	190

2.2. Stress Measurement

2.2.1. Stress Measurement System

A stress measurement system was designed to measure the stress exerted on the transplanting device, as shown in Figure 2. The stress measurement system comprised strain gauges, a data acquisition system (TG009E, HBK, Darmstadt, Germany), and a laptop. The strain data measured using the strain gauges were transmitted to a data acquisition unit and recorded on a laptop. Two types of strain gauges were used to obtain strain data for the transplanting device links. The uniaxial strain gauge is suitable for measuring the strain in a singular direction, thus rendering it appropriate for scenarios in which a primary loading direction is evident, such as axial bars or links (KFGS-5-350-C1-11 L10M3R, KYOWA, Tokyo, Japan). Additionally, a rosette strain gauge comprising three strain gauges positioned at 45° intervals was used (KFGS-1-350-D17-11 L5M3S; KYOWA, Tokyo, Japan). The rosette strain gauge is optimal for measuring areas where the main loading direction is uncertain because it encompasses three strain gauges positioned at distinct angles. The transplanting device of the four-bar link type comprises hopper and link components. The primary loading direction on the link is well-defined; therefore, a uniaxial strain gauge is selected and affixed to the link. In contrast, the hopper section has an area with an unclear main loading direction. Consequently, a rosette-type strain gauge is chosen as the strain measurement tool for the hopper.

Figure 3 shows the locations at which 13 uniaxial strain gauges (S1–S2 and S5–S15) and two rosette-strain gauges (S3 and S4) were installed for this study. Strain gauges were strategically positioned in different sections of the transplanting device, such as on a link, where they were installed both in the middle and at the ends. This placement was implemented to enable the measurement of strains across various components of the transplanting device. The two uniaxial strain gauges (S1 and S2) were attached to the end of the transplanting hopper. The rosette strain gauges (S3 and S4) were attached to the curved upper part on both sides of the transplanting hopper, respectively. The uniaxial strain gauges of S5, S6, and S7 were attached to the left, middle left, and middle right sides of the sub-link F, respectively. The strain gauges S6 and S7 were both situated in the middle of sub-link F, with S6 positioned closer to the left end and S7 closer to the right end. This placement was selected to capture a more comprehensive measurement of stress in sub-link F. The uniaxial strain gauges of S8, S9, and S10 were attached to the left, middle, and right sides of link B, respectively. Two strain gauges (S11 and S12) were installed on the left and right sides of link A. The uniaxial strain gauge S13 was placed at the right side of sub-link F, the strain gauge S14 was placed at middle side link D, and the strain gauge S15 was attached to middle side link C. The effect of the measurement system on the stress exerted on the transplanting device was ignored.

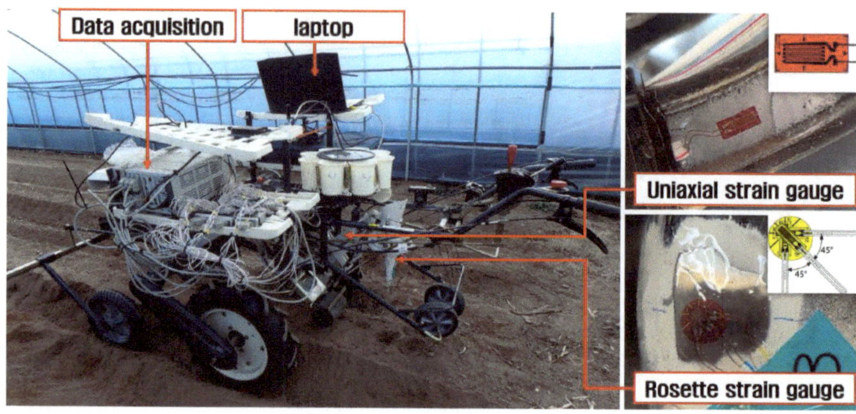

Figure 2. Shape of stress measurement system.

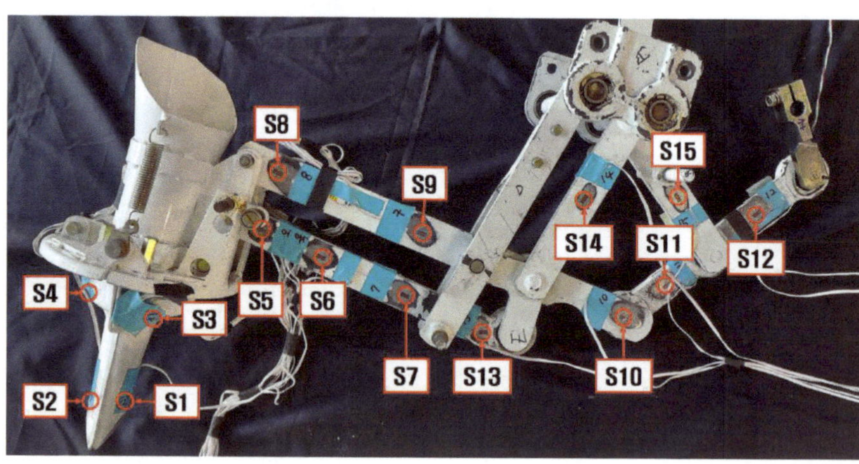

Figure 3. Installation location of strain gauges on the four-bar link-type transplanting device.

The specifications of the strain gauges and data acquisition module are shown in Tables 3 and 4, respectively.

Table 3. Specifications of strain gauges used.

Gauge	Item	Specification
uniaxial strain gauge	model/manufacture/nation	KYOWA KFGS-5-350-C1-11 L10M3R/KYOWA/Tokyo, Japan
	gauge factor (%)	2.12 ± 1.0
	gauge length (mm)	5
	gauge resistance (%)	351.2 Ω ± 0.4
rosette strain gauge	model/manufacture/nation	KYOWA KFGS-1-350-D17-11 L5M3S/KYOWA/Tokyo, Japan
	gauge factor (%)	2.11 ± 1.0
	gauge length (mm)	1
	gauge resistance (%)	350.0 Ω ± 0.7

Table 4. The specifications of the data acquisition module used.

Item	Specification
model/manufacture/nation	TG009E/HBK/Darmstadt, Germany
length/width/height (mm)	177/161/386
weight (kg)	5
number of channels	16
sampling rate (Hz)	Up to 320

2.2.2. Operating Condition

The field test took place in a field featuring consistent soil conditions, situated at coordinates 37°56′24.0″ N and 127°46′59.1″ E. The location has an elevation of 111.00 m above sea level and is in Sinbuk-eup, Chuncheon, within Gangwon Province, South Korea. The test bed was measured as 45 m (length) × 0.6 m (width) × 0.3 m (height). Before the experiments, the soil was tilled using a plow and a rotavator to account for the actual operating conditions of the vegetable transplanter. The vegetable transplanter was operated at an engine rotational speed of 750 rpm at a planting distance of 0.35 m and a planting depth of 0.07 m. The test was performed in triplicate and the data were analyzed using the average as a representative value.

2.2.3. Analysis of Measured Data

The strain data acquired during field tests were converted to stress values. This conversion was contingent upon the specific type of strain gauge used, thus resulting in two distinct categories of strain data. The stress was calculated by multiplying the strain data derived from the uniaxial strain gauge by the modulus of elasticity, as shown in Equation (1). By contrast, the rosette strain gauge was used to measure strains along three different axes. By utilizing the strain values recorded for each direction, significant stress values such as the maximum and minimum principal stresses and von Mises stress can be calculated using Equations (2)–(5) [11].

$$\sigma = \mathrm{E} \times \varepsilon, \tag{1}$$

$$\sigma_1 = \frac{\mathrm{E}}{2(1-\mathrm{v}^2)}\left[(1+\mathrm{v})(\varepsilon_a+\varepsilon_c) + (1-\mathrm{v}) \times \sqrt{2\left\{(\varepsilon_a-\varepsilon_b)^2+(\varepsilon_b-\varepsilon_c)^2\right\}}\right], \tag{2}$$

$$\sigma_2 = \frac{\mathrm{E}}{2(1-\mathrm{v}^2)}\left[(1+\mathrm{v})(\varepsilon_a+\varepsilon_c) - (1-\mathrm{v}) \times \sqrt{2\left\{(\varepsilon_a-\varepsilon_b)^2+(\varepsilon_b-\varepsilon_c)^2\right\}}\right], \tag{3}$$

$$\sigma_v = \sqrt{\sigma_1^2 - \sigma_1\sigma_2 + \sigma_2^2}, \tag{4}$$

$$\tau_{max} = \frac{E}{2(1+v)} \times \sqrt{2\left\{(\varepsilon_a - \varepsilon_b)^2 + (\varepsilon_b - \varepsilon_c)^2\right\}}, \quad (5)$$

where

σ	= calculated axial stress (Pa)
σ_1	= maximum principal stress (Pa)
σ_2	= minimum principal stress (Pa)
σ_v	= von Misses stress (Pa)
ε	= measured strain for the components of transplanting device
E	= modulus of elasticity (Pa)
v	= Poisson's ratio
τ_{max}	= maximum shear stress (Pa)
ε_a	= strain measured by rosette strain gauge in horizontal direction
ε_b	= strain measured by rosette strain gauge in 45° direction
ε_c	= strain measured by rosette strain gauge in vertical direction

2.3. Stress Simulation

2.3.1. Dynamic Simulation Model

A dynamic simulation to derive the stress exerted on the four-bar link-type transplanting device was performed using commercial software (Recurdyn V9R4, Functionbay, Seongnam, Republic of Korea). This software is typically used in studies for predicting forces or loads within diverse multibody systems comprising rigid and flexible components. Figure 4 illustrates a 3D model of the four-bar link-type transplanting device. A 3D model of the transplanting device was created based on its actual dimensions and material properties.

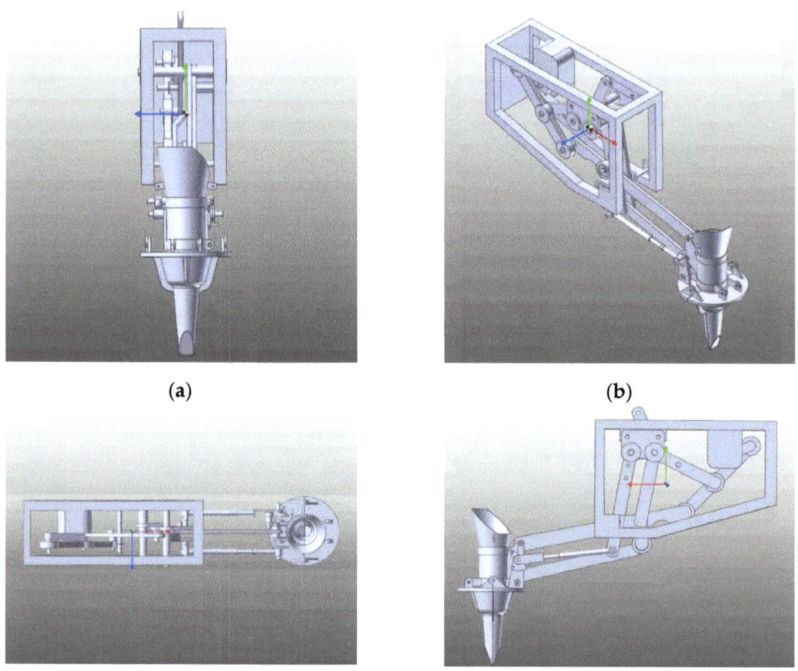

Figure 4. Three-dimensional model of four-bar link-type transplanting device: (**a**) Front view; (**b**) Isometric view; (**c**) Top view; (**d**) Side view.

2.3.2. Simulation Condition

The simulation model for deriving the stress exerted by the contact between the transplanting hopper and ground is shown in Figure 5. The simulation was conducted at an engine rotational speed of 750 rpm and a planting distance of 0.35 m, i.e., the condition at which the highest stress was recorded in the experiment. In the stress simulation, the effect of vibration from the engine, the transmission, etc. was not considered. The gravitational acceleration was set to 9.81 m/s^2 to act vertically downward. A simulation was set to derive the stress exerted on the links and transplanting hopper when it contacts the soil during the time when the four-bar link-type transplanting device and the transplanter operate together. The conditions of the interaction between the hopper and ground are listed in Table 5.

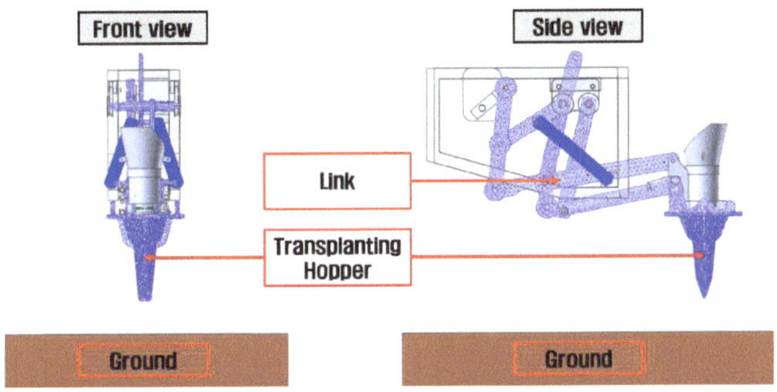

Figure 5. Stress simulation model for the four-bar link-type transplanting device.

Table 5. Material properties used in simulation.

	Item	Value
interaction between hopper and ground	stiffness coefficient	35
	damping coefficient	0.03
	dynamic friction coefficient	1.0

2.4. Verification of Stress Simulation

To verify the simulation model, the stress values derived from the field test and simulation were compared and analyzed. The validation process involved a comparison between the maximum stress at 15 specific points on the transplanting device links and the transplanting hopper (Figure 3). The stress data obtained from the experiment, which validated the simulation results, were the averages of the maximum stress values (peak stress) derived from three repeated tests. The root mean square error (RMSE) was calculated to determine the disparity between the maximum stress values derived from the simulation results and those obtained from the field test measurements. As the RMSE decreases, the disparity between the maximum stress in the experiment and the simulation becomes smaller [21]. The RMSE is determined by Equation (6).

$$RMSE = \sqrt{\frac{1}{n} \times \sum_{i=1}^{n} (y_i - \hat{y}_i)^2} \quad (6)$$

where,

RMSE = root mean square error

y_i = measured maximum stress in the experiment

$\hat{y}_i$ = measured maximum stress in the stress simulation

n = number of stress measurement locations

2.5. Maximum Stress and Static Safety Factor

The simulation results indicated stress across all regions of the transplanting device. These results can be used to determine the position and magnitude of the highest stress point. The maximum stress from the simulation may differ among the 15 locations where the strain gauges were applied during the field test.

The static safety factor is commonly used in engineering design and analyzed to ensure that structures and materials can withstand the stresses and loads that they will encounter during their expected service life [22]. In order to derive the static safety factor, it is necessary to identify both the yield strength that a material can endure and the maximum stress exerted on the transplanting device during operation [23]. The static safety factor is determined by the ratio of the yield strength and the maximum stress incurred. If the static safety factor exceeds 1.0, it can be judged that a safe design has been achieved. On the other hand, if the static safety factor is less than 1.0, it indicates that the transplanting device of the four-bar link-type vegetable transplanter can be damaged or destroyed by static load. Prior to constructing a machine, specifically during the machine tool design stage, calculation of a static safety factor is performed. If the result of the static safety factor calculated by simulation is less than 1.0, design improvement is necessary. This process aims to optimize the design, ensuring its viability and suitability for production. The static safety factor can be calculated by Equations (7) and (8) [11]. In addition, considering the static safety factors, it is essential to address dynamic loads when designing a machine. This is particularly important when the results of static safety factors are only marginally higher than 1.0, as dynamic loads can pose greater damage.

$$\mathrm{SF} = \frac{S_y}{\sigma_{max}} \tag{7}$$

$$\mathrm{SF} = \frac{S_y}{\sigma_{v\_max}} \tag{8}$$

where

SF = static safety factor
S_y = yield strength (Pa)
σ_{max} = maximum axial stress (Pa)
σ_{v_max} = maximum von Mises stress (Pa)

3. Results and Discussion

3.1. Verification of Stress Simulation

The maximum stress values obtained from the experimental and simulation results are presented in Figure 6 and Table 6.

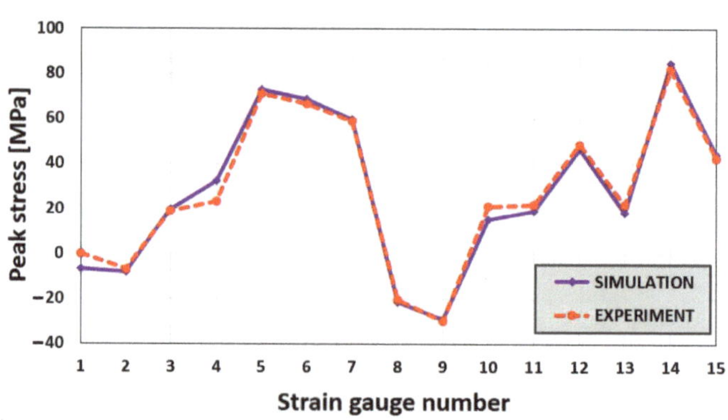

Figure 6. Comparison of maximum stress between experiment and simulation.

Table 6. Maximum stress values for 15 locations of the four-bar link-type transplanting device.

Strain Gauge Number	Measured Maximum Stress (MPa)	Maximum Stress Derived by Simulation (MPa)	RMSE
S1	−6.30	−6.74	
S2	−6.97	−8.33	
S3	19.11	19.88	
S4	23.28	32.35	
S5	71.27	72.72	
S6	66.67	68.82	
S7	58.99	59.47	
S8	−20.32	−21.34	3.3117
S9	−29.81	−29.17	
S10	21.10	15.17	
S11	21.98	19.12	
S12	48.59	46.83	
S13	21.94	18.21	
S14	81.81	84.71	
S15	42.65	44.21	

When considering the absolute value of the measured maximum stress, the maximum stress of the S14 location was the highest at 81.81 MPa in the experiment and 84.71 MPa in the simulation, respectively. Then, the maximum stress in strain gauge S5 was 71.27 MPa (experiment) and 72.72 MPa (simulation), which was the second, followed by strain gauge S5, S6, S7, S12, and S15. In the case of the location S1, the maximum stress was the lowest at −6.30 MPa in the experiment and −6.74 MPa in the simulation, respectively.

Based on the verification results obtained from the 15 distinct points, one can deduce that the simulation findings aligned with the experimental results. Nevertheless, significant disparities were observed at certain points, particularly at strain gauges 4 and 10. Conversely, strain gauges 5, 6, 7, and 14, which manifested greater stress levels than the other positions, exhibited a substantial level of agreement.

The RMSE between the maximum stress data from the simulation and the maximum stress data from the field test was 3.3117 MPa, which signifies an insignificant variance between the anticipated and recorded values. This result can be regarded as modest, particularly when considering the typical range of maximum stress, i.e., from −29.81 MPa to 81.81 MPa. In addition, the normalized RMSE for the four-bar link arrangement was 0.0297, which approached zero. This figure shows a precise model with a strong correlation between the projected and actual measurements.

The differences between the simulated and measured results can be attributed to various factors, including the omission of frictional forces between joints in the simulation and disparities in the assembly of components when comparing the simulated scenarios to real-world conditions. Additionally, the effects of vibrations generated by the operating engine were not incorporated in the simulation. Moreover, the load input from the main body, which was facilitated by the connection to the machine frame, was not considered in the simulation. Despite the significant deviations at specific points, 12 out of 15 points indicated consistency. Consequently, one can infer that the stress values derived from the simulation aligned adequately with the stress levels measured during the field test.

3.2. Maximum Stress and Static Safety Factor Based on Simulation

Figure 7 shows the stress simulation results for the four-bar link type transplanting device. The maximum stress value was 150.82 MPa, which was recorded within sub-link E and served as a connection among links G, D, and F. The position of sub-link E, i.e., at the structural center of the entire transplanting device, resulted in a high stress owing to forces from the ground and engine, thus resulting in greater stress in this region compared with other regions.

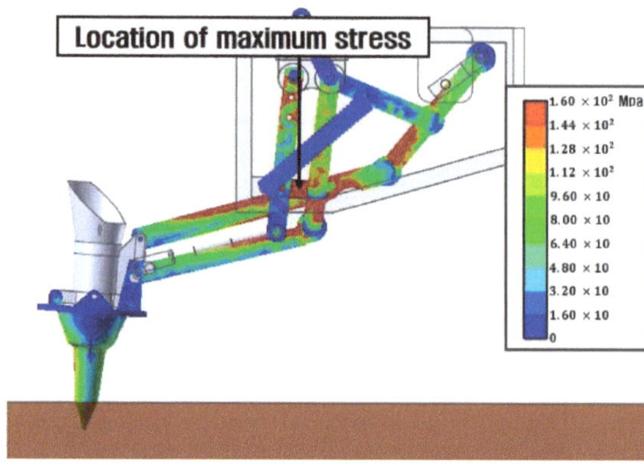

Figure 7. Location of maximum stress on a four-bar link-type transplanting device.

Based on the maximum stress data from the simulation results, the static safety factor for the four-bar link type was 1.39, which was deficient owing to its proximity to 1. Several strategies can be applied to augment the static safety factor of the transplanting equipment. One potential approach involves substituting an existing material with one that possesses a higher yield strength, thereby contributing to an elevated level of safety. Alternatively, employing AISI 1030, which possesses a yield strength of 344.7 MPa, can increase the static safety factor for the four-bar link connection to 2.29 [18].

When selecting a material, one must consider not only the yield strength, but also its specific attributes. Additionally, the economic viability of the material must be prioritized because its price significantly affects the production cost of a machine. This consideration is essential for ensuring that the machine design maintains affordability while satisfying the required criteria.

4. Conclusions

In this study, stress measurements and dynamic simulations were conducted to determine the stress level of the four-bar link-type transplanting devices of a semiautomatic vegetable transplanter.

The main results of this study were as follows. The maximum stress obtained from the simulation results for the four-bar link-type transplanting device correlated with the measurement results, although significant differences were observed for strain gauges numbers 4 and 10. This study's outcomes demonstrate the feasibility of employing simulation for stress measurement, yielding precise results. Based on the simulation results, the maximum stress of the four-bar link type at the sub-link E was 150.82 MPa. The sub-link E, located at the structural midpoint of the entire transplanting device, experienced elevated stress due to forces originating from both the ground and engine. Consequently, this area exhibited higher stress levels compared to other areas. The static safety factor obtained from the simulation of the four-bar link type was 1.39. The transplanting device design is deemed secure, as indicated by the static safety factor value surpassing 1.0. Even so, this static safety factor is still relatively low so it is still necessary to increase the safety level of four-bar link-type transplanting devices. To heighten safety, various avenues can be explored, including replacing the current material with a stronger alternative or adjusting the design of susceptible components in terms of shape or size. The findings from this study demonstrate that employing simulation techniques for stress assessment in machine components can expedite and economize stress measurements for safety analysis. Furthermore, the findings of this study can serve as foundational data to establish the process and guidelines of design for four-bar link-type transplanting devices. As a future study, based on the results of this study, a kinematic analysis of the four-bar link-type transplanting device will be performed to establish a design process that can improve durability and economic efficiency while satisfying the appropriate planting trajectory.

Author Contributions: Data curation and formal analysis, S.M., S.-J.H., J.-H.K. and M.-K.J.; writing—original draft, S.M. and S.-J.H.; writing—review and editing, J.-S.N. All authors have read and agreed to the published version of the manuscript.

Funding: This work was supported by Korea Institute of Planning and Evaluation for Technology in Food, Agriculture and Forestry (IPET) through Machinery Mechanization Technology Development Program for Field Farming Program, funded by Ministry of Agriculture, Food and Rural Affairs (MAFRA) (RS-2023-00236724).

Institutional Review Board Statement: Not applicable.

Data Availability Statement: Data are contained within the article.

Conflicts of Interest: The authors declare no conflicts of interest.

References

1. Hwang, S.-J.; Park, J.-H.; Lee, J.-Y.; Shim, S.-B.; Nam, J.-S. Optimization of main link lengths of transplanting device of semi-automatic vegetable transplanter. *Agronomy* **2020**, *10*, 1938. [CrossRef]
2. Yang, Q.; Huang, G.; Shi, X.; He, M.; Ahmad, I.; Zhao, X.; Addy, M. Design of a control system for a mini-automatic transplanting machine of plug seedling. *Comput. Electron. Agric.* **2020**, *169*, 105226. [CrossRef]
3. Tsuga, K. Development of fully automatic vegetable transplanter. *Jarq Jpn. Agric. Res. Q.* **2000**, *34*, 21–28.
4. Jo, J.S.; Okyere, F.G.; Jo, J.M.; Kim, H.T. A study on improving the performance of the planting device of a vegetable transplanter. *J. Biosyst. Eng.* **2018**, *43*, 202–210. [CrossRef]
5. Park, J.-H.; Hwang, S.-J.; Nam, J.-S. Operational characteristics of a domestic commercial semi-automatic vegetable transplanter. *J. Agric. Life Sci.* **2018**, *52*, 127–138. [CrossRef]
6. Min, Y.B.; Kang, J.K.; Ryu, C.S. Development of onion transplanter: Analysis of a transplanting locus on the type of transplanting devices for a vegetable transplanter. *J. Agric. Life Sci.* **2015**, *49*, 289–294. [CrossRef]
7. Kumar, G.V.P.; Raheman, H. Vegetable transplanters for use in developing countries—A review. *Int. J. Veg. Sci.* **2008**, *14*, 232–255. [CrossRef]
8. Tian, S.; Qiu, L.; Kondo, N.; Yuan, T. Development of automatic transplanter for plug seedling. *IFAC Proc.* **2010**, *43*, 79–82. [CrossRef]
9. Kim, W.-S.; Kim, Y.-S.; Kim, Y.-J.; Choi, C.-H.; Inoue, E.; Okayasu, T. Analysis of the load of a transplanter PTO shaft based on the planting distance. *J. Fac. Agric. Kyushu Univ.* **2018**, *63*, 97–102. [CrossRef]
10. Kim, W.-S.; Kim, Y.-S.; Kim, T.-J.; Nam, K.-C.; Kim, T.-B.; Han, T.-H.; Im, R.-G.; Kim, Y.-H.; Kim, Y.-J. Effects of planting distance and depth on PTO load spectrum of a small riding-type transplanter. *Int. J. Agric. Biol. Eng.* **2020**, *13*, 57–63. [CrossRef]
11. Sri, M.; Hwang, S.-J.; Nam, J.-S. Experimental safety analysis for transplanting device of the 4-bar link type semi-automatic vegetable transplanter. *Agronomy* **2022**, *12*, 1890. [CrossRef]
12. Radhakrishnan, V.M. Multiaxial fatigue—An overview. *Sadhana* **1995**, *20*, 103–122. [CrossRef]
13. Paraforos, D.S.; Griepentrog, H.W.; Vougioukas, S.G. Country road and field surface profiles acquisition, modelling and synthetic realisation for evaluating fatigue life of agricultural machinery. *J. Terramech.* **2016**, *63*, 1–12. [CrossRef]
14. Yurdem, H.; Degirmencioglu, A.; Cakir, E.; Gulsoylu, E. Measurement of strains induced on a three-bottom moldboard plough under load and comparisons with finite element simulations. *Measurement* **2019**, *136*, 594–602. [CrossRef]
15. Kešner, A.; Chotěborský, R.; Linda, M.; Hromasová, M.; Katinas, E.; Sutanto, H. Stress distribution on a soil tillage machine frame segment with a chisel shank simulated using discrete element and finite element methods and validate by experiment. *Biosyst. Eng.* **2021**, *209*, 125–138. [CrossRef]
16. Shao, X.; Song, Z.; Yin, Y.; Xie, B.; Liao, P. Statistical distribution modelling and parameter identification of the dynamic stress spectrum of a tractor front driven axle. *Biosyst. Eng.* **2021**, *205*, 152–163. [CrossRef]
17. Paulson, I.W.P.; Dolovich, A.T.; Noble, S.D. Development of a dynamic simulation model of a towed seeding implement. *J. Terramech.* **2018**, *75*, 25–35. [CrossRef]
18. Tekeste, M.Z.; Balvanz, L.R.; Hatfield, J.L.; Ghorbani, S. Discrete element modeling of cultivator sweep-to-soil interaction: Worn and hardened edges effects on soil-tool forces and soil flow. *J. Terramech.* **2019**, *82*, 1–11. [CrossRef]
19. Xie, Z.; Shi, W.; Tian, Q.; Zheng, Y.; Tan, L. Fatigue life assessment and damage investigation of centrifugal pump runner. *Eng. Fail. Anal.* **2021**, *124*, 105256. [CrossRef]
20. Juvinall, R.C.; Marshek, K.M. *Machine Component Design*, 5th ed.; John Wiley & Sons, Inc.: Hoboken, NJ, USA, 2020; pp. 338–340.
21. Harwell, M. A strategy for using bias and RMSE as outcomes in Monte Carlo studies in statistics. *J. Mod. Appl. Stat. Methods* **2019**, *17*, 5. [CrossRef]

22. Fadier, E.; la Garza, C.D. Safety design: Towards a new philosophy. *Saf. Sci.* **2006**, *44*, 55–73. [CrossRef]
23. Bluff, E. Safety in machinery design and construction: Performance for substantive safety outcomes. *Saf. Sci.* **2014**, *66*, 27–35. [CrossRef]

agriculture

MDPI

Article

Stress Simulation on Cam-Type Transplanting Device of Semiautomatic Vegetable Transplanter

Sri Markumningsih [1,†], Seok-Joon Hwang [2,3,†], Jeong-Hun Kim [2,3], Moon-Kyeong Jang [2,3] and Ju-Seok Nam [2,3,*]

1 Department of Agricultural and Biosystems Engineering, Faculty of Agricultural Technology, Universitas Gadjah Mada, Yogyakarta 55281, Indonesia; sri_markumningsih@ugm.ac.id
2 Department of Biosystems Engineering, Kangwon National University, Chuncheon 24341, Republic of Korea; human51@kangwon.ac.kr (S.-J.H.); jhkim1995@kangwon.ac.kr (J.-H.K.); moon2842@kangwon.ac.kr (M.-K.J.)
3 Interdisciplinary Program in Smart Agriculture, Kangwon National University, Chuncheon 24341, Republic of Korea
* Correspondence: njsg1218@kangwon.ac.kr; Tel.: +82-33-250-6497
† These authors contributed equally to this work.

Abstract: Stress measurements play a crucial role in safety analyses of transplanting devices. Strain gauges for stress measurements during field tests can be expensive and time-consuming. The aim of this study was to investigate the stress on the transplanting device of a cam-type semiautomatic vegetable transplanter using a simulation method. A three-dimensional simulation model was established, considering the dimensions and material properties of the transplanting device. The stress distribution and maximum stress values were obtained through simulations. The maximum stress values at 15 points within the transplanting device determined via the simulation were compared with the experimental stress data to verify the stress simulation model. The results show that the maximum stress obtained from the simulation correlated with that of the measured results, although differences were observed at different locations, particularly at strain gauge positions 11 and 13. Based on the simulation results, the maximum stress occurs at the upper link of the cam-type transplanting device, reaching a magnitude of 201.21 MPa, and the static safety factor is 1.04.

Keywords: cam type; static safety factor; stress simulation; transplanting device; vegetable transplanter

Citation: Markumningsih, S.; Hwang, S.-J.; Kim, J.-H.; Jang, M.-K.; Nam, J.-S. Stress Simulation on Cam-Type Transplanting Device of Semiautomatic Vegetable Transplanter. *Agriculture* **2023**, *13*, 2230. https://doi.org/10.3390/agriculture13122230

Academic Editors: Chung-Liang Chang and Mustafa Ucgul

Received: 24 October 2023
Revised: 28 November 2023
Accepted: 29 November 2023
Published: 1 December 2023

1. Introduction

The modernization of agricultural practices has led to the development of specialized machinery to optimize various aspects of crop cultivation. Among these innovations, the vegetable transplanter is a pivotal advancement in addressing challenges, such as labor shortages and rising production costs in vegetable farming [1,2]. A vegetable transplanter is a machine used to plant vegetable seedlings into the ground. Based on the method for extracting the seedling and its placement within the seedling cylinder, the vegetable transplanter is categorized into two types as semiautomatic and fully automatic. In the case of a fully automatic vegetable transplanter, the seedlings are automatically supplied and put into the seedling cylinder. Conversely, in a semiautomatic vegetable transplanter, the operator manually supplies the seedling and places them within the seedling cylinder, necessitating human intervention in the seedling handling process [3,4].

The primary component of a vegetable transplanter is the transplanting device, which is responsible for planting seedlings in the soil. There are several types of vegetable transplanter that have been developed based on the design of the transplanting device, namely wheel, rotary, four-bar-link, and cam types [5–7]. Among the various developed types, the cam mechanism is widely used. The cam type employs a cam that opens and closes the hopper by contacting the bearing. This type is particularly suited for small cultivation areas because of its simple design and user-friendly operation [8–10].

The transplanting device is subjected to significant mechanical stresses during its operational cycles. These stresses, arising from the demanding conditions of field operations, can potentially lead to material fatigue, diminished performance, and safety hazards within the transplanting service [11–13]. Sri et al. [14] conducted load and safety analyses to ensure the safety of the transplanting device of a cam-type semiautomatic vegetable transplanter. The safety analysis involved measuring the strain at several points on the transplanting device using a strain gauge. Subsequently, the measured strain data were converted into stress data to calculate the static safety factor and fatigue life. The experiments were carried out using four different engine speeds and twelve planting distances. The findings revealed that as the engine speed increased or the planting distance decreased, the stress on the transplanting device tended to increase. The cam-type transplanting device demonstrated static safety factors exceeding 1.0 across all measurement points and under various working conditions. At the upper section of the hopper, the minimum fatigue life was determined to be 66,416 h. This is considered an ample lifespan, especially when taking into account the annual usage time of 25.5 h in Korea.

The initial step in the load and safety analysis of a machine component is stress determinative [15,16]. The prevalent technique for measuring stress in machine parts involves using sensors, such as strain gauges and load cells in field tests [17–19]. The sensor can measure stress exclusively at a specific location. Numerous sensors must be installed to assess the stress levels at multiple points. Consequently, the use of sensors for measuring stress can be expensive and time-consuming [20,21]. A potential solution to this problem is to conduct stress simulations.

Stress simulation entails the utilization of software or computer programs to emulate and analyze the stress experienced by various components of a machine. Numerous studies have been conducted on stress simulations of various types of agricultural machinery. Plouffe et al. [22] investigated the influence of various components and adjustments on the performance of a moldboard plow operating on clay soil by combining modeling applied with the finite element method (FEM) and incorporating experimental observations. Makange et al. [23] performed a FEM analysis on a nine-tine cultivator to recognize potential weaknesses within the shovel element under varying loads and speeds in medium-black soil. The findings indicated that the highest and lowest principal stresses registered at 5.1726 and 0.20944 megapascals (MPa), respectively, along with a total deformation of 0.076953 mm. Importantly, the maximum stress remained below the yield point, indicating that the deformation did not lead to failure in the tine. Similarly, Yurdem et al. [24] conducted their research on a three-bottom moldboard plow to assess the stresses through field measurements using strain gauges affixed to various sections of the moldboard frame. These measurements were then verified with the outcomes derived from finite element simulations. The study concluded that the decreased thickness of the moldboard frame did not lead to undue stresses, and the observed strains closely matched the simulated data. Kesner et al. [25] established a computational model of a tillage machine to analyze these loads. The result showed that the experimental stress measurements aligned well with the simulation data obtained from the model. Consequently, the methods employed in this study can be applied in designing and refining tillage machinery. Islam et al. [26] analyzed the stress resisted by the gear mechanism within the picking device of an automatic pepper transplanter to determine optimal materials and dimensions and predict the fatigue life based on damage assessment. Both the crank and cam gear sets underwent testing using finite element analysis simulation and stress analysis theory based on the American Gear Manufacturers Association standard. These tests were performed with various materials and dimensions. The findings from this research act as a valuable reference for designing the picking device gears with optimal material characteristics, ensuring the recommended service life of pepper-planting equipment. However, from all the research that has been carried out, a virtual model for stress simulations that specifically focuses on cam-type transplanting devices has not yet been developed, to the best of our knowledge.

The aim of this study was to investigate the stress distribution and analyze the safety of cam-type transplanting devices for semiautomatic vegetable transplanters using simulations. The specific objectives were to (1) establish a three-dimensional model of the transplanting device using commercial dynamic simulation software, (2) conduct a simulation-based analysis of stress distribution, maximum stress levels, and the static safety factor associated with the transplanting device, and (3) validate the precision of the established virtual model by contrasting simulation results with experimental data. The results of this study have the potential to provide valuable insights for the manufacturers and designers of transplanting devices, enhancing safety, performance, and reliability. In addition, the results of this study can be used as basic data to establish the design process or design guidelines for cam-type planting devices.

2. Materials and Methods

2.1. Transplanting Device of Cam-Type Vegetable Transplanter [7]

The shape and specifications of the cam-type vegetable transplanter used in this study are presented in Figure 1 and Table 1, respectively. The cam-type vegetable transplanter consisted of an engine for supplying power, a transmission for transferring power to the wheel and transplanting device, a control panel for adjusting plant spacing and depth, a seedling cylinder designed for seedling placement, and a transplanting device for planting the seedlings supplied from the seedling cylinder into the soil.

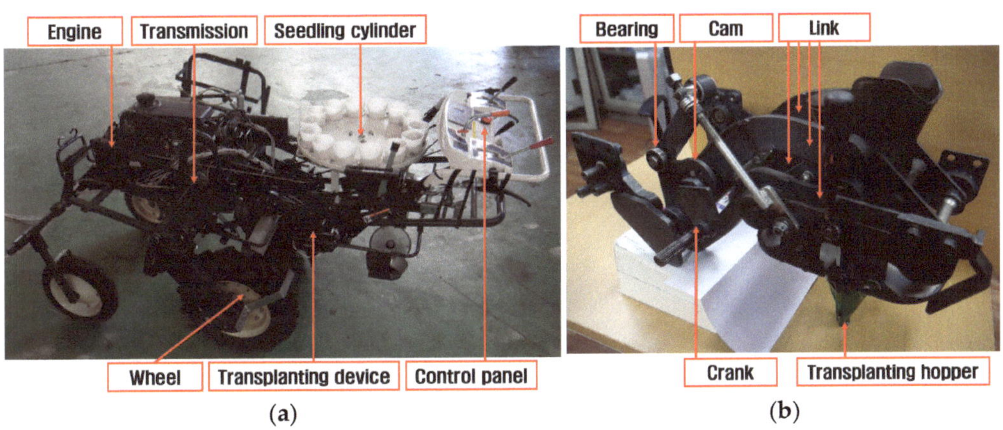

Figure 1. Shape of cam-type vegetable transplanter used in this study: (**a**) overall view; (**b**) transplanting device.

Table 1. Specifications for cam-type vegetable transplanter used in this study.

Item/Parameter		Specification
model/manufacturer/country		KP-100KR/KUBOTA/Osaka, Japan
length/width/height (mm)		2150/1360/1130
weight (kg)		280
engine	rated power (kW)	2.6
	rated speed (rpm)	1550
planting distance (mm)		350–900
maximum working speed (m/s)		0.57
working efficiency (h/m^2)		0.0015–0.0025

The transplanter with the cam-type transplanting device operates as follows: A user determines the travel speed and the spacing in the row of the transplanter. Then the seedlings are supplied to the seedling cylinders manually. The transplanter moves and plants the seedlings into the ground by the motion of the transplanting device. That motion makes the transplanting hopper of the cam-type transplanting device move up and down in a certain trajectory. When the hopper is at the top, it is located below one of the seedling cylinders, the seedling cylinder opens and drops the seedling into the transplanting hopper. When the transplanting hopper comes to the lower end, it is located at a certain depth in the ground. At that moment, the seedlings are planted into the ground after the opening of the transplanting hopper. The row spacing suitable for the target crop can be set by the control panel. The seedlings planted in the ground are covered with soil, and the transplanting work is complete. During the transplanting process, the mechanical operations of the transplanter generate loads on the transplanting device. The main components of the transplanting device in this study were links, bearings, a cam, a crank, and a transplanting hopper. The links played a role in determining the trajectory of the transplanting hopper. The crank supplied the power transmitted from the engine and transmission to the transplanting device. The shape of the cam had an influence on the behavior of the planting hopper. Therefore, the cycle of opening and closing the planting hopper was determined mechanically due to the contact between the bearing and the cam. The material properties of the transplanting device are listed in Table 2.

Table 2. Mechanical properties of transplanting device (steel alloy 1020).

Property	Specification
density, ρ (kg/m^3)	7.85×10^3
modulus of elasticity, *E* (GPa)	207
Poisson's ratio, ν	0.3
yield strength, S_y (MPa)	210
yield strength in shear, S_{sy} (MPa)	105
ultimate strength, S_{ut} (MPa)	380
fatigue strength of 10^6 cycles, S_n (MPa)	190

2.2. Stress Measurement

2.2.1. Stress Measurement System

A stress measurement system was constructed to measure the stress exerted on the transplanting device, as shown in Figure 2. The stress measurement system consisted of strain gauges, a data acquisition system (TG009E, HBK, Darmstadt, Germany), and a laptop. The strain data measured using the strain gauges were transmitted to a data acquisition unit and recorded on a laptop. Two types of strain gauges were used to obtain strain data for the transplanting device links. One was a uniaxial strain gauge (KFGS-5-350-C1-11 L10M3R, KYOWA, Tokyo, Japan), which is suitable for measuring the strain in a singular direction, making it well suited for scenarios where a primary loading direction is evident, such as in axial bars or links. The second was a rosette strain gauge (KFGS-1-350-D17-11 L5M3S, KYOWA, Tokyo, Japan) that featured three strain gauges positioned at 45° intervals. The rosette strain gauge is optimal for measuring areas where the main loading direction is unknown because it encompasses three strain gauges positioned at distinct angles. Figure 3 shows the installation locations of the 13 uniaxial strain gauges (S1–S2 and S5–S15) and 2 rosette strain gauges (S3 and S4). Two uniaxial strain gauges (S1 and S2) were attached to the end of the transplanting hopper. The rosette strain gauges (S3 and S4) were attached to the curved upper part on either side of the transplanting hopper. The uniaxial strain gauges of S5, S6, and S7 were attached to the left, center, and right sides of the left upper link, respectively. Uniaxial strain gauges (S10, S11, and S12) were attached on the right upper link in the same positions as those on the left upper link. Two uniaxial strain gauges (S8 and S9) were attached to the left bottom link, and three uniaxial strain gauges (S13, S14, and S15) were attached to the bottom right link. The specifications of the strain gauges and data acquisition are listed in Tables 3 and 4, respectively.

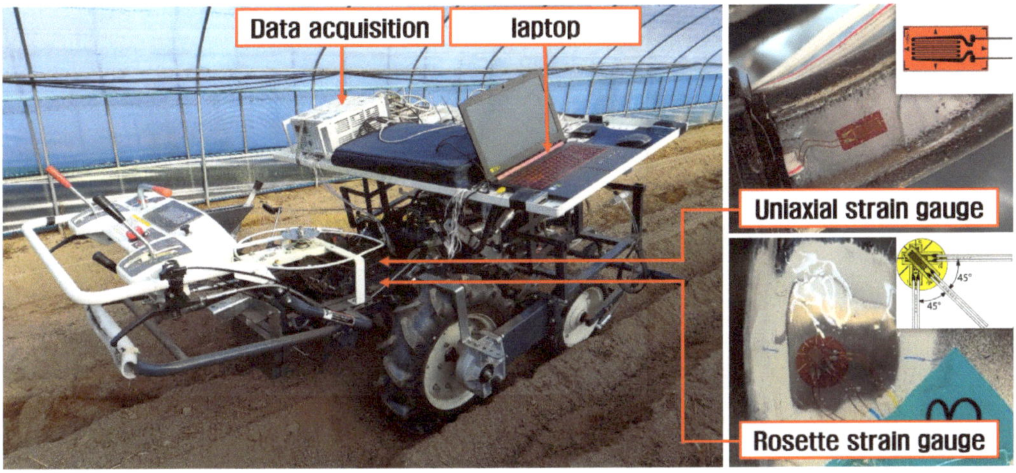

Figure 2. Shape of stress measurement system.

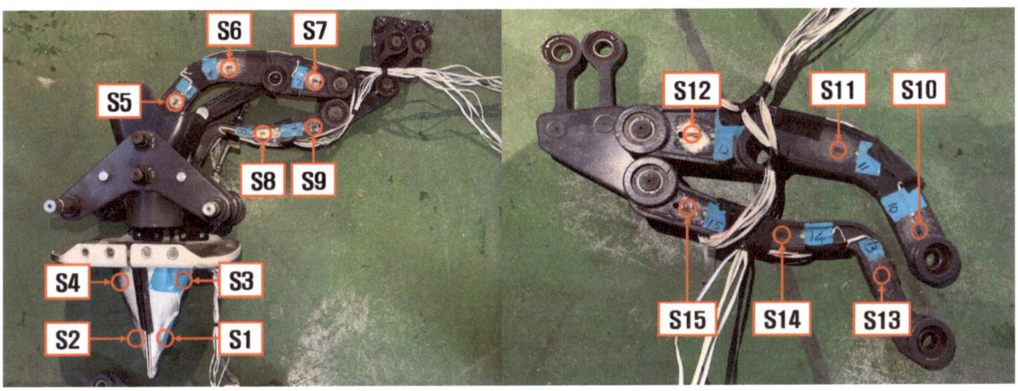

Figure 3. Installation location of strain gauges on transplanting device.

Table 3. Specifications for strain gauges.

Item/Parameter		Specification
uniaxial strain gauge	model/manufacturer/country	KYOWA KFGS-5-350-C1-11 L10M3R/KYOWA/Tokyo, Japan
	gauge factor (%)	2.12 ± 1.0
	gauge length (mm)	5
	gauge resistance (%)	351.2 ± 0.4 Ω
rosette strain gauge	model/manufacturer/country	KYOWA KFGS-1-350-D17-11 L5M3S/KYOWA/Tokyo, Japan
	gauge factor (%)	2.11 ± 1.0
	gauge length (mm)	1
	gauge resistance (%)	350.0 ± 0.7 Ω

2.2.2. Working Conditions

The field test took place in a field featuring consistent soil conditions, situated at coordinates 37°56′24.0″ N and 127°46′59.1″ E. The location has an elevation of 111.00 m above sea level and is located in Sinbuk-eup, Chuncheon, within Gangwon Province, South Korea. The length, width, and depth of the test bed were 45, 0.6, and 0.3 m, respectively. Prior to the experiments, the soil was tilled using a plow and a rotavator, considering the actual working conditions of the vegetable transplanter. Therefore, the soil of the transplanting ridge is extremely soft. And the variation of planting depth is less than the hardpan depth of the test bed. The vegetable transplanter worked at an engine speed of 1550 RPM with a set planting distance of 0.5 m and a set planting depth of 0.07 m. The test was performed in triplicate, and the data were analyzed using the average as a representative value.

Table 4. Specifications of data acquisition used.

Item/Parameter	Specification
model/manufacturer/country	TG009E/HBK/Darmstadt, Germany
length/width/height (mm)	177/161/386
weight (kg)	5
number of channels	16
sampling rate (Hz)	Up to 320

2.2.3. Analysis Data

The strain data acquired during the field tests were converted to stress values. This conversion depended on the specific type of strain gauge used, resulting in two distinct categories of strain data. The stress was calculated by multiplying the strain data derived from the uniaxial strain gauge by the modulus of elasticity, as expressed by Equation (1). In contrast, the rosette strain gauge was capable of measuring strains along three different axes. By using the strain values recorded for each direction, significant stress values, such as the maximum and minimum principal stresses and the von Mises stress, can be calculated using Equations (2)–(5) [14]:

$$\sigma = E \times \varepsilon \tag{1}$$

$$\sigma_1 = \frac{E}{2(1-v^2)}\left[(1+v)(\varepsilon_a+\varepsilon_c) + (1-v) \times \sqrt{2\left\{(\varepsilon_a-\varepsilon_b)^2 + (\varepsilon_b-\varepsilon_c)^2\right\}}\right] \tag{2}$$

$$\sigma_2 = \frac{E}{2(1-v^2)}\left[(1+v)(\varepsilon_a+\varepsilon_c) - (1-v) \times \sqrt{2\left\{(\varepsilon_a-\varepsilon_b)^2 + (\varepsilon_b-\varepsilon_c)^2\right\}}\right] \tag{3}$$

$$\sigma_v = \sqrt{\sigma_1^2 - \sigma_1\sigma_2 + \sigma_2^2} \tag{4}$$

$$\tau_{max} = \frac{E}{2(1+v)} \times \sqrt{2\left\{(\varepsilon_a-\varepsilon_b)^2 + (\varepsilon_b-\varepsilon_c)^2\right\}} \tag{5}$$

where σ is the calculated axial stress (Pa), σ_1 is the maximum principal stress (Pa), σ_2 is the minimum principal stress (Pa), σ_v is the von Mises stress (Pa), ε is the measured strain for the components of the transplanting device, E is the modulus of elasticity (Pa), ν is Poisson's ratio, τ_{max} is the maximum shear stress (Pa), ε_a is the measured strain in the horizontal direction in the rosette strain, ε_b is the measured strain in the 45° direction in the rosette strain, and ε_c is the measured strain in the vertical direction in the rosette strain.

2.3. Stress Simulation

2.3.1. Simulation Model

A dynamic simulation was performed using commercial software (Recurdyn V9R4, Functionbay, Seongnam, Republic of Korea) to derive the stress exerted on the transplanting device. This software is widely used in studies on predicting forces and loads within diverse multibody systems comprising rigid and flexible components. Figure 4 depicts a 3D model of the cam-type transplanting device. A 3D model of the transplantation device was developed, considering its actual dimensions and material properties.

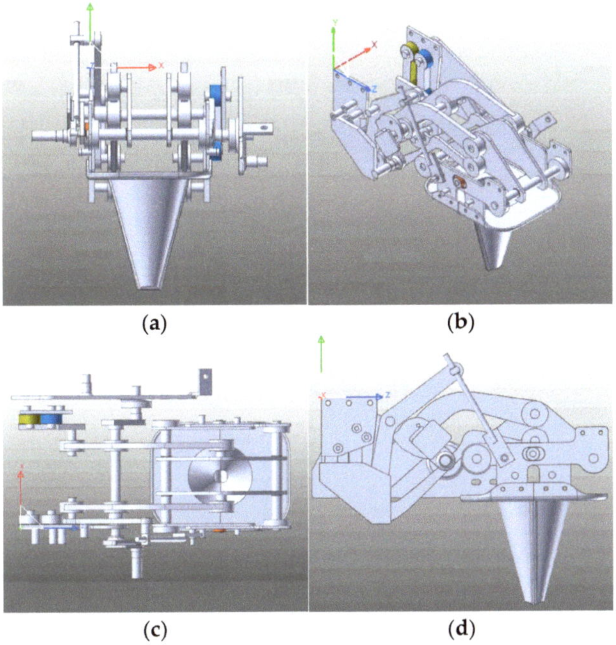

Figure 4. Three-dimensional model of cam-type transplanting device: (**a**) front view; (**b**) isometric view; (**c**) top view; (**d**) side view.

2.3.2. Simulation Condition

The simulation model for determining the stress exerted by the contact between the transplanting hopper and the ground is shown in Figure 5. The simulation conditions were set as an engine speed of 1550 rpm, a planting distance of 0.5 m, and a planting depth of 0.07 m, which yielded the maximum stress during the experiments. Therefore, in the simulation, the components of the transplanting device were operated by the behavior of the rotated cam with the 1550 rpm angular speed. Then a mesh was applied to the 3D model to derive the stress exerted on the transplanting device. The average size of the mesh for stress analysis was 1~5 mm, and the mesh shape was set as a tetrahedron. The behavior and influence of all parts, except the planting device, were ignored to minimize the analysis time.

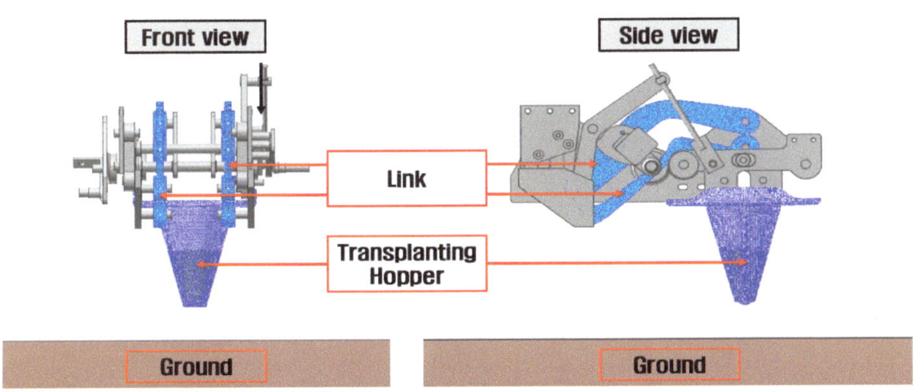

Figure 5. View of stress simulation model for transplanting device.

In the stress simulation, the effect of the vibration from the engine, the transmission, etc., was not considered. The gravitational acceleration was set to 9.81 m/s^2 to act vertically downward. A simulation was set to derive the stress generated in the links and transplanting hopper when the hopper contacted the soil during transplanting and the cam-type transplanting device operated at the same time. The conditions of the interaction between the hopper and the ground are listed in Table 5. The determination of interaction conditions between the hopper and the ground involved a multi-step process. Initially, an exploration of relevant references was conducted to establish the scope of interaction conditions applicable to the hopper–ground interface. Subsequently, a simulation was executed within this identified range, iteratively refining the parameters until optimal results were achieved.

Table 5. Material properties for simulation.

Parameter		Value
Interaction between hopper and ground	Stiffness coefficient	35
	Damping coefficient	0.03
	Dynamic friction coefficient	1.0

2.3.3. Verification of Stress Simulation

The stress values derived from the field test and simulation were compared and analyzed to verify the simulation model. The validation process involved performing a comparison between the maximum stress at 15 specific points on the transplanting device links and the transplanting hopper (Figure 3). The stress data obtained from the experiments, which validated the simulation results, were the averages of the maximum stress values (peak stress) derived from three repeated tests.

2.3.4. Maximum Stress and Static Safety Factor Based on Simulation

Unlike the field test results, which only indicated stress values at specific points where the strain gauges were positioned, the simulation results could indicate stress values across all components of the transplanting device. The simulation results identified the precise point with the highest stress level, which might be in contrast to the 15 locations where the strain gauges were positioned during the field experiments. This maximum stress value was subsequently used to compute the static safety factor. The static safety factor is a numerical value that represents the degree of safety for the machinery or structure. It can be determined as a ratio by comparing the yield strength that can sustain machinery or a structure with the maximum stress that the material or structure is expected to experience during normal use [27]. If the yield strength surpasses the measured maximum stress, thereby resulting in a static safety factor exceeding 1.0, it can be determined that a safe design has been implemented. A higher static safety factor indicates that the system is more resistant to failure and is considered safe. Conversely, if the static safety factor is less than 1.0, the design is deemed unsafe, indicating that the part may malfunction or fail due to inadequate yield strength in comparison to the maximum stress exerted on it [28]. The static safety factor can be calculated by Equations (6) and (7) [14]:

$$SF = \frac{S_y}{\sigma_{max}} \tag{6}$$

$$SF = \frac{S_y}{\sigma_{v\_max}} \tag{7}$$

where SF is the static safety factor, S_y is the yield strength (Pa), σ_{max} is the maximum axial stress (Pa), and σ_{v_max} is the maximum von Mises stress (Pa).

3. Results and Discussion

3.1. Verification of Stress Simulation

Figure 6 and Table 6 present the maximum stress values at 15 installation locations of the strain gauges obtained from the experiment and simulation.

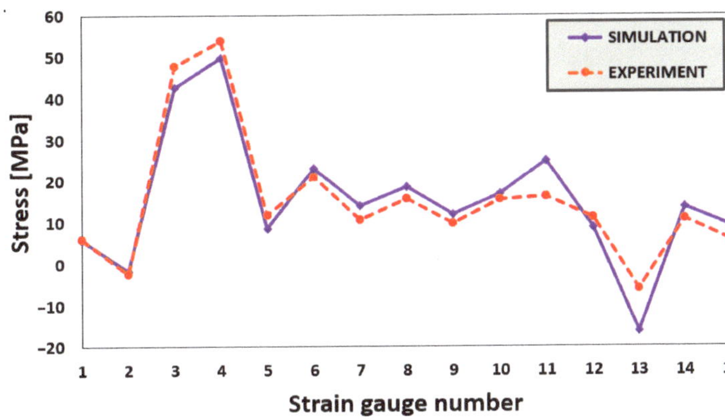

Figure 6. Comparison of maximum stress between experiment and simulation.

Table 6. Maximum stress values for 15 locations of cam-type transplanting device.

Strain Gauge Number	Measured Maximum Stress (MPa)	Maximum Stress Derived via Simulation (MPa)	RMSE
S1	5.94	5.70	
S2	−2.47	−1.84	
S3	47.53	42.54	
S4	53.69	49.64	
S5	11.75	8.38	
S6	20.81	22.95	
S7	10.42	13.96	
S8	15.68	18.53	4.4274
S9	9.66	11.89	
S10	15.51	16.78	
S11	16.04	24.85	
S12	11.06	8.66	
S13	−6.16	−16.36	
S14	10.65	13.48	
S15	5.63	9.05	

When considering the absolute value of the measured maximum stress, the maximum stress of the S4 location was highest at 53.69 Mpa in the experiment and 49.64 Mpa in the simulation. The maximum stress of the S3 location followed next at 47.53 Mpa (experiment) and 42.54 Mpa (simulation). In the case of location S2, the maximum stress was the lowest at −2.47 Mpa in the experiment and −1.84 Mpa in the simulation. The validation results show consistency between the highest stresses recorded using the simulation and field tests. However, significant differences were observed within the 15 points, such as at gauge positions 11 and 13. In contrast, there was good agreement between the experimental and simulated data for strain gauges 3 and 4, representing the highest stress values.

The root-mean-square error (RMSE) is a widely used statistical parameter for evaluating the predictive accuracy and fit quality between predicted and observed values. RMSE is calculated using the following formula:

$$RMSE = \sqrt{\frac{1}{n}\sum_{i=1}^{n}(P_i - O_i)^2}$$

where P_i is the predicted value for the i observation in the data set, O_i is the observed value for the i observation in the data set, and n is the sample size.

The RMSE calculation results for simulations on the cam type were 4.4274 MPa. This value is significantly high compared with the standard range of the maximum stress, which spanned from −6.16 to 53.69 MPa. A technique for clarifying the adequacy of the RMSE is to calculate the normalized RMSE. Normalized RMSE is calculated using the following formula:

$$Normalized\ RMSE = RMSE/(\max\ value - \min\ value)$$

For the cam type, the normalized RMSE is 0.074. This proximity to zero signifies that the model yields predictive outcomes that closely correspond to the actual measurement results.

Various factors contributed to the discrepancies between the simulation results and experimental test data. These factors include the omission of frictional effects between the joints within the simulation, inconsistencies in the assembly of components compared to actual conditions, disregard for vibrations induced by the operation of the engine, and exclusion of the load input from the main body via its connection to the machine frame. Despite the observed deviations at a few points, 13 of the 15 examined points demonstrated consistency. Consequently, the stress values obtained from the simulation experimental measurements are in good agreement.

3.2. *Maximum Stress and Static Safety Factor Based on Simulation*

Figure 7 shows the stress simulation results for the cam-type vegetable transplanter. Figure 7 highlights the position with the maximum stress among all cam-type structures, as indicated by the simulation results. A maximum stress with a magnitude of 201.21 MPa was observed at the location of the maximum stress when the hopper entered the ground and it was in the lowest position. This specific location was positioned below the location where strain gauge S6 was attached. This component functioned as the linkage between the frame and the hopper and showed a curved configuration. Because of the rotational force exerted by the engine and the pressure originating from the interaction between the hopper and the ground, this particular area was subjected to a significantly higher force than the other elements.

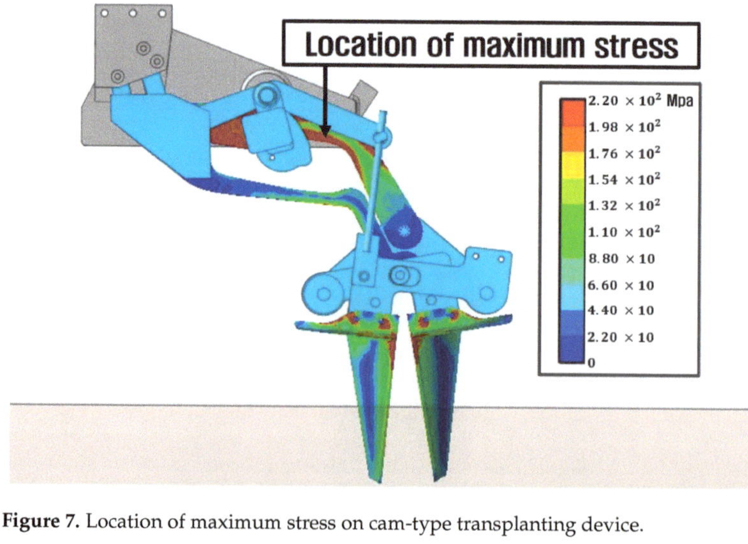

Figure 7. Location of maximum stress on cam-type transplanting device.

The maximum stress derived from the simulation significantly exceeded the highest stress value obtained from the field test measurements, which was only 53.69 MPa at strain gauge S4. This difference existed because stress measurements via field tests have constraints; specifically, stress can only be gauged at predetermined points where the strain gauges have been positioned.

Based on the maximum stress values obtained from the simulation, the static safety factor of the transplanting device of the cam-type vegetable transplanter was 1.04. The simulation results indicate that the static safety factor for the cam-type vegetable transplanter is significantly low, with values almost equal to 1. Several approaches can be considered to improve the static safety factor of transplanting devices. A potential method involves replacing the existing material with a higher-yield-strength material, which would contribute to increasing the safety level. If the material used to construct the transplanting devices is replaced with AISI 1040, which has a yield strength of 413.7 MPa, the static safety factor increases to 2.06. This value is safer than the previously reported static safety factor (1.04) [21]. When selecting a material, it is crucial to consider not only the yield strength but also the characteristics of the material and its economic viability.

4. Conclusions

In this study, the stress distribution in cam-type transplanting devices of a semiautomatic vegetable transplanter was investigated through dynamic simulation. A dynamic simulation model was developed based on the exact size and material properties of a real transplanting device. The simulation was carried out at an engine speed of 1550 rpm and planting distance of 0.5 m. The field tests were conducted to measure stress on the transplanting device using 15 strain gauges installed at certain locations. Then, the stress data from the simulation results were compared with the stress measurements obtained from field tests to validate the accuracy of the simulation method. The maximum stress values obtained across the entire structures of both types of devices in the simulation were used to compute the static safety factor.

The main findings of this study are as follows. The highest stress derived from the simulation results for the cam-type device correlated with the measured results, although discrepancies were observed, particularly at strain gauge positions 11 and 13. Based on the simulation results, the most significant stress occurred at the upper link of the transplanting device, reaching a magnitude of 201.21 MPa. This part served as the connection between the frame and the hopper, displaying a curved shape. Due to the engine's rotational force and the pressure generated from the interaction between the hopper and the ground, this specific area experienced relative greater force compared to the other components. The static safety factor calculated from the simulation was 1.04. The static safety factor value obtained from the simulation is relatively low; it is close to 1.0, indicating that the design of the transplanting device is still susceptible to stress. To enhance the safety level, there are several options available, such as substituting the current material with one possessing higher strength or modifying the design of susceptible components to a different shape or size. Furthermore, the findings of this study can serve as foundational data to establish the guidelines and process of designing cam-style planting devices. Based on the results of this study, a kinematic analysis of the cam-type transplanting device will be performed in future to establish a design process that can improve durability and economic efficiency while satisfying the appropriate planting trajectory.

Author Contributions: Data curation and formal analysis, S.M., S.-J.H., J.-H.K. and M.-K.J.; writing—original draft, S.M. and S.-J.H.; writing—review and editing, J.-S.N. All authors have read and agreed to the published version of the manuscript.

Funding: This work was supported by the Korea Institute of Planning and Evaluation for Technology in Food, Agriculture and Forestry (IPET) through the Machinery Mechanization Technology Development Program for Field Farming Program, funded by the Ministry of Agriculture, Food and Rural Affairs (MAFRA) (RS-2023-00235957).

Institutional Review Board Statement: Not applicable.

Data Availability Statement: Data are contained within the article.

Conflicts of Interest: The authors declare no conflict of interest.

References

1. Khadatkar, A.; Mathur, S.M.; Dubey, K.; BhusanaBabu, V. Development of embedded automatic transplanting system in seedling transplanters for precision agriculture. *Artif. Intell. Agric.* **2021**, *5*, 175–184. [CrossRef]
2. Javidan, S.M.; Mohamadzamani, D. Design, construction, and evaluation of automated seeder with ultrasonic sensors for row detection. *J. Biosyst. Eng.* **2021**, *46*, 365–374. [CrossRef]
3. Tian, S.; Qiu, L.; Kondo, N.; Yuan, T. Development of automatic transplanter for plug seedling. *IFAC Proc. Vol.* **2010**, *43*, 79–82. [CrossRef]
4. Moon, S.; Min, Y.; Park, J.-C. Analysis of working capacity of a hand-fed transplanter. *J. Bio-Environ. Control* **1997**, *6*, 159–167.
5. Tsuga, K. Full automatic vegetable transplanter. *J. Jpn. Soc. Agric. Mach.* **1997**, *59*, 109–110. [CrossRef]
6. Min, Y.B.; Kang, J.K.; Ryu, C.S. Development of onion transplanter: Analysis of a transplanting locus on the type of transplanting devices for a vegetable transplanter. *J. Agric. Life Sci.* **2015**, *49*, 289–294. [CrossRef]
7. Kumar, G.V.P.; Raheman, H. Vegetable transplanters for use in developing countries—A review. *Int. J. Veg. Sci.* **2008**, *14*, 232–255. [CrossRef]
8. Park, J.-H.; Hwang, S.-J.; Nam, J.-S. Operational characteristics of a domestic commercial semi-automatic vegetable transplanter. *J. Agric. Life Sci.* **2018**, *52*, 127–138. [CrossRef]
9. Hwang, S.-J.; Park, J.-H.; Lee, J.-Y.; Shim, S.-B.; Nam, J.-S. Optimization of main link lengths of transplanting device of semi-automatic vegetable transplanter. *Agronomy* **2020**, *10*, 1938. [CrossRef]
10. Yang, Q.; Huang, G.; Shi, X.; He, M.; Ahmad, I.; Zhao, X.; Addy, M. Design of a control system for a mini-automatic transplanting machine of plug seedling. *Comput. Electron. Agric.* **2020**, *169*, 105226. [CrossRef]
11. Shao, X.; Song, Z.; Yin, Y.; Xie, B.; Liao, P. Statistical distribution modelling and parameter identification of the dynamic stress spectrum of a tractor front driven axle. *Biosyst. Eng.* **2021**, *205*, 152–163. [CrossRef]
12. Xie, Z.; Shi, W.; Tian, Q.; Zheng, Y.; Tan, L. Fatigue life assessment and damage investigation of centrifugal pump runner. *Eng. Fail. Anal.* **2021**, *124*, 105256. [CrossRef]
13. Siddique, M.A.A.; Kim, W.S.; Baek, S.Y.; Kim, Y.J.; Park, S.U.; Choi, C.H.; Choi, Y.S. Analysis of the axle load of a rice transplanter according to gear selection. *J. Drive Control* **2020**, *17*, 125–132. [CrossRef]
14. Sri, M.; Hwang, S.-J.; Nam, J.-S. Experimental safety analysis for transplanting device of the cam type semi-automatic vegetable transplanter. *J. Terramechanics* **2022**, *103*, 19–32. [CrossRef]
15. Tekeste, M.Z.; Balvanz, L.R.; Hatfield, J.L.; Ghorbani, S. Discrete element modeling of cultivator sweep-to-soil interaction: Worn and hardened edges effects on soil-tool forces and soil flow. *J. Terramechanics* **2019**, *82*, 1–11. [CrossRef]
16. Paulson, I.W.P.; Dolovich, A.T.; Noble, S.D. Development of a dynamic simulation model of a towed seeding implement. *J. Terramechanics* **2018**, *75*, 25–35. [CrossRef]
17. Shao, Y.; Liu, Y.; Xuan, G.; Hu, Z.; Han, X.; Wang, Y.; Chen, B.; Wang, W. Design and test of multifunctional vegetable transplanting machine. *IFAC PapersOnLine* **2021**, *52*, 92–97. [CrossRef]
18. Swe, K.M.; Islam, M.N.; Chowdhury, M.; Ali, M.; Wing, S.; Jun, H.-J.; Lee, S.-H.; Chung, S.-O.; Kim, D.-G. Theoretical analysis of power requirement of a four-row tractor-mounted Chinese cabbage collector. *J. Biosyst. Eng.* **2021**, *46*, 139–150. [CrossRef]
19. Radhakrishnan, V.M. Multiaxial fatigue—An overview. *Sadhana* **1995**, *20*, 103–122. [CrossRef]
20. Paraforos, D.S.; Griepentrog, H.W.; Vougioukas, S.G.; Kortenbruck, D. Fatigue life assessment of a four-rotor swather based on rainflow cycle counting. *Biosyst. Eng.* **2014**, *127*, 1–10. [CrossRef]
21. Juvinall, R.C.; Marshek, K.M. *Machine Component Design*, 5th ed.; John Wiley & Sons: Hoboken, NJ, USA, 2020; pp. 338–340.
22. Yurdem, H.; Degirmencioglu, A.; Cakir, E.; Gulsoylu, E. Measurement of strains induced on a three-bottom moldboard plough under load and comparisons with finite element simulations. *Measurement* **2019**, *136*, 594–602. [CrossRef]
23. Plouffe, C.; Laguë, C.; Tessier, S.; Richard, M.J.; McLaughlin, N.B. Moldboard plow performance in a clay soil: Simulations and experiment. *Trans. ASAE* **1999**, *42*, 1531–1540. [CrossRef]
24. Makange, N.R.; Parmar, R.P.; Tiwari, V.K. Stress analysis on tyne of cultivator using finite element method. *Trends Biosci.* **2015**, *8*, 3919–3923.
25. Kešner, A.; Chotěborský, R.; Linda, M.; Hromasová, M.; Katinas, E.; Sutanto, H. Stress distribution on a soil tillage machine frame segment with a chisel shank simulated using discrete element and finite element methods and validate by experiment. *Biosyst. Eng.* **2021**, *209*, 125–138. [CrossRef]
26. Islam, M.N.; Iqbal, M.Z.; Chowdhury, M.; Ali, M.; Shafik, K.; Kabir, M.S.N.; Lee, D.-H.; Chung, S.-O. Stress and fatigue analysis of picking device gears for a 2.6 kW automatic pepper transplanter. *Appl. Sci.* **2021**, *11*, 2241. [CrossRef]

27. Bluff, E. Safety in machinery design and construction: Performance for substantive safety outcomes. *Saf. Sci.* **2014**, *66*, 27–35. [CrossRef]
28. Knabner, D.; Hauschild, S.; Suchý, L.; Vetter, S.; Leidich, E.; Hasse, A. Calculation method for the fail-safe design of steel-steel contacts subject to fretting fatigue based on a worst-case assumption. *Int. J. Fatigue* **2022**, *165*, 107217. [CrossRef]

agriculture

MDPI

Article

Precision Location-Aware and Intelligent Scheduling System for Monorail Transporters in Mountain Orchards

Shilei Lyu [1,2,3], Qiafeng Li [1], Zhen Li [1,2,3], Hengmao Liang [1], Jiayu Chen [1], Yuanyuan Liu [1] and Huixian Huang [1,*]

1 College of Electronic Engineering (College of Artificial Intelligence), South China Agricultural University, Guangzhou 510642, China; lvshilei@scau.edu.cn (S.L.); 17688015023@163.com (Q.L.); lizhen@scau.edu.cn (Z.L.); lianghm@scau.edu.cn (H.L.); 20213163195@stu.scau.edu.cn (J.C.); yuanyuan_liu5@163.com (Y.L.)
2 Pazhou Lab, Guangzhou 510330, China
3 Division of Citrus Machinery China, Agriculture Research System of MOF and MARA, Guangzhou 510642, China
* Correspondence: huanghuixian@scau.edu.cn; Tel.: +86-1511-202-6883

Abstract: This study addressed the issue of the real-time monitoring and control of the transporter in a mountain orchard terrain characterized by varying topography, closed canopy, shade, and other environmental factors. This study involved independent research and the development of a series of electric monorail transporters. First, the application requirements of "Where is the monorail transporter?" were examined, and an accurate location-aware method based on high-frequency radio frequency identification (RFID) technology was proposed. In addition, a location-aware hardware system based on STM32 + RFID + LoRa was designed to determine the position of the monorail transporter on a rail. Second, regarding the application requirements of "Where is the monorail transporter going?", a multimode control gateway system based on Raspberry Pi + LoRa + 5G was designed. An Android mobile terminal can obtain operational information about the transport plane in real time through the gateway system and remotely control its operation. The track-changing branch structure enables multimachine autonomous intelligent avoidance. Based on the experimental results of monorail transporter positioning in mountain orchards under various typical terrains, such as flat surfaces, turning paths, and uphill/downhill slopes, the road section average relative error of the 7ZDGS–250-type monorail transporter was 1.27% when the distance between benchmark positioning tags was set at 10 m on both flat and turning roads, and that of the 7ZDGS–300-type monorail transporter was 1.35% when the distance between benchmark positioning tags was set at 6 m uphill/downhill. The road section relative error of the 7ZDGS–250-type monorail transporter was 21.18%, and that of the 7ZDGS–300-type monorail transporter was 9.96%. In addition, the experimental results of monorail transporter communication control showed that the combination of the multimode control gateway control system and track-changing branch structure can achieve multimachine cooperation and autonomous avoidance function, ensuring that multiple monorail transporters can operate simultaneously without collision. The findings of this study establish the communication link of "monorail transporter-gateway system-control terminal" and form a precise positioning and real-time control scheme applicable to the operating environment of monorail transporters, thereby improving the intelligence and safety of mountain orchard monorail transporters.

Keywords: mountain orchard; monorail transporter; positioning system; gateway system; RFID; Raspberry Pi; LoRa

Citation: Lyu, S.; Li, Q.; Li, Z.; Liang, H.; Chen, J.; Liu, Y.; Huang, H. Precision Location-Aware and Intelligent Scheduling System for Monorail Transporters in Mountain Orchards. *Agriculture* **2023**, *13*, 2094. https://doi.org/10.3390/agriculture13112094

Academic Editors: Chung-Liang Chang and Mustafa Ucgul

Received: 8 October 2023
Revised: 1 November 2023
Accepted: 2 November 2023
Published: 3 November 2023

1. Introduction

Hilly and mountainous areas are essential production bases for grain and unique agricultural products. In China, for example, the amount of arable land in 2022 reached 168,695 kilo hectares [1], with hilly and mountainous areas accounting for one-third of it [2]. However, these areas have varied terrains and rugged, narrow roads, making it difficult for most agricultural machinery and equipment designed for flatlands to pass through and function effectively. The problems of "no machine available" and "no machine good" are prominent [3]. This is particularly notable in the transportation of supplies during production, where the traditional approach relies on labor or animal transport, leading to high labor intensity, high transport cost, low operational efficiency, and increased safety hazards [4]. The monorail transporter has a smooth operation and strong climbing ability, and is the mainstream machine for mechanized transport operations in mountain orchards [5]. The transporter can be divided into double track [6] and single track [7] based on the layout of the tracks and the condition of the power drive. The transporter mainly includes traction [8], internal combustion engine-driven [9], or electric [10] power drives.

Monorail transporters have received significant attention in the industry because of their advantageous features, such as flexible track-laying and a small turning radius [11]. In previous studies, the author's team has independently developed a series of electrically operated monorail transporters [12]. Among them, the 7ZDGS–250-type monorail transporter can carry a maximum load of 250 kg, travel at a speed of 0.53 m/s, and climb slopes with a maximum angle of 30° while, the 7ZDGS–300-type monorail transporter can handle a maximum load of 300 kg, travel at a speed of 0.94 m/s, and climb slopes with a maximum angle of 36°. The mechanical theory and technical application of monorail transporters have progressively matured. However, with the increasing scale of track laying, the monorail transporter faces challenges during the driving process, such as changes in terrain within mountain orchards, closed canopies, and severe shade. These environmental factors make real-time monitoring and control difficult. Thus, the transporter should be equipped with autonomous operation and navigation capabilities, aligning with the intelligent agriculture development trend [13]. To achieve this goal, two fundamental questions must be addressed: "Where is the monorail transporter?" and "Where is the monorail transporter going?"

The primary objective of addressing the "Where is the monorail transporter?" problem is to determine the real-time on-track position of the monorail transporter. Currently, the universal positioning system for agricultural equipment mainly adopts satellite navigation [14,15], light detection and ranging (LiDAR) [16,17], and machine vision [18]. However, the positioning signal propagation of satellite navigation is affected by terrain elevation changes, tree canopy shading, and other factors, resulting in low positioning accuracy, making it challenging to apply directly to a mountain orchard operating environment. LiDAR and machine vision fall under the simultaneous localization and mapping (SLAM) field category. These technologies require scanning in advance to build a map model, which is then combined with environmental parameters to achieve positioning. However, the dynamic nature of ground vegetation and tree canopy in the orchard production process makes it difficult to maintain a static map model. In addition, the cost of implementing this technology is high, making it unsuitable for mountainous orchard environments.

The primary objective of addressing the "Where is the monorail transporter going?" problem is to effectively communicate with and control the monorail transporter in real time, which can be categorized in the context of IoT in agriculture. Common IoT communication technologies [19] mainly use Wi-Fi [20], Zigbee [21], LoRa [22], Bluetooth [23], and 5G [24]. Agricultural equipment communication requires low power consumption, low cost, and short response time. Compared to other communication technologies, LoRa communication has comprehensive range coverage, low application cost, convenient networking, and can generally support thousands of nodes. Through the communication experiment in the orchard environment with a closed canopy, the measured communication distance between two Lora communication modules can reach 400 m without relay nodes. Therefore, LoRa is more suitable for the operating environment of mountain orchards.

In this study, a comprehensive analysis is performed on a self-developed series of electric monorail transporters. This study focused on two main aspects. First, we propose an accurate positioning method based on high-frequency radio frequency identification (HF

RFID) technology. This solution solves the problem of "Where is the monorail transporter?" through the design of a location-aware hardware system based on STM32 + RFID + LoRa to determine the monorail transporter's on-track position in real time. Second, a multimodal control gateway system was designed based on Raspberry Pi + LoRa + 5G, allowing an Android mobile terminal to obtain real-time information about the operation of the monorail transporter and remotely control its operation status through the gateway system, thus solving the problem of "where is the monorail transporter going?". In addition, this study combines the track-changing branch structure to enable the autonomous and intelligent avoidance of multitransporters.

This study is organized as follows: Section 2 describes the RFID location-aware method and feasibility analysis of the monorail transporter, as well as the design of the location-aware hardware system; Section 3 describes the hardware design of the multimode control gateway system, software control process, autonomous avoidance strategy of the monorail transporter, and software control process of the mobile client. The experiments on positioning and communication control of the monorail transporter of the mountain orchard and the analysis of their results are elaborated in Section 4; and finally, Section 5 summarizes the study.

2. Location-Aware Methods and Systems for Monorail Transporter

2.1. Accurate Location-Aware Method Based on High-Frequency Radio Frequency Identification Technology

The operating frequency of HF RFID is 13.56 MHz. Reader and tag communication is characterized by a fast data transmission rate and a close effective reading distance [25]. This study used this property to deploy double readers (R_1 and R_2) in the nose of a monorail transporter with distance D_R and benchmark positioning tag sets (T_a, T_b and etc.) on the trackside with distance D_T (Figure 1).

The process of determining the on-track position of the monorail transporter is described as follows: when the monorail transporter passes the benchmark positioning tag, the double readers sequentially identify the benchmark positioning tag T_n ($n = a, b, \ldots$) to obtain the on-track position of the monorail transporter, P_n ($n = a, b, \ldots$), and the time difference of identifying the benchmark positioning tag T_n, Δt. The monorail transporter's instantaneous velocity V_n' ($n = a, b, \ldots$) was calculated using Equation (1) to characterize the monorail transporter's average velocity V_n ($n = a, b$) between the two benchmark positioning tags T_n and T_{n+1}.

$$V_n' = \frac{D_R}{\Delta t} = V_n \quad (1)$$

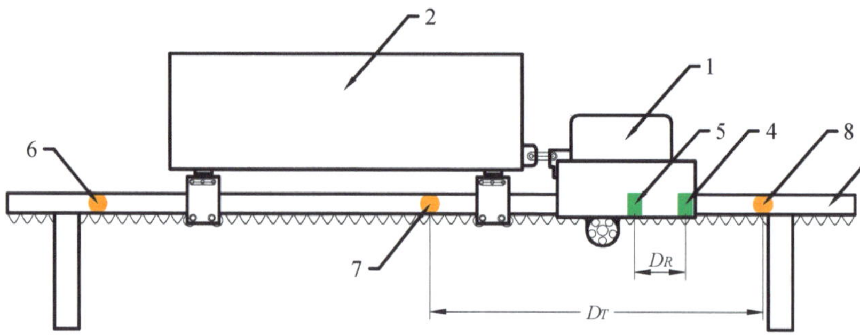

Figure 1. Accurate positioning method based on HF RFID technology. 1. The head of the monorail transporter. 2. The rack of the monorail transporter. 3. Track. 4. Reader R_1. 5. Reader R_2. 6. Tag T_{n-1}. 7. Tag T_n. 8. Tag T_{n+1}.

Double readers identified the benchmark positioning tag Tn and then timed it to obtain the traveling time t_n ($n = a, b, \ldots$) of the monorail transporter between the two benchmark positioning tags T_n and T_{n+1}, and Equation (2) was used to calculate the traveling distance

S_n ($n = a, b, \ldots$) of the monorail transporter between the two benchmark positioning tags T_n and T_{n+1}. Equation (3) was used to calculate the monorail transporter's real-time on-track position P.

$$S_n = V_n * t_n \tag{2}$$

$$P = P_n + S_n \tag{3}$$

2.2. Feasibility Analysis of Radio Frequency Identification Location Methods for Monorail Transporter

The first prerequisite for the feasibility of the monorail transporter RFID positioning method is that double readers can correctly read the baseline positioning tag data while the monorail transporter is traveling, and the smaller the distance D_R between R_1 and R_2, the more accurate the monorail transporter's instantaneous traveling speed V_n'. However, because HF RFID uses a fixed communication band, the near-neighboring R_1 and R_2 inevitably encounter the reader collision interference problem when reading the same tag data. At this point, the tag fails to communicate because it cannot correctly parse the readers' query signal [26]. In this study, a double-reader interlocking anti-interference strategy was designed to address this problem: The first reader along the traveling direction of the monorail transporter is activated in priority, and the second reader is closed and activated immediately when the reader correctly reads the tag data. Only one reader is activated at any time by controlling the double readers to take turns starting and stopping. The key to the double-reader interlocking anti-jamming strategy is the existence of a start initialization process for reader start-up, which requires some data processing time, T_{data}. Thus, a lower limit value for the distance d between R_1 and R_2 must exist to satisfy this time requirement.

To verify the above ideas, an RFID model (MFRC522, ZLG, Guangzhou, China) was identified that supports the ISO 14443A/MIFARE protocol [27], with an adequate reading distance of 6–10 cm and, communication modes such as Serial Peripheral Interface (SPI), inter-integrated circuit, and universal asynchronous receiver/transmitter. Second, the double readers were placed relative to each other at a distance of 12 cm (Figure 2), and 20 matching tags were randomly selected; each tag was placed at the center of the double readers, and the time difference in reading tag data was recorded by starting and stopping the reader in turn, which was repeated 20 times. Approximately 400 sets of time-difference data were obtained, with the statistical results of 3.19 $\pm$ 0.01 ms, which can characterize the data processing time T_{data} to some extent. If the monorail transporter's maximum traveling speed is 0.94 m/s, and the total value of the reading time difference is 3.20 ms, the distance D_R should be approximately larger than 0.30 cm. Considering the complexity of the operating environment of the actual application, the distance D_R was located at 5 cm in this study, which is an order of magnitude higher than the lower limit value to avoid the problem of RFID collision interference to the greatest extent possible.

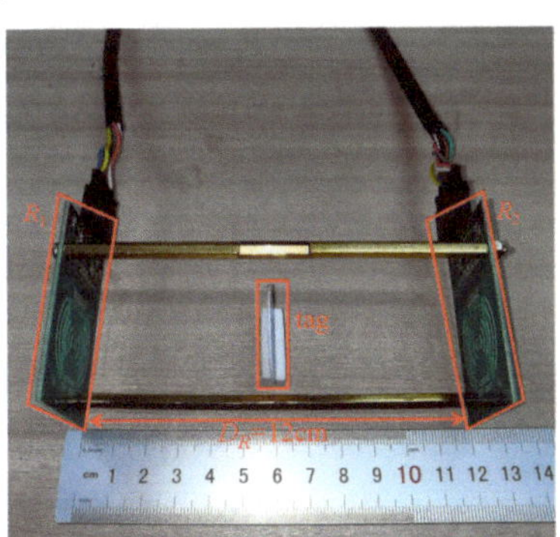

Figure 2. Double readers interlock anti-interference idea verification.

2.3. Design of Location-Aware Hardware System Based on STM32 + RFID + LoRa

For the application requirement analysis of "Where is the monorail transporter?", the transport location-aware hardware system should have the functions capable of location-aware, calculating, and transmitting on-track position information. Therefore, the control logic of the location-aware hardware system designed in this study used the HF RFID to sense the benchmark positioning tag's position information deployed on the trackside, an embedded STM32 microcontroller to process the RFID data to obtain the on-track position information of the monorail transporter, and a low power LoRa communication module to realize the information transmission function. The design of the location-aware hardware system based on STM32 + RFID + LoRa is shown in Figure 3, and the fabricated circuit board is shown in Figure 4.

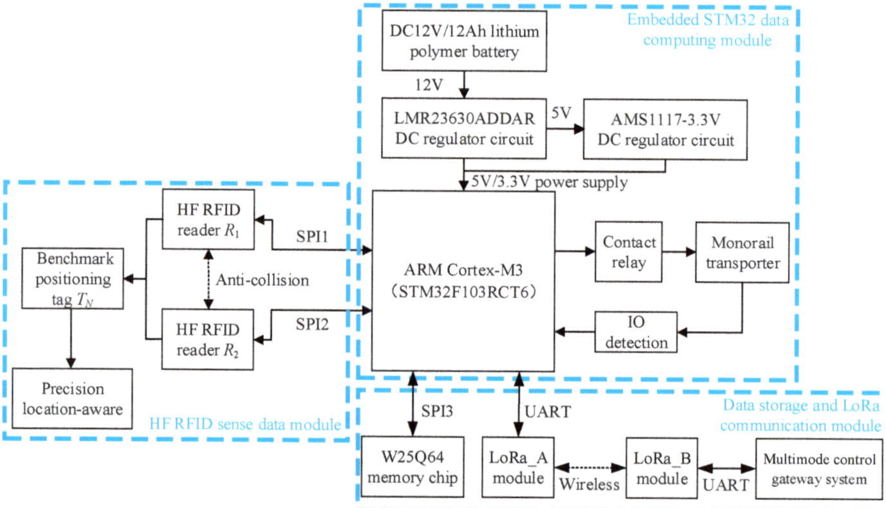

Figure 3. Framework of the location-aware hardware system.

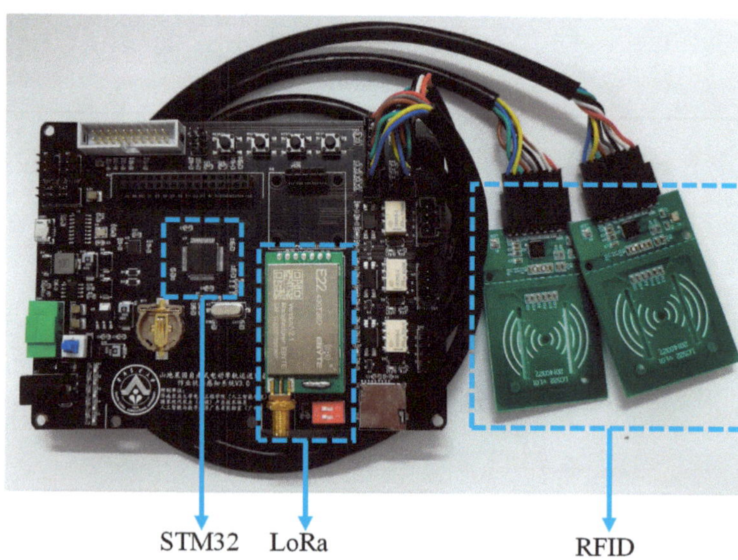

Figure 4. Circuit board of the location-aware hardware system.

The main functional modules of the location-aware hardware system are as follows:

- HF RFID sense data module: The system uses an MFRC522-type RFID reader with a matching passive tag (for performance parameters, see Section 2.1). Readers R_1 and R_2 use the interlocking anti-interference strategy to solve the RFID collision interference problem and combine it with a hardware timer to obtain double readers to read the same benchmark positioning tag's data time difference; double readers through the SPI channel to provide real-time sensing data for the main control chip.
- Embedded STM32 data computing module: The main control chip of the system is a 32-bit microcontroller STM32F103RCT6 (ST Microelectronics, Muar, Malaysia) based on an ARM Cortex-M3 core that supports multiple communication methods and timer configurations, and operates at a clock frequency of 72 MHz. The power supply uses a Direct Current (DC) 12V/12Ah lithium polymer battery and provides 5/3.3 V working voltage through a DC regulator circuit. The main control chip determines the position of the monorail transporter on the track in real time based on RFID real-time sensing data using the proposed positioning method, controls the motor of the monorail transporter using the contact relay (HFD4/5, Hongfa, Xiamen, China), and monitors the motor operating status in real time through the I/O ports.
- Data storage and LoRa communication module: The main control chip of the system reads the benchmark positioning tag data in the flash memory (W25Q64, Winbond, Taiwan, China) through the SPI channel and stores the determined on-track position of the monorail transporter in real time, addressing the problem of data loss by power-down. In addition, the system uses a LoRa serial communication module (E22-400TBH-01, Ebyte, Chengdu, China) to upload the measured on-track monorail transporter positions, traveling speeds, historical paths, and other operational statuses to the multimode control gateway system. It receives real-time monorail transporter forward/backward, stop, and other control commands from the gateway system.

3. Multimode Control Gateway System for Monorail Transporter

3.1. Design of Multimode Control Gateway System

For the application requirement analysis of "Where is the monorail transporter going?" the multimode control gateway system should have the functions of remotely obtaining the operation status of the monorail transporter, issuing control instructions in real time based on the operation tasks, and realizing the safe scheduling of multiple machines. The monorail transporter operation status is sent to the gateway system in real time after being determined by the location-aware hardware system. In contrast, the mobile client releases the monorail transporter control instructions subscribed by the gateway system to the cloud server. The architecture of the multimode control gateway system designed in this study is shown in Figure 5, and the design diagram and system object are shown in Figure 6.

The main functional modules of the multimode control gateway system are as follows:

- Raspberry Pi 4B central control module: The central control module of the gateway system adopts the Raspberry Pi 4B board (Raspberry Pi, UK) based on the BCM2711 chip, equipped with a 64-bit 4-core processor with a central frequency of 1.5 GHz, and Python based on the transplanted Linux operating system to complete the development of communication functions. The power supply is made of a high-energy-density DC12V/12Ah lithium polymer battery. It adopts an isolated DC–DC small power step-down power module (DM41-20W1205B1, Ebyt, Chengdu, China) to effectively and consistently suppress the spike voltage and output 5 V working voltage.
- LoRa communication module: The gateway system LoRa communication module is consistent with the configuration of the location-aware hardware system, and the working frequency band is 433.125 MHz. Based on the coordinated operation of multiple transporters, after polling by the gateway system, the location-aware hardware system of each monorail transporter uploads its operation status information using the LoRa peer-to-peer transmission mode, and the gateway system releases

monorail transporter control and scheduling commands using the LoRa peer-to-peer transmission mode.

- 5G communication module: The gateway system adopts a 5G module based on the RG500U-CN module (Quectel, Hefei, China), which can automatically adapt to 5G NSA and SA dual-mode networks and is also compatible with 4G/3G. The Gigabit Ethernet port (RJ-45) of the controller module connects to the 5G communication module to access the Internet through a CAT-6 cable, and the system uses the Message Queuing Telemetry Transport [28] (MQTT) protocol to facilitate information interaction with the cloud server.
- MQTT cloud server: The controller module uses Mosquitto to deploy the MQTT cloud server, which is responsible for forwarding the communication data between the mobile client and multiple equipment. In addition, it adopts Phddns [29] intranet penetration to map the intranet ports to the cloud, converting the private IP address of the intranet into the legal public IP address, realizing the domain name-based Internet access of the LAN application, and applying the MQTT protocol to realize the information interaction with the mobile client.

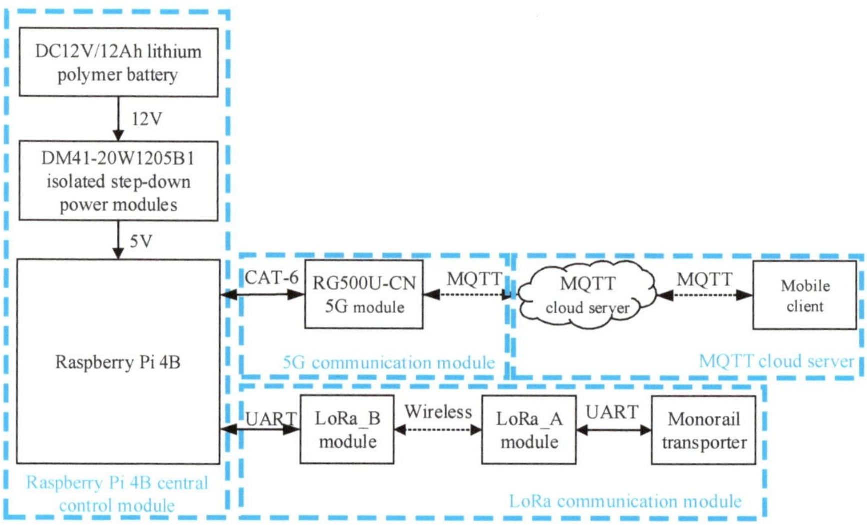

Figure 5. Multimode control gateway system architecture.

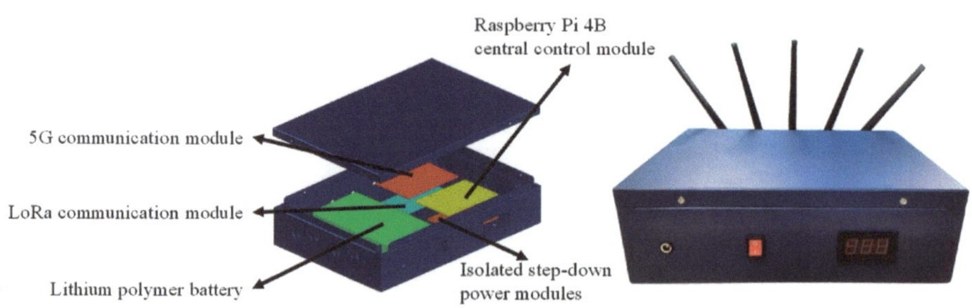

Figure 6. Multimode control gateway system design and object.

3.2. Multimode Control Gateway System Workflow

Figure 7 shows the workflow diagram of the gateway system. After the gateway system starts running, it first performs the start operation, creates the equipment dictionary for storing the job status information of multiple equipment, and creates the queues used for process communication, which are the LoRa data downstream queue, LoRa data upstream queue, MQTT data downstream queue, and MQTT data upstream queue. Secondly, based on the Linux system, Python language for multitasking programming, creating two task processes and two sub-threads, namely LoRa process, MQTT process, polling thread, and scheduling thread. The processes and threads work together in the system to obtain transmitter operation status information and execute the scheduling commands, ensuring efficient operation of the transmitter.

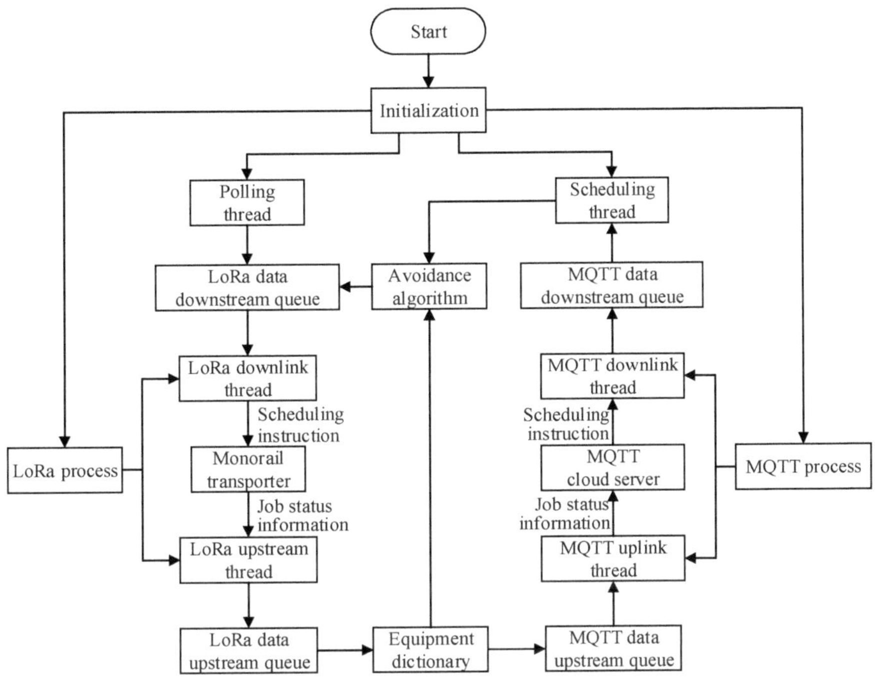

Figure 7. Workflow diagram of multimodal control gateway system for multitasking programming using Python based on Linux system.

Two subthreads are created in the LoRa process: the LoRa downlink and uplink threads. The LoRa downstream thread reads the scheduling commands in the LoRa data downstream queue in real time and sends them to the monorail transporter point-to-point using the LoRa communication module. The LoRa upstream thread reads the data in the First Input First Output (FIFO) data buffer of the LoRa communication module in real time and writes the data into the LoRa data upstream queue. Two subthreads are created in the MQTT process: the MQTT downlink and uplink threads. The MQTT downlink thread subscribes to the scheduling commands of the MQTT cloud server in real time and writes them into the MQTT data downlink queue. The MQTT uplink thread reads the job status information of the monorail transporter in the MQTT data uplink queue and publishes the data to the MQTT cloud server.

The polling thread sends commands to the LoRa data downstream queue at regular intervals to poll the monorail transporter job status information in the self-organizing network. Then, it waits to read the monorail transporter job status information into the LoRa data upstream queue and writes it into the equipment dictionary and MQTT data upstream queue. It sends the instruction to the avoidance algorithm, which performs avoidance path planning by combining the operational status information of each monorail transporter in the equipment dictionary. The equipment priority obtains the avoidance scheduling instruction and writes the scheduling instruction to the LoRa data downstream queue, while also monitoring and updating the subsequent operation status information of each monorail transporter and scheduling instruction in real time.

3.3. *Transporters Avoidance Strategies Combined with the Track-Changing Branch Structure*

To ensure the efficient coordination and safe operation of multiple monorail transporters, this study proposed a monorail transporter avoidance strategy combined with a track-changing branch structure [30], in which the track-changing branch structure was designed independently by the author's team in previous work. It can communicate with the multimode control gateway system in real time. The design and physical objects are shown in Figure 8.

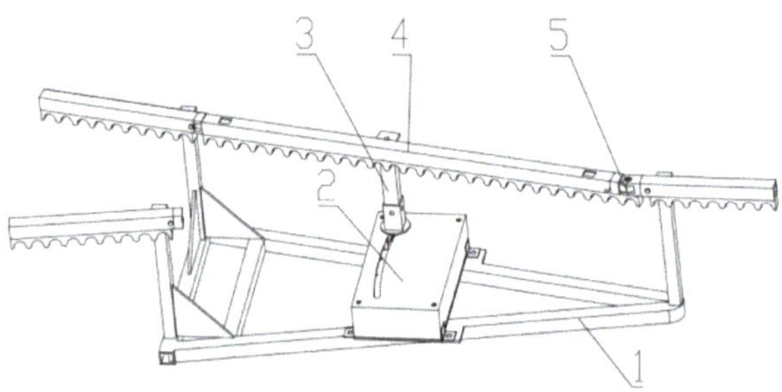

1. Track change bracket. 2. Control box. 3. Conversion support columns. 4. Conversion track. 5. Track-switching connector.

(**a**) Track-changing branch structure design

(**b**) Track-changing branch structure objects

Figure 8. Track-changing branch structure of monorail transporter.

The avoidance strategy of the monorail transporter based on the link of the "monorail transporter-gateway system-track-changing branch structure" is summarized as follows:

1. The avoidance waiting area was set up at each branch of the track-changing branch structure (Figure 9);
2. The gateway system detects the operation status information of each monorail transporter in real time and calculates a safe avoidance control strategy using the avoidance algorithm based on the operation priority of the monorail transporters when there is a risk of collision between the detected monorail transporters;
3. Using the results of the avoidance strategy, the gateway system controls the track-changing branch structure to change the track direction, ensuring that monorail transporters with low operational priority go to the nearest avoidance waiting area and give the right of way of the track to monorail transporters with high operational priority;
4. During the avoidance process, the gateway system continuously detects the operation status information of each monorail transporter to ensure that all monorail transporters cooperate to avoid conflicts and optimize the avoidance path. If a new risk of collision emerges, the gateway system will devise a new avoidance plan for the monorail transporters to avoid the collision.

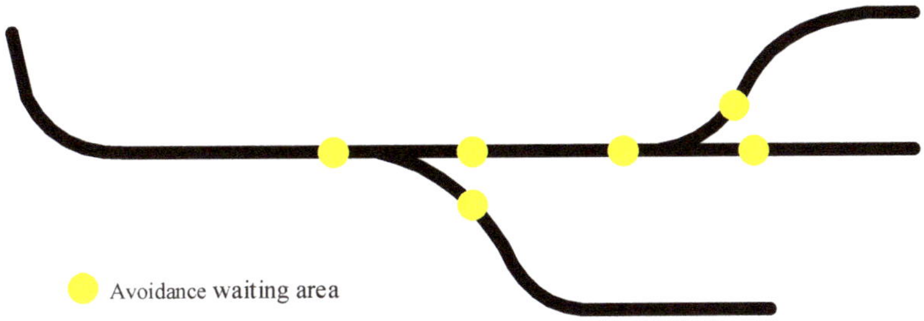

Figure 9. Schematic diagram of avoidance waiting areas set up at each branch of the track-changing branch structure.

3.4. Mobile Client Workflow

The mobile client APP was built on the basis of Android Studio and Java language to implement the function; the workflow is shown in Figure 10, and the operation interface is shown in Figure 11. First, the initialization procedure establishes a connection with the MQTT cloud server. Following a successful link, distinct MQTT uplink and downlink threads are generated for subscribing to and transmitting data. The MQTT uplink thread subscribes to the real-time operation status information of the monorail transporter from the MQTT cloud server, analyzes and processes the information, and updates it to the APP interface, thereby allowing for the monitoring of the operation status information of the monorail transporter. By clicking the button in the interface, the MQTT uplink thread sends the corresponding scheduling command to the MQTT cloud server, thus indirectly sending scheduling commands to the monorail transporter.

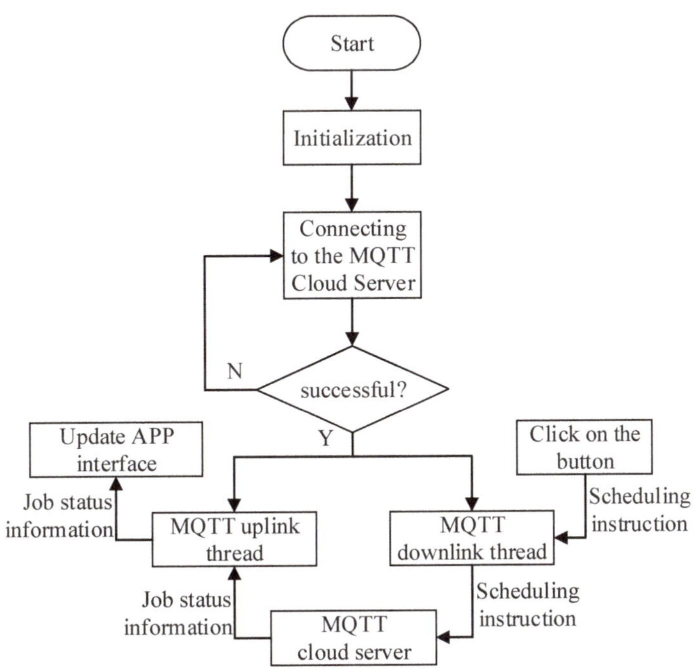

Figure 10. Workflow diagram of a mobile client app based on Android Studio.

Figure 11. Mobile client interface. Numbers 1 to 12: the benchmark positioning tag number.

4. Experimental Results and Analyses

4.1. Experiments on the Precision of Monorail Transporter Operation State Perception Sensing

In this study, a field experiment was performed in April 2023 in a mountainous (lychee) orchard of the Changsheng fruit industry family farm (Yangxi County, Yangjiang City, Guangdong, China) to validate the performance of the location-aware hardware system based on STM32 + RFID + LoRa, as well as to investigate the deployment mode of the HF RFID benchmark positioning tag set and the influencing law of the track road condition on the sensing accuracy of the monorail transporter's operation state. The experimental site covers three typical terrains, including flat, turning, and sloping (uphill/downhill), with an average gradient of the sloping road being 19.22°. In each typical terrain, a test rail with a length of 12 m was set up, the RFID benchmark positioning tag sets numbered 1–25 were deployed at 0.5 m intervals. The site road conditions are shown in Figure 12. To ensure the comprehensiveness of the experiments, two types of monorail transporters, the 7ZDGS–250 and 7ZDGS–300, were used. In addition, the Hall positioning and satellite navigation positioning methods were used in this study to perform side-by-side comparison experiments. In addition, the Hall positioning and the GPS/Beidou positioning methods were used to perform cross-comparison experiments. The Hall positioning method detects the permanent magnet on the motor rotor shaft through the magnetic Hall switch module (NJK-5002-NKK) and achieves the positioning function by realistically counting the number of motor rotation turns, while the GPS/Beidou positioning method achieves positioning by obtaining the latitude and longitude information of the monorail transporter through the WGS-84 coordinate system through the GPS/Beidou dual-mode positioning module (BH-ATGM332D).

Figure 12. The site road conditions.

4.1.1. Experiments on the Performance of Location-Aware Hardware System

The experimental program for the performance of the location-aware hardware system is described as follows:

1. The 7ZDGS–250-type monorail transporter was used, and the flat road condition was selected to start the experiment. Using the No. 1 tag as the starting point, the monorail transporter was started at 2 m from the starting point so that it reached the starting point at a uniform speed.
2. When the monorail transporter passes the starting point, the location-aware hardware system calculates the instantaneous traveling speed (V_n') of the monorail transporter based on the accurate positioning method of high-frequency RFID technology.
3. When the monorail transporter passes the tag No. 2–25 (T_n), the location-aware hardware system calculates the traveling distance (S_n) of the monorail transporter based on the speed obtained in step 2 and the timing of the timer (T_n).
4. After completing the above steps, the monorail transporter type was changed, and steps 1 to 3 were repeated to obtain the experimental data of different monorail transporter types.
5. Finally, the road condition was changed, and steps 1 to 4 were repeated to study the effect of different road conditions on the experimental results.

Figure 13 shows the experimental results. The road section error gradually increases as the tag distance D_T increases, whereas the road section relative error remains stable. The size of the road section relative error of the RFID positioning for the same type of monorail transporter is uphill > downhill > turning > flat road. When going downhill, the road section relative error of 7ZDGS–250-type transport was smaller than that of the 7ZDGS–300-type monorail transporter. In contrast, when going uphill, the road section relative error of the 7ZDGS–250-type monorail transporter was larger than that of the

7ZDGS–300-type monorail transporter, mainly because the power of the 7ZDGS–250-type monorail transporter is low under the uphill condition, causing the cumulative error and the road section relative error of the 7ZDGS–250 type monorail transporter to increase.

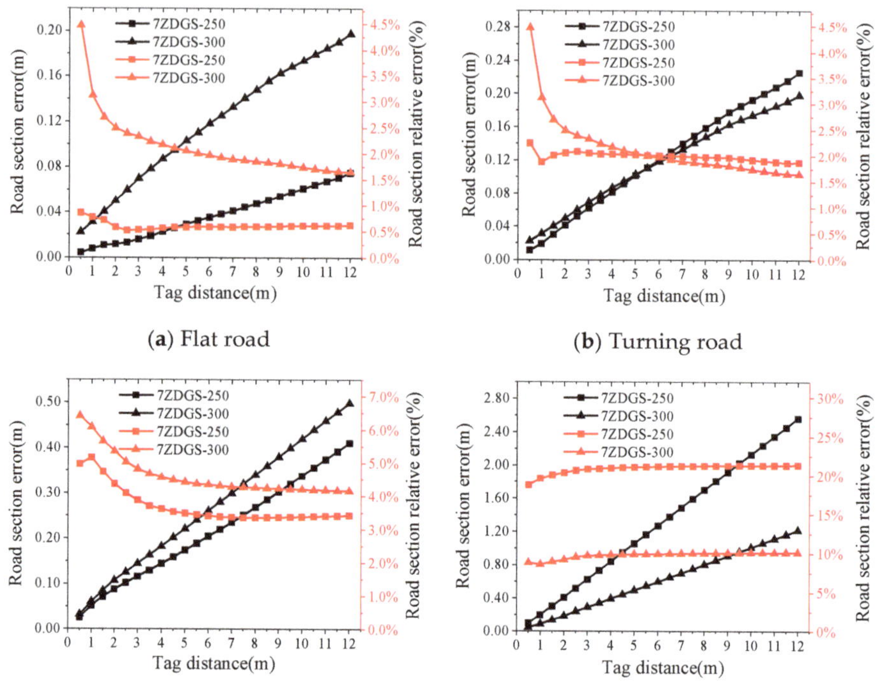

(**a**) Flat road (**b**) Turning road

(**c**) Downhill road (**d**) Uphill road

Figure 13. Curves for experiments on the performance of location-aware hardware systems.

Based on the experimental results, the following tag distance selection strategies are recommended to ensure the efficiency of farm work:

1. Flat road and turning road: The tag distance can be set at 10 m to control the road section error within 20 cm and the road section relative error within 2%. Under this tag distance, the road section average relative error of the 7ZDGS–300-type monorail transporter is 1.35%, and the road section average relative error of the 7ZDGS–250-type monorail transporter is 1.27%.
2. Sloping road: Because the length of the monorail transporter used in this study is 3.42 m, it is recommended that the tag distance should be greater than 3.42 m. In addition, the uphill condition of the road section relative error was greater than that of the downhill condition of the road section relative error to ensure the safety of multimachine cooperation. Therefore, it is recommended that the difference between the tag distance and length of the monorail transporter should be twice greater than the uphill section of the road section relative error and that the tag distance be set at 6 m.

4.1.2. Verification Experiments of the Location-Aware Hardware System

The accurate positioning method based on the HF RFID technology characterizes the average driving speed of the monorail transporter under road conditions by the instantaneous driving speed at the starting point to obtain the monorail transporter's real-time on-track position. Validation experiments were performed to verify the applicability of the location-aware hardware system based on this method and to determine whether different starting points under the same road condition affect the positioning accuracy of the location-aware hardware system. The 7ZDGS–300-type monorail transporter was used to start at 2 m from the starting point and traveled 10 m on a flat road with tags No. 1, 2, 3, 4, and 5 as the starting points. The experimental data are shown in Table 1.

Table 1. Experimental data for the verification of the location-aware hardware system.

Starting Point	Difference between the Road Section Error at This Point and That at No. 1 (m)	Difference between the Road Section Error at This Point and That at No. 1.
No. 2	0.006	0.06%
No. 3	0.008	0.08%
No. 4	0.026	0.26%
No. 5	0.045	0.45%
Average	0.021	0.21%

The experimental results show that when the location-aware hardware system is turned on at different starting points of the same road section, the difference between the road section relative error for the starting points of tags No. 2, No. 3, No. 4, and No. 5 and the road section relative error for the starting point of tag No. 1 ranges from 0.21% on average only. Therefore, the location-aware hardware system has minimal influence on its positioning accuracy when it is turned on at different starting points of the same road section, and it also shows that the location-aware hardware system has good universality and robustness.

4.1.3. Cross-Comparison Experiments—The Hall Positioning Method

The experimental program for the comparison of the Hall positioning method is described as follows:

1. The 7ZDGS–250-type monorail transporter was used, and the flat road condition was selected to start the experiment. Using tag No. 1 as the starting point, the monorail transporter was positioned at 2 m from the starting point so that it reached the starting point at a uniform speed.
2. When the monorail transporter passes the starting point, the location-aware hardware system activates the Hall positioning method and counts the motor shaft of the monorail transporter in real time;
3. When the monorail transporter passes tag Nos. 2–25 (T_n), the number of motor rotations was obtained at the corresponding tag (n) and the traveling distance of the monorail transporter (H_n) using Equation (4),

$$H_n = \frac{2\pi rn}{a}, \tag{4}$$

   where r is the gear radius with a value of 0.135 m, and a is the gear ratio of 10.4166:1.
4. After completing the above steps, the monorail transporter type was changed, and steps 1 to 3 were repeated to obtain experimental data for different monorail transporter types.
5. Finally, the road conditions were changed, and steps 1 to 4 were repeated to study the effect of different road conditions on the experimental results.

The experimental results show that the RFID positioning method has a lower road section relative error than the road section relative error of the Hall positioning method for the 7ZDGS–250- and 7ZDGS–300-type monorail transporters on flat and turning roads (Figure 14). The relative errors were approximately 4.90% and 8.59%, and 6.08% and 51.72%, respectively, when the tag distance was set at 10 m. Furthermore, when the tag distance was set to 6 m, the road section relative error of the RFID positioning method was lower than that of the Hall positioning method for both the 7ZDGS–250- and 7ZDGS–300-type monorail transporters, by approximately 45.25% and 27.48%, and 23.43% and 16.13%, respectively, for both downhill and uphill road conditions.

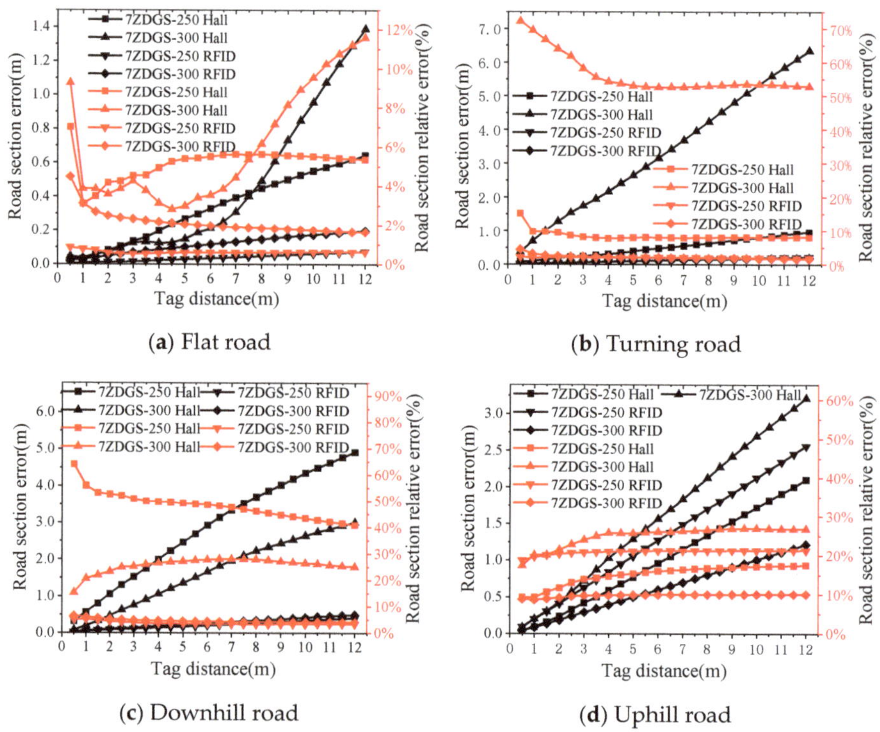

Figure 14. Curves for cross-comparison experiments using the Hall positioning method.

The results of the cross-comparison experiments show that the RFID positioning method exhibits superior positioning accuracy under different road conditions, and the road section relative error of the Hall positioning method was relatively increased. This phenomenon may be because the magnetic Hall switch module works on the principle of magnetic pole induction, and the permanent magnet affects the counting accuracy of the magnetic Hall switch module under high-speed rotation and oscillation, leading to a larger road section relative error.

4.1.4. Cross-Comparison Experiments—The GPS/Beidou Positioning Method

The experimental program for comparing the GPS/Beidou positioning methods is described as follows:

1. The 7ZDGS–300-type monorail transporter was used, and the flat road condition was selected to start the experiment. Using tag No. 1 as the starting point, the monorail transporter was positioned at 2 m from the starting point so that it reached the starting point at uniform speed.
2. When the monorail transporter passes the starting point, the location-aware hardware system starts the GPS/BeiDou positioning method to obtain the latitude and longitude coordinates (x_1, y_1) of the starting point
3. When the monorail transporter passes tag No. 2–25 (T_n), the location-aware hardware system acquires the latitude and longitude coordinates (x_m, y_m) at the correspond-

ing tag and obtains the traveling distance (G_n) of the monorail transporter using Equation (5),

$$G_n = Rarccos(\cos y_1 \cos y_2 \cos(x_2 - x_1) + \sin y_1 \sin y_2), \tag{5}$$

where R is the radius of the earth, which is approximately 6,371,004 m.

4. The road conditions were changed, and steps 1 to 3 were repeated to study the effect of different road conditions on the experimental results.

When the tag distance was set at 10 m, the GPS/BeiDou positioning effect of the 7ZDGS–300 monorail transporter at the slope was significantly poor, with a road section error of 12.77 m and a road section relative error of 127.69%. The cause of this phenomenon may be because the slope was located on the hillside, and the canopy of the fruit trees above it was heavily covered, resulting in a weak acceptance strength of the GPS/BeiDou satellite signals, which significantly affects the positioning accuracy. Based on the data of the flat road and turning road for cross-comparison with RFID positioning, the experimental results are shown in Figure 15. When the tag distance was set at 10 m, the road section relative error of the RFID positioning method was better than the road section relative error of the GPS/Beidou positioning method, by approximately 11.94% and 14.215%, respectively. Based on these experimental results, it is recommended that the GPS/BeiDou positioning methods be used with caution in the positioning task at ramps, particularly in the case of severe fruit tree canopy cover, where the GPS/BeiDou positioning effect may be significant. In contrast, RFID positioning methods showed better positioning accuracy in all road conditions.

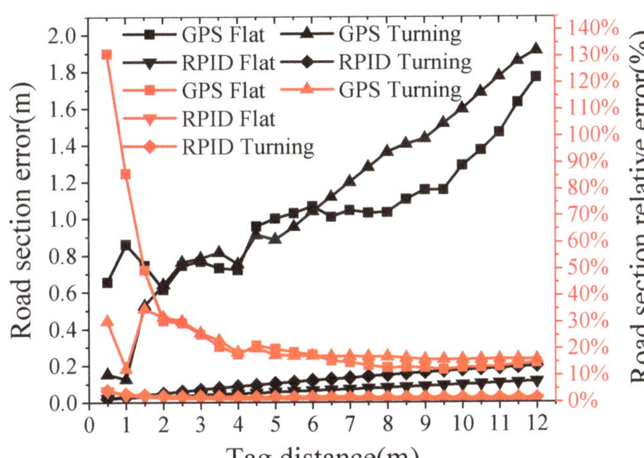

Figure 15. Curves for cross-comparison experiments using the GPS/Beidou positioning method.

4.2. Communication and Control Experiments for the Multimode Control Gateway System for the Monorail Transporter

In this study, a self-assembling gateway system network was built to verify the communication function of the gateway system and the avoidance strategy of the monorail transporter. The terminals have a 7ZDGS–300-type monorail transporter (Transporter A), a 7ZDGS–250-type monorail transporter (Transporter B), and a track-changing branch structure, with the initial state of the track-changing branch structure on the left. Transporter A's operation priority was higher than Transporter B's. The benchmark positioning tags were deployed at a distance of 10 m for multitransporter communication and control experiments based on two typical avoidance scenarios.

The first avoidance scenario is that Transporter A and B were at tag No. 2 and 3 of the main branch of the track-changing branch structure, respectively, and the destination of Transporter A was tag No. 9 of the left branch of the track-changing branch structure. A schematic of the avoidance results and field experiments is shown in Figure 16. The gateway system controls the track-changing branch structure to switch the track to the right, then controls Transporter B to travel to avoidance waiting areas at tag No. 10 of the right branch, controls the track-changing branch structure to switch the track to the left, and finally controls Transporter A to travel to tag No. 9 of the left branch.

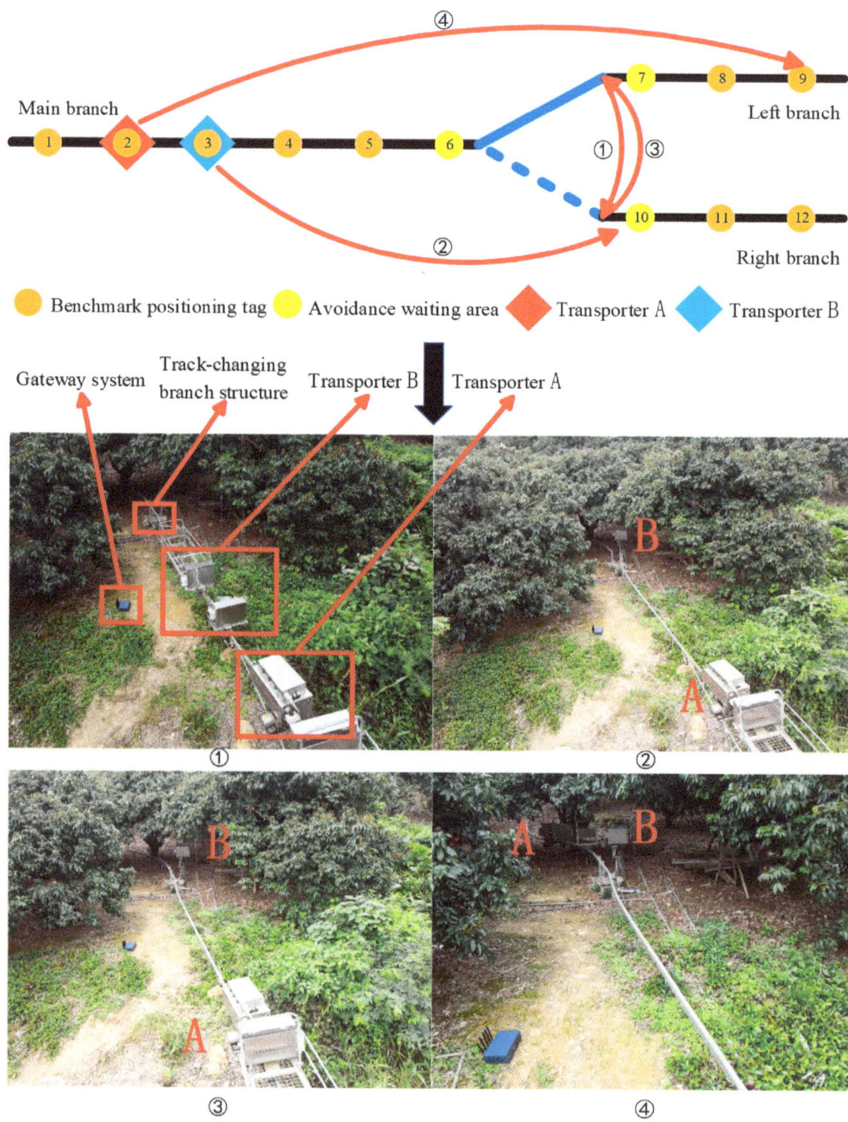

Figure 16. Schematic of the field experiment for the first avoidance results.

The second avoidance scenario involves Transporters A and B being at tag Nos. 2 and 3 of the main track-changing branch structure, respectively, and the destination of Transporter A being tag No. 12 of the right branch of the track-changing branch structure. A schematic of the avoidance results and the field experiment are shown in Figure 17, where the gateway system controls Transporter B to travel to tag No. 7 of the left branch trunk road, then controls the track-changing branch structure to switch the track to the right side, and finally controls Transporter A to travel to tag No. 12 of the right branch trunk road. The experimental results show that the gateway system equipped with the autonomous avoidance strategy of monorail transporters can detect the operation status information of multiple monorail transporters in real time and complete the avoidance operation based on the avoidance strategy algorithm.

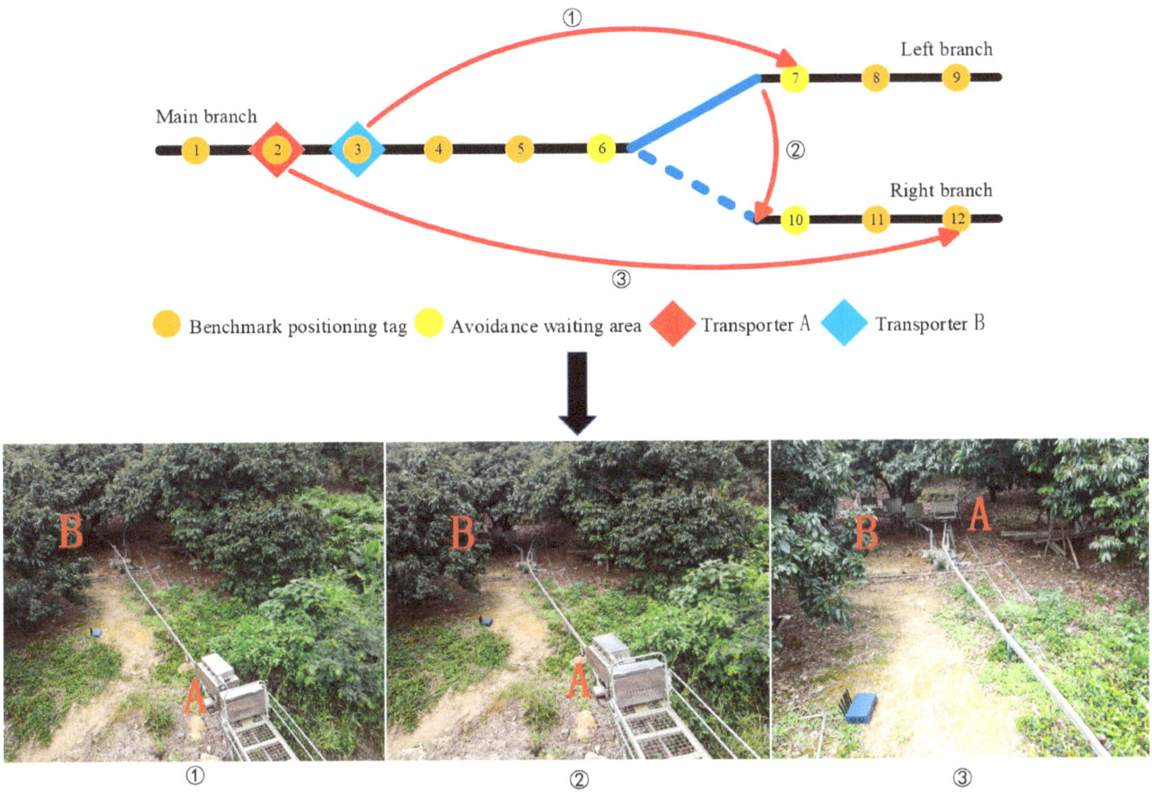

Figure 17. Schematic and field experiment for the second avoidance results.

5. Conclusions

The location-aware hardware system based on STM32 + RFID + LoRa designed in this study shows good feasibility and stability in monorail transport. This strategy characterizes the average driving speed of a monorail transporter under road conditions by the instantaneous driving speed at the starting point. It accurately the monorail transporter's on-track position in real time. Based on the performance of the location-aware hardware system, a tag distance selection strategy was proposed; the distance of the benchmark positioning tags was set at 10 m under flat and turning road conditions. Currently, the average relative error of road sections of the 7ZDGS–300-type monorail transporter is 1.35%, while the average section relative error of the 7ZDGS–250-type monorail transporter is 1.27%. The 6 m distance of the reference positioning tags in the uphill/downhill benchmark effectively ensures that the monorail transporter avoids collision. This strategy provides practical guidance to improve the efficiency of orchard transport and optimizes the location-aware hardware system to enhance its positioning accuracy and reliability.

The effect of different starting points on the positioning accuracy of the location-aware hardware system was also verified in this study, and the results showed that the effect was small and negligible on the positioning accuracy of the location-aware hardware system. In addition, the superiority of the RFID positioning method under different road conditions was verified using a cross-comparison experiment with other positioning methods. In addition, the RFID positioning method shows better positioning accuracy and robustness than the Hall and GPS/Beidou positioning methods.

In this study, an avoidance strategy for monorail transporters combined with a track-changing branch structure was proposed, which significantly allows for the simultaneous operation of multiple monorail transporters to achieve the efficient and safe completion of operations on the track. This not only improves the flexibility and reliability of the monorail transporters but also provides technical support for cooperative work and the intelligent management of the monorail transporter fleet in real agricultural scenarios.

Some tasks in this study require further improvement, such as the poor positioning accuracy of the location-aware hardware system in uphill/downhill conditions and the avoidance strategy. In future research, the performance and parameters of the location-aware hardware system will be evaluated and compared in detail to provide reliable data support for further optimizing and improving the accuracy of the location-aware hardware system. In addition, the avoidance strategy can be optimized by integrating deep learning technology to improve the system's adaptive ability in complex situations. This would help the system better cope with the challenges commonly encountered in practical applications.

At the same time, the coverage of the multimodal control gateway system is affected by the actual orchard environment. When deploying the system, it is possible to increase the coverage area of the gateway system by adding relay nodes or using dual LoRa communication modules on the gateway system to handle different tasks, reduce data conflicts, and enhance the scalability of the multimodal control gateway system. In addition, in the future, various agricultural machinery can be carried on the monorail transport for collaborative operation, such as spray, which can achieve self-spray at a fixed time, improve the efficiency and effect of spray, make the monorail transport more relevant in other aspects, and promote the development of intelligent agriculture.

Author Contributions: Conceptualization, S.L. and Z.L.; methodology, S.L. and Q.L.; software, Q.L.; validation, Q.L., J.C. and Y.L.; formal analysis, S.L. and Q.L.; data curation, Q.L., H.L. and H.H.; writing—original draft preparation, S.L. and Q.L.; writing—review and editing, Z.L.; visualization, Q.L. and H.L.; supervision, S.L., Z.L. and H.H.; funding acquisition, S.L. and Z.L. All authors have read and agreed to the published version of the manuscript.

Funding: This research was funded by National Natural Science Foundation of China (32271997, 62241105), the General Program of Guangdong Natural Science Foundation (2021A1515010923), Special Projects for Key Fields of Colleges and Universities in Guangdong Province (2020ZDZX3061), the Key Technologies R&D Program of Guangdong Province (2023B0202100001), and the China Agriculture Research System of MOF and MARA (CARS-26).

Institutional Review Board Statement: Not applicable.

Data Availability Statement: Not applicable.

Acknowledgments: The authors would like to thank the anonymous reviewers for their criticisms and suggestions. We would also like to thank Tiansheng Hong for research data support.

Conflicts of Interest: The authors declare that the research was conducted in the absence of any commercial or financial relationships that could be construed as a potential conflict of interest.

References

1. National Bureau of Statistics of the People's Republic of China. *China Statistical Yearbook-2022*; China Statistics Press: Beijing, China, 2022.
2. Zhang, Z. Some Important Problems and Measures of Farmland Construction Suitable for Mechanization in Hilly and Mountainous Areas during the 14th Five-year Plan Period. *Chin. Rural. Econ.* **2020**, *11*, 13–28.
3. Zheng, Y.; Jiang, S.; Chen, B.; Lyu, H.; Wan, C. Review on Technology and Equipment of Mechanization in Hilly Orchard. *Trans. Chin. Soc. Agric. Mach.* **2023**, *51*, 40–42.
4. Zou, B.; Liu, F.; Zhang, Z.; Hong, T.; Wu, W.; Lai, S. Mechanization of Mountain Orchards: Development Bottleneck and Foreign Experiences. *J. Agric. Mech. Res.* **2019**, *41*, 254–260.

5. Morinaga, K.; Sumikawa, O.; Kawamoto, O.; Yoshikawa, H.; Seiji Nakao, S.; Shimazaki, M.; Kusaba, S.; Hoshi, N. New technologies and systems for high quality citrus fruit production, labor-saving and orchard construction in mountain areas of Japan. *J. Mt. Sci.* **2005**, *2*, 59–67. [CrossRef]
6. Ouyang, Y.; Hong, T.; Su, J.; Xu, N.; Ni, X.; Yang, C. Design and experiment for rope brake device of mountain orchard traction double-track transporter. *Trans. Chin. Soc. Agric. Eng.* **2014**, *30*, 22–29.
7. Zhang, H.; Ren, L.; Song, Y. Design and research of meshing mechanism of orchard monorail transporter. *J. Phys. Conf. Ser.* **2020**, *1550*, 042009. [CrossRef]
8. Ouyang, Y.; Hong, T.; Su, J.; Xu, N.; Li, Z.; Jiao, F.; Chen, J. Design of wire rope ranging device for mountain orchard traction double-track cargo vehicle. *J. Huazhong Agric. Univ.* **2014**, *33*, 123–129.
9. Li, J.; Li, S.; Zhang, Y.; Wu, H.; Liu, M.; Gao, Z. Design of hydraulic drive system for mountain orchard transporter. *J. Anhui Agric. Univ.* **2021**, *48*, 143–149.
10. Sun, S.; Wang, B. Low-energy Mountain Transportation System with PRT Rail Transit Technology. *J. Landsc. Res.* **2020**, *26*, 15–17.
11. Zou, S.; Liu, S.; Li, M.; Chen, Z.; Zhang, X. Research Progress of Mountain Orchard Rail Conveyor. *Mod. Agric. Equip.* **2021**, *42*, 9–13.
12. Li, Z.; Lyu, S.; Hong, T.; Xue, X.; Yang, Z.; Dai, Q.; Chen, S.; Song, S.; Wu, W.; Li, J.; et al. A Kind of Mountain Orchard Self-propelled Electric Monorail Transportation Equipment and Control Method. CN114013461A, 2 August 2022.
13. Xie, B.; Jin, Y.; Faheem, M.; Gao, W.; Liu, J.; Jiang, H.; Cai, L.; Li, Y. Research progress of autonomous navigation technology for multi-agricultural scenes. *Comput. Electron. Agric.* **2023**, *211*, 107963. [CrossRef]
14. Yang, L.; Wang, X.; Li, Y.; Xie, Z.; Xu, Y.; Han, R.; Wu, C. Identifying Working Trajectories of the Wheat Harvester In-Field Based on K-Means Algorithm. *Agriculture* **2022**, *12*, 1837. [CrossRef]
15. Radočaj, D.; Plaščak, I.; Heffer, G.; Jurišić, M. A Low-Cost Global Navigation Satellite System Positioning Accuracy Assessment Method for Agricultural Machinery. *Appl. Sci.* **2022**, *12*, 693. [CrossRef]
16. Iberraken, D.; Gaurier, F.; Roux, J.C.; Chaballier, C.; Lenain, R. Autonomous Vineyard Tracking Using a Four-Wheel-Steering Mobile Robot and a 2D LiDAR. *AgriEngineering* **2022**, *4*, 826–846. [CrossRef]
17. Hu, L.; Wang, Z.; Wang, P.; He, J.; Jiao, J.; Wang, C.; Li, M. Agricultural robot positioning system based on laser sensing. *Trans. Chin. Soc. Agric. Eng.* **2023**, *39*, 1–7, (In Chinese with English abstract).
18. Bi, S.; Wang, Y. Inter-line Pose Estimation and Fruit Tree Location Method for Orchard Robot. *Trans. Chin. Soc. Agric. Mach.* **2021**, *52*, 16–26+39.
19. Andrés, V.H.; Gareth, T.C.E.; Liisa, A.P.; Ole, G.; Claus, A.G.S. Internet of Things in arable farming: Implementation, applications, challenges and potential. *Biosyst. Eng.* **2020**, *191*, 60–84.
20. Giordano, S.; Seitanidis, L.; Ojo, M.; Adami, D.; Vignoli, F. IoT Solutions for Crop Protection Against Wild Animal Attacks. In Proceedings of the 2018 IEEE International Conference on Environmental Engineering (EE), Milan, Italy, 12–14 March 2018.
21. Wang, X.D.; Xie, W.; Song, X.Y.; Wan, T.; Liu, A. Grassland Ecological Protection Monitoring and Management Application Based on ZigBee Wireless Sensor Network. *Math. Probl. Eng.* **2022**, *2022*, 2623183. [CrossRef]
22. Sinha, R.S.; Wei, Y.; Hwang, S.H. A survey on LPWA technology: LoRa and NB-IoT. *ICT Express* **2017**, *3*, 14–21. [CrossRef]
23. Maddikunta, P.K.R.; Hakak, S.; Alazab, M.; Bhattacharya, S.; Gadekallu, T.R.; Khan, W.Z.; Pham, Q.-V. Unmanned Aerial Vehicles in Smart Agriculture: Applications, Requirements, and Challenges. *IEEE Sens. J.* **2021**, *21*, 17608–17619. [CrossRef]
24. Wilson, A.H.; Adelaida, O.B.; Andrés, S.B.; Geovanny, R.I.; Alexis, B.U.; Dora, C.P.; Francisco, M.A.C.; Juan, A.M.L.; Alejandro, C.P.; Francisco, M.A. Precision Agriculture and Sensor Systems Applications in Colombia through 5G Networks. *Sensors* **2022**, *22*, 7295.
25. Luis, R.G.; Loredana, L. The role of RFID in agriculture: Applications, limitations and challenges. *Comput. Electron. Agric.* **2011**, *79*, 42–50.
26. Golsorkhtabaramiri, M.; Tahmasbi, M.; Ansari, S. A Distributed Mobile Reader Collision Avoidance Protocol for Dense RFID Networks. *Wirel. Pers. Commun.* **2022**, *125*, 2719–2735. [CrossRef]
27. Aerts, W.; Mulder, E.D.; Preneel, B. Dependence of RFID reader antenna design on read out distance. *IEEE Trans. Antennas Propag.* **2008**, *56*, 3829–3837. [CrossRef]
28. Silva, L.M.D.; Menezes, H.B.D.B.; Luccas, M.D.S.L.; Mailer, C.; Pinto, A.S.R.; Boava, A.; Rodrigues, M.; Ferrão, I.G.; Estrella, J.C.; Branco, K.R.L.J.C. Development of an Efficiency Platform Based on MQTT for UAV Controlling and DoS Attack Detection. *Sensors* **2022**, *22*, 6567. [CrossRef]
29. Zhang, X.; Qu, T.; Tang, P.; Liu, Y. A Method of Realizing External Network Access to Intranet for Embedded Web Server. In Proceedings of the 2022 2nd International Conference on Electronic Information Technology and Smart Agriculture (ICEITSA), Huaihua, China, 9–11 December 2022.
30. Li, Z.; Chen, J.; Lyu, S.; Xue, X.; Chen, S.; Song, S.; Wu, W.; Sun, D.; Hong, T.; Yang, Z.; et al. A Self-Propelled Monorail Transportation Track-Changing Branch Structure for Mountain Orchards. CN114803335A, 29 July 2022.

agriculture

MDPI

Article

Design and Experiment of Uniform Seed Device for Wide-Width Seeder of Wheat after Rice Stubble

Weiwen Luo [1], Xulei Chen [1], Mingyang Qin [1], Kai Guo [1], Jie Ling [1], Fengwei Gu [2,*] and Zhichao Hu [1,*]

[1] Nanjing Institute of Agricultural Mechanization, Ministry of Agriculture and Rural Affairs, Nanjing 210014, China; 82101211053@caas.cn (W.L.); njscxltg@163.com (X.C.); njsqmytg@163.com (M.Q.); njsgktg@163.com (K.G.); njsljtg@163.com (J.L.)

[2] Graduate School of Chinese Academy of Agricultural Sciences, Beijing 100083, China

* Correspondence: gufengwei@caas.cn (F.G.); huzhichao@caas.cn (Z.H.)

Abstract: When wide-width sowing wheat after rice stubble (WRS) in a rice-wheat rotation area, there is a problem of poor uniform of seed distribution. To solve the problem, this study designed the seed distribution plate (SDP) structure and optimized its critical structure parameters. Firstly, combined with the operating principles of the wide-width seeder and the agricultural standards for WRS, the main structural parameters affecting seed movement were determined by a theoretical analysis of seed grain dynamics and SDP structure. Secondly, the operational performance of six different structures of SDP under different structural parameters was compared using discrete element simulation technology. The structure of SDP most suitable for WRS wide-width seeding and the value ranges of key structural parameters that have a significant impact on the coefficient of the variation of seed lateral uniformity (CVLU) were determined. Finally, the pattern and mechanism of the influence of key structural parameters of SDP on the CVLU were analyzed. The optimum parameter combination was obtained and a field validation test was conducted on this. The results showed that the anti-arc ridge and arc bottom structure (S6) is more suitable for the agronomy standards of WRS wide-width seeding. The chord length of ridge, installation inclination, angle between the chord and tangent of the end of ridge line (ACT), span, and bottom curve radius are determined as the key structural parameters affecting the CVLU, and there is a lower CVLU (42.8%) when the ACT is 13°. The primary and secondary order of the influence of each factor on CVLU is the chord length of the ridge, span, installation inclination, and bottom curve radius. The corresponding parameter values after optimization are 140 mm, 40°, 75 mm and 50 mm, respectively. A field test was conducted on the SDP after optimizing parameters, and the CVLU was 30.27%, which was significantly lower than the CVLU before optimization.

Keywords: agricultural equipment; structural design; DEM; sowing; seed distribution plate

Citation: Luo, W.; Chen, X.; Qin, M.; Guo, K.; Ling, J.; Gu, F.; Hu, Z. Design and Experiment of Uniform Seed Device for Wide-Width Seeder of Wheat after Rice Stubble. *Agriculture* **2023**, *13*, 2173. https://doi.org/10.3390/agriculture13112173

Academic Editors: Mustafa Ucgul and Chung-Liang Chang

Received: 10 October 2023
Revised: 10 November 2023
Accepted: 15 November 2023
Published: 20 November 2023

1. Introduction

The rice-wheat rotation agronomic model with two crops per year is widely practiced in the middle and lower Plain of the Yangtze River in China. The rice-wheat rotation agronomic model is characterized by sowing wheat immediately after the rice harvest, that is, wheat after rice stubble (WRS). The annual planting area of WRS is about five million hectares, accounting for about 20% of the total wheat planting area in China [1,2]. So, the promotion and development of WRS are crucial to stabilize the total yield of wheat in China. The WRS sowing time is from late October to early November each year. The traditional sowing technology is "straw crushing–burying and returning to field-rotary ploughing–sowing and fertilizing–suppression" [3,4]. Under this background, there are many constraints to the mechanized sowing of WRS [5]. The major points of constraints are as follows: (a) With the postponement of rice harvesting, the sowing period of WRS is further shortened. However, the mechanized sowing process of WRS is complicated. (b)

The rice straw content in the field is huge, and the rice straw on the seed strip cannot be handled well, resulting in poor sowing quality. (c) The paddy soil is wet and sticky due to high moisture content, which reduces the smoothness of mechanized sowing. Combined with the agronomic conditions of WRS production in the middle and lower plain of the Yangtze River in China, the agronomic model of straw inter-row mulching and wide-width sowing of WRS was proposed. This technical model can solve the above-mentioned WRS sowing problems. However, the poor uniformity of wheat sown in the wide seed strips is the main reason that the popularization of this technology at present is restricted.

Many agricultural experts have studied the mechanized strip-sowing technology of grain. Wang [6,7] designed a noncontact self-suction wheat shooting device, analyzed the working principle of this device, and designed the dimensions of key components. On this basis, by analyzing the accelerating process of wheat seeds inside the seed discharger and the seed casting process, the seeding performance of the device was investigated, to obtain the factors that affect the wheat seeding effect [8]. Wang et al. [9–11] proposed a non-contact pneumatic WRS seeding technology. After analyzing the injection speed required for qualified seeding under soil conditions with different water contents through bench tests, a WRS seeding device with high-pressure accelerating airflow was designed; subsequently, the characteristics of the airflow field and the influence relationship of various structural parameters on sowing quality were analyzed based on CFD techniques, and the key parameters of the seeding device were optimized. Xing et al. [12] designed a high-speed strip seeding device. Through CFD-DEM coupled simulation, the aerodynamic characteristics and distribution law of the gas-phase flow field inside the device, as well as the influence law of the structural parameters on the operation quality, were analyzed. Wang et al. [13] analyzed the influence mechanism of fluted force-feed seeder parameters on sowing uniformity, and optimized the structural parameters of the feeding mechanism of the wheat planter. Through a theoretical analysis, numerical simulation and actual experiment, Li [14], Xi [15], and Han [16] analyzed and optimized the effects of structure and operation parameters of a rotary-tillage sowing device, wide strip seed guiding device and non-tube sowing device on sowing uniformity and sowing depth. Compared with traditional seeders, it is shown that these optimized devices have better operating performance and economic benefits. These devices are mainly used on land without straw, and the seeds are distributed in a narrow strip shape. Lei [17,18] and Tang [19] simulated the motion characteristics of seeds in a pneumatic seed metering device based on CFD-DEM simulation technology. The effects of the seed metering tube, seed distribution device structure and air flow on the sowing performance were analyzed by simulation. The optimized parameters of the device structure and airflow were determined to improve the seeding quality. Liu [20] analyzed the effects of factors (rotational speed, adjusting plate height and eyelet length) on the operational performance of the seed wide dispensing device with eyelet wheeled through experiments, and comprehensively optimized the performance of the dispensing device [21]. Niu [22] proposed a technical solution for WRS seeding with a straw covering on the surface after sowing, and studied the operational performance of the whole machine through a theoretical analysis and field test. Hu [23] proposed the technical scheme of combining "seed-fertilizer-seed" type wide seeding and belt rotary tillage, and conducted theoretical design and field experiments on key components such as furrow openers and floating soil covering plates. The sowing width of this sowing technique is about 70 mm, and the qualified rate of sowing depth is about 85%. Taking the sowing uniformity as the index, Zhu [24] analyzed the effect relationship between the spherical radius, installation angle and span on the index, and optimized the performance of the wide seed distribution plate by an orthogonal rotary combination test. These studies mainly focus on the performance analysis and optimization of hole-sowing and narrow strip-sowing techniques. However, there are few studies on the uniform seed distribution mechanism and structure optimization of WRS wide-width sowing.

The purpose of this study is to design and optimize a wide-width seed uniform distribution device for WRS to solve the problem of the poor sowing uniformity of wide-width (240 mm) wheat. This research is based on an agronomic model of straw inter-row mulching and wide-width sowing. The first research task is to simulate the working process of different structures of seed uniform distribution device with a wide width to obtain the optimal structural form and the corresponding key parameter range. The second research task is to conduct a comprehensive optimization of the structural parameters of the seed uniform distribution device, to obtain the optimal combination of structural parameters. By achieving this goal, mechanical technology and agronomy of WRS are integrated, thus promoting the advancement of a full-process mechanized production technology of WRS.

2. Material and Methods

2.1. WRS Agronomic Model with Wide Width

An appropriate sowing period is one of the crucial guarantees for high wheat yield [25]. However, with the postponement of rice harvest time, the WRS sowing period is further delayed. WRS sowing with a wide width can optimize the wheat population structure and help stabilize the yield when sowing dates are delayed.

Combined with the agronomic conditions of WRS production in the middle and lower plain of the Yangtze River in China, the agronomic model of straw inter-row mulching and the wide-width sowing of WRS was proposed. The practice shows that the yield by this technology is basically the same as that sown by conventional methods, but the operation procedure is more simplified, the operation smoothness is higher, and the efficiency is higher [26]. This sowing method has the following characteristics: (a) We complete all processes at one time, including rice straw crushing, seed strip straw removal, wide sowing, and suppression. (b) The width of the seed strip is 240 mm, and the row spacing is 300 mm. (c) We fertilize and rotary till the seed strip, which allows for both belt arrangement and reduced soil disturbance. (d) After the operation, seeds are evenly distributed in the seed strip of 240 mm, and all the crushed straw is placed in a gap of 300 mm. As shown in Figure 1, the operation effect of the technique with straw inter-row mulching and wide-width sowing WRS is shown.

(a) (b)

Figure 1. Operation effect of technique with straw inter-row mulching and wide-width sowing WRS: (**a**) untreated rice straw land after rice harvest; (**b**) field after operation.

2.2. Structure and Principle

The equipment with straw inter-row mulching and wide-width sowing mainly performs three processes: seed strip manufacturing, wide-width sowing, and suppression. The overall structure of the equipment is shown in Figure 2, which is mainly composed of a straw-crushing device, straw diversion device, seed strip rotary tillage device, and seed uniform distribution device with a wide width and press wheel.

The technical principle of seed strip manufacturing is as follows: the high-speed rotating straw-crushing device crushes rice straw and throws it to the straw diversion device; the crushed straw sprayed at high speed is coupled with the wedge-shaped straw diversion device for movement so that the straw originally located in the seed strip is forced to slide into the gap between rows. The technical principle of wide-width sowing is as follows: when the strip tillage device rotates the seed strip without straw, the soil is thrown up to a certain height; the seed uniform distribution device performs sowing between the thrown soil particles and the ground surface; most of the soil thrown up crosses over the seed uniform distribution device and falls back to the ground surface by the blocking plate of the strip tillage device, to realize the mulching of seeds; and there is press wheel compaction of the ground after sowing. After operation, 240 mm seed strip and 300 mm crushed straw strip were alternately arranged longitudinally in the field, as shown in Figure 1b.

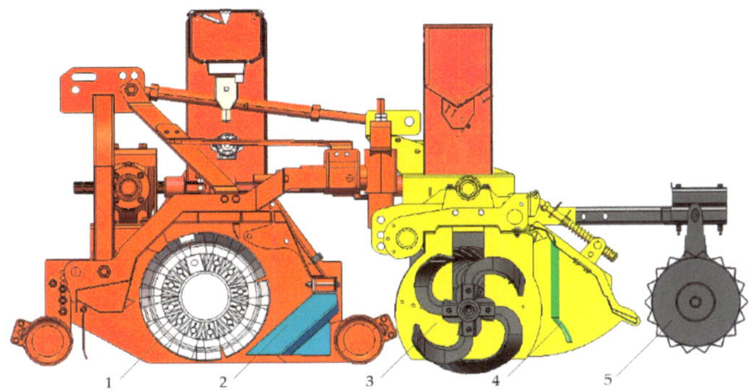

Figure 2. The overall structure of a wide-width planter with crushed straw inter-row mulching: 1. Straw-crushing device; 2. straw diversion device; 3. seed strip rotary tillage device; 4. seed uniform distribution device with wide width; 5. press wheel.

The seed uniform distribution device with a wide width is the key component of the WRS wide-width sowing system. The seed uniform distribution device mainly consists of a feed tube, seed distribution plate (SDP), seed-retaining plate, and soil-retaining plate. The seed discharger, seed transporting tube, and seed equalizer are connected up and down sequentially, as shown in Figure 3. The seed uniform distribution device is installed between the strip rotary cutter roller and the rear cover. The installation of the seed uniform distribution device needs a certain height from the ground, which can be adjusted according to the demand of sowing depth. Each seed strip is equipped with two seed uniform distribution devices, each with a sowing width of 120 mm. The operation principle of the seed uniform distribution device is as follows: when sowing, the wheat seeds pass through the seed transporting tube to SDP. Wheat seeds fall uniformly on the seed strip after collision and slip on SDP. While controlling the trajectory of the seed, the seed-retaining plate and soil-retaining plate can prevent the thrown-up soil from falling into SDP to avoid clogging.

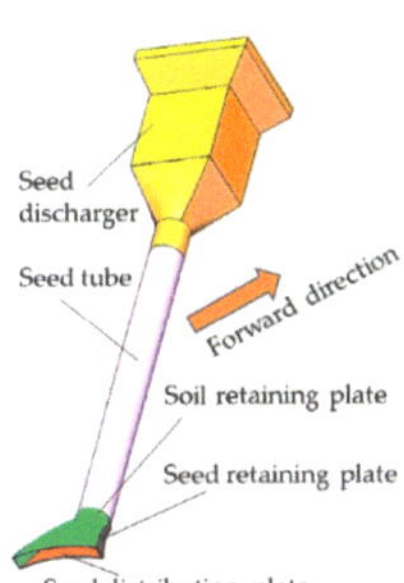

Figure 3. Structure of seed uniform distribution device with wide width.

2.3. Analysis of the Relationship between Seed Distribution Plate Structure and Particle Motion

A mathematical structural model of SDP is shown in Figure 4. The SDP is a convex circular arc shape, which is formed by the movement of the bottom arc CD along the ridge arc AB. Point A is located at the midpoint of the arc CD and point B is the highest point of the SDP. The distance of AB is the chord length of ridge, which is l. Line EQ is the mid-perpendicular of chord AB, and Q is the center position of arc AB. Points C and D are the two boundary points of the SDP at the bottom, and the distance of CD is the span of the plate, which is $2b$.

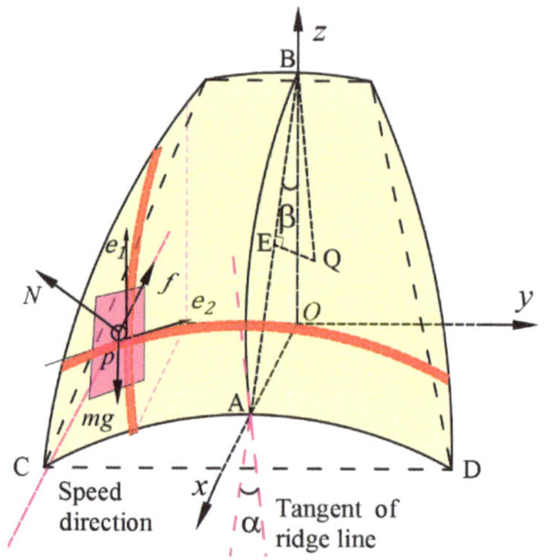

Figure 4. Mathematical structural model of the seed distribution plate.

Through theoretical analysis, it can be observed that the coordinates of point A are $(l \cdot \sin\beta,\ 0,\ 0)$, the coordinates of point B are $(0,\ 0,\ l \cdot cos\beta)$ and the coordinates of the midpoint E on the string are $(\frac{l \cdot \sin\beta}{2},\ 0,\ \frac{l \cdot cos\beta}{2})$.

The slope of the line EQ is as follows:

$$k = -tan\beta \tag{1}$$

where β is the installation inclination.

So, the Equation of line EQ is as follows:

$$z = -\tan\beta \cdot x + \frac{l}{2\cos\beta} \tag{2}$$

Assume that the center coordinate corresponding to the arc ridge is Q $(m, 0, n)$, and the radius is R; then, the following relationship holds:

$$\begin{cases} (x-m)^2 + (z-n)^2 = R^2 \\ z' = -\frac{x-m}{z-n} \end{cases} \tag{3}$$

The slope corresponding to the tangent of the ridge line at point A is shown in Equation (3).

$$z' = \frac{l \cdot sin\beta - m}{n} = cot(\beta - \alpha) \tag{4}$$

where α is the angle between the chord and tangent of the end of the ridge line (ACT).

Moreover, point Q (m, 0, n) is on the line EQ, so through the simultaneous Equations (2)–(4), the parameter values (m, n, and R) of the ridge can be obtained.

$$\begin{cases} m = \frac{2l\sin\beta\cos\beta\tan(\beta-\alpha)-1}{2\cos\beta\tan(\beta-\alpha)-2\sin\beta} \\ n = -\frac{2l\sin^2\beta\cos\beta\tan(\beta-\alpha)-\sin\beta}{2\cos^2\beta\tan(\beta-\alpha)-\sin 2\beta} + \frac{l}{2\cos\beta} \end{cases} \tag{5}$$

$$R = \sqrt{(m - x_A)^2 + (\mathrm{n} - y_A)^2} = \sqrt{(m - l\sin\beta)^2 + n^2} \tag{6}$$

Taking any point P ($\mathrm{p_x}$, $\mathrm{p_y}$, $\mathrm{p_z}$) on the SDP surface as the research object, at this point the following relationship holds.

$$\begin{cases} y^2 + (z + r - h)^2 = r^2 \quad (-b \le y \le b) \\ h = \sqrt{R^2 - (x_p - m)^2} + n + r \end{cases} \tag{7}$$

where r is the bottom curve radius, mm.

The tangents of the two curves at point P are both on the tangent plane of the SDP surface at point P, and the slopes of the two tangents are shown in Equation (8).

$$\begin{cases} k_1 = -\frac{x_p - m}{z_p - n} \\ k_2 = -\frac{y_p}{z_p + r - h} \end{cases} \tag{8}$$

where k_1 and k_2 are the slopes of two tangents, respectively. These two tangents respectively belong to the section curves parallel to the ridge arc and the bottom arc.

These direction vectors corresponding to the two tangents are shown in Equation (9). Then, the normal vector of the tangent plane of the SDP surface at that point is shown in Equation (10).

$$\begin{cases} \vec{e}_1 = (\frac{z_p - k_1 x_p}{k_1})\vec{\mathrm{i}} + (z_p - k_1 x_p)\vec{k} \\ \vec{e}_2 = (\frac{z_p - k_2 y_p}{k_2})\vec{\mathrm{j}} + (z_p - k_2 y_p)\vec{k} \end{cases} \tag{9}$$

$$\vec{n} = \frac{1}{k_2}\vec{i} + \frac{1}{k_1}\vec{j} + \frac{1}{k_1 k_2}\vec{k} \tag{10}$$

Ignore air resistance and discretize the point particle for study. The dynamic differential equation along the coordinate axis is established for the particles at any point P.

$$\begin{cases} \frac{k_1 \cdot N}{\sqrt{{k_1}^2 + {k_2}^2 + {k_1}^2{k_2}^2}} + f_x = m\frac{d^2x}{dt^2} \\ \frac{k_2 \cdot N}{\sqrt{{k_1}^2 + {k_2}^2 + {k_1}^2{k_2}^2}} + f_y = m\frac{d^2y}{dt^2} \\ mg + \frac{N}{\sqrt{{k_1}^2 + {k_2}^2 + {k_1}^2{k_2}^2}} + f_y = m\frac{d^2z}{dt^2} \end{cases} \tag{11}$$

It can be observed from Equation (11) that the mechanical relationship of particles on SDP is closely related to the curve's shape in longitudinal and transverse positions. Therefore, when the initial velocity of seeds and the installation position of SDP are determined, the distribution of seeds on the ground is mainly affected by the structural parameters of SDP. Equations (1)–(7) shows that the structure of SDP is determined by the chord length of the ridge, installation inclination, the angle between the chord and tangent of the end of the ridge (ACT), span and bottom curve radius.

2.4. Simulation System Settings

Experiments and data statistics can be efficiently carried out through discrete element numerical simulation methods. The discrete element simulation system of this experiment includes a wheat particle model, device model, and soil plate used to host seeding and statistical data. By actually measuring the three-dimensional dimensions of wheat, a wheat particle model with a length of 6.9 mm and a maximum height of 3.4 mm was composed of spheres. The particle distribution ranged from 0.95 to 1.05 [27]. The simplified model of the device was imported into EDEM (2019) software, and then a flat plate with a length of 2000 mm and a width of 200 mm was built under it. The simulation platform of the device–seed-statistic plate is built as shown in Figure 5. Considering the convenience of manufacturing, the device is made of PLA polymer material. The discrete element parameters refer to the related literature and measurements [28–30], as shown in Table 1. The operating speed of wheat seeders is generally 0.8~1.2 m/s. In this study, the operating speed is set to 1 m/s. The wheat-seeding dose in the rice-wheat rotation area is generally 180~270 kg/hm^2. In this study, the wheat-seeding dosage was set to 225 kg/hm^2, that is, the wheat production speed of the particle factory is set to 6 g/s.

(**a**) (**b**)

Figure 5. Using EDEM software post-processing module for data statistics: (**a**) DEM model of wheat; (**b**) test and statistical method of CVLU.

Table 1. Material property parameters of the simulation model.

Item	Wheat Particle	PLA	Soil Board
Density/(kg/m^3)	1350	1060	2650
Shear modulus/Pa	5.1×10^7	8.9×10^8	1.0×10^6
Poisson ratio	0.29	0.4	0.3
Rolling friction coefficient (Interaction with wheat)	0.08	0.05	0.3
Static friction coefficient (Interaction with wheat)	0.58	0.4	0.58
Restitution coefficient (Interaction with wheat)	0.50	0.6	0.52

2.5. Evaluation Index Calculation Method

The experiments described in Section 2.4 and Section 2.5 are implemented through numerical simulations of the discrete element method. The test results can be easily counted directly through the post-processing module of EDEM software. The data need to be counted under the stable operating; thus, the middle position of the simulated seed strip was used as the counting area. The section of the long 500 mm is randomly selected in the area where the operation is stable. Considering the length of wheat particles and the unity of the statistics of field trials, the statistical area was divided into 6 rows for the CVLU statistics. The size of each test cell is 20 × 100 mm, and the number of seeds in each cell is counted. The CVLU statistical method is shown in Equation (12). Figure 6 shows the statistical diagram of CVLU.

$$\begin{cases} Q = \sum\limits_{i=1}^{N} \sum\limits_{j=1}^{M} Q_{ij}/MN \\ Y = \frac{1}{Q}\sqrt{\sum\limits_{i=1}^{N} (Q_{ij} - Q)^2/(N-1)} \times 100\% \end{cases} \quad (12)$$

where Q is the average number of seeds in each column of grid cells; Q_{ij} is the number of seeds in the ijth statistical cell; M denotes the number of grids per row; N represents the number of grids per column; Y is the CVLU value, %.

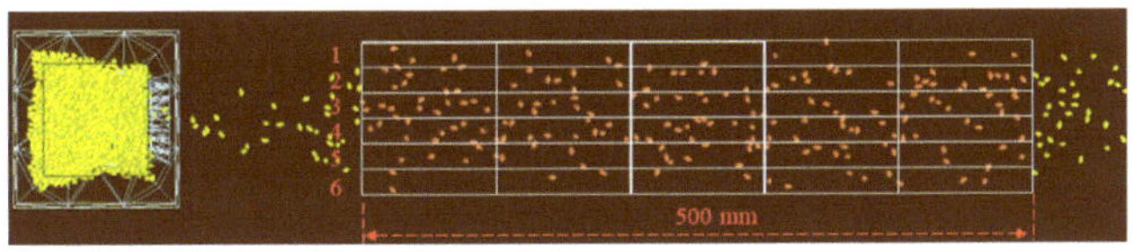

Figure 6. Schematic diagram of data statistics in the sampling area.

When conducting sampling for CVLU in field trials, steel plates are inserted into the soil according to the edge line of the cell shown in Figure 6, and all the soil in the steel plate is dug to count the number of seeds in each cell. In the statistics of sowing depth, taking the symmetrical plane of the SDP as the datum plane, 5 seed depths were randomly measured in each group of symmetrical rows, and the average value was calculated.

2.6. The Effect of SDP Structure on the Seed Uniform Distribution Performance

The SDP structure is the main factor affecting the wheat distribution uniformity in the seed belt. Considering the installation limitation of SDP and agronomic conditions, six structures of SDP were designed to analyze the effect of SDP structure on seeding performance, as shown in Table 2. The corresponding structural characteristic parameters of SDP include the chord length of ridge, installation inclination, ACT, span, and bottom curve radius. These parameters of SDP were changed to analyze the effect of the structure of SDP on the seed uniform distribution performance. The variation of these parameters of SDP is shown in Table 3, and when one of the parameters is varied, the others are at an intermediate level.

Table 2. The different structures of the seed distribution plate.

Cross Section/Ridge	Line	Arc	Ant-Arc
Line	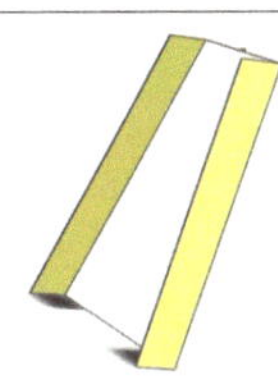 S1	 S2	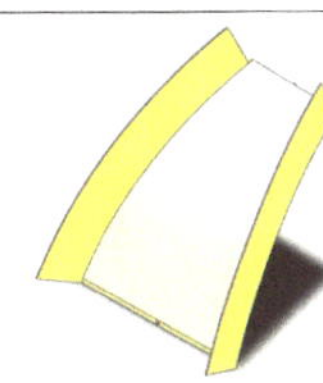 S3
Arc	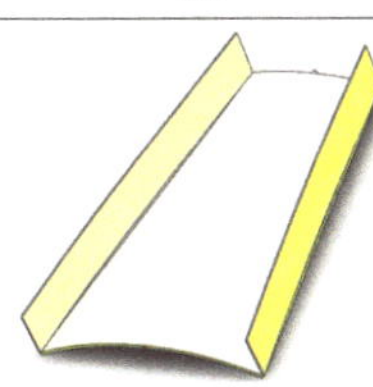 S4	 S5	 S6

2.7. Comprehensive Optimization of Crucial Structural Parameters of SDP

Combined with Section 2.3 and the analysis result of Section 2.4, a comprehensive optimization simulation test was conducted on the crucial parameters that affect seed uniform distribution performance, including chord length of ridge, installation inclination, span, and bottom curve radius. Considering the actual sowing situation in the field, the seeds are regarded as evenly distributed longitudinally. At the same time, the SDP will hinder the movement of the soil, which will have a certain impact on the smoothness of the operation. Therefore, the coefficient of the variation of seed lateral uniformity (CVLU) through SDP is used as an evaluation indicator. A four-factor central combination test was implemented. Its test scheme and results are shown in Table 4.

Table 3. Structural parameters and test levels for performance testing.

	Symbols	Parameters		Test Levels				
Ridge parameters	A	The ridge length of the ridge line	mm	50	75	100	125	150
	B	Installation inclination	°	20	25	30	35	40
	C	The angle between chord and tangent of the end of ridgeline	°	5	10	15	20	25
Bottom parameters	D	Span	mm	50	65	80	95	110
	E	Bottom transverse radius	mm	50	80	110	140	170

Table 4. The scheme and result for comprehensive optimization test.

Test Number	A/mm	B/°	D/mm	E/mm	Y/%
1	75	36.5	98.75	110	42.18
2	75	29.5	76.25	110	56.67
3	100	33	87.5	90	39.04
4	50	33	87.5	90	51.65
5	100	33	87.5	90	35.91
6	75	36.5	76.25	70	42.51
7	100	26	87.5	90	46.93
8	125	36.5	76.25	110	37.38
9	125	36.5	98.75	70	32.72
10	100	33	65	90	50.63
11	125	36.5	98.75	110	33.32
12	75	36.5	98.75	70	40.8
13	75	36.5	76.25	110	50.96
14	100	33	87.5	90	35.4
15	75	29.5	76.25	70	48.5
16	100	33	87.5	90	38.83
17	125	29.5	76.25	70	43.87
18	100	33	87.5	90	37.73
19	75	29.5	98.75	110	39.87
20	150	33	87.5	90	40.85
21	100	33	87.5	50	31.94
22	100	33	87.5	130	42.78
23	100	40	87.5	90	33.36
24	75	29.5	98.75	70	39.32
25	125	29.5	76.25	110	45.44
26	100	33	110	90	42.59
27	125	29.5	98.75	110	39.26
28	125	29.5	98.75	70	37.11
29	100	33	87.5	90	37.72
30	125	36.5	76.25	70	32.89

2.8. Verification Test

To verify the accuracy of the results of optimizing SDP by numerical simulation, the optimized SDP was compared with simulations and bench tests under different operating parameters. The wheat seed variety used is Ningmai No. 31, and the sowing dosage is 180, 225 and 270 kg/hm^2, respectively. The forward speed of the bench test is set to 0.8, 1.0 and 1.2 m/s, respectively. After the bench test is completed, the seeds on the surface are collected according to the zones shown in Figure 6. Each level test was repeated three times and the average value was taken. In addition, to further verify the field operation performance and smoothness of the optimized SDP, a field verification test was conducted at the rice-wheat rotation test base of the Nanjing Institute of Agricultural Mechanization,

Ministry of Agriculture and Rural Affairs. The soil type was clayey, and no treatment was done in the field after the rice harvest. The average height of the rice straw is ≥400 mm. A WRS wide-width planter with crushed straw inter-row mulching was used as the test platform, and an optimized SDP was configured after its rotary tillage device to conduct the field trial, as shown in Figure 7. The forward speed and seeding dosage are set to be the same as those of the bench test, and the other working parameters of the machine were matched at the same time; that is, the rotational speed of the crushing device was 2000 r/min, and the rotational speed of the rotary tillage device was 300 r/min. Each level test was repeated three times and the average value was taken. The performance of the SDP is evaluated according to the method of Section 2.5.

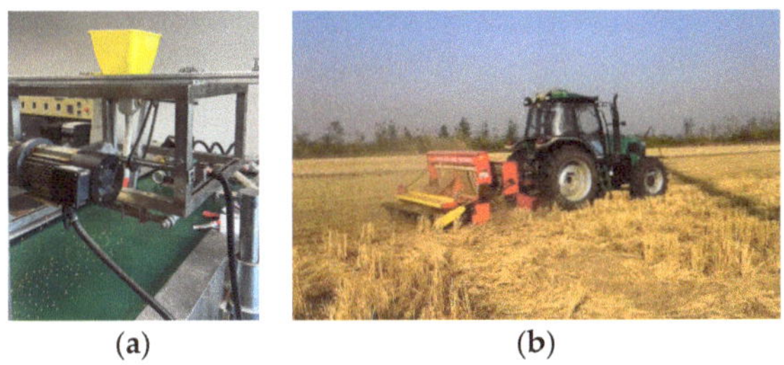

Figure 7. Performance verification test of SDP: (**a**) bench test to verify the accuracy of simulation results; (**b**) field performance verification test.

3. Results and Discussion

3.1. The Effect Law of the SDP Structure Parameter on the CVLU

3.1.1. The Effect of Ridge Parameters on the CVLU

Figure 8a shows the variation trend of the CVLU of SDP at different chord lengths of ridge. S1 and S4 have a similar change rule. With the increase in the chord length of the ridge, the CVLU shows a trend of decreasing and then increasing, and there is a minimum value (51.28% and 47.20%) when the chord length of the ridge is 130 mm. S2 and S5 have a similar change rule. With the increase in the chord length of the ridge, the CVLU shows a tendency to increase and then decrease, and it reaches the maximum value (84.74% and 76.87%) when the chord length of the ridge is 90 mm. S3 and S6 have a similar pattern of change, and the CVLU has a larger value at both smaller and larger chord lengths of the ridge. When the chord length of the ridge is 70–130 mm, the CVLU shows a fluctuating change trend. Overall, the CVLU order from low to high is S6, S3, S4, S1, S5 and S2. In other words, the seed distribution performance of the anti-arc ridge type is the best, the linear type is the second, and the arc is the worst.

Figure 8b shows the variation trend of the CVLU of SDP at different installation inclinations. With the increase in the installation inclination, the corresponding variation rule of CVLU of each structure is as follows. S1-S4, S2-S5 and S3-S6 have similar variation rules, respectively, and the CVLU value from small to large is sorted as S6, S3, S4, S2, S5 and S1. Overall, with the increase in the installation inclination, the CVLU shows a decreasing trend. When the installation inclination is small, the CVLU decreases greatly, but when the installation inclination is greater than 30°, the CVLU decreases slowly with the increase in the angle.

Figure 8c shows the variation trend of the CVLU of SDP at a different ACT. SDP with a straight ridge does not have the characteristic parameter of the ACT, so no test was conducted. As the ACT increases, S3 and S6 show a trend of first decreasing and then increasing, but S2 and S5 show a trend of first increasing and then decreasing. When the ACT ranges from 5° to 25°, each structure has a relatively stable change pattern. Overall, the ranking of the CVLU performance of SDP is S6, S3, S5 and S2. Compared with other parameters, the amplitude of the CVLU with the ACT is smaller. Therefore, the curve of the ACT–CVLU in the range of 5° to 20° was fitted to establish a mathematical model.

The abscissa corresponding to the extreme value of the fitting curve is the ACT value corresponding to the minimum value of the CVLU, which is 13.35 degrees. Therefore, the ACT value was set to 13°, and the corresponding CVLU is 42.98%. When the ACT is too large, the flat surface at the bottom end of the curved SDP is increased, the seed contact time with the SDP surface is increased, and the CVLU of S2 and S5 are decreased under the combined effect of seed rotation and lateral slip. When the ACT is too large, the seeds are more likely to detach from the lower half of the plate and fly directly out, so that the CVLU of S3 and S6 increase.

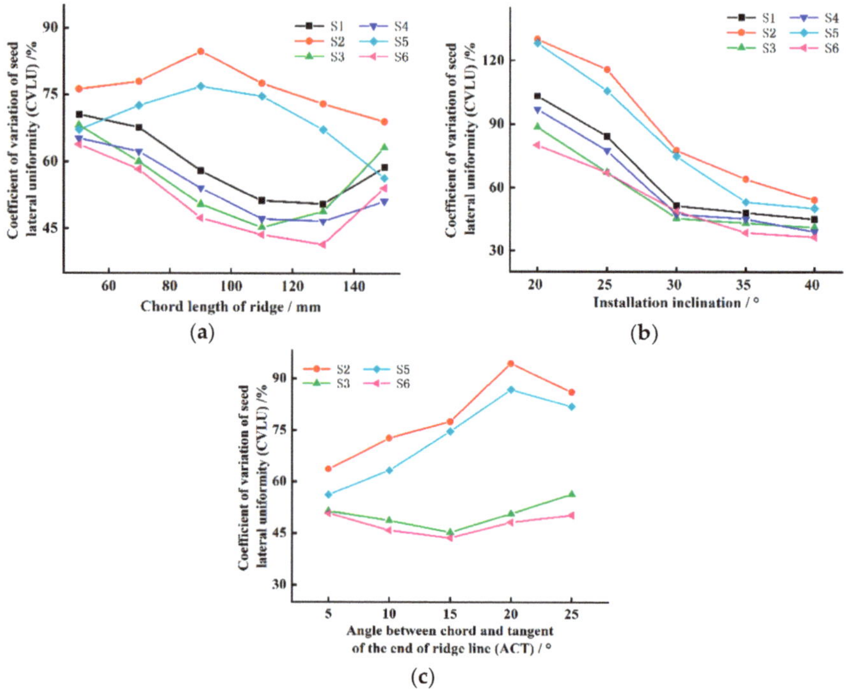

Figure 8. The effect of ridge parameters on CVLU: (**a**) the effect of the chord length of the ridge on CVLU; (**b**) the effect of installation inclination on CVLU; (**c**) the effect of ACT on CVLU.

3.1.2. The Effect of Bottom Parameters on the CVLU

Figure 9a shows the variation trend of the CVLU of SDP at different span values. With the increase in the span value, the variation rules of the corresponding CVLU of each structure are as follows. S1-S2-S3 and S4-S5-S6 have similar patterns of change, respectively, but the CVLU of S2 and S5 are significantly higher than that of other structures. With the increase in span, the CVLU of S1, S2, and S3 all show the trend of first rapid decrease and then slow decrease, and the cut-off point of the change of the curve decreasing trend is at the span of 80 mm. S1, S2, and S3 obtain the minimum value at the span of 110 mm, which is 49.80%, 68.68% and 43.05%, respectively. With the increase in span, the CVLU of S4, S5, and S6 showed a rapid decrease followed by a slow increase with minimum values of 52.69%, 72.18%, and 42.05%, respectively. The variation trend of the CVLU curve of SDP with different structures is mainly affected by the shape of the bottom, while the variation range of the value is mainly affected by the shape of the ridge.

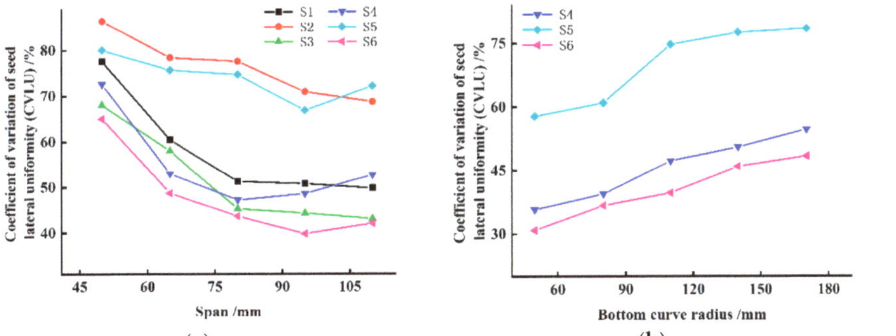

Figure 9. The effect of ridge parameters on CVLU: (**a**) the effect of span on CVLU; (**b**) the effect of bottom transverse radius on CVLU.

Figure 9b shows the variation trend of the CVLU of SDP at different bottom curve radii. The ridge shapes of S1, S2, and S3 are linear, and the tests for analyzing the influence law of the bottom transverse radius on CVLU were carried out only for S4, S5, and s6. The CVLU increased for all test groups as the bottom transverse radius increased. S4 and S6 have closer CVLU values and similar curve patterns. However, the CVLU of S5 is much larger than the other groups and the trend of the curve change is different. The CVLU value from small to large is sorted as S6, S4, S5. In addition, when the bottom transverse radius is greater than 110 mm, the CVLU increases slowly as the radius increases.

3.2. Results Analysis of Response Surface Test

3.2.1. Results of Variance Analysis

The simulation results based on the center combination test are shown in Table 4. Design-expert 8.0 software is used to analyze the variance of the statistical data in Table 4. The analysis results are shown in Table 5. We establish a regression fitting model between CVLU and the four factors, as shown in Equation (13).

$$Y = 311.299 - 0.642A - 2.652B - 4.508D + 0.547E - 0.015AB \\ +0.004AD + 0.039BD - 0.005DE + 0.003A^2 + 0.017D^2 \quad (13)$$

Table 5. Model and variance analysis results of CVLU.

Source	Sum of Squares	df	Mean Square	F-Value	p-Value	Significance
Model	1063.09	10	106.31	27.71	<0.0001	***
A	269.47	1	269.47	70.23	<0.0001	***
B	172.91	1	172.91	45.07	<0.0001	***
D	202.54	1	202.54	52.79	<0.0001	***
E	100.21	1	100.21	26.12	<0.0001	***
AB	28.78	1	28.78	7.50	0.0130	**
AD	23.28	1	23.28	6.07	0.0235	**
BD	36.60	1	36.60	9.54	0.0060	***
DE	20.25	1	20.25	5.28	0.0331	**
A^2	110.95	1	110.95	28.92	<0.0001	***
D^2	121.29	1	121.29	31.61	<0.0001	***
Lack of Fit	61.74	14	4.41	1.98	0.2328	
Pure Error	11.16	5	2.23			
Cor Total	1135.99	29				

Note: *** means extremely significant ($p < 0.01$); ** means significant ($0.01 < p < 0.05$).

It can be observed from Table 5 that the regression model is extremely significant ($p < 0.01$), and the lack of fit item is not significant ($p = 0.2328 > 0.1$). Meanwhile, the model regression coefficient $R^2 = 0.94$, and the adjusted $R^2 = 0.90$. It shows that the model is well-fitted and high reliability. Comparing the F-value and *p*-value of each factor term, the following conclusions were obtained. The order of the main effect relationships of the four factors is A > C > B > D. The extremely significant items are ordered from large to small as A, C, B, C^2, A^2, D, BC. The significant items ordered from large to small are AB, AC, and CD.

3.2.2. Analysis of Interaction Effects of Factors

The ANOVA table shows that the interaction terms AB, AC, BC and CD have different degrees of significant effect on CVLU. To more intuitively display the impact of interaction terms on CVLU, response surface plots were drawn, as shown in Figure 10.

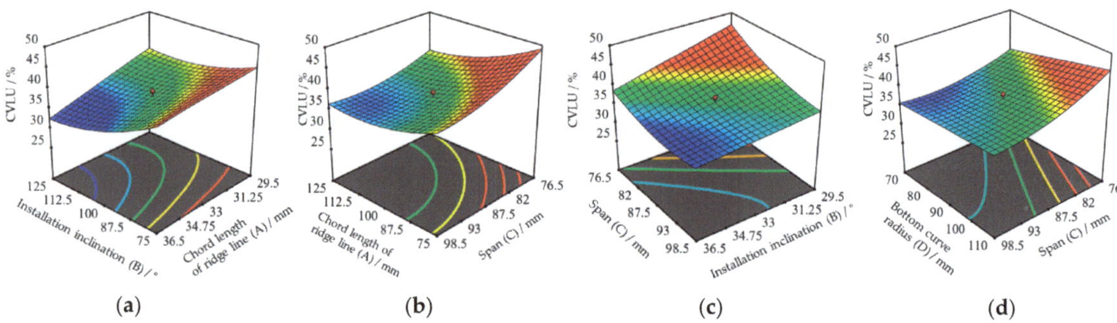

Figure 10. Effect of interaction between factors on CVLU. (**a**) Y = f (A, B, 0, 0); (**b**) Y = f (A, 0, C, 0); (**c**) Y = f (0, B, C, 0); (**d**) Y = f (0, 0, C, D).

The interaction effect of AB item is shown in Figure 10a. When the installation inclination is low, the CVLU shows a quadratic trend of decreasing and then increasing with the increase in the chord length of the ridge. Moreover, when the installation inclination is larger, CVLU shows a large decrease followed by a slow increase as the chord length of the ridge increases. When the chord length of the ridge is larger or smaller, the CVLU shows a decreasing trend with increasing the installation inclination, but the decrease is larger when the chord length of the ridge is larger.

The interaction effect of AC item is shown in Figure 10b. When the chord length of the ridge is smaller or larger, the CVLU shows a tendency to decrease and then increase with increasing the span, but the decrease in CVLU is greater when the chord length of the ridge is smaller. When the span is smaller or larger, the CVLU shows a quadratic trend of decreasing and then increasing with the increase in the chord length of the ridge, but the change of the CVLU is greater when the span is smaller.

The interaction effect of the BC item is shown in Figure 10c. When the installation inclination is small, the CVLU shows a trend of rapid decrease with the increase in span. When the installation inclination is large, the CVLU shows a quadratic trend of first decreasing and then slowly increasing as the span increases. When the span is small or large, the CVLU decreases with the increase in the installation inclination, and the amplitude of change is larger when the span is small.

The interactive effect of the CD item is shown in Figure 10d. When the span is small or large, the CVLU decreases as the bottom curve radius increases, but when the span is small, the CVLU decreases more. When the bottom curve radius is small or large, the CVLU shows a quadratic trend that decreases rapidly and then increases slowly as the span increases.

3.2.3. Acquisition and Verification of Optimal Parameter Combination

To optimize the performance of SDP, the parameters are comprehensively optimized based on the above analysis. According to the actual working conditions of the seeder, operational requirements, and relevant theoretical analysis, the objective function and constraints of comprehensive optimization are established, as shown in Equation (13). Design-expert software is used to optimize and solve the four parameters in the objective function. Combined with the agronomic demand of the WRS sowing and installation process, the best parameter combination after optimization was obtained as a chord length of a ridge of 140 mm, installation inclination of 40°, span of 75 mm and bottom curve radius of 50 mm, corresponding to a CVLU of 21.42%.

$$\begin{cases} \min Y(A,B,C,D) \\ s.t. \begin{cases} 50\ \text{mm} \leq A \leq 150\ \text{mm} \\ 26\ ^\circ \leq B \leq 40\ ^\circ \\ 64\ \text{mm} \leq C \leq 110\ \text{mm} \\ 50\ \text{mm} \leq D \leq 130\ \text{mm} \end{cases} \end{cases} \quad (14)$$

3.3. Results Analysis of Verification Tests

The data statistical of the field test is shown in Figure 11. By comparing the results of simulation tests, bench tests and field tests under different operating parameters, the following conclusions can be drawn. It can be observed from Figure 12a that when the forward speed is 0.8 m/s and the seeding dosage is 270 kg/hm^2, the CVLU of the simulation, bench and field tests all reach the minimum value, which are 16.47%, 18.77% and 22.62%, respectively; when the forward speed is 1.2 m/s and the seeding dosage is 180 kg/hm^2, the CVLU of the simulation, bench and field tests all reach the maximum value, which are 25.89%, 29.28% and 33.28%, respectively; as the forward speed increases or the dosage decreases, CVLU gradually increases, but the effect of dosage on CVLU is greater than that of the forward speed. The possible reason for this is that with the increase in forward speed, the wheat mass flow through the SDP increases, and the stacking between wheat particles increases, resulting in a decrease in distribution uniformity. With the increase in sowing dosage, the number of seeds in each test cell increases, which reduces the CVLU value. Under the conditions of different forward speeds and sowing dosage, the results of the simulation test and the bench test at each factor level are relatively close, but the simulation test has better uniformity than the bench test. This is mainly because the three-dimensional dimensions of the wheat particles in the simulation test are the same, but there is a small deviation in the actual wheat size. What is more, in the simulation test, wheat particles remain relatively stationary with the receiving plate after contact with the particles, while in the actual test, there are different degrees of rebounds due to the difference in phase angle and falling velocity. Overall, the comparison between the bench test and the simulation test proves the accuracy of the optimization results through simulation.

Figure 11. Data statistics of field trials.

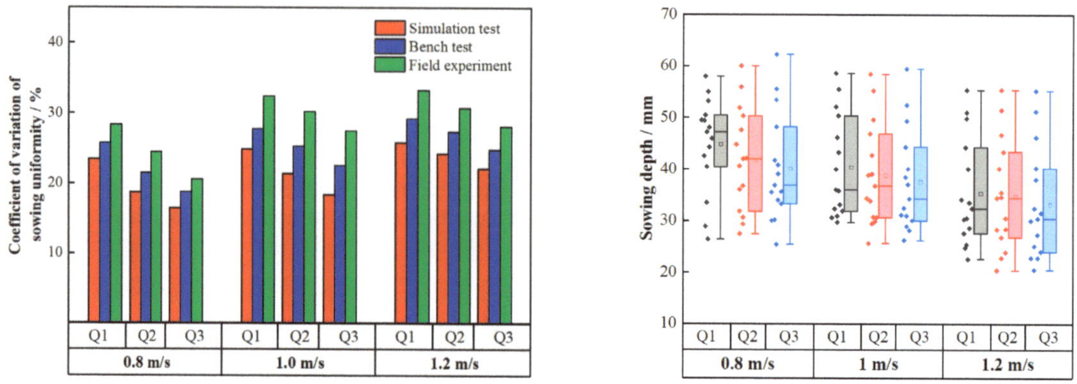

(a) (b)

Figure 12. Statistical results of actual verification tests: (**a**) Comparison of the results of CVLU at different levels; (**b**) comparison of the results of sowing depth at different levels. Note: Q1, Q2, and Q3 mean 180, 225, and 270 kg/hm^2, respectively.

It can be observed from Figure 12b that when the forward speed is 0.8 m/s and the seeding dosage is 180 kg/hm^2, the average sowing depth has a maximum value of 44.6 mm, while when the forward speed is 1.2 m/s and the dosage is 270 kg/hm^2, the average sowing depth has a minimum value of 33.2 mm. With the increase in forward speed or sowing dosage, the average sowing depth decreased gradually, and the influence of the forward speed on sowing depth is greater than that of the sowing dosage. The possible reason for this is that the absolute velocity of soil particles throwing backward decreases with the increase in forward speed, which leads to the weakening of soil movement ability, thus reducing the ability of soil to cross the SDP. With the increase in sowing dosage, the collision between soil particles and seeds increased, which led to a decrease in the spanning ability of some soil particles, so the sowing depth decreased slightly. In addition, a phenomenon was noticed; that is, the consistency of the sowing depth was more variable than that of the ordinary furrow sowing method. This is because there is no trench opener in the seeding device, and the soil thrown by rotary tillage directly covers the seeds across the SDP; thus, the consistency of the sowing depth is not as consistent as the conventional operation mode. However, due to the special planting environment of WRS sowing, agronomic experts have reached a consensus that even if wheat seeds are not completely covered by the soil, they can be regarded as qualified; that is, there is no strict requirement for the consistency of sowing depth. From this perspective, the uniform seed device meets agronomic requirements for WRS sowing in full rice straw and stubble.

In addition, under the condition that the forward speed is 1 m/s and the sowing dosage is 225 kg/hm^2, the seed uniformity performance of the SDP before and after optimization was compared. The results of the field test showed that the CVLU was 30.27%. This has a certain variability from the theoretical value (21.42%), which is mainly due to the poor leveling of the site ground and the interference of soil particles on the movement of seed grains. However, comparing the results of the field test before optimization (63%), the CVLU was significantly reduced after optimization, which indicates the accuracy of the parameter optimization.

4. Discussions

The influence mechanism of the chord length of ridge on CVLU may be as follows. With the change in the shape of the ridge, the contact time between the seed and the arc surface is changed. The shorter contact time makes the seeds not uniformly dispersed, and the longer contact time makes the seeds too dispersed to both sides of the seed belt. The influence mechanism of installation inclination on CVLU may be as follows. The installation inclination affects the collision position between the seed and the SDP surface, the coupling time, and the velocity of the seed. When the installation inclination is small, the collision position is at the lower end of the SDP surface, the coupling time with the surface is short, and the seed movement speed is large, which leads to a CVLU. On the contrary, the CVLU is small. The possible reasons why the ACT affects performance are as follows. When other parameters remain unchanged, as ACT increases, the corresponding ridge curvature also increases. As a result, the seed collecting ability of SDP with a curved ridge is strengthened, while the ability of SDP with an anti-arc is weakened. The influence mechanism of the span on CVLU may be as follows. The smaller span results in a smaller lateral movement distance for the seeds, which causes the seed falling point to concentrate in the center of the seed strip. For larger spans, under the action of the bottom arc curve, the seeds are easily dispersed to the boundary of the SDP. The reason for the influence of the bottom curve radius on the CVLU may be that as the bottom curve radius increases, the plate surface curvature decreases, which increases the lateral sliding of seeds in varying degrees.

The influence of the installation inclination on CVLU is similar to that of reference [4]. However, the research results of reference [4] show that the uniform seed distribution performance is better when the arc radius is 141.26 mm and the installation angle is 35.53°, which is different from the results of this study. The main reason is that the structure and manufacturing requirements of the uniform seed device are different. In addition, by comparing the working effect of the same type of strip-sowing device without a trench opener, it can be found that the uniform seeding device in this study has good seed uniformity in wide seed strips [14,15]. Compared with the strip-sowing method using the trench opener, it can be found that the uniform seed distribution device can reduce the operation smoothness problem caused by the dense arrangement of the trench opener [16,18,26]. In addition, this sowing technique is more suitable for sowing wheat in fields with high straw content, because this technique can cover the rice straw neatly in the gap between the seed strips [16,20].

To take into account the operation indexes such as sowing efficiency, economic benefit and sowing depth, the seeder operating parameters, such as rotary ploughing knife speed, are designed to be fixed. At present, only the influence of the structural parameters of the SDP on the operation performance is considered under the fixed operating parameters. However, these operating parameters, which are designed as fixed values, may also have a certain impact on the sowing uniformity. For example, the machine vibration caused by these parameters affects the performance of sowing uniformity. Future research can further evaluate the impact of machine operating parameters on sowing uniformity. The specific method is to evaluate the effects of different forward speeds, rotary tillage speeds, sowing dosage, and wheat varieties on sowing uniformity, emergence rate, and yield through bench tests or field experiments. Moreover, we learn how to adjust the structural parameters of the SDP to adapt to the operation of different crops and different soil conditions.

In addition, this study did not consider the effects of machine operation parameters, field straw content, wheat varieties, soil inclination, and other factors on operation quality. However, it is found that the factors such as land inclination have a certain influence on the sowing uniformity. It is necessary to comprehensively study the effects of soil parameters, seed parameters, and machine operation parameters on sowing performance, to further develop a real-time control device with soil inclination and machine operation parameters as independent variables.

5. Conclusions

To solve the problem of heap soil around the opener due to the sticky and wet nature of soil, the structural design and parameter optimization of SDP were carried out through a theoretical analysis, structural comparative analysis test, parameter comprehensive optimization test, and field verification test. The specific conclusions are as follows.

1. Combined with agronomic standards, the structural design and theoretical analysis of the SDP were carried out. The key factors (chord length of ridge, installation inclination, ACT, span, and bottom curve radius) affecting CVLU value are identified.
2. Through simulation tests, six structures of SDP were compared, and the structure of S6 was determined to be the optimal structural model. The influence of key parameters on performance was analyzed. Four factors (chord length of ridge, installation inclination, span, and bottom curve radius) were determined as factors for a comprehensive optimization. In addition, the ACT was determined to be 13°, and the corresponding CVLU is 42.98%.
3. Through comprehensive optimization experiments, the influence of parameters on the CVLU is analyzed, and parameters are comprehensively optimized. The order of the main effect relationships of the four factors is A > C > B > D. The best parameter combination after optimization was obtained as the chord length of a ridge of 140 mm, installation inclination of 40°, span of 75 mm and' bottom curve radius of 50 mm. The corresponding theoretical CVLU value is 21.42%, and the corresponding field test value is 30.27%.

Author Contributions: Conceptualization, W.L. and Z.H.; methodology, F.G. and K.G.; software, J.L. and X.C.; validation, W.L. and X.C.; formal analysis, W.L. and F.G.; investigation, K.G. and X.C.; resources, M.Q.; data curation, W.L.; writing—original draft preparation, W.L.; writing—review and editing, W.L.; visualization, J.L.; supervision, M.Q.; project administration, Z.H.; funding acquisition, F.G. All authors have read and agreed to the published version of the manuscript.

Funding: This work was financially supported by the Natural Science Foundation of Jiangsu Province (Grant No. BK 20221187).

Institutional Review Board Statement: Not applicable.

Data Availability Statement: The data presented in this study are available in the article.

Conflicts of Interest: The authors declare no conflict of interest.

References

1. Cao, W. Study on Mechanization Production Pattern and Efficiency of Rice-Wheat Double Cropping in Jiangsu Province. Ph.D. Thesis, China Agricultural University, Beijing, China, 2015.
2. Singh, A.; Phogat, V.K.; Dahiya, R.; Batra, S.D. Impact of Long-Term Zero till Wheat on Soil Physical Properties and Wheat Productivity under Rice–Wheat Cropping System. *Soil Tillage Res.* **2014**, *140*, 98–105. [CrossRef]
3. Ahmad, F.; Weimin, D.; Qishou, D.; Rehim, A.; Jabran, K. Comparative Performance of Various Disc-Type Furrow Openers in No-Till Paddy Field Conditions. *Sustainability* **2017**, *9*, 1143. [CrossRef]
4. Ding, J.; Li, F.; Le, T.; Xu, D.; Zhu, M.; Li, C.; Zhu, X.; Guo, W. Tillage and Seeding Strategies for Wheat Optimizing Production in Harvested Rice Fields with High Soil Moisture. *Sci. Rep.* **2021**, *11*, 119. [CrossRef] [PubMed]
5. He, J.; Li, H.; Chen, H.; Lu, C.; Wang, Q. Research Progress of Conservation Tillage Technology and Machine. *Trans. Chin. Soc. Agric. Mach.* **2018**, *49*, 1–19. [CrossRef]
6. Wang, Y.; Li, H.; Hu, H.; He, J.; Wang, Q.; Lu, C.; Liu, P.; Yang, Q.; He, D.; Jiang, S. A Noncontact Self-Suction Wheat Shooting Device for Sustainable Agriculture: A Preliminary Research. *Comput. Electron. Agric.* **2022**, *197*, 106927. [CrossRef]
7. Wang, Y.; Li, H.; He, J.; Wang, Q.; Lu, C.; Liu, P.; Yang, Q. Design and Experiment of Wheat Mechanical Shooting Seed-Metering Device. *Trans. Chin. Soc. Agric. Mach.* **2020**, *51*, 73–84. [CrossRef]
8. Wang, Y.; Li, H.; He, J.; Wang, Q.; Lu, C.; Liu, P.; Yang, Q. Parameters Optimization and Experiment of Mechanical Wheat Shooting Seed-Metering Device. *Trans. Chin. Soc. Agric. Eng.* **2020**, *36*, 1–10. [CrossRef]
9. Wang, C.; Li, H.; He, J.; Wang, Q.; Lu, C.; Yang, H. Optimization Design of a Pneumatic Wheat-Shooting Device Based on Numerical Simulation and Field Test in Rice–Wheat Rotation Areas. *Agriculture* **2022**, *12*, 56. [CrossRef]
10. Wang, C.; Lu, C.; Li, H.; He, J.; Wang, Q.; Cheng, X. Preliminary Bench Experiment Study on Working Parameters of Pneumatic Seeding Mechanism for Wheat in Rice-Wheat Rotation Areas. *Int. J. Agric. Biol. Eng.* **2020**, *13*, 66–72. [CrossRef]
11. Wang, C.; Li, H.; Wang, J.; He, J.; Wang, Q.; Lu, C. CFD Simulation and Optimization of a Pneumatic Wheat Seeding Device. *IEEE Access* **2020**, *8*, 214007–214018. [CrossRef]
12. An, X.; Cheng, X.; Wang, X.; Han, Y.; Li, H.; Liu, L.; Liu, M.; Liu, M.; Zhang, X. Design and Experimental Testing of a Centrifugal Wheat Strip Seeding Device. *Agriculture* **2023**, *13*, 1883. [CrossRef]
13. Wang, J.; Sun, W.; Simionescu, P.A.; Ju, Y. Optimization of the Fluted Force-Feed Seeder Meter with the Helical Roller Using the Discrete Element Method and Response Surface Analysis. *Agriculture* **2023**, *13*, 1400. [CrossRef]

14. Li, C.; Tang, Y.; McHugh, A.D.; Wu, X.; Liu, M.; Li, M.; Xiong, T.; Ling, D.; Tang, Q.; Liao, M.; et al. Development and Performance Evaluation of a Wet-Resistant Strip-till Seeder for Sowing Wheat Following Rice. *Biosyst. Eng.* **2022**, *220*, 146–158. [CrossRef]
15. Xi, X.; Gao, W.; Gu, C.; Shi, Y.; Han, L.; Zhang, Y.; Zhang, B.; Zhang, R. Optimisation of No-Tube Seeding and Its Application in Rice Planting. *Biosyst. Eng.* **2021**, *210*, 115–128. [CrossRef]
16. Han, J.; Du, J.; Gu, X.; Liu, L.; Li, Z.; Liu, C. Design and test of the separate and combined double row wide strip seedguiding device for wheat. *Trans. Chin. Soc. Agric. Eng.* **2023**, *39*, 35–46. [CrossRef]
17. Lei, X.; Liao, Y.; Liao, Q. Simulation of Seed Motion in Seed Feeding Device with DEM-CFD Coupling Approach for Rapeseed and Wheat. *Comput. Electron. Agric.* **2016**, *131*, 29–39. [CrossRef]
18. Lei, X.; Liao, Y.; Zhang, Q.; Wang, L.; Liao, Q. Numerical Simulation of Seed Motion Characteristics of Distribution Head for Rapeseed and Wheat. *Comput. Electron. Agric.* **2018**, *150*, 98–109. [CrossRef]
19. Tang, H.; Xu, F.; Xu, C.; Zhao, J.; Wang, Y.-J. The Influence of a Seed Drop Tube of the Inside-Filling Air-Blowing Precision Seed-Metering Device on Seeding Quality. *Comput. Electron. Agric.* **2023**, *204*, 107555. [CrossRef]
20. Liu, C.; Wei, D.; Du, X.; Jiang, M.; Song, J.; Zhang, F. Design and Test of Wide Seedling Strip Wheat Precision Hook-hole Type Seed-metering Device. *Trans. Chin. Soc. Agric. Mach.* **2019**, *50*, 75–84.
21. Jiang, M.; Liu, C.; Wei, D.; Du, X.; Cai, P.; Song, J. Design and test of wide seedling strip wheat precision planter. *Trans. Chin. Soc. Agric. Mach.* **2019**, *50*, 53–62. [CrossRef]
22. Niu, Q.; Wang, Q.; Chen, L.; Li, H.; He, J.; Li, W. Design and Experiment on Straw Post-covering Wheat Planter. *Trans. Chin. Soc. Agric. Mach.* **2017**, *48*, 52–59.
23. Hu, H.; Li, H.; Li, C.; Wang, Q.; He, J.; Li, W.; Zhang, X. Design and Experiment of Broad Width and Precision Minimal Tillage Wheat Planter in Rice Stubble Field. *Trans. Chin. Soc. Agric. Eng.* **2016**, *32*, 24–32. [CrossRef]
24. Zhu, Q.; Wu, G.; Chen, L.; Zhao, C.; Meng, Z.; Shi, J. Structural design and optimization of seed separated plate of wheat wide-boundary sowing device. *Trans. Chin. Soc. Agric. Eng.* **2019**, *35*, 1–11. [CrossRef]
25. Shi, Y.; Chu, J.; Yin, L.; He, M.; Deng, S.; Zhang, L.; Sun, X.; Tian, Q.; Dai, X. Wide-range sowing improving yield and nitrogen use efficiency of wheat sown at different dates. *Trans. Chin. Soc. Agric. Eng.* **2018**, *34*, 127–133. [CrossRef]
26. Luo, w.; Gu, F.; Wu, F.; Xu, H.; Chen, Y.; Xu, H. Design and Experiment of Wheat Planter with Straw Crushing and Inter-Furrow Collecting Mulching under Full Amount Of Straw and Root Stubble Cropland. *Trans. Chin. Soc. Agric. Mach.* **2019**, *50*, 42–52. [CrossRef]
27. Sun, K.; Yu, J.; Liang, L.; Wang, Y.; Yan, D.; Zhou, L.; Yu, Y. A DEM-Based General Modelling Method and Experimental Verification for Wheat Seeds. *Powder Technol.* **2022**, *401*, 117353. [CrossRef]
28. Lu, C.; Gao, Z.; Li, H.; He, J.; Wang, Q.; Wei, X.; Wang, X.; Jiang, S.; Xu, J.; He, D. An Ellipsoid Modelling Method for Discrete Element Simulation of Wheat Seeds. *Biosyst. Eng.* **2023**, *226*, 1–15. [CrossRef]
29. Liu, F.; Zhang, J.; Li, B.; Chen, J. Calibration of Parameters of Wheat Required in Discrete Element Method Simulation Based on Repose Angle of Particle Heap. *Trans. Chin. Soc. Agric. Eng.* **2016**, *32*, 247–253. [CrossRef]
30. Lei, X. Design and Working Mechanism of Air-Assisted Centralized Metering Device for Rapeseed and Wheat. Ph.D. Thesis, Huazhong Agricultural University, Wuhan, China, 2017.

agriculture

Article

Safety Analysis of Fastening Device of Agricultural By-Product Collector in Various Ground Conditions

Jeong-Hun Kim [1,2], Markumningsih Sri [3], Seok-Joon Hwang [1,2], Moon-Kyeong Jang [1,2], Seung-Jun Kim [1,2], Yun-Jeong Yang [1,2] and Ju-Seok Nam [1,2,*]

1 Department of Biosystems Engineering, College of Agricultural and Life Science, Kangwon National University, Chuncheon 24341, Republic of Korea; jhkim1995@kangwon.ac.kr (J.-H.K.); human51@naver.com (S.-J.H.); moon2842@kangwon.ac.kr (M.-K.J.); gracenav4@naver.com (S.-J.K.); yyj000501@naver.com (Y.-J.Y.)
2 Interdisciplinary Program in Smart Agriculture, Kangwon National University, Chuncheon 24341, Republic of Korea
3 Department of Agricultural and Biosystems Engineering, Faculty of Agricultural Technology, Universitas Gadjah Mada, Yogyakarta 55281, Indonesia; sri_markumningsih@ugm.ac.id
* Correspondence: njsg1218@kangwon.ac.kr; Tel.: +82-33-250-6497

Abstract: In this study, to evaluate the safety of the fastening device, which is a vulnerable part of the agricultural by-product collector, the stress in fastening devices was measured, and the operational and driving safety were analyzed by deriving the static safety factor and fatigue life. The position with the maximum stress in fastening devices was identified through structural analysis simulation, and a stress measurement system was constructed using strain gauges. Test conditions for stress measurement were classified into three operating conditions (collection operation, driving with the loading part lifted to the highest point, and driving with the loading part lifted to the lowest point) and three soil conditions (even pavement, sloped pavement, and farmland). A process for deriving the fatigue life based on the measured stress was constructed by applying the rain-flow counting method, Goodman equation, and Palmgren–Miner's rule via commercial software. From the stress measurement results, the collection operation exhibited the highest maximum stress, followed by driving with the loading part lifted to the highest point and driving with the loading part lifted to the lowest point. Under all conditions, the static safety factor of the fastening devices was found to be higher than 1.0 (1.16–1.33). The fatigue life of the fastening devices was also found to be longer than the service life of Korean agricultural machinery under all operating conditions. Therefore, the fastening devices are expected to operate safely under generated static and dynamic loads. The agricultural by-product collector can perform agricultural work and drive stably and is expected to contribute to reducing unnecessary labor force for Korean farms.

Keywords: agricultural by-product collector; rain-flow counting; static safety factor; fastening device; fatigue life

Citation: Kim, J.-H.; Sri, M.; Hwang, S.-J.; Jang, M.-K.; Kim, S.-J.; Yang, Y.-J.; Nam, J.-S. Safety Analysis of Fastening Device of Agricultural By-Product Collector in Various Ground Conditions. *Agriculture* **2023**, *13*, 2064. https://doi.org/10.3390/agriculture13112064

Academic Editors: Chung-Liang Chang and Mustafa Ucgul

Received: 15 September 2023
Revised: 25 October 2023
Accepted: 25 October 2023
Published: 27 October 2023

1. Introduction

Recently, the necessity of developing alternative energy sources has been emphasized worldwide owing to the lack of fossil energy sources and reinforced environmental regulations on emissions [1]. Accordingly, attention has been paid to biomass, which uses agricultural by-products as fuel, as an alternative energy source in Korea [2], and studies have been conducted to produce and process biomass [3–6]. There are, however, few studies on agricultural machinery that can directly collect or process agricultural by-products, such as pruned branches. In this regard, Hwang et al. designed an agricultural by-product collector for pruned branches of fruit trees, which have the highest potential generation and potential energy among agricultural by-products produced in Korea [7]. Owing to the nature of the cultivation environment of fruit trees, large loads can arise because of irregular road surface conditions and the vibration of the operating part during the operation and driving of the agricultural by-product collector [8]. Such loads may cause deformation and cracking in the collector, which may lead to damage to the vulnerable part. Therefore, safety analysis of agricultural by-product collectors is required to increase work safety and prevent damage and failure.

Fatigue failure refers to a fracture due to the accumulation of damage caused by the application of repeated stress to the material over an extended period of time [9–11]. Fatigue failure is mainly caused by cracking. Cracking, which is the first stage of fatigue failure, is caused by local plastic deformation that occurs inside the material. Cracks grow with repeated loads, and they eventually lead to the fracture of the material. In this instance, the number of repetitions or time of the load until the fracture of the material under cyclic loads is referred to as fatigue life. In fact, fatigue has mostly been reported as the cause of damage to machines and structures, and it causes human casualties and property damage [12]. To minimize such damage, it is necessary to predict and design the fatigue life of material as a reference [13]. This is important in the design of agricultural machinery. To accurately determine the fatigue life of agricultural machinery, it is necessary to derive the load and stress frequency that occur during actual agricultural work and driving. There are various techniques for deriving stress frequency. Among them, the representative technique is rain-flow counting. It is possible to derive the stress range, mean stress, and number of stress cycles using rain-flow counting and to predict fatigue life using the S-N curve—a characteristic of the material—and Palmgren–Miner's linear cumulative damage rule. Kulkarni et al. (2016) assessed fatigue life to analyze the effect of an increase in the mass of the switched reluctance motor (SRM), which is used in small vehicles, on suspension design [14]. Kim et al. (2018) measured the stress generated in the transplanter PTO axis according to the planting distance during transplanting and derived fatigue life by applying rain-flow counting [15]. Choi et al. (2020) conducted fatigue analysis on the tunnel boring machine (TBM) cutter head, which is the representative equipment of mechanized tunnel construction, to predict its fatigue life under cyclic load conditions [16]. Bohm and Kowalsk (2020) developed a fatigue test model for the case in which aluminum alloys are subjected to torsional loads with a certain amplitude through rain-flow counting and Palmgren–Miner's linear cumulative damage rule [17]. Han et al. (2022) derived the load acting on the differential gear under the application of braking loads, such as a sudden stop or start, using a tractor front-axle analysis model to evaluate safety [18]. As described, the fatigue life of components is derived to evaluate safety during the design of various machines, including agricultural machinery.

In this study, the stress in the fastening devices of the agricultural by-product collector was measured under various operating and soil conditions, and safety was evaluated by deriving the static safety factor and fatigue life. The stress measurement system was constructed by installation of a strain gauge in the location where the maximum stress generated on the fastening device. The static safety factor was derived based on the measured maximum stress. In addition, the dynamic safety and fatigue life were derived by applying the rain-flow counting, Goodman equation, S-N curve, and Palmgren–Miner's linear cumulative damage rule. As a result, it is judged that the application of the agricultural by-product collector to Korean farms will contribute to reducing farmers' labor force and increasing convenience.

2. Materials and Methods

2.1. Agricultural by-Product Collector [7]

In this study, the agricultural by-product collector consists of a collecting part, a transferring part, a loading part, and a driving part, as shown in Figure 1. The collecting part adopts the longitudinal axis rotation method to collect agricultural by-products in the middle using two rotating collection brushes. The transferring part transports the collected agricultural by-products from the ground to the loading part through a conveyor belt. The maximum capacity of the loading part is 100 kg, and the cargo box can be moved up and down through the application of a lifting device [19]. The size of the cargo box was set to 900 × 1100 × 450 mm (W × L × H). Tracked wheels were applied to the agricultural by-product collector, so that it could travel on soft soil in a stable manner.

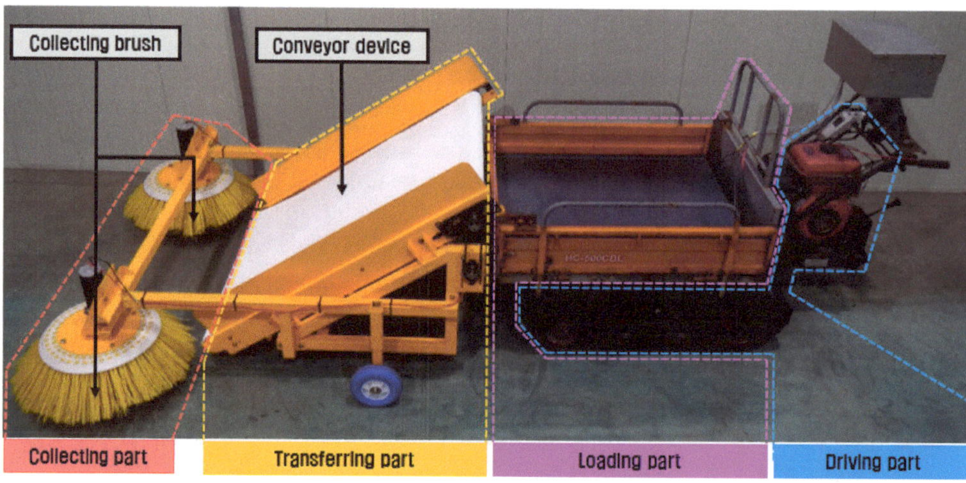

Figure 1. Shape of the agricultural by-product collector.

In the agricultural by-product collector, the transferring and the loading parts are combined by inserting two cantilever-type fastening devices connected to the transferring part frame into the hollow pipes located on the bottom frame of the loading part (Figure 2). The fastening devices were divided into the left (L) and right (R) devices, as viewed from the driving part toward the collecting part. The fastening devices must bear the self-weight of the collecting–transferring parts when the parts are raised from the ground, and they are subjected to variable loads during the collection operation due to the vibration caused by the stone, gravel, and obstacles on the ground. Considering that the fastening devices have structurally vulnerable cantilever geometry, their safety must be secured for the safe operation of the agricultural by-product collector.

Figure 2. Shape of the fastening device.

2.2. Stress Measurement

2.2.1. Measurement System Configuration [20]

To evaluate the safety of the fastening devices, it is necessary to identify the position with the maximum stress and measure the stress at that position during the operation of the agricultural by-product collector. Kim et al. (2022) conducted structural analysis through commercial dynamics simulation and derived the position with the maximum stress in the fastening device [19]. The position was the upper part of the fastening devices, 175 mm away from the transferring part frame. Therefore, in this study, strain gauges were installed in the position to measure the strain during the operation of the agricultural by-product collector (Figure 3). The measurement system was composed of strain gauges for measuring the axial vertical stress, a data acquisition system (DAQ), and a laptop for real-time display and storage of the measurement results (Figure 4). Also, the stress value measured through the data acquisition system shows an accuracy of 0.05–0.15% from the actual value.

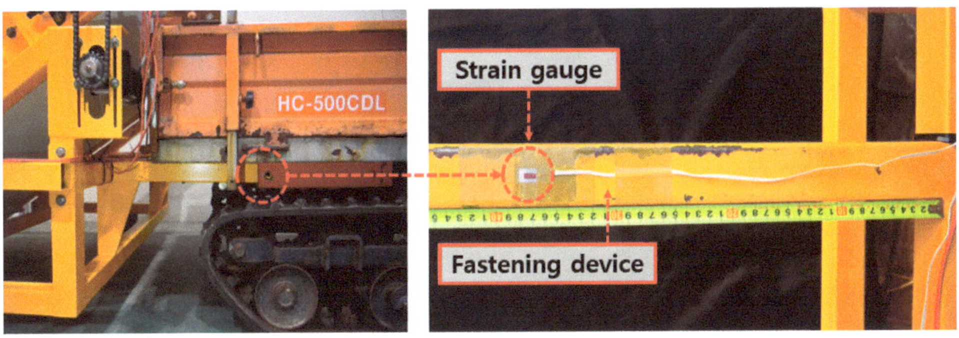

Figure 3. Attachment location of strain gauge.

Figure 4. Configuration of measurement system.

2.2.2. Test Conditions

The actual operating conditions of the agricultural by-product collector can be divided into collecting pruned branches of fruit trees and driving to the farmland. Therefore, the following three conditions were set as the operating conditions to measure the stress of the fastening devices: collection operation, driving with the loading part lifted to the highest point, and driving with the loading part lifted to the lowest point (Figure 5).

(a) (b) (c)

Figure 5. Driving conditions of agricultural by-product collector. (**a**) Collection operation. (**b**) Driving with the loading part lift-ed to the highest point. (**c**) Driving with the loading part lifted to the lowest point.

Considering the operating environment of the agricultural by-product collector, the following three soil conditions were selected: even pavement, sloped pavement, and farmland (Figure 6). The average ground slope of orchards in Korea ranges from 0 to 5°; however, a small number of orchards have a slope of approximately 16° to increase the drainage capacity of the ground [21]. Therefore, the slope of the sloped pavement was set to 16°, which is the harshest condition. The stress according to the three types of operation was measured under each soil condition, and the operating distance per test was set to 30 m. Under each condition, the average value of three repeated tests was used as the representative value. The driving speed of the agricultural by-product collector was fixed at 1.0 m/s by referring to a previous study [7].

(a) (b) (c)

Figure 6. Soil conditions for stress measurement. (**a**) Even pavement. (**b**) Sloped pavement. (**c**) Farmland.

The soil from farmland was analyzed by soil sampling and classified as sand composed of clay (2.18%), silt (4.41%), and sand (93.41%). The average water content was found to be 14.86% after measuring at six random locations on the farmland through the oven dry method.

2.3. Safety Analysis Method

2.3.1. Static Safety Factor

In mechanical design, uncertainties of material strength, machining precision, and workload must be considered, so that the designed product can have higher safety than the required performance or capability. In this instance, the safety of a machine under static loads is evaluated through the static safety factor. An increase in the safety factor involves an increase in cost and weight, while a decrease in the safety factor leads to higher risk. Therefore, it is necessary to set a proper static safety factor in the design of machine products, considering both the cost added by increasing the safety factor and the risk of damage. The stress-based static safety factor is defined as the ratio of the allowable strength of the material to the maximum stress acting on it. The stress acting on the material can be divided into shear stress, vertical stress, and equivalent stress. The yield strength of the material is generally applied as the allowable strength. Therefore, the static safety factor can be calculated as the ratio of the shear yield strength to the maximum shear stress for shear stress and as the ratio of the tensile yield strength to the maximum vertical stress/maximum equivalent stress for vertical and equivalent stress, respectively. In general, when the static safety factor exceeds 1.0, it can be said to be safe under static loads. If the maximum stress of a specific area is higher than the allowable strength of the material, the static safety factor becomes less than 1.0, and static damage may occur in the area, thereby causing failure or malfunction [22]. Since the fastening devices are mainly subjected to the uniaxial vertical stress caused by bending as a cantilever type, the static safety factor was derived using the tensile yield strength and Equation (1).

$$\text{S.F.} = \frac{S_y}{\sigma_{max}} \tag{1}$$

where

S.F. = Static safety factor;
S_y = Tensile yield strength of the material;
σ_{max} = Measured maximum normal stress during the operation.

2.3.2. Fatigue Life

Even a very lower load compared to the load that causes static damage may lead to the fracture of the material if applied repeatedly, due to the accumulation of damage. Fatigue life is a criterion for determining safety under repeated loads, and it means the number or time of load cycles until the fracture of the material due to the accumulation of damage. Fatigue life must be longer than the required service life for the stable operation of the system. Fatigue life is derived by applying rain-flow counting, the Goodman equation, and Palmgren–Miner's linear cumulative damage rule to the measured stress data in the time domain. The strain measured through the uniaxial strain gauge can be converted into vertical stress using Equation (2).

$$\sigma = E \times \varepsilon \tag{2}$$

where

σ = Measured normal stress;
E = Modulus of elasticity;
ε = Measured normal strain.

To derive fatigue life, it is necessary to convert the measured stress data in the time domain into the frequency domain and count the number of stresses of a certain amplitude. This is referred to as cycle counting. Cycle counting methods include level crossing counting, peak counting, and rain-flow counting. Among them, rain-flow counting has been most widely used [23]. It is possible to obtain information on the stress amplitude, mean stress, and number of cycles by applying rain-flow counting to the stress data in the time domain. A life cycle must be derived by applying stress data that correspond to each level to the S-N curve of the material. To this end, the stress composed of the amplitude and average must be converted into the equivalent completely reversed stress with an average of zero. The equivalent completely reversed stress can be obtained through the Goodman equation, as shown in Equation (3).

$$\sigma_{eq} = \frac{S_u \times \sigma_a}{S_u - \sigma_m} \quad (3)$$

where

σ_{eq} = Equivalent completely reversed stress;
S_u = Ultimate tensile strength of the material;
σ_a = Measured stress amplitude;
Σ_m = Measured mean stress.

When the equivalent completely reversed stress is substituted into the S-N curve of the material, it is possible to obtain the life cycle corresponding to the completely reversed stress. The total damage sum can be calculated using Equation (4) by applying the number of actually applied cycles of the stress and life cycle to Palmgren–Miner's linear cumulative damage rule. The linear cumulative damage rule derives the total damage sum by adding the partial damage caused by all applied loads under the assumption that fatigue damage is linearly accumulated, and it assumes that the fatigue damage of the material occurs when the total damage sum becomes 1.0 [24]. The fatigue life based on the total damage sum (obtained via Equation (4)) can be calculated using Equation (5).

$$D_t = \sum_{i=1}^{k} \frac{n_i}{N_i} \quad (4)$$

where

D_t = Total damage sum;
n_i = Number of actually applied cycles for ith stress;
N_i = Life cycles for ith stress.

$$L_f = \frac{1}{D_t} \times t \quad (5)$$

where

L_f = Fatigue life;
D_t = Total damage sum;
t = Working time, which generates damage sum.

2.3.3. Process of Deriving Fatigue Life

Fatigue life was derived from the measured stress data in the time domain using nCode (nCode, Version 19.0.0, HBM Prencia, Southfield, MI, USA), a commercial software program. The process was constructed by applying the fatigue life calculation method described in Equations (3)–(5), so that the load spectrum and fatigue life can be derived by applying rain-flow counting, the Goodman equation, and Palmgren–Miner's linear cumulative damage rule when the stress data in the time domain are entered (Figure 7). The S-N curve of the fastening device material was selected based on the properties of the material and the library of the nCode software (Prenscia, nCode Book of Fatigue theory). The material was Steel UML UTS300. The properties and S-N curve of the material are presented in Table 1 and Figure 8.

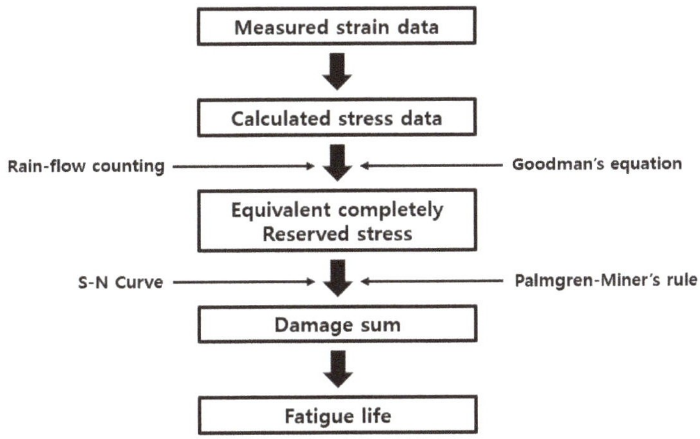

Figure 7. Theoretical analysis of fatigue life.

Table 1. Material properties of fastening device.

Item	Material Properties
Material	Steel UML UTS300
Yield strength (MPa)	231
Ultimate strength (MPa)	300
Elastic modulus (MPa)	2.07×10^5

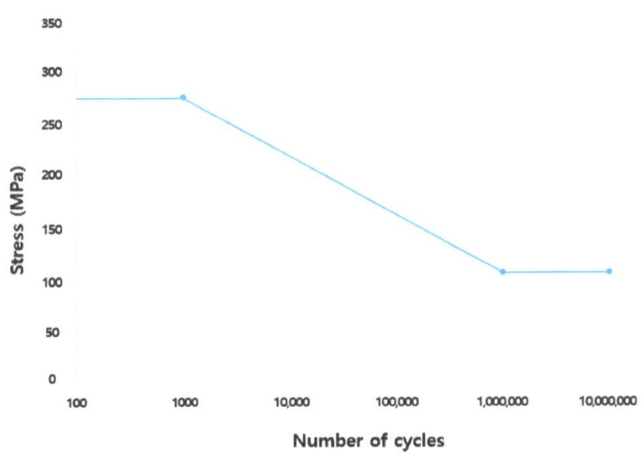

Figure 8. S-N curve for Steel UML UTS300.

3. Results and Discussion

3.1. Stress Measurement Results

The highest maximum stress in the fastening devices was measured during the collection operation on the farmland. The measured stress graphs in the time domain are shown in Figure 9. On farmland, the maximum stress exerted on each fastening device was 192.95–199.37 MPa (left) and 196.29–185.13 MPa (right). In case of the sloped pavement condition, the generated maximum stress was 187.60–180.79 MPa (left) and 181.96–175.98 MPa (right). The maximum stresses of 184.64–176.18 MPa (left) and 179.51–173.48 MPa (right) were measured in the even pavement condition.

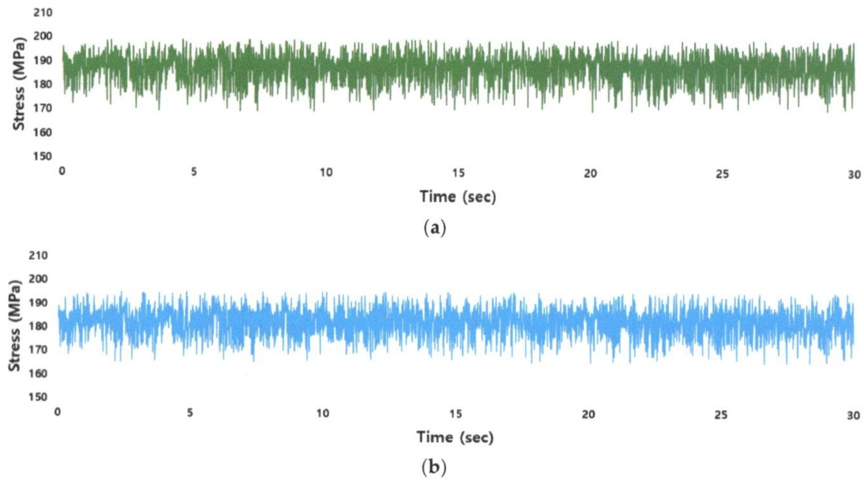

Figure 9. Time-series stress date of each fastening device during the collection operation on the farmland. (**a**) Fastening device (L). (**b**) Fastening device (R).

Regarding soil conditions, farmland exhibited the highest maximum stress, followed by sloped pavement and even pavement, in that order. It appears that high workloads were induced on the farmland because high vibration and impacts were generated in the fastening devices by the irregular road surface and obstacles. In addition, there was a relatively higher stress on the sloped pavement than on the even pavement. This appears to be because the load in the direction of the slope was additionally exerted on the fastening devices by the slope. Regarding the operating conditions, the collection operation exhibited the highest maximum stress, followed by the driving with the loading part lifted to the highest point and driving with the loading part lifted to the lowest point, in that order, under all soil conditions (Figure 10). It was found that the maximum stress during the collection operation was higher than that during driving with the loading part lifted to the highest point because the vibration caused by the collection operation had a larger impact than the moment load by the weight of the collecting and transferring parts.

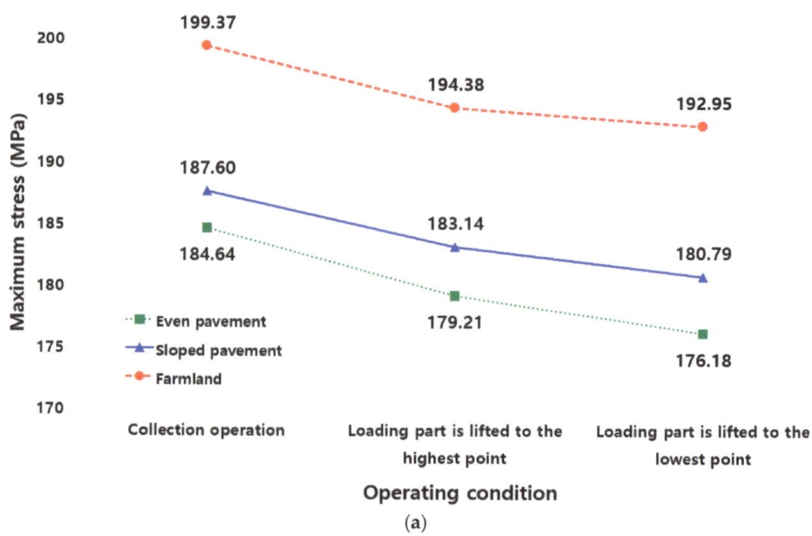

Figure 10. *Cont.*

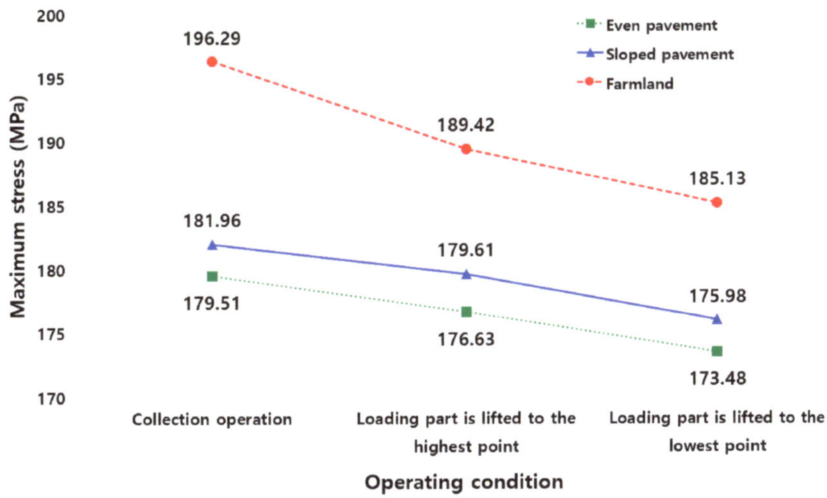

(**b**)

Figure 10. Measured maximum stress under each condition. (**a**) Fastening device (L). (**b**) Fastening device (R).

3.2. Results of Deriving Fastening Device Safety

3.2.1. Static Safety Factor

Based on the maximum stress derived under the test condition and the yield strength of the fastening device material, the static safety was derived. Table 2 summarizes the results of the static safety factor obtained via Equation (1). The static safety was found to be 1.16–1.29 for the collection operation, 1.19–1.31 for the driving with the loading part lifted to the highest point, and 1.21–1.33 for the driving with the loading part lifted to the lowest point. The operating condition with the lowest static safety factor was the collection operation on the farmland. In this case, the static safety factors were 1.16 and 1.18 for the left and right fastening devices, respectively. The static safety factor exceeded 1.0 under all test conditions, indicating that the fastening devices will operate safely, without damage under static loads.

Table 2. Static safety factor of the fastening device.

Driving Condition	Soil Condition	Fastening Device	
		L	R
Collection operation	Even pavement	1.25	1.29
	Sloped pavement	1.23	1.27
	Farmland	1.16	1.18
Loading part lifted to the highest point	Even pavement	1.29	1.31
	Sloped pavement	1.26	1.28
	Farmland	1.19	1.22
Loading part lifted to the lowest point	Even pavement	1.29	1.33
	Sloped pavement	1.29	1.33
	Farmland	1.21	1.25

3.2.2. Fatigue Life

The stress amplitude, mean stress, and number of cycles were derived by applying rain-flow counting to the measured time-series stress data under each test condition. The rain-flow counting results for the farmland-collection operation, which causes the highest stress, are shown in Figure 11. Under all operating and soil conditions, the mean stress and stress amplitude for the fastening devices were found to be 155–200 MPa and 0–35 MPa, respectively. Stress conditions resulting in the maximum damage to the fastening devices are presented in Table 3. Under all test conditions, there was just one cycle that caused maximum damage, which was more than 70% of the total damage. This indicates that certain maximum-load conditions that occur during the operation and driving have decisive effects on damage and fatigue life.

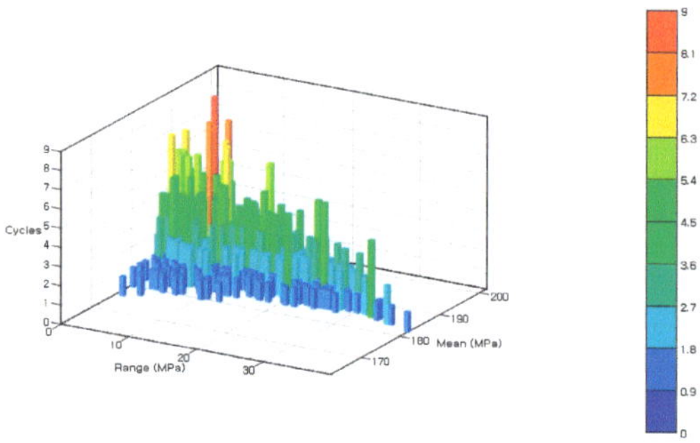

(a)

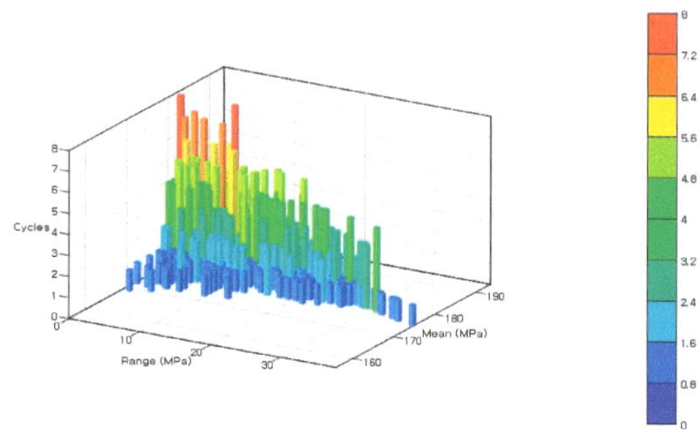

(b)

Figure 11. Rain-flow counting result during the collection operation on the farmland. (**a**) Fastening device (L). (**b**) Fastening device (R).

Table 3. Maximum-damage condition for each fastening device.

Driving Condition	Soil Condition	Fastening Device	Mean Stress (MPa)	Stress Amplitude (MPa)	Equivalent Stress (MPa)	Damage	Number of Cycle
Collection operation	Even pavement	L	171.95	14.55	34.08	9.39×10^{-8}	1
		R	166.54	15.05	33.84	6.64×10^{-8}	
	Sloped pavement	L	168.34	18.95	43.18	3.17×10^{-7}	
		R	167.29	16.50	37.29	1.24×10^{-7}	
	Farmland	L	182.05	19.73	50.17	1.93×10^{-6}	
		R	174.78	18.76	44.94	5.91×10^{-7}	
Loading part lifted to the highest point	Even pavement	L	167.26	13.14	29.70	3.14×10^{-8}	1
		R	161.76	11.41	24.76	7.99×10^{-9}	
	Sloped pavement	L	173.01	12.02	28.40	3.35×10^{-8}	
		R	169.96	10.69	24.65	1.19×10^{-8}	
	Farmland	L	196.34	20.67	50.13	1.27×10^{-6}	
		R	171.96	18.39	43.09	3.85×10^{-7}	
Loading part lifted to the lowest point	Even pavement	L	161.31	11.69	25.29	8.88×10^{-9}	1
		R	159.71	10.85	23.19	4.90×10^{-9}	
	Sloped pavement	L	164.57	12.49	27.67	1.79×10^{-8}	
		R	160.97	11.72	25.28	8.72×10^{-9}	
	Farmland	L	171.35	20.48	47.75	6.89×10^{-7}	
		R	170.91	15.96	37.08	1.47×10^{-7}	

The load spectrum during the farmland-collection operation, which causes the highest stress, is shown in Figure 12. During the operation on the farmland, the number of cycles due to stresses below 10 MPa was less than 50%. It, however, exceeded 70% for operation on both the sloped and even pavements. In other words, the number of cycles due to stresses below 10 MPa increased with decreasing maximum stress, confirming that the farmland has the highest high-load occurrence rate, followed by sloped pavement and even pavement, in that order.

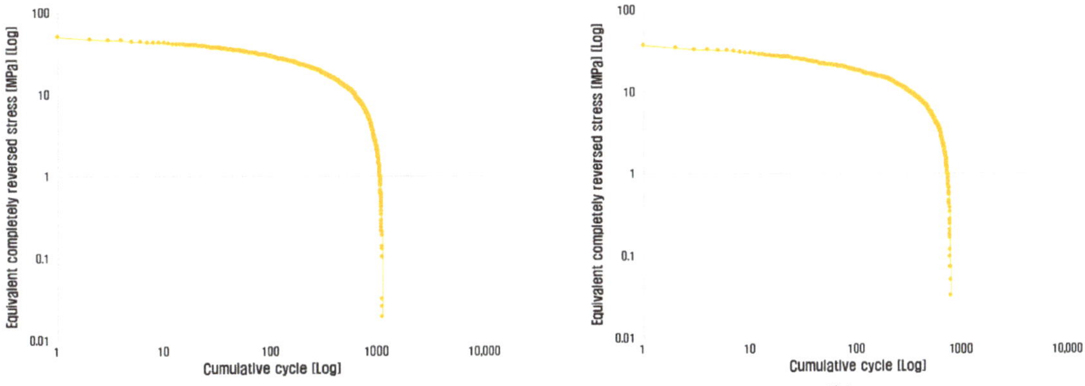

Figure 12. Load spectrum during the collection operation on the farmland. (**a**) Fastening device (L). (**b**) Fastening device (R).

Table 4 summarizes the total damage and fatigue life under each operation condition. Fatigue life was found to be shortest during the farmland-collection operation. The maximum stress tended to increase with decreasing fatigue life. In a literature review, the average annual pruning time of orchards in Korea was found to be 38.5 h [25]. Under the assumption that the agricultural by-product collector is continuously used during that time, the fatigue life of the fastening devices was found to be 621–2167 y for the collection operation, 1060–5041 y for driving with the loading part lifted to the highest point, and 2285–7205 y for driving with the loading part lifted to the lowest point. Considering that the average service life of agricultural machinery is 9 y [26], fastening devices of the agricultural by-product collector are expected to operate safely during their required service life, without damage under variable loads.

Table 4. Total damage and fatigue life of the fastening device.

Driving Condition	Soil Condition	Total Damage		Fatigue Life (hours)		Fatigue Life Considering 38.5 h of Annual Working Time (y)	
		Fastening Device (L)	Fastening Device (R)	Fastening Device (L)	Fastening Device (R)	Fastening Device (L)	Fastening Device (R)
Collection operation	Even pavement	8.094×10^{-7}	6.102×10^{-7}	1.277×10^{6}	1.932×10^{6}	33,168	50,181
	Sloped pavement	1.974×10^{-6}	1.251×10^{-6}	5.114×10^{5}	1.100×10^{6}	13,283	28,571
	Farmland	4.328×10^{-5}	1.212×10^{-5}	2.391×10^{4}	8.346×10^{4}	621	2167
Loading part lifted to the highest point	Even pavement	3.160×10^{-7}	1.358×10^{-7}	4.901×10^{6}	1.164×10^{7}	127,298	302,337
	Sloped pavement	2.072×10^{-7}	8.651×10^{-8}	3.209×10^{6}	7.907×10^{6}	83,350	205,376
	Farmland	2.645×10^{-5}	6.043×10^{-6}	4.084×10^{4}	1.941×10^{5}	1060	5041
Loading part lifted to the lowest point	Even pavement	3.464×10^{-7}	1.163×10^{-7}	1.175×10^{7}	1.986×10^{7}	305,194	511,168
	Sloped pavement	1.328×10^{-8}	6.503×10^{-9}	3.416×10^{6}	8.936×10^{6}	88,727	232,103
	Farmland	4.123×10^{-6}	1.382×10^{-6}	8.799×10^{4}	2.774×10^{5}	2285	7205

4. Conclusions

To analyze the safety of the agricultural by-product collector, the stress in the fastening devices—the vulnerable part of the collector—was measured, and the static safety factor, as well as fatigue life, was derived in this study. A stress measurement system based on strain gauges was constructed to measure the stress at the position with the maximum stress.

Three operating conditions (collection operation, driving with the loading part lifted to the highest point, and driving with the loading part lifted to the lowest point) and three soil conditions (even pavement, sloped pavement, and farmland) were selected as test conditions for stress measurement. Regarding soil conditions, farmland showed the highest maximum stress, followed by sloped pavement and even pavement. For the farmland, it appears that high stress was due to the vibration and impacts caused by the irregular road surface and obstacles. There was relatively higher stress on the sloped pavement compared with that on the even pavement, owing to the applied load in the direction of the slope by the slope. Regarding operating conditions, the collection operation exhibited the highest maximum stress, followed by driving with the loading part lifted to the highest point and driving with the loading part lifted to the lowest point. This indicates that the vibration caused by the collection operation has a significant impact on the stress of the fastening devices. In addition, when compared with the results derived from previous studies, the soil conditions were in the order of farmland, hard flat ground, sloped pavement, and even pavement. Based on the yield strength of the fastening device material and the maximum stress measured under each test condition, the static safety factor of the fastening devices was derived. The safety factor by operating condition was found to be 1.16–1.29 for the collection operation, 1.19–1.31 for driving with the loading part lifted to the highest point, and 1.21–1.33 for driving with the loading part lifted to the lowest point. It was judged that agricultural machinery is safe under static loads if the static safety factor of its vulnerable parts is higher than 1.0 during design by referring to a previous study. The static safety factor of the fastening devices exceeded 1.0 under all test conditions, indicating that the fastening devices will operate safely, without damage under static loads. The fatigue life of the fastening devices was found to be shortest during the collection operation on the farmland. Considering that the average annual pruning time of orchards in Korea is 38.5 y, the minimum fatigue life of the fastening devices is 621 y. Considering that the fatigue life of the fastening devices exceeds the average service life of agricultural machinery under all test conditions, the fastening devices are expected to operate safely during the required service life. The fatigue life of the fastening devices was much longer than their service life; however, the static safety factor was found to be less than 1.5, indicating that the fastening devices need to be designed with a focus on the static damage caused by the maximum load rather than the fatigue failure caused by variable loads. As future work, it is necessary to measure the stress generated in the fastening device and evaluate its safety when collecting agricultural by-products using the agricultural by-product collector in various agricultural fields.

Author Contributions: Theoretical and simulation analysis, J.-H.K., M.S., S.-J.H., M.-K.J., S.-J.K. and Y.-J.Y.; writing—original draft preparation, J.-H.K.; writing—review and editing, J.-S.N. All authors have read and agreed to the published version of the manuscript.

Funding: This work was supported by the Korea Institute of Planning and Evaluation for Technology in Food, Agriculture and Forestry (IPET) through the Machinery Mechanization Technology Development Program for Field Farming Program, funded by the Ministry of Agriculture, Food and Rural Affairs (MAFRA)(RS-2023-00235957).

Institutional Review Board Statement: Not applicable.

Data Availability Statement: Not applicable.

Conflicts of Interest: The authors declare no conflict of interest.

References

1. Jung, J.W.; Son, S.H.; Jang, Y.H.; Ryou, K.H.; Jeong, J.H.; Lee, S.Y.; Lee, S.M. *Structural Changes in the Global Energy Market and Diversification Policy in Korea's Energy Cooperation with the Middle East*; Korea Institute for International Economic Policy: Sejong, Republic of Korea, 2020.
2. Park, W.K.; Park, N.B.; Shin, J.D.; Hong, S.G.; Kwon, S.I. Estimation of Biomass Resource Conversion Factor and Potential Production in Agricultural Sector. *Korean J. Environ. Agric.* **2011**, *30*, 252–260. [CrossRef]

3. Cha, J.Y.; Ali, M.; Hong, Y.S.; Yu, B.K.; Lee, S.H.; Seonwoo, H.; Kim, H.J. Development of a Pelletizing System of Fermented TMR for Pig Feeding. *J. Biosyst. Eng.* **2018**, *43*, 119–127. [CrossRef]
4. Patel, D.K.; Kim, M.H.; Lim, K.T. Synthesis and Characterization of Eggshell-Derived Hydroxyapatite Bioceramics. *J. Biosyst. Eng.* **2019**, *44*, 128–133. [CrossRef]
5. Alahakoon, A.; Karunarathna, A.; Dharmakeerthi, R.; Silva, F. Design and Development of a Double-chamber Down Draft(DcDD) Pyrolyzer for Biochar Production from Rice Husk. *J. Biosyst. Eng.* **2022**, *47*, 458–467. [CrossRef]
6. Park, S.Y.; Kim, S.J.; Oh, K.C.; Jeon, Y.K.; Kim, Y.M.; Cho, A.Y.; Lee, D.H.; Jang, C.S.; Kim, D.H. Biochar from Agro-byproducts for Use as a Soil Amendment and Solid Biofuel. *J. Biosyst. Eng.* **2023**, *48*, 93–103. [CrossRef]
7. Hwang, S.J.; Kim, J.H.; Nam, J.S. Factorial Experiment for the Collecting Device of and Agricultural By-product Collector. *J. Biosyst. Eng.* **2020**, *45*, 422–431. [CrossRef]
8. Han, G.H.; Kim, D.K.; Ahn, D.V.; Park, Y.J. Comparative Analysis of Tractor Ride Vibration According to Suspension System Configuration. *J. Biosyst. Eng.* **2023**, *48*, 69–78. [CrossRef]
9. Vincent, M.K.; Varghese, V.; Sukumaran, S. Fabrication and Analysis of Fatigue Testing Machine. *Int. J. Eng. Sci.* **2016**, *5*, 15–19.
10. Jeong, S.U. Bending Fatigue Characteristics and Stress Analysis of STS304 Steel. *J. Inst. Mar. Ind.* **2007**, *20*, 79–84.
11. Kang, M.S.; Koo, J.M.; Seok, C.S.; Park, J.S. Fatigue Characteristic of High Impact Polystyrene(HR-1360). *Korean Soc. Mech. Eng. A* **2010**, *34*, 763–769. [CrossRef]
12. Yeom, H.H.; Jung, Y.C.; Kim, C.Y.; Kang, K.Y.; Lee, M.G.; Hong, M.S.; Jeon, Y.H. Ultrasonic Fatigue Test for a High Strength Steel Plate. *J. Korean Soc. Manuf. Technol. Eng.* **2015**, *24*, 589–593. [CrossRef]
13. Murakami, M.; Takai, T.; Wada, K.; Matsunaga, H. Essential structure of S-N curve: Prediction of fatigue life and fatigue limit of defective materials and nature scatter. *Int. J. Fatigue* **2021**, *146*, 106–138. [CrossRef]
14. Kulkarni, A.; Ranjha, S.A.; Kapoor, A. Fatigue analysis of a suspension far and in-wheel electric vehicle. *Eng. Fail. Anal.* **2016**, *68*, 150–158. [CrossRef]
15. Kim, W.S.; Kim, Y.S.; Kim, Y.J. Analysis of the Load of a Transplanter PTO Shaft Based on the Planting Distance. *J. Fac. Agric. Kyushu Univ.* **2018**, *63*, 97–102. [CrossRef]
16. Choi, S.W.; Kang, T.H.; Lee, C.H.; Chang, S.H. Fatigue Analysis for Structural Stability Review of TBM Cutterhead. *J. Korean Tunn. Undergr. Sp. Assoc.* **2020**, *22*, 529–541. [CrossRef]
17. Bohm, M.; Kowalski, M. Fatigue Life Estimation of Explosive Cladded Transition Joints with the Use of the Spectral Method for the Case of a Random sea state. *Mar. Struct.* **2020**, *71*, 102739. [CrossRef]
18. Han, H.W.; Oh, J.S.; Yoo, H.J.; Shim, S.B. Strength Evaluation of the Tractor Differential Case Under Braking Load. *J. Agric. Life Environ. Sci.* **2022**, *34*, 9–16. [CrossRef]
19. Kim, S.J.; Kim, J.H.; Nam, J.S. Performance Test of Agricultural By-product Collecting Monitoring System. *Precis. Agric. Sci. Technol.* **2022**, *4*, 71–80. [CrossRef]
20. Kim, J.H.; Sri, M.; Nam, J.S. Safety Evaluation on a Fastening Device of an Agricultural By-Product Collector for Hard Flat Ground Driving. *Agriculture* **2022**, *12*, 1071. [CrossRef]
21. Noh, J.S.; Cho, Y.J.; Kim, T.S.; Lee, D.H.; Lee, K.S. Driving Running Stability of Orchard Aerial Lift by Using Recurdyn that is Analysis Tool. *Proc. Korean Soc. Agric. Mach. Conf.* **2012**, *17*, 153–159.
22. Budynas, R.G.; Nisbett, J.K. *Shigley's Mechanical Engineering Design*, 9th ed.; Mcgraw-Hill: New York, NY, USA, 2011; pp. 10–15.
23. Park, J.J.; Seo, S.M.; Seo, M.W.; Choi, Y.S.; Kim, Y.J. Development of Material Properties Measurement and Fatigue Life Evaluation System. *Korean Soc. Mech. Eng. A* **1994**, *18*, 1465–1473.
24. Lee, W.S.; Lee, H.W. A Study on the Prediction of the Fatigue Life of a Lug through the Finite Element Analysis. *J. Korean Soc. Precis. Eng.* **1998**, *15*, 88–95.
25. Lee, Y.H.; Lee, J.H.; Lee, K.S.; Kim, K.R.; Lee, S.J. Ergonomic Risk Factors Related to Musculoskeletal Symptoms in the Vineyard Workers. *J. Korean Soc. Occup. Environ. Hyg.* **2008**, *18*, 122–132.
26. Lee, J.M.; Kim, Y.Y.; Hwang, S.J. *A Study of Useful Life for Agricultural Machinery*; National Institute of Agricultural Sciences: Wanju, Republic of Korea, 2017.

agriculture

MDPI

Article

Simulation of Mouldboard Plough Soil Cutting Based on Smooth Particle Hydrodynamics Method and FEM–SPH Coupling Method

Xiaoming Jin, Fangping Ma, Di Wang * and Zhengtao Zhu

Department of Mechanical Engineering, Xinjiang University, Urumqi 830017, China; 107552103920@stu.xju.edu.cn (X.J.)
* Correspondence: wangdi1210@126.com; Tel.: +86-188-1926-2328

Abstract: In the field of agricultural machinery, various empirical field tests are performed to measure the tillage force for precision tillage. However, the field test performance is costly and time-consuming, and there are many constraints on weather and field soil conditions; the utilization of simulation studies is required to overcome these shortcomings. As a result, the SPH method and the coupled FEM-SPH method are used in this paper to investigate the mouldboard plough–soil interaction. In this paper, the finite element software LS-DYNA was used to build the SPH model and the FEM-SPH coupling model of soil cutting, as well as to investigate the change in cutting resistance during the soil cutting process. The simulation results are compared with those of the experiments, and the curves of the simulation and experiment are in good agreement, which verifies the reliability of the model. The validated simulation model was used to investigate the effects of the cutting speed, depth of cut, inclination angle, and lifting angle of the mouldboard plough on cutting resistance. The simulation studies show that the SPH model takes 5 h and 2 min to compute, while the FEM-SPH coupled model takes 38 min; obviously, the computational efficiency of the FEM-SPH coupled model is higher. The relative errors between the SPH model and the experiment are 2.17% and 3.65%, respectively. The relative errors between the FEM-SPH coupled model and the experiment are 5.96% and 10.67%, respectively. Obviously, the SPH model has a higher computational accuracy. The average cutting resistances predicted by the SPH model and the FEM-SPH coupled model, respectively, were 349.48 N and 306.25 N; these resistances are useful for precision tillage. The cutting resistance increases with the increase in cutting speed and is quadratic; the cutting resistance increases with the increase in cutting depth and is quadratic; the horizontal cutting resistance and the combined cutting resistance increase with the increase in inclination angle, while the vertical cutting resistance remains essentially constant with the increase in inclination angle; the horizontal cutting resistance and combined cutting resistance increase as the lifting angle increases, while the vertical cutting resistance decreases as the lifting angle increases.

Keywords: mouldboard plough; soil cutting; cutting resistance; SPH method; FEM-SPH coupling method

Citation: Jin, X.; Ma, F.; Wang, D.; Zhu, Z. Simulation of Mouldboard Plough Soil Cutting Based on Smooth Particle Hydrodynamics Method and FEM–SPH Coupling Method. *Agriculture* **2023**, *13*, 1847. https://doi.org/10.3390/agriculture13091847

Academic Editors: Chung-Liang Chang and Mustafa Ucgul

Received: 15 August 2023
Revised: 16 September 2023
Accepted: 18 September 2023
Published: 21 September 2023

1. Introduction

The mouldboard plough is still today's most widely used soil tillage tool [1,2]. Since Roman times, the mouldboard plough has been widely used as the dominant tillage tool for the following purposes: soil turning, burying litter and crop residues, laying the foundation for seedbeds, and making the soil loose and airy [3]. However, ploughing with a plough is a highly energy-consuming process [4,5], and the high energy consumption of agricultural machinery results in the emission of large amounts of exhaust gases, which cause pollution [6]. Precision agriculture is a technology-enabled farming management approach that observes, measures, and analyzes the needs of individual fields. Precision tillage involves measuring within-field soil strength variations and making applications to the field accordingly rather than applying an equal amount of cutting force during tillage operations in each field. This approach significantly reduces tillage energy consumption [7]. As a result, it is necessary to investigate the mouldboard soil tillage process and its mechanical properties to precisely till the soil and thus reduce energy consumption.

In recent years, many researchers have investigated the soil–tool interaction and the mechanical properties. Given that tillage force is primarily related to soil properties, tillage speed, tillage depth, and tool geometry [8], scholars have investigated the effects of soil properties [9,10], tillage speed [11,12], tillage depth [13,14], and tool geometry [15,16] on tillage force using field experiments and other methods. Some researchers have also developed tillage prediction models [17,18]. However, developing a fast and accurate tillage prediction model remains difficult because of differences in soil types, working conditions, and other factors.

Experimental and analytical methods are the two approaches to studying soil–tool interactions [19]. The experimental method is intuitive and reliable. However, the experimental method requires significant labor, material, financial resources, and time, and it is also easily affected by seasonality [20–22]. Although analytical methods can evaluate cutting forces and soil behavior such as soil displacement and velocity, the solutions of the analytical models are primarily based on the concept of equilibrium limits under quasi-static conditions and predetermined assumptions regarding the damaged surface, and the shape of the tillage tool must be simplified or of regular geometry [23,24]. The numerical simulation method has been developed as an efficient calculation method in recent years. Using numerical simulation methods to study the soil cutting process saves time and resources. It eliminates the interference of chance factors in the experimental research process and thus represents the modern design development direction [25–27].

The finite element method (FEM) [28–30], the discrete element method (DEM) [19,31–33], and computational fluid dynamics (CFD) [34–36] are the main numerical simulation methods used by scholars in the study of soil–tool interaction. In the classical finite element method (FEM), the nodes are fixed in the material, and the elements deform with the material. Large soil deformation, however, frequently results in severe mesh distortion, which significantly impacts the accuracy of the simulation results and can even cause the calculation to stop [37,38]. DEM is a good tool for simulating particle materials such as soil, but the stress–strain concept does not exist directly and cannot be calculated directly. DEM simulations require a rigorous and sophisticated method to calibrate the parameters representing soil mechanical properties. Furthermore, the large computational volume of the discrete element method (DEM) makes the computation very expensive [39,40]. Because the computational domain is fixed in space, the Eulerian description is used to prevent element deformation in computational fluid dynamics (CFD). Because the mesh is fixed in space, it is inapplicable when the material boundary is deformable. CFD simulation of the unconfined deformation of soil-free surfaces during cutting is therefore difficult [41].

The recently developed SPH method, also known as the smooth particle hydrodynamics method, is a meshless pure Lagrangian method [42]. Originally developed to solve astrophysical problems, the SPH method has since been extended to solve fluid flow, elastic–plastic flow, and brittle body damage [43,44]. In recent years, the SPH method has also been used for soil cutting. Man Hu et al. [45] investigated the cutting process and soil–tool interaction in non-cohesive and cohesive soils using an SPH model of soil–tool interaction based on an elastic–plastic structure, and the predicted cutting forces during tillage were in good agreement with the experimental results. As a result, using the SPH method to study soil–tool interactions is both practical and informative.

Few people have yet to apply the SPH method to simulate mouldboard plough–soil interactions. Thus, the following are the aims and objectives of the current work: (1) The mouldboard plough–soil interaction is simulated using the SPH method utilizing the finite element software LS-DYNA. (2) A coupled FEM-SPH model is developed to increase computational efficiency. (3) The simulation results are compared with those of the experiments [46] to verify the model's reliability. (4) The effects of cutting speed, depth, inclination angle, and lifting angle on cutting resistance are investigated using the validated simulation model.

2. Materials and Methods

2.1. Brief Introduction to SPH

The SPH method is a meshless pure Lagrangian method with interpolation theory at its core [47,48], and its essence is to discretize a continuum with physical quantities such as mass and velocity into interacting particles and then to solve for the discrete individual particles to obtain the overall information of the continuum [49,50]. Any macroscopic variable function of a continuous variable field is represented as an integral form by smooth approximate interpolation, which can be expressed as:

$$u(x) = \int_{\Omega} u(\overline{x})\delta(x - \overline{x})d\overline{x} \tag{1}$$

in which $u(x)$ is a continuous function, such as displacement, stress and so on, and $\delta(x - \overline{x})$ is a delta function, which can be expressed as:

$$\delta(x - \overline{x}) = \begin{cases} 1 & x = \overline{x} \\ 0 & x \neq \overline{x} \end{cases} \tag{2}$$

If a strongly peaked function $W(x - \overline{x}, h)$ is employed to replace the delta function $\delta(x - \overline{x})$, then Equation (1) can be rewritten as:

$$\langle u(x) \rangle = \int_{\Omega} u(\overline{x})W(x - \overline{x}, h)d\overline{x} \tag{3}$$

in which the strongly peaked function $W(x - \overline{x}, h)$ is called the smooth kernel function; when $h \to 0$, the strongly peaked function tends to $\delta(x - \overline{x})$. The most commonly used smooth kernel in SPH is the cubic B-sample, defined as:

$$\theta(\mu) = \begin{cases} (1 - 1.5\mu^2 + 0.75\mu^3)C & (|\mu| \leq 1) \\ 0.25C(2 - \mu)^3 & (1 < |m| < 2) \\ 0 & (2 \leq |\mu|) \end{cases} \tag{4}$$

2.2. FEM–SPH Coupling Method

The methods in LS-DYNA that can be used to treat sliding and impact along the interface include the kinematic constraint method, the penalty method, the distributed parameter method, and others. As shown in Figure 1a, the SPH particle and the finite element in the coupling model established in this paper are divided into two parts: attachment coupling and contact coupling, with attachment coupling achieved using the kinematic constraint algorithm and contact coupling achieved using the penalty algorithm. All of the preceding algorithms are based on a master–slave scheme, in which the interfaces are defined as master and slave surfaces, and the nodes that lie within these surfaces are defined as master and slave nodes, respectively [51]. In general, it finds the nearest point in the master surface to each slave node to determine the penetration, unless the point is at the intersection of two master segments. As shown in Figure 1b, it is assumed that a master segment has been found for the slave node n_s and that n_s has not been determined as being located at the intersection of two master segments. The contact point is then defined as the nearest point on the master segment to the slave node n_s.

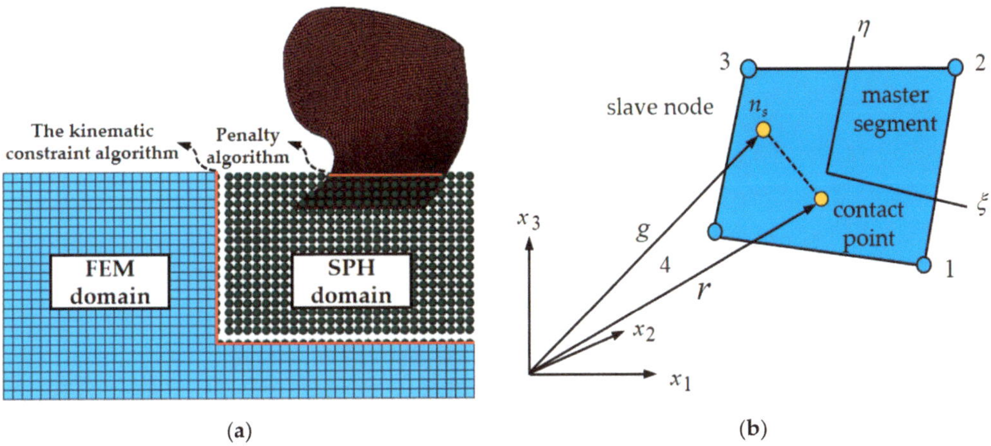

Figure 1. FEM–SPH coupling: (**a**) the attachment and contact algorithm between FEM and SPH; (**b**) location of the contact point when n_s lies above master segment.

A kinematic constraint algorithm is used to achieve the attachment between the SPH particles and the finite element, which takes place at their interface [51]. During the computation, the slave nodes (SPH particles) are constrained to move on the master surface (finite element) after the impact, and the orthogonal projection keeps the relative position of the interpolated contact points on the master segment constant at each step of the computation. The tied interfaces are independently checked and updated at each time step. First, the nodal forces and masses of each slave node are distributed to the master node, and the mass increment and force increment of the master node define the segment containing the contact points, which can be calculated using Equations (5) and (6):

$$\Delta M_m^i = \phi_i(\xi_c, \eta_c) M_s \tag{5}$$

$$\Delta f_m^i = \phi_i(\xi_c, \eta_c) f_s \tag{6}$$

where $\phi_i(\xi_c, \eta_c)$ is the interpolation function, and i is the number of master nodes; the force and mass increments calculated by Equations (5) and (6) will then be added to the force and mass vectors of the master surface. After completing the summation for all slave nodes, the acceleration of the jth node of the master surface a_i^j can be calculated. Using a_i^j, Equation (7) can be used to calculate the acceleration a_{is} of each slave node, updating the slave nodes' velocity and displacement.

$$a_{is} = \sum_{j=1}^{4} \phi_j(\xi_c, \eta_c) a_i^j \tag{7}$$

The penalty method is used to achieve contact between the finite element plough and the SPH area of the soil [51]. During the calculation, the contact will only occur if the penetration is positive; so, each slave node n_s will be checked for penetration l through the master surface.

$$l = n \times [g - r(\xi_c, \eta_c)] < 0 \tag{8}$$

where n is the vector normal to the master segment at contact point (ξ_c, η_c). g and r are the position vectors drawn to the slave and master nodes, respectively.

If penetration l occurs $(l < 0)$, an interface force vector f_s, which is proportional to the magnitude of the penetration, is applied between the slave node and its corresponding contact point.

$$f_s = -l k_i n_i \tag{9}$$

Then, the interface force applied to the four nodes ($i = 1, 2, 3, 4$) of the master segment is expressed as:

$$f_m^i = \phi_i(\xi_c, \eta_c) f_s \tag{10}$$

where $\phi_i(\xi_c, \eta_c)$ is the interpolation function.

$$k_i = \frac{f_{si} K_i A_i^2}{V_i} \tag{11}$$

where k_i is the stiffness factor for master segment, f_{si} is the scale factor, K_i is the bulk modulus, A_i is the surface area of the element, and V_i is the volume.

2.3. Soil Constitutive Law

Over 300 material models are available in the LS-DYNA material library, but only a few of them have proven to be effective in describing flow and sliding properties. Remarkably, the elastic–plastic material MAT_010 (MAT_ELASTIC_PLASTIC_HYDRO) is suitable for simulating a wide range of materials, including fluid materials with large deformations [52]. As a result, in this paper, the elastic–plastic material MAT_010 is used to simulate soil. The type of soil is clay loam, and the parameters for the plough soil cutting simulation are shown in Table 1.

Table 1. Simulation parameters of soil–plough coupling system [46,53].

Parameters	Value	Parameters	Value
Density of soil/(kg/m^3)	2600	Static friction coefficient between soil and plough	0.6
Shear modulus of soil/Pa	1.2×10^6	Dynamic friction coefficient between soil and plough	0.1
Poisson's ratio of soil	0.38	Static friction coefficient between two soil particles	0.422
Density of plough/(kg/m^3)	7970	Dynamic friction coefficient between two soil particles	0.282
Young's modulus of plough/Pa	2.10×10^{11}	Radius of particle/(mm)	5
Poisson's ratio of plough	0.32	Gravity/(m/s^2)	9.806

If the effective stress (ES) and effective plastic strain (EPS) are not defined in MAT_010, the yield stress and plastic hardening modulus are taken from SIG0 and EH. In this case, the bilinear stress–strain curve shown in Figure 2 is obtained using hardening parameters. The yield strength is calculated as:

$$\sigma_y = \sigma_0 + E_h \bar{\varepsilon}^p + (a_1 + p a_2)\max[p, 0] \tag{12}$$

where p is the positive compression pressure, and E_h is the plastic hardening modulus defined by Young's modulus E and the tangent modulus E_t, which are defined as follows:

$$E_h = \frac{E_t E}{E - E_t} \tag{13}$$

If the effective stress (ES) and effective plastic strain (EPS) are specified, a curve similar to the one shown in Figure 2 can be defined. The effective stress is defined by the deviatoric stress tensor s_{ij} as:

$$\bar{\sigma} = \left(\frac{3}{2} s_{ij} s_{ij}\right)^{1/2} \tag{14}$$

and effective plastic strain by:

$$\bar{\varepsilon}^p = \int_0^t \left(\frac{2}{3} D_{ij}^p D_{ij}^p\right)^{1/2} dt \tag{15}$$

where t is the time, and D^p_{ij} is the plastic component of the deformation tensor rate. In this case, ignoring the plastic hardening modulus, the yield stress is given as:

$$\sigma_y = f(\bar{\varepsilon}^p) \tag{16}$$

where the value of $f(\bar{\varepsilon}^p)$ is obtained by interpolation of the data curve.

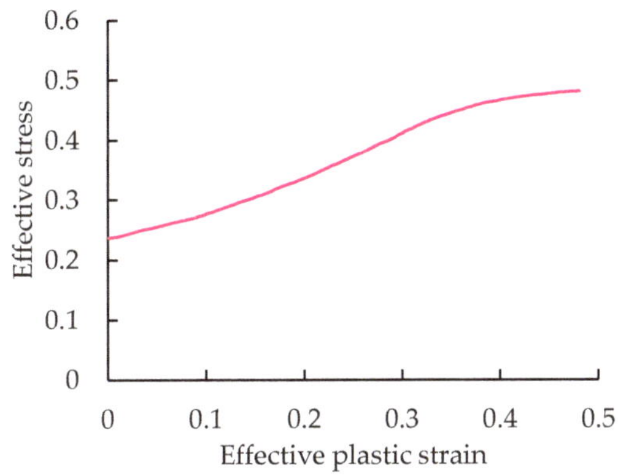

Figure 2. Effective stress versus effective plastic strain curve.

3. Numerical Simulation

3.1. Establishment of the Finite Element Model of the Plough

In this paper, the plough chosen for the tractor CF700 tillage system in the experiment of Jinming Zhang et al. [46] is used as the research object. The numerical simulation method is used to study the entire process of soil cutting by the plough, and the numerical simulation results are compared with those of the experiment for analysis. This paper only studies the working condition of a single plough in the soil to simplify the model and improve calculation efficiency.

The plough is modeled utilizing 3D modeling software using the horizontal straight element line method, and its geometric model is shown in Figure 3a. Considering the irregularity of the plough's surface, the geometric model of the plough is meshed, and the finite element mesh is free-meshed with a mesh size of 10 mm. According to the literature [46,53], the plough chosen for this study is made of 65 Mn steel with a material density of 7970 kg/m^3, a modulus of elasticity of 2.10×10^{11} Pa, and a Poisson's ratio of 0.32. The model was imported into LS-DYNA; the element type was set to Solid; the material model was set to MAT_RIGID; and the constraints were set to move in the y and z directions and rotate in the x, y, and z directions. Figure 3b depicts the plough's finite element model.

3.2. Establishment of SPH Model, FEM–SPH Coupling Model

3.2.1. SPH Model

There are two methods in LS-DYNA for generating SPH soil particles: one is to convert the established finite element mesh into SPH particles; this method is mostly used for modeling objects with irregular and complex structures. The other is to generate SPH particles directly; this is the best way to model objects with simple structures. The direct SPH particle generation method is used in this paper for soil modeling. First, the established finite element mesh model of the plough was imported into the LS-PrePost. The rectangular soil model was generated after setting the particle generation approach, the number of particles in each direction, the filling rate, the particle density, and other parameters in the Mesh_Sph Generation, with a soil size of $3 \times 2 \times 0.6$ m and particle number of 132,000. Second, parameters such as particle type and radius were set in Settings_General Settings.

Finally, Element Tools_Transform was used to adjust the cutting position of the plough based on the plough's cutting mode and the boundary condition processing requirements. Figure 4 depicts the SPH model of the plough soil cutting.

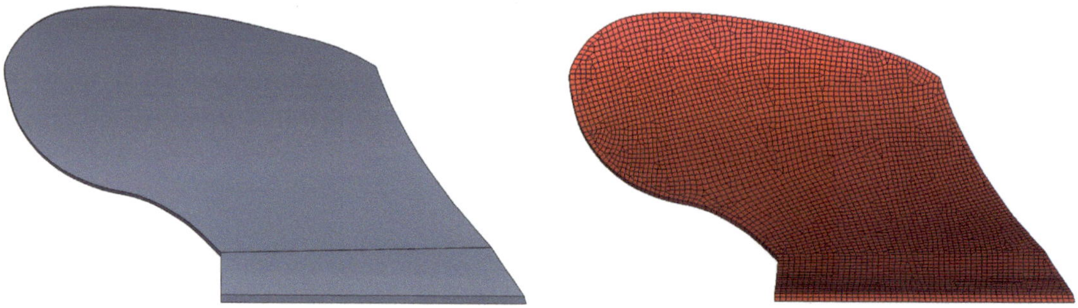

(**a**) (**b**)

Figure 3. (**a**) Geometric model of the plough; (**b**) finite element mesh of plough.

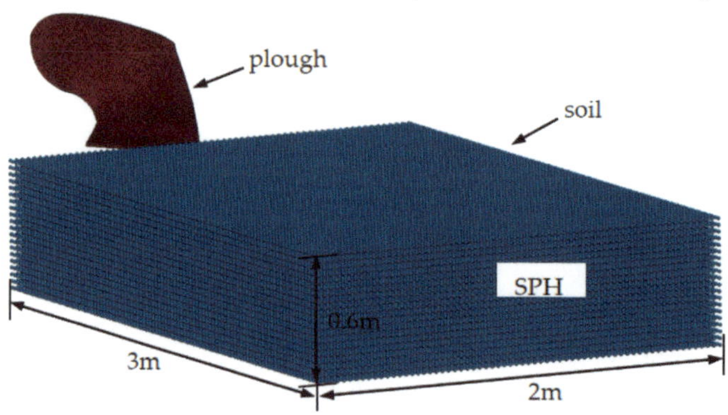

Figure 4. SPH model of plough soil cutting.

3.2.2. FEM-SPH Coupling Model

The soil in the coupled model comprises two parts: the finite element mesh and the SPH particles, with the SPH particles obtained through finite element mesh transformation. First, the established finite element mesh model of the plough is imported into the LS-PrePost, and Mesh_ShapeM generates a finite element mesh with a soil size of $3 \times 2 \times 0.6$ m and an element size of 0.024 m. Second, by using Mesh_Sph Generation, the finite element mesh of the middle part of the soil is transformed into SPH particles with a soil size of $3 \times 0.96 \times 0.46$ m. Finally, in Settings_General Settings, parameters such as particle type and particle radius are set, and the cutting position of the plough is adjusted using Element Tools_Transform. Figure 5 depicts the FEM–SPH plough soil cutting model.

3.2.3. Boundary Conditions Imposed

The plough's depth into the soil during the soil cutting process is 0.1 m, and it moves forward with the tractor. To reduce invalid cutting time, the plough should be as close to the soil edge as possible. In the numerical simulation, the following boundary conditions are imposed: ① Select the numerical algorithm as the SPH algorithm. ② The plough's forward speed is 1 m/s, and the cutting depth is 0.1 m. ③ The soil material was chosen as MAT_010 (MAT_ELASTIC_PLASTIC_HYDRO). ④ Using SPC, set the soil bottom, with the left and right sides as fixed constraints, and apply gravity to the plough and the soil. ⑤ Set the contact between the plough and the soil as AUTOMATIC_NODES_TO_SURFACE. Set the contact between FEM and SPH soil in the coupled model as a TIED contact (TIED_NODES_TO_SURFACE). ⑥ Set the solution time to 4 s. Complete the other keyword settings in the LS-PrePost and generate the K file to be submitted to the LS-DYNA solver for solution.

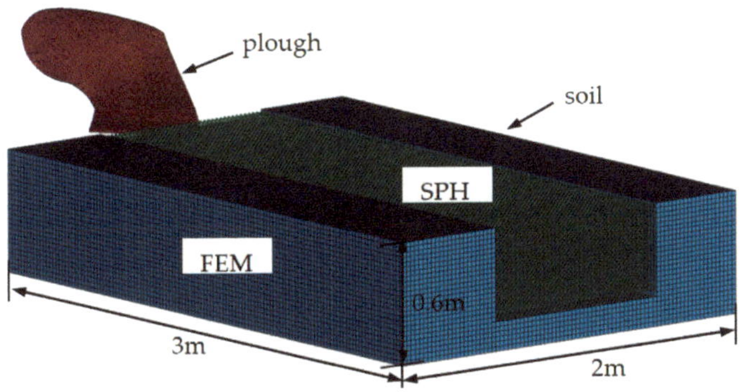

Figure 5. FEM-SPH model of plough soil cutting.

3.3. Parameter Study

To investigate the effect of cutting speed and depth on cutting resistance, the cutting resistance corresponding to the different cutting speeds (1 m/s, 2 m/s, 3 m/s, 4 m/s, and 5 m/s) under the different cutting depth conditions (0.1 m, 0.15 m, 0.2 m, 0.25 m, and 0.3 m) of the plough was simulated using the simulation model.

The inclination angle α and lifting angle β of the plough are depicted in Figure 6 and are as given in the study in [54]. The cutting resistance at different cutting depths was simulated using a simulation model to investigate the effect of plough inclination and lifting angles on cutting resistance. For the inclination angle, the cutting resistance was simulated for different depths (0.15 m, 0.2 m, and 0.25 m) as the inclination angle was increased from 30° to 75°. For the lifting angle, the cutting resistance was simulated for different depths (0.15 m, 0.2 m, and 0.25 m) as the lifting angle was increased from 15° to 60°. In the simulation, the plough's cutting speed was 1 m/s.

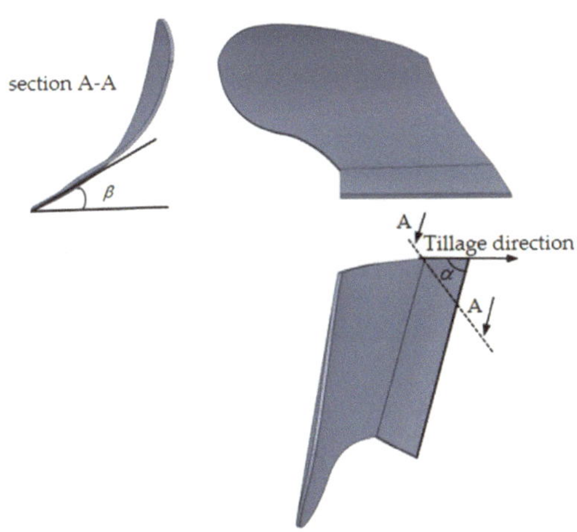

Figure 6. Inclination angle and lifting angle of the plough.

4. Results and Discussion

4.1. Cutting Process and Cutting Resistance Analysis

Figure 7 depicts the soil cutting process and the cutting resistance using the SPH model and the coupled FEM-SPH model, which clearly shows the trajectory and displacement changes of the soil during the cutting process. When the plough moves into contact with the soil, the soil is destroyed due to shearing and squeezing (Figure 7a,d); as the plough cuts entirely into the soil, the contact area between the plough and the soil increases, and the soil is gradually destroyed to a greater extent (Figure 7b,e). When the plough moves forward to the edge of the soil and then leaves the soil, the soil remains in motion due to inertia, causing the plough to cut the soil continuously (Figure 7c,f). The soil cutting processes in both simulation models are very stable, and both can reflect the soil being turned up and falling. The soil cutting process is consistent with the actual situation, indirectly confirming the simulation model's reliability.

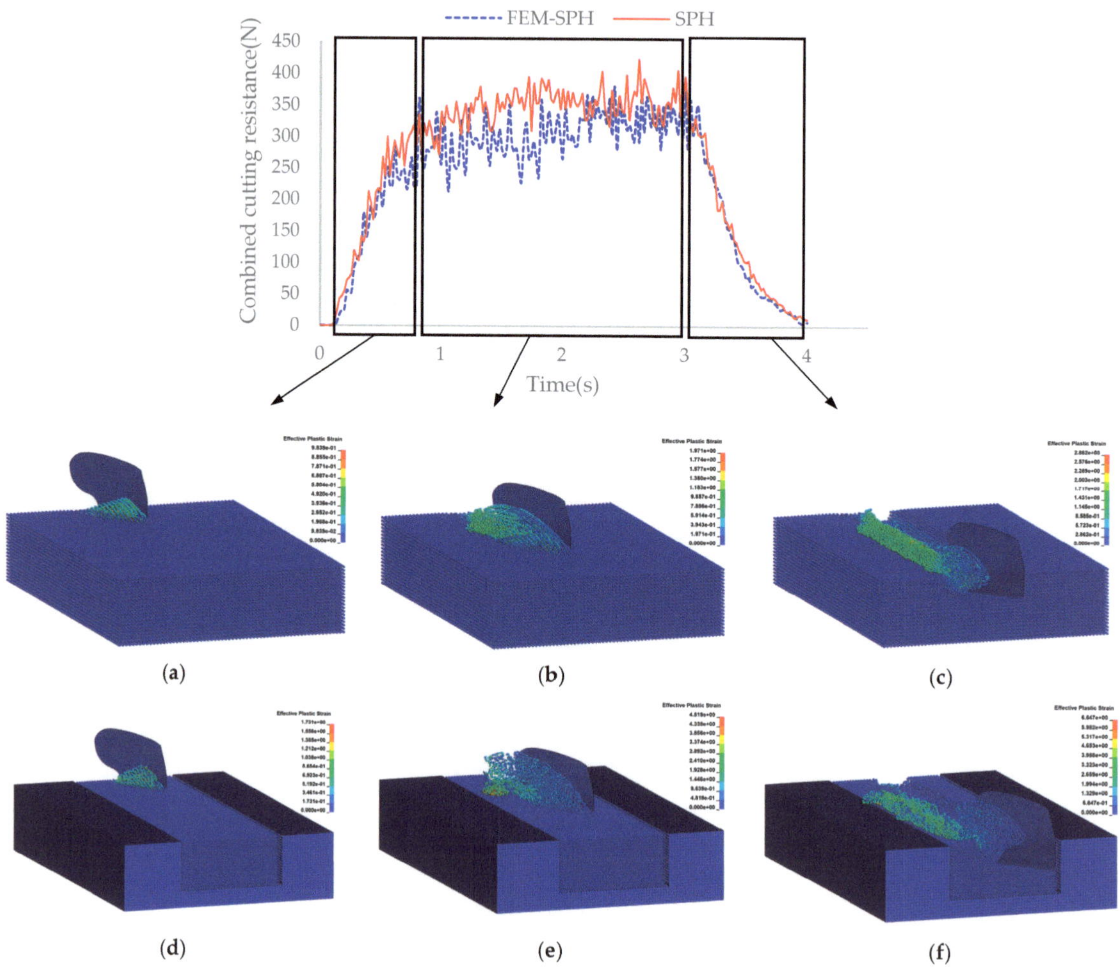

Figure 7. Cutting process and cutting resistance of soil based on SPH model and coupled FEM-SPH model: SPH model: (**a**) t = 0.74 s, (**b**) t = 2.02 s, (**c**) t = 3.78 s; FEM-SPH model: (**d**) t = 0.74 s, (**e**) t = 2.02 s, (**f**) t = 3.78 s.

The cutting resistance of a plough is the reaction force of the soil on the plough during the cutting process. Loading the rcforc file into the LS-DYNA displays the plough's cutting resistance. Figure 7 shows the combined cutting resistance curves obtained from the simulations of the SPH model and the coupled FEM-SPH model. As shown in the figure, there is no initial contact between the plough and the soil, and the cutting resistance value is 0. The cutting resistance increases gradually as the plough contacts and cuts into the soil.

At 0.8 s, the plough completely cuts into the soil, and the cutting resistance reaches and fluctuates near its maximum value. The plough starts cutting out of the soil at 3 s, and the cutting resistance gradually decreases. At 4 s, the plough completely removes the soil, and the cutting resistance value is 0.

The plough cut completely into the soil at 0.8 s and began to cut out of the soil at 3 s. The cutting resistance data from the two numerical simulation models were extracted from 0.8 s to 3 s. After analysis and calculation, the average cutting resistances of the SPH and coupled FEM-SPH models were 349.48 N and 306.25 N, respectively.

4.2. Stress Variation of Soil

Figure 8 depicts the change in soil effective stress during the plough soil cutting process using the coupled FEM-SPH model. The effective stress on the soil increases as the contact area between the plough and the soil increases during the soil cutting process. When the soil is completely destroyed, the effective stress reaches a stable value. During the soil cutting process, the maximum effective stress on the soil is 4000 Pa. As can be seen from the entire cutting process, the soil is cut relatively steadily, and the effective stress on the soil varies less.

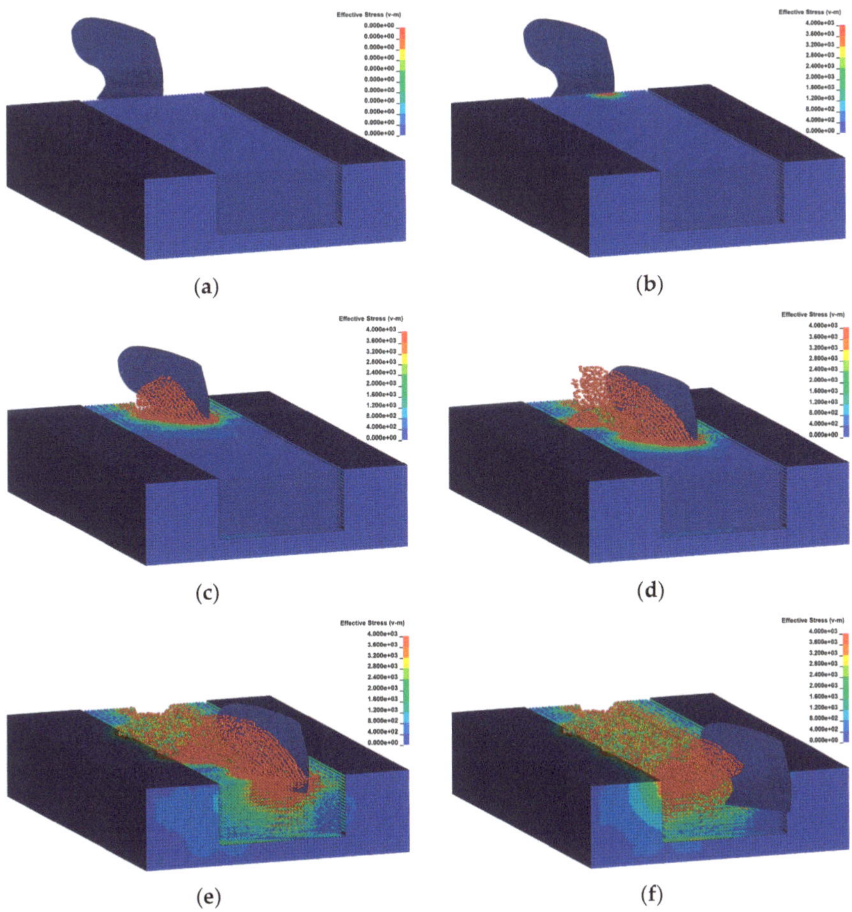

Figure 8. Stress variation in soil cutting process (FEM-SPH model): (**a**) t = 0 s; (**b**) t = 0.16 s; (**c**) t = 0.98 s; (**d**) t = 1.96 s; (**e**) t = 3.12 s; (**f**) t = 4 s.

4.3. Model Validation

To validate the reliability of the simulation model, the simulation results of the SPH model and the coupled FEM-SPH model were compared with those of the experiment and the DEM simulation [46]. Figure 9 depicts the simulation and experimental cutting resistance curves at 1 m/s cutting speed and 0.1 m depth of cut (Figure 9a shows the horizontal cutting resistance of the plough, and Figure 9b shows the vertical cutting resistance of the plough.). Figure 9 shows that the cutting resistance of the two numerical simulation models is basically consistent with that of the experiment, proving the model's reliability.

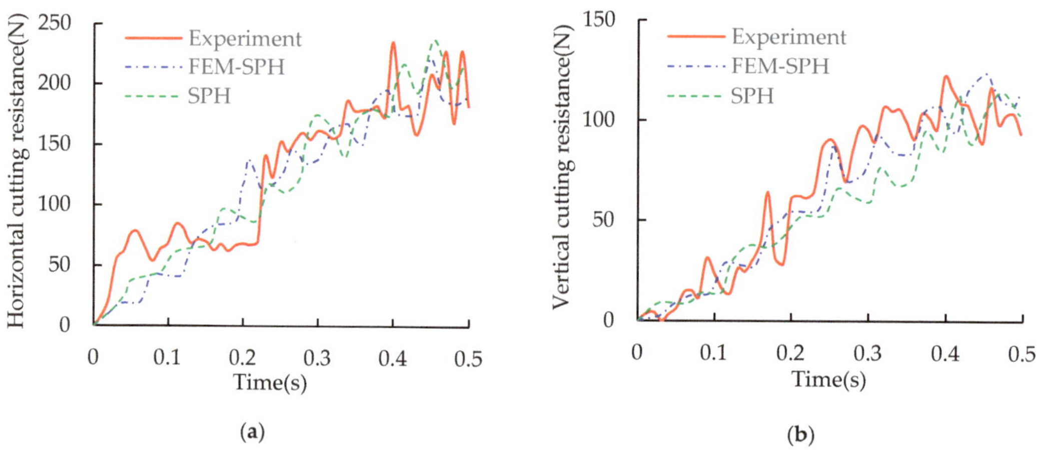

Figure 9. Simulated and experimental cutting resistance: (**a**) horizontal; (**b**) vertical.

When the plough makes contact with the soil, the cutting resistance in the horizontal and vertical directions increases; it stabilizes after a while and fluctuates up and down near the stable value. The following can be derived from both the simulation model and the literature: The SPH model's average cutting resistances in the horizontal and vertical directions are 124.97 N and 62.50 N, respectively. The coupled FEM-SPH model's average cutting resistances in the horizontal and vertical directions are 130.07 N and 72.62 N, respectively. The experiment's average cutting resistances in the horizontal and vertical directions are 122.32 N and 64.87 N, respectively. The DEM model's average cutting resistances in the horizontal and vertical directions are 121.05 N and 62.64 N, respectively. They are calculated as follows: The relative errors between the SPH model and the experiment are 2.17% and 3.65%, respectively. The relative errors between the coupled FEM–SPH model and the experiment are 5.96% and 10.67%, respectively. The relative errors between the DEM model and the experiment are 1.04% and 3.43%, respectively.

Regarding computational efficiency and accuracy, the SPH model and the coupled FEM-SPH model have advantages and disadvantages. Regarding computational efficiency, the SPH model takes 5 h and 2 min to compute using the same computing device, while the FEM-SPH coupled model takes 38 min. The FEM-SPH coupled model is more computationally efficient than the SPH model. In terms of computational accuracy, it is obvious from the above error analysis that the SPH model has a higher computational accuracy than the coupled FEM-SPH model. The DEM model has the lowest error and the highest accuracy. However, the large computational volume of DEM may render it inefficient in terms of computational efficiency.

Figure 10 depicts the linear regression plots of cutting resistance in the horizontal and vertical directions for both the experiment and the simulation. As shown in the figure, both simulation models predicted a cutting resistance with good regression results. Numerically, the SPH model has the best regression results for cutting resistance in the horizontal direction, followed by the FEM-SPH model in the vertical direction, then the SPH model in the vertical direction and, finally, the FEM-SPH model in the horizontal direction.

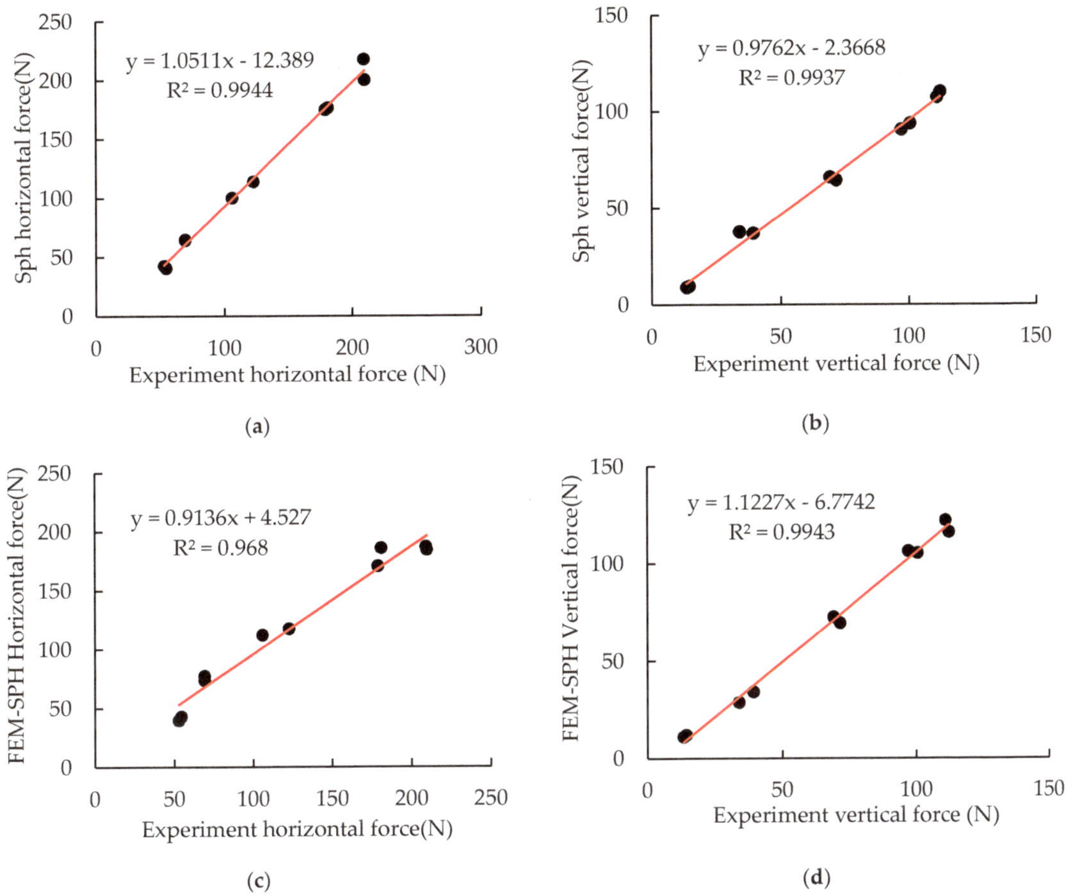

Figure 10. The linear regression plots for the estimated against measured: (**a**,**c**) horizontal force; (**b**,**d**) vertical force.

4.4. Effect of Cutting Speed and Depth on Cutting Resistance

Figure 11 depicts the variation in cutting resistance as the plough's cutting speed increases from 1 m/s to 5 m/s under various cutting depth (0.1 m, 0.15 m, 0.2 m, 0.25 m, and 0.3 m) operating conditions (Figure 11a depicts the horizontal cutting resistance; Figure 11b depicts the vertical cutting resistance; and Figure 11c depicts the combined cutting resistance). Fitting the data into the graph results in the following: The cutting resistance increases with the increase in cutting speed and has a quadratic relationship under different cutting depth conditions. The cutting resistance increases with the depth of the cut and has a quadratic relationship under different cutting speeds. According to the study in [55], the relationship between cutting resistance and cutting speed and depth can be expressed as:

$$f = av^2 + b \tag{17}$$

$$f = pd^2 + qd \tag{18}$$

where *a* and *b* are empirical functions of the depth *d*, and *p* and *q* are empirical functions of the velocity *v*, which depends on specific soil properties and geometric parameters of the plough. Obviously, cutting resistance is a quadratic function of cutting speed and depth. Therefore, the results predicted by the simulations in this study are consistent with those of the above studies.

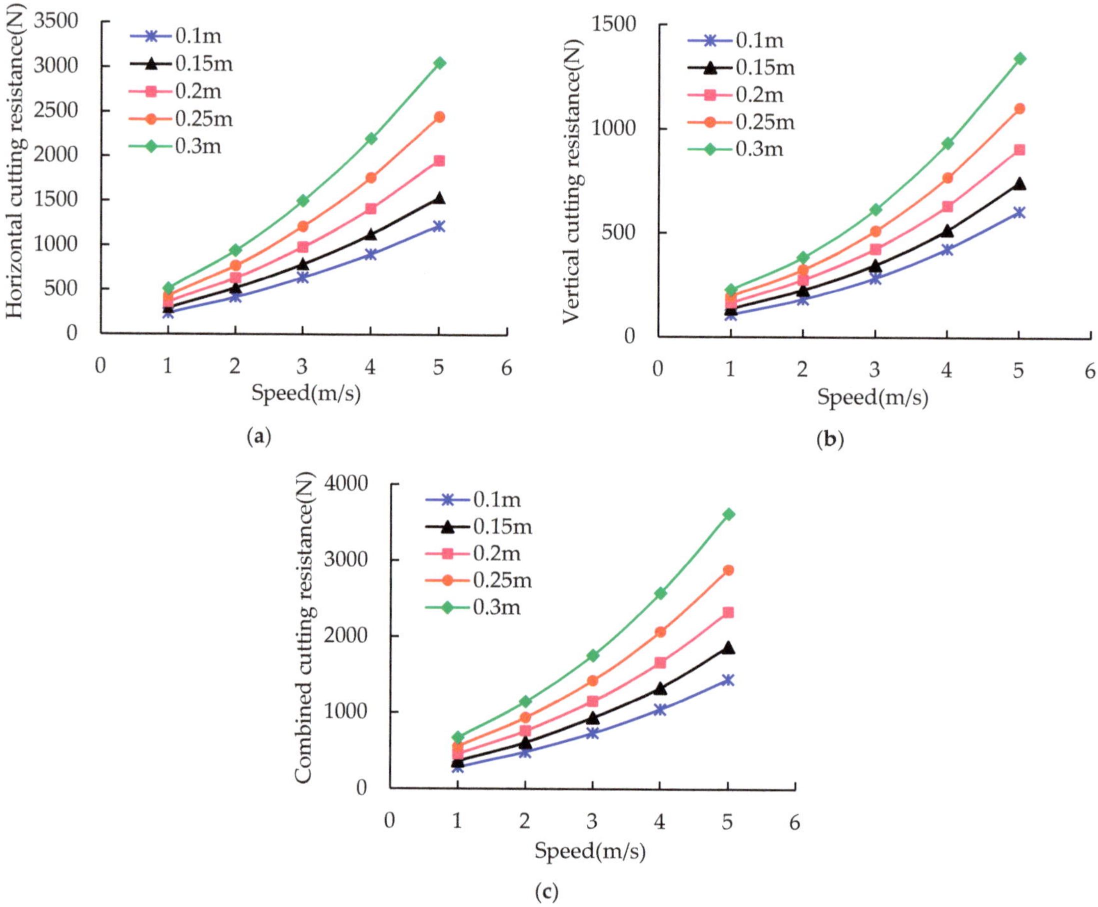

Figure 11. Relationship between cutting resistance and cutting speed: (**a**) horizontal; (**b**) vertical; (**c**) combined cutting resistance.

4.5. Effect of Inclination Angle α and Lifting Angle β of the Plough on Cutting Resistance

Figure 12 depicts the variation in cutting resistance as the inclination angle increases from 30° to 75° at a cutting speed of 1 m/s and different depths of cut (0.15 m, 0.2 m, and 0.25 m) (Figure 12a depicts the horizontal cutting resistance; Figure 12b depicts the vertical cutting resistance; and Figure 12c depicts the combined cutting resistance). It can be seen that under different cutting depth operating conditions, the horizontal cutting resistance and combined cutting resistance increase with the increase in inclination angle α, whereas the vertical cutting resistance remains essentially constant with the increase in inclination angle α.

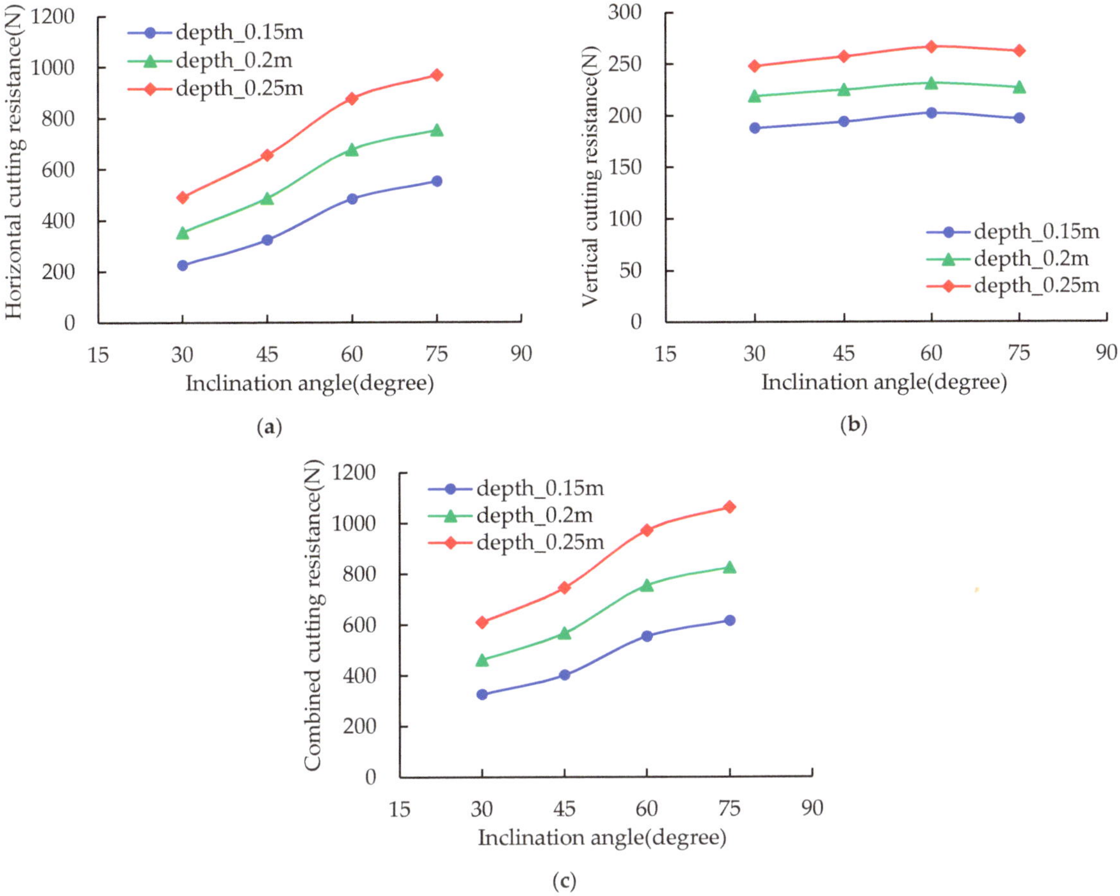

Figure 12. Relationship between cutting resistance and inclination angle: (**a**) horizontal; (**b**) vertical; (**c**) combined cutting resistance.

Figure 13 depicts the variation in cutting resistance as the lifting angle increases from 15° to 60° at a cutting speed of 1 m/s and different depths of cut (0.15 m, 0.2 m, and 0.25 m) (Figure 13a depicts the horizontal cutting resistance; Figure 13b depicts the vertical cutting resistance; and Figure 13c depicts the combined cutting resistance). It can be seen that, under different cutting depth operating conditions, horizontal cutting resistance and combined cutting resistance increase with the increase in lifting angle β, whereas vertical cutting resistance decreases with the increase in lifting angle β.

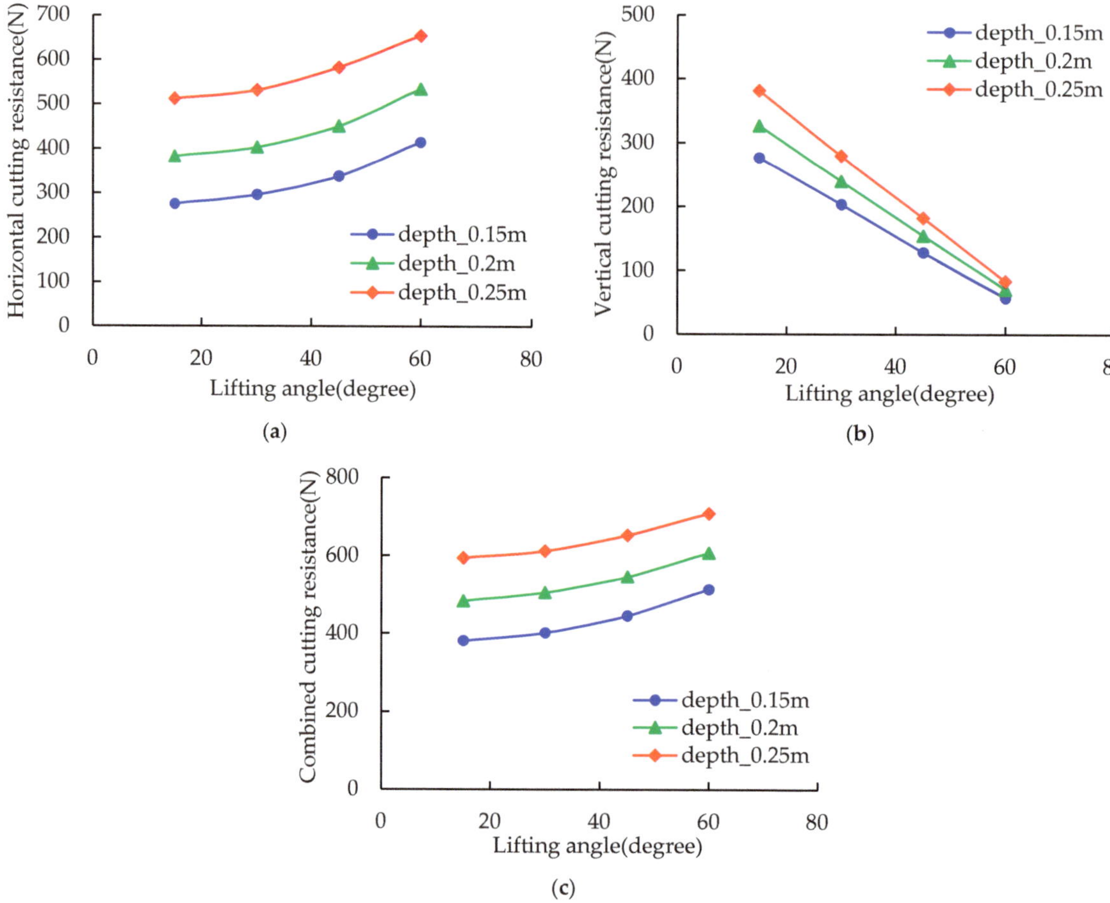

Figure 13. Relationship between cutting resistance and lifting angle: (**a**) horizontal; (**b**) vertical; (**c**) combined cutting resistance.

5. Conclusions

The mouldboard plough-soil interaction was simulated using the SPH method and the FEM-SPH coupling method. The cutting resistance obtained from the two simulation models was compared with that of the experiment in the literature, and the curves of the simulation and experiment were in good agreement, verifying the reliability of the SPH model and the FEM-SPH coupled model. As a result, the two simulation models can be used as a quick and accurate tool to predict cutting resistance during soil cutting.

The SPH model and the FEM-SPH coupled model can both accurately simulate the soil cutting process, but each has advantages and disadvantages in terms of computational efficiency and accuracy. The SPH model takes 5 h and 2 min to compute, while the FEM-SPH coupled model takes 38 min; obviously, the computational efficiency of the FEM-SPH coupled model is higher. The relative errors between the SPH model and the experiment are 2.17% and 3.65%, respectively. The relative errors between the FEM-SPH coupled model and the experiment are 5.96% and 10.67%, respectively. Obviously, the SPH model has a higher computational accuracy.

The maximum effective stress on the soil during the entire cutting process was 4000 Pa, and the maximum effective stress was primarily concentrated on the soil in contact with the mouldboard plough. The average cutting resistances predicted by the SPH model and the FEM-SPH coupled model, respectively, were 349.48 N and 306.25 N; these resistances are useful for precision tillage.

The validated simulation model was used to investigate the effects of cutting speed, depth of cut, inclination angle, and lifting angle of the mouldboard plough on cutting resistance. Simulation studies have shown that the cutting resistance increases with the increase in cutting speed and is quadratic; the cutting resistance increases with the increase in cutting depth and is quadratic; the horizontal cutting resistance and the combined cutting resistance increase with the increase in inclination angle, while the vertical cutting resistance remains essentially constant with the increase in inclination angle; the horizontal cutting resistance and combined cutting resistance increase as the lifting angle increases, while the vertical cutting resistance decreases as the lifting angle increases.

Author Contributions: Conceptualization, X.J. and D.W.; methodology, X.J.; software, X.J., F.M. and Z.Z.; validation, X.J., F.M. and D.W.; formal analysis, X.J.; investigation, X.J.; resources, X.J.; data curation, X.J.; writing—original draft preparation, X.J.; writing—review and editing, D.W.; visualization, X.J., F.M. and Z.Z.; supervision, D.W.; project administration, D.W.; funding acquisition, D.W. All authors have read and agreed to the published version of the manuscript.

Funding: This research was funded by the Natural Science Foundation of the Xinjiang Uygur Autonomous Region (No. 2021D01C115).

Institutional Review Board Statement: Not applicable.

Data Availability Statement: Not applicable.

Conflicts of Interest: The authors declare no conflict of interest.

References

1. Naderi-Boldaji, M.; Karparvarfard, S.H.; Azimi-Nejadian, H. Investigation of the predictability of mouldboard plough draught from soil mechanical strength (cone index vs. shear strength) using finite element modelling. *J. Terramech.* **2023**, *108*, 21–31. [CrossRef]
2. Renton, M.; Flower, K.C. Occasional mouldboard ploughing slows evolution of resistance and reduces long-term weed populations in no-till systems. *Agric. Syst.* **2015**, *139*, 66–75. [CrossRef]
3. Saunders, C. *Optimising the Performance of Shallow High-Speed Mouldboard Ploughs*; Cranfield University: Cranfield, UK, 2002.
4. Bulgakov, V.; Pascuzzi, S.; Adamchuk, V.; Ivanovs, S.; Pylypaka, S. A theoretical study of the limit path of the movement of a layer of soil along the plough mouldboard. *Soil Tillage Res.* **2019**, *195*, 11. [CrossRef]
5. Moitzi, G.; Haas, M.; Wagentristl, H.; Boxberger, J.; Gronauer, A. Energy consumption in cultivating and ploughing with traction improvement system and consideration of the rear furrow wheel-load in ploughing. *Soil Tillage Res.* **2013**, *134*, 56–60. [CrossRef]
6. Nabavi-Pelesaraei, A.; Rafiee, S.; Mohtasebi, S.S.; Hosseinzadeh-Bandbafha, H.; Chau, K.W. Energy consumption enhancement and environmental life cycle assessment in paddy production using optimization techniques. *J. Clean. Prod.* **2017**, *162*, 571–586. [CrossRef]
7. Lieve, V.W. *Precision-Agriculture and the Future of Farming in Europe*; European Maritime Safety Agency: Lisbon, Portugal, 2016.
8. Naderloo, L.; Alimadani, R.; Akram, A.; Javadikia, P.; Khanghah, H.Z. Tillage depth and forward speed effects on draft of three primary tillage implements in clay loam soil. *J. Food Agric. Environ.* **2009**, *7*, 382–385.
9. Abu-Hamdeh, N.H. The Effect of Tillage Treatments on Soil Water Holding Capacity and on Soil Physical Properties. In Proceedings of the ISCO 13th International Soil Conservation Organization Conference, Brisbane, Australia, 4–8 July 2004.
10. Guan, C.; Fu, J.; Xu, L.; Jiang, X.; Wang, S.; Cui, Z. Study on the reduction of soil adhesion and tillage force of bionic cutter teeth in secondary soil crushing. *Biosyst. Eng.* **2022**, *213*, 133–147. [CrossRef]
11. Kichler, C.M.; Fulton, J.P.; Raper, R.L.; McDonald, T.P.; Zech, W.C. Effects of transmission gear selection on tractor performance and fuel costs during deep tillage operations. *Soil Tillage Res.* **2011**, *113*, 105–111. [CrossRef]
12. Salahloo, M.; Alasti, B.M.; Mardani, A.; Abbasgholipour, M. Effect of Forward Speed, Working Depth and Overlay Parameters of Cultivator Tillage on Power Consumption and Draft Force. *Iran. J. Biosyst. Eng.* **2021**, *51*, 749–756.
13. Rashidi, M.; Najjarzadeh, I.; Jaberinasab, B.; Emadi, S.M.; Fayyazi, M. Effect of Soil Moisture Content, Tillage Depth and Operation Speed on Draft Force of Moldboard Plow. *Middle East J. Sci. Res.* **2013**, *16*, 245–249.
14. Kim, Y.S.; Kim, T.J.; Kim, Y.J.; Lee, S.D.; Park, S.U.; Kim, W.S. Development of a Real-Time Tillage Depth Measurement System for Agricultural Tractors: Application to the Effect Analysis of Tillage Depth on Draft Force during Plow Tillage. *Sensors* **2020**, *20*, 912. [CrossRef] [PubMed]
15. Arvidsson, J.; Keller, T.; Gustafsson, K. Specific draught for mouldboard plough, chisel plough and disc harrow at different water contents. *Soil Tillage Res.* **2004**, *79*, 221–231. [CrossRef]

16. Hoseinian, S.H.; Hemmat, A.; Esehaghbeygi, A.; Shahgoli, G.; Baghbanan, A. Development of a dual sideway-share subsurface tillage implement: Part 2. Effect of tool geometry on tillage forces and soil disturbance characteristics. *Soil Tillage Res.* **2022**, *215*, 105200. [CrossRef]
17. Ibrahmi, A.; Bentaher, H.; Maalej, A. Soil-blade orientation effect on tillage forces determined by 3D finite element models. *Span. J. Agric. Res.* **2014**, *12*, 941–951. [CrossRef]
18. Sadek, A.A.; Chen, Y.; Zeng, Z. Draft force prediction for a high-speed disc implement using discrete element modelling. *Biosyst. Eng.* **2021**, *202*, 133–141. [CrossRef]
19. Wang, X.; Li, P.; He, J.; Wei, W.; Huang, Y. Discrete element simulations and experiments of soil-winged subsoiler interaction. *Int. J. Agric. Biol. Eng.* **2021**, *14*, 50–62. [CrossRef]
20. Kesner, A.; Choteborsky, R.; Linda, M.; Hromasova, M.; Katinas, E.; Sutanto, H. Stress distribution on a soil tillage machine frame segment with a chisel shank simulated using discrete element and finite element methods and validate by experiment. *Biosyst. Eng.* **2021**, *209*, 125–138. [CrossRef]
21. Kim, Y.S.; Siddique, M.A.; Kim, W.S.; Kim, Y.J.; Lee, S.D.; Lee, D.K.; Hwang, S.J.; Nam, J.S.; Park, S.U.; Lim, R.G. DEM simulation for draft force predition of moldboard plow according to the tillage depth in cohesive soil. *Comput. Electron. Agric.* **2021**, *189*, 18. [CrossRef]
22. Makange, N.R.; Ji, C.Y.; Torotwa, I. Prediction of cutting forces and soil behavior with discrete element simulation. *Comput. Electron. Agric.* **2020**, *179*, 11. [CrossRef]
23. Bal, A.R.L.; Dang, T.S.; Meschke, G. A 3D particle finite element model for the simulation of soft soil excavation using hypoplasticity. *Comput. Part. Mech.* **2020**, *7*, 151–172. [CrossRef]
24. Bazarov, D.; Toshtemirov, S.; Mustafayev, S.; Xo'jayev, A.; Mamatkulov, I.; Boboev, F. Technology and machine parameters for preparing the soil for sowing cotton. *E3S Web Conf.* **2021**, *264*, 04046. [CrossRef]
25. Chen, P.; Tao, W.; Zhu, L.; Wu, Q.M.; Zhang, J.W.; Dong, S.J.; Moray, P.J. Effect of varying remote cylinder speeds on plough-breast performances in alternative shifting tillage. *Comput. Electron. Agric.* **2021**, *181*, 11. [CrossRef]
26. Gelybo, G.; Barcza, Z.; Dencso, M.; Potyo, I.; Kasa, I.; Horel, A.; Pokovai, K.; Birkas, M.; Kern, A.; Hollos, R.; et al. Effect of tillage and crop type on soil respiration in a long-term field experiment on chernozem soil under temperate climate. *Soil Tillage Res.* **2022**, *216*, 13. [CrossRef]
27. Krauss, M.; Wiesmeier, M.; Don, A.; Cuperus, F.; Gattinger, A.; Gruber, S.; Haagsma, W.K.; Peigne, J.; Palazzoli, M.C.; Schulz, F.; et al. Reduced tillage in organic farming affects soil organic carbon stocks in temperate Europe. *Soil Tillage Res.* **2022**, *216*, 11. [CrossRef]
28. Yurdem, H.; Degirmencioglu, A.; Cakir, E.; Gulsoylu, E. Measurement of strains induced on a three-bottom moldboard plough under load and comparisons with finite element simulations. *Measurement* **2019**, *136*, 594–602. [CrossRef]
29. Nazemosadat, S.M.R.; Ghanbarian, D.; Naderi-Boldaji, M.; Nematollahi, M.A. Structural analysis of a mounted moldboard plow using the finite element simulation method. *Span. J. Agric. Res.* **2022**, *20*, 14. [CrossRef]
30. Azimi-Nejadian, H.; Karparvarfard, S.H.; Naderi-Boldaji, M.; Rahmanian-Koushkaki, H. Combined finite element and statistical models for predicting force components on a cylindrical mouldboard plough. *Biosyst. Eng.* **2019**, *186*, 168–181. [CrossRef]
31. Song, W.; Jiang, X.H.; Li, L.K.; Ren, L.L.; Tong, J. Increasing the width of disturbance of plough pan with bionic inspired subsoilers. *Soil Tillage Res.* **2022**, *220*, 14. [CrossRef]
32. Aikins, K.A.; Ucgul, M.; Barr, J.B.; Awuah, E.; Antille, D.L.; Jensen, T.A.; Desbiolles, J.M.A. Review of Discrete Element Method Simulations of Soil Tillage and Furrow Opening. *Agriculture* **2023**, *13*, 29. [CrossRef]
33. Azimi-Nejadian, H.; Karparvarfard, S.H.; Naderi-Boldaji, M. Weed seed burial as affected by mouldboard design parameters, ploughing depth and speed: DEM simulations and experimental validation. *Biosyst. Eng.* **2022**, *216*, 79–92. [CrossRef]
34. Zhu, L.; Cheng, X.; Peng, S.S.; Qi, Y.Y.; Zhang, W.F.; Jiang, R.; Yin, C.L. Three dimensional computational fluid dynamic interaction between soil and plowbreast of horizontally reversal plow. *Comput. Electron. Agric.* **2016**, *123*, 1–9. [CrossRef]
35. Wei, M.; Zhu, L.; Luo, F.; Zhang, J.W.; Dong, X.W.; Jen, T.C. Share-soil interaction load and wear at various tillage conditions of a horizontally reversible plough. *Comput. Electron. Agric.* **2019**, *162*, 21–30. [CrossRef]
36. Guo, Y.; Yu, X. Analysis of surface erosion of cohesionless soils using a three-dimensional coupled computational fluid dynamics—Discrete element method (CFD-DEM) model. *Can. Geotech. J.* **2019**, *56*, 687–698. [CrossRef]
37. Saunders, C.; Ucgul, M.; Godwin, R.J. Discrete element method (DEM) simulation to improve performance of a mouldboard skimmer. *Soil Tillage Res.* **2021**, *205*, 13. [CrossRef]
38. Ibrahmi, A.; Bentaher, H.; Hbaieb, M.; Maalej, A.; Mouazen, A.M. Study the effect of tool geometry and operational conditions on mouldboard plough forces and energy requirement: Part 1. Finite element simulation. *Comput. Electron. Agric.* **2015**, *117*, 258–267. [CrossRef]
39. Nadykto, V.; Kyurchev, V.; Bulgakov, V.; Findura, P.; Karaiev, O. Influence of the Plough with Tekrone Mouldboards and Landsides on Ploughing Parameters. *Acta Technol. Agric.* **2020**, *23*, 40–45. [CrossRef]
40. Wang, X.Z.; Zhang, Q.K.; Huang, Y.X.; Ji, J.T. An efficient method for determining DEM parameters of a loose cohesive soil modelled using hysteretic spring and linear cohesion contact models. *Biosyst. Eng.* **2022**, *215*, 283–294. [CrossRef]
41. Zhang, L.B.; Cai, Z.X.; Liu, H.F. A novel approach for simulation of soil-tool interaction based on an arbitrary Lagrangian-Eulerian description. *Soil Tillage Res.* **2018**, *178*, 41–49. [CrossRef]

42. Zhang, J.H.; Wang, J.; Wang, T. Research on the Coupled Water-Soil SPH Algorithm Improvement and Application Based on Two-phase Mixture Theory. *Int. J. Comput. Methods* **2022**, *19*, 23. [CrossRef]
43. Lucy, L.B. A numerical approach to the testing of the fission hypothesis. *Astron. J.* **1977**, *82*, 1013–1024. [CrossRef]
44. Gingold, R.A.; Monaghan, J.J. Smoothed particle hydrodynamics: Theory and application to non-spherical stars. *Mon. Not. R. Astron. Soc.* **1977**, *181*, 375–389. [CrossRef]
45. Hu, M.; Gao, T.; Dong, X.; Tan, Q.; Yi, C.; Wu, F.; Bao, A. Simulation of soil-tool interaction using smoothed particle hydrodynamics (SPH). *Soil Tillage Res.* **2023**, *229*, 105671. [CrossRef]
46. Zhang, J.M.; Yao, H.P.; Chen, L.Z.; Zheng, E.L.; Zhu, Y.; Xue, J.L. Vibration characteristics analysis and suspension parameter optimization of tractor/implement system with front axle suspension under ploughing operation condition. *J. Terramech.* **2022**, *102*, 49–64. [CrossRef]
47. Wang, H.B.; Yan, F.; Zhang, L.W.; Zhang, W.; Li, X.M.; Wang, S.Q.; Wang, S. Mechanism and flow process of debris avalanche in mining waste dump based on improved SPH simulation. *Eng. Fail. Anal.* **2022**, *138*, 22. [CrossRef]
48. He, H.Z.; Karsai, A.; Liu, B.Y.; Hammond, F.L.; Goldman, D.I.; Arson, C. Simulation of compound anchor intrusion in dry sand by a hybrid FEM plus SPH method. *Comput. Geotech.* **2023**, *154*, 16. [CrossRef]
49. Feng, D.; Imin, R. A kernel derivative free SPH method. *Results Appl. Math.* **2023**, *17*, 100355. [CrossRef]
50. Zhan, L.; Peng, C.; Zhang, B.Y.; Wu, W. A SPH framework for dynamic interaction between soil and rigid body system with hybrid contact method. *Int. J. Numer. Anal. Methods Geomech.* **2020**, *44*, 1446–1471. [CrossRef]
51. Wang, W.; Wu, Y.J.; Wu, H.; Yang, C.Z.; Feng, Q.S. Numerical analysis of dynamic compaction using FEM-SPH coupling method. *Soil Dyn. Earthq. Eng.* **2021**, *140*, 11. [CrossRef]
52. Feng, S.J.; Gao, H.Y.; Gao, L.; Zhang, L.M.; Chen, H.X. Numerical modeling of interactions between a flow slide and buildings considering the destruction process. *Landslides* **2019**, *16*, 1903–1919. [CrossRef]
53. Li, Z.X.; Chen, J.S.; Wang, X.N.; Shen, X.J.; Cen, Y.M.; Chen, J.; Chu, Y.J.; Han, Y.J. Microstructure distribution and bending fracture mechanism of 65Mn steel in the laser surface treatment. *Mater. Sci. Eng. A-Struct. Mater. Prop. Microstruct. Process.* **2022**, *850*, 8. [CrossRef]
54. Bentaher, H.; Ibrahmi, A.; Hamza, E.; Hbaieb, M.; Kantchev, G.; Maalej, A.; Arnold, W. Finite element simulation of moldboard-soil interaction. *Soil Tillage Res.* **2013**, *134*, 11–16. [CrossRef]
55. Godwin, R.J.; O'Dogherty, M.J.; Saunders, C.; Balafoutis, A.T. A force prediction model for mouldboard ploughs incorporating the effects of soil characteristic properties, plough geometric factors and ploughing speed. *Biosyst. Eng.* **2007**, *97*, 117–129. [CrossRef]

agriculture

MDPI

Article

Impact of Application Rate and Spray Nozzle on Droplet Distribution on Watermelon Crops Using an Unmanned Aerial Vehicle

Luis Felipe Oliveira Ribeiro [1,†] and Edney Leandro da Vitória [1,2,*,†]

1 Department of Agricultural and Biological Sciences (DCAB), Federal University of Espirito Santo (UFES), São Mateus CEP 29932-540, ES, Brazil; luis.f.ribeiro@edu.ufes.br
2 Graduate Program in Tropical Agriculture (PPGAT), Federal University of Espirito Santo, São Mateus CEP 29932-540, ES, Brazil
* Correspondence: edney.vitoria@ufes.br; Tel.: +55-27-99943-0907
† The authors contributed equally to this work.

Abstract: Watermelon is one of the most commonly grown vegetable crops worldwide due to the economic and nutritional importance of its fruits. The yield and quality of watermelon fruits are affected by constant attacks from pests, diseases, and weeds throughout all phenological stages of the crop. Labor shortages and unevenness of pesticide applications using backpack and tractor sprayers are significant challenges. The objective of this study was to evaluate the effect of different spray nozzles (XR110015 and MGA60015) and application rates (8, 12, and 16 L ha^{-1}) on droplet distribution on different targets in watermelon plants using an unmanned aerial vehicle. Water-sensitive papers were used as targets to analyze the droplet coverage, deposition, density, and volume median diameter. Data were collected from targets placed on the leaf adaxial and abaxial sides, fruit, apical bud, and stem of each plant. The mean droplet coverage and density increased as the application rate was increased, with no significant interaction between the factors or statistical difference between spray nozzles, except for the leaf abaxial side. No significant differences were found for the variables analyzed at application rates of 12 and 16 L ha^{-1}, whereas significant differences were observed at 8 L ha^{-1}. The use of unmanned aerial vehicles in watermelon crops is efficient; however, further studies should be conducted to evaluate their effectiveness in pest control and compare them with other application methods.

Keywords: *Citrullus lanatus* L.; remotely piloted aircraft; UAV; cucurbits; droplet distribution; application technology

Citation: Ribeiro, L.F.O.; Vitória, E.L.d. Impact of Application Rate and Spray Nozzle on Droplet Distribution on Watermelon Crops Using an Unmanned Aerial Vehicle. *Agriculture* **2024**, *14*, 1351. https://doi.org/10.3390/agriculture14081351

Academic Editors: Mustafa Ucgul and Chung-Liang Chang

Received: 12 August 2023
Revised: 18 September 2023
Accepted: 22 September 2023
Published: 13 August 2024

1. Introduction

Watermelon (*Citrullus lanatus* L.) is one of the most commonly grown vegetables in the world due to its economic importance and nutritional benefits. The fruit contains more than 90% water, considerable levels of vitamins A and C, a high lycopene content, and is widely consumed fresh [1]. More than 100 million Mg of watermelon are produced annually in the world, with an export value of EUR 2.02 billion in 2021 [2]. The largest watermelon-producing countries are China, Turkey, Iran, and Brazil [3].

Watermelon yield and quality are significantly impacted by constant attacks of pests, diseases, and weeds in all phenological stages of the crop [4,5]. Major pests affecting watermelon crops include underground pests such as the cutworm caterpillar (*Agrotis ipsilon*) and aerial pests such as aphids (*Aphis gossypii*), whitefly (*Bemisia tabaci*), and thrips (*Frankliniella schultzei*). Major diseases include anthracnose (*Colletotrichum orbiculare*), powdery mildew (*Sphaerotheca fuliginea*), downy mildew (*Pseudoperonospora cubensis*), and fusarium wilt (*Fusarium oxysporum* f. *niveum*).

Conventional methods for pest and disease control in watermelon crops are based on ground pesticide applications using backpack or tractor sprayers. However, these operations are sometimes inefficient and costly since watermelon has a creeping growth with countless herbaceous branches and large leaf areas covering the crop area, making it impossible for machinery and agricultural equipment to move between crop rows.

The use of new technologies to spray droplets with active ingredients on targets with greater efficiency is essential to achieve application efficiency. Despite unmanned aerial vehicles (UAVs) having been a suitable alternative for applications of pesticides and foliar fertilizers, there are some gaps or little information on droplet deposition quality for some crops, in which this technology is little explored. UAVs are used with pre-programmed flight plans and controlled by an operator at a ground station [6]. UAVs have advantages over conventional sprayers, including a low application rate, greater water savings, and a lower risk of applicator contamination; in addition, they are not limited by specific take-off and landing sites, topography, turning space, crop pattern, and canopy architecture, making them suitable for mountainous areas, muddy fields, and creeping crops with high leaf area indexes [7–11].

The spray nozzle, application rate, application route, application range, flight height, and flight speed are significant factors for UAV spraying [12]. They affect droplet deposition and in-flight planning, as the droplet spectrum and uniformity vary according to the canopy of plants. Research on UAV application technology is common for perennial (apple trees [13], peach [14], chestnut [15], arabica coffee [16], conilon coffee [17], vine [18], palm [19]) and annual (rice [20], corn [21], cotton [22], and weeds [23]) crops. Contrastingly, the droplet distribution and spray penetration in cucurbits vary, and some operational parameters used in the aforementioned studies are not applicable due to the irregular plant foliage and unpredictable branch growth direction.

Previous studies have investigated the impact of UAV application parameters on different crops. Spray nozzles and application rates significantly affected droplet deposition on different layers of papaya [24] and wheat [25] plants. The optimum spraying parameters for sugarcane crops were 15.0 L ha^{-1}, with a 3.0 m flight height and 4.0 m s^{-1} flight speed [26]. An application route parallel to the planting line maximized droplet deposition on grapevine plants while minimizing endo-drift losses [18]. A flight height of 2.5 m and a flight speed of 3.0 m s^{-1} increased droplet density on the leaf abaxial side of pumpkin plants [27]. Despite these findings, research on UAVs for creeping crops remains limited.

Therefore, the objective of this work was to fill theoretical gaps in the technical feasibility of using UAVs in watermelon crops. Three main gaps justify the present study:

(i) Research advances related to UAV operational parameters in the application of pesticides and fertilizers for annual (wheat, rice, corn, and cotton) and perennial (fruit trees and shrubs) crops are lacking for vegetable crops.

(ii) Fruit ripening is the most important stage of watermelon crops, when the control of the incidence of pests and diseases that occur in different plant parts (leaf, fruit, flower, and apical bud) hindering the final quality of the fruit is difficult. Therefore, despite studies that have shown the effects of different operational parameters (operational flight height and speed, application rate, spray nozzle, and flight route), they are not applicable to watermelon crops and should be investigated using different application rates and spray nozzles to test the application efficiency.

(iii) Watermelon crops present important specific characteristics that are not found in other studied crops, including the growing area, which is covered by primary and secondary herbaceous branches; plants covered with trichomes; fruit weights ranging from less than 1 kg to more than 20 kg (depending on the crop); and fruits initially developed below the leaf layer, which hinders the penetration of droplets to lower targets using conventional equipment.

Despite the economic importance of watermelon crops and the need for pest control during the crop cycle and the lack of information on UAV-based application technology in vegetable crops, droplet-deposition processes using UAVs in watermelon crops have never been investigated. Thus, the following hypotheses were tested to address these knowledge gaps: (a) the interaction between application rate and spray nozzles is significant for the optimal application configuration in watermelon crops; (b) droplet distribution by a UAV on watermelon crops has high potential to control fungal diseases and pests, depending on the target to be reached; (c) high application rates result in more uniform applications on watermelon crops.

Consequently, the objective of this study was to evaluate the effect of different spray nozzles and application rates on droplet distribution in different parts of watermelon plants using a UAV.

2. Materials and Methods

2.1. Characterization of the Area and Crop

The experiment was implemented, monitored, and evaluated at the Experimental Farm of the Northern University Center of Espírito Santo, Federal University of Espírito Santo, in São Mateus, ES, Brazil (18°40′25″ S, 40°51′23″ W) (Figure 1). The soil of the experimental area was classified as a Typic Hapludult of sandy loam texture. The climate of the region was classified as Aw, hot and humid, with a dry season in autumn–winter and a rainy season in spring–summer, according to the Köppen classification [28].

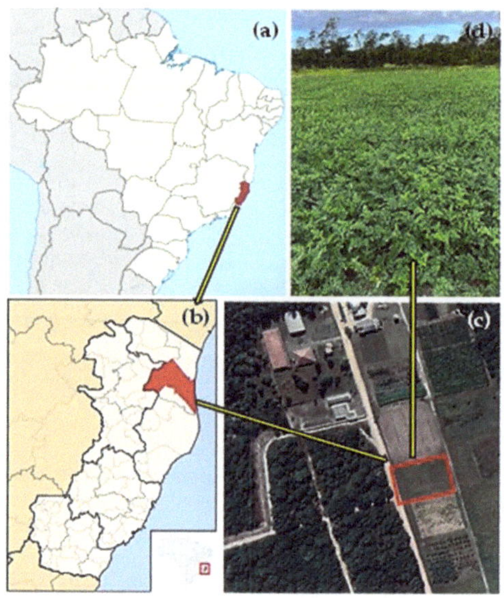

Figure 1. (**a**) Location of the state of Espirito Santo in Brazil, (**b**) city of São Mateus in the state of Espirito Santo, (**c**) Experimental Farm of the Federal University of Espirito Santo, (**d**) experimental area.

The experiment was conducted ins an experimental area of 2500 m^2 (50 × 50 m), grown with watermelon (Liverpool hybrid; Syngenta®, São Paulo, Brazil), planted with spacing of 3.0 × 0.8 m, totaling a stand of 1042 plants in the experimental area.

2.2. Characterization of the Unmanned Aerial Vehicle

The unmanned aerial vehicle (UAV) was a DJI Agras, model T10 (DJI, SZ DJI Technology, Shenzhen, China) (Figure 2).

The UAV has a solution tank capacity of 10 L, which was previously adjusted and calibrated before application. In addition to the product tank, the UAV is equipped with a water pump, piping circuit for liquid circulation, electronic control, and valves. The spray nozzles of the UAV were placed equidistantly and perpendicularly to the aircraft axis, 1.0 m apart. The main specifications of the UAV are listed in Table 1.

Figure 2. Unmanned aerial vehicle used in the experiment.

Table 1. Specifications of the DJI Agras UAV model T10.

Operating efficiency per hour	15 acres
Number of rotors	4
Maximum operational flight speed	0 to 7 m s^{-1}
Maximum level flight speed	4 to 10 m s^{-1} (with strong GNSS signals)
Maximum bearable wind speed	0 to 8 m s^{-1}
Tank capacity	10 L
Maximum effective spray width	3 to 5.5 m
Stationary flight duration	0 to 17 min
Maximum spraying flow	1.81 L/min^{-1}
Number of nozzles	4

Before the implementation of the experiment, tests were carried out at different moments to define optimal flight operating parameters to be used as reference, since there is no scientific data on these parameters in the literature for cucurbits. In this sense, four plants were randomly selected within the experimental area, and five water-sensitive papers were placed on the plants using metal clips. Preliminary application tests were carried out at different flight heights (2, 3, and 4 m), speeds (3 and 5 m s^{-1}), and application ranges (3.5 and 5.0 m). Water-sensitive papers were then digitized in a wireless system and carefully analyzed to establish optimal parameters based on the visual distribution of the water-sensitive paper and the lowest risk of dripping into the soil. In this sense, the application rate and spray nozzles were fixed as experimental treatments, according to the proposed objective of the present study. After the preliminary tests, the defined parameters were average flight height of 3.0 m above the canopy of watermelon plants, flight speed of 5.0 m s^{-1}, average application range of 5.0 m, and application route perpendicular to the planting line. These parameters were kept constant and used for all treatments.

2.3. Experimental Design

The experiment was conducted in a randomized block design in a 3 $\times$ 2 factorial arrangement. The treatments consisted of different application rates (8.0, 12.0, and 16.0 L ha^{-1}) and spray nozzles (XR110015 flat jet nozzle (Teejet®, Cotia, Brazil) and MGA60015 empty cone nozzle (Magnojet®, Ibaiti, Brazil) (Table 2). The treatments were applied four times; each replication corresponded to a day of application, i.e., the variation of the blocks was the different days of application.

Table 2. Experimental treatments.

Treatments	Spray Nozzle	Application Rate (L ha)$^{-1}$	Flow Rate (L min)$^{-1}$ [a]	Droplet Classification [b]
T1	MGA60015	8.0	1.18	Very thin
T2		12.0	1.78	
T3		16.0	2.38	
T4	XR110015	8.0	1.18	Thin
T5		12.0	1.78	
T6		16.0	2.38	

[a] Flow rate corresponding to the application rate at 3 bar pressure; [b] droplet classification according to the nozzle manufacturer.

The experimental area was 2500 m^2. The treatments were applied in the total area, and all treatments were applied on the same day (block). The border of the area, 10.0 m from the edges, was disregarded to avoid the edge effect and allow some distance for the UAV sprayer to be activated.

The experimental unit for data sampling was a central area of 1500 m^2 (50 × 30 m), where most vigorous plants developed. Each treatment was applied to four target plants that jointly composed the experimental unit. A strip of 10.0 m was left between targeted plants. Each treatment had four plots, totaling 24 experimental units. The applications were carried out during the fruit-formation stage, at maximum crop development, and when the experimental area had full canopy cover. Figure 3 illustrates the experimental design used.

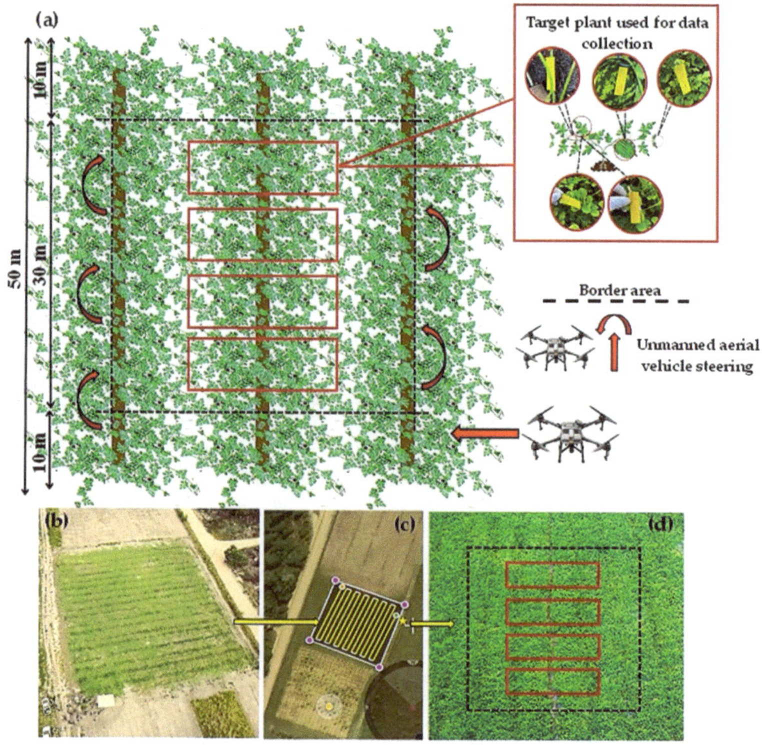

Figure 3. (**a**) Experimental design, (**b**) aerial photography of the experimental area, (**c**) unmanned aerial vehicle application route, and (**d**) target area used for data collection.

2.4. Variables Related to Application Efficiency

In all treatments, the sprayed solutions were composed of water mixed with brilliant blue dye (MarcAzul®, Porto Alegre, Brazil) at a rate of 1.25 mL L^{-1} and non-siliconized adjuvant based on balanced polymers, specific for low-volume aerial applications (0.2% v v^{-1}) (Helper Air®, ICL, São Paulo, Brazil).

Water-sensitive paper with dimensions of 76 × 26 mm was used for collecting data on droplet coverage, deposition, density, and volume median diameter. The targets were placed on different parts of each plant used for data collection: leaf adaxial and abaxial sides, fruit, apical bud, and stem. The productive branch with the highest vigor of each target plant was previously selected. Water-sensitive papers were fixed using wooden clips on the leaf adaxial (Figure 4a) and abaxial (Figure 4b) sides and using double-sided tape on fruits (Figure 4c), apical buds (Figure 4d), and stems (Figure 4e); therefore, 5 water-sensitive papers were placed on each target plant, totaling 20 water-sensitive papers per experimental plot and 80 water-sensitive papers per treatment. Despite its limitations in detecting very fine droplets, water-sensitive paper is still employed because of its ease of use in obtaining droplet spectrum data.

(a) (b) (c) (d) (e)

Figure 4. Water-sensitive papers fixed on watermelon plant parts: (**a**) leaf adaxial side, (**b**) leaf abaxial side, (**c**) fruit, (**d**) apical bud, and (**e**) stem.

The water-sensitive papers were removed five minutes after application of each treatment to allow the solution to evaporate, maintaining only the dye. They were then packed in labeled kraft paper bags and analyzed for droplet characterization on the same day. The analyses were performed at the Laboratory of Mechanization and Agricultural Defensives (LMDA) of the Northern University Center of Espírito Santo, Federal University of Espírito Santo, São Mateus, ES, Brazil.

A wireless system (DropScope®; SprayX, São Carlos, Brazil) was used to scan the water-sensitive papers for data acquisition and analysis. This system is composed of application programs and a wireless digital microscope with a digital image sensor of more than 2500 dpi, which allows the estimation of partially overlapping droplets of approximately 35 µm (Figure 5). Recent studies have demonstrated the reliability of the data obtained by this system in analyses of spectrum of droplets sprayed by UAV [24,29,30]. After scanning the water-sensitive papers, data were generated, and the following parameters were evaluated: droplet coverage (%), density (droplets cm^{-2}), deposition (µL cm^{-2}), and volume median diameter (µm).

2.5. Climate Conditions

The applications were carried out in late afternoon. Wind speed, air humidity, and temperature were recorded during the applications (Table 3) using an automatic weather station Davis®; Vantage Pro2 wireless K6152 (Davis Instruments, Hayward, CA, USA) installed at approximately 100 m from the experimental area. The applications were carried out considering the methodology described by the International Organization for Standardization (Standard 22866) [31], which recommends temperature during applications between 5 and 35 °C, a maximum of 10% of wind speed measurements below 1.0 m s^{-1}, and wind direction within 90° ± 30° in relation to the application line.

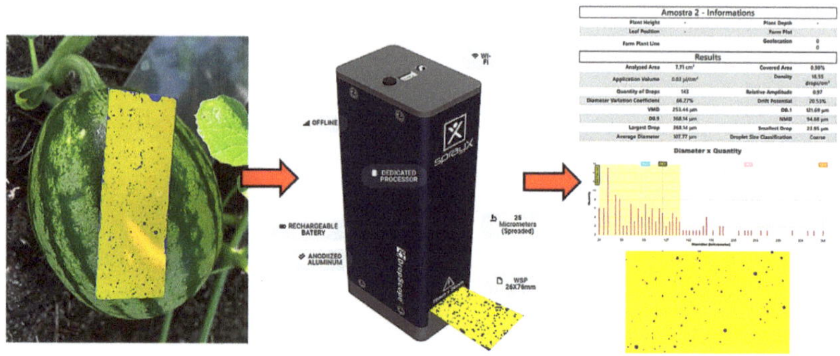

Figure 5. Wireless system (DropScope®; SprayX, São Carlos, Brazil) used for scanning the water-sensitive papers.

Table 3. Climate conditions at the time of applications.

Application Date	Time	Temperature (°C)		Relative Humidity (%)		Wind Speed (m s)$^{-1}$		Wind Direction
		Minimum	Maximum	Minimum	Maximum	Minimum	Maximum	
30 May 2023	16:00	26.0	27.3	60.0	65.0	3.1	6.7	East-northeast.
	17:00	24.8	26.0	63.0	69.0	2.7	6.3	
	18:00	24.0	24.8	69.0	75.0	1.8	4.5	
	19:00	23.0	24.0	75.0	85.0	0.9	3.6	
1 June 2023	16:00	26.8	28.8	68.0	77.0	1.2	3.2	
	17:00	25.5	26.8	77.0	82.0	1.1	3.9	
	18:00	24.2	25.6	82.0	87.0	1.9	4.6	
	19:00	23.3	24.2	87.0	91.0	0.8	3.7	
5 June 2023	16:00	25.9	27.1	55.0	61.0	1.30	3.6	
	17:00	24.1	25.9	60.0	70.0	1.7	3.1	
	18:00	21.8	24.1	70.0	79.0	1.5	1.4	
	19:00	21.1	21.8	79.0	82.0	0.4	2.3	
7 June 2023	16:00	25.3	26.6	63.0	72.0	1.3	5.4	
	17:00	23.9	25.4	71.0	79.0	2.1	3.1	
	18:00	22.5	23.9	79.0	85.0	1.8	1.2	
	19:00	21.2	22.5	84.0	89.0	0.4	2.1	

2.6. Statistical Analysis

The Shapiro–Wilk test was used to assess the normality of residuals. Data related to application efficiency were subjected to analysis of variance. When the effects of factors or their interactions were significant, means were compared using Tukey's test.

3. Results

The results of the analysis of variance for variables related to application efficiency indicated statistical differences in the application rates (Table 4). The interaction between the factors (spray nozzle and application rate) was not significant for any of the variables. Thus, each factor was analyzed separately within the proposed targets.

3.1. Droplet Coverage

The application rate was significant for droplet coverage on all watermelon plant targets, except for the leaf abaxial side (Table 4). The spray nozzle effect and the interaction effect between factors were not significant for any target, indicating that all treatments were statistically similar. The normality of residuals was confirmed by the Shapiro–Wilk test, with values ranging from 0.667 to 0.990. Normality tests yield important assumptions before analyses for greater accuracy of the data obtained.

Table 4. Analysis of variance (Anova) and normality of residuals for droplet coverage on different watermelon plant parts.

	Droplet coverage (%)									
	Analysis of variance									
	Leaf adaxial side		**Leaf abaxial side**		**Fruit**		**Apical Bud**		**Stem**	
Factor	*p*-value	CV (%)	*p*-value	CV (%)	*p*-value	CV (%)	*p*-value	CV (%)	*p*-value	CV (%)
T	<0.01 **		0.80 ^{ns}		<0.001 ***		<0.01 **		<0.001 ***	
P	0.36 ^{ns}	39.78	0.12 ^{ns}	117.68	0.21 ^{ns}	23.43	0.66 ^{ns}	30.11	0.054 ^{ns}	38.08
T × P	0.80 ^{ns}		0.80 ^{ns}		0.44 ^{ns}		0.98 ^{ns}		0.91 ^{ns}	
	Normality of residuals									
W=	0.753 ^{ns}		0.897 ^{ns}		0.667 ^{ns}		0.990 ^{ns}		0.986 ^{ns}	

P = spray nozzles; T = application rate; T × P = interaction between spray nozzle and application rate; ** = significant at $p < 0.01$; *** = significant at $p < 0.001$; ^{ns} = not significant; CV (%) = coefficient of variation of Anova; W = normality of residuals by the Shapiro–Wilk test.

The means found for the spray nozzle factor within the targets were significantly similar to Tukey's test and confirmed by an Anova (Table 5). However, the XR10015 nozzle presented the highest means in the leaf adaxial side (LAD), fruit (FT), and stem (ST), with a difference in droplet coverage of 0.25, 0.10, and 0.13 percentual points, equivalent to 14.45%, 12.20%, and 28.90%, respectively, compared to the MGA60015 nozzle. On the leaf abaxial side, the MGA60015 nozzle presented the highest mean value, with a difference equivalent to 56.52%, 2.3-fold higher compared to the XR110015 nozzle.

Table 5. Droplet coverage (%) on watermelon plant parts (leaf adaxial side—LAD, leaf abaxial side—LAB, fruit—FT, apical bud—AB, and stem—ST) using different spray nozzles and application rates. Mean values are followed by standard deviation.

Targets	Spray Nozzles	
	MGA60015	XR110015
LAD	1.48 ± 0.70 a	1.73 ± 0.86 a
LAB	0.46 ± 0.42 a	0.20 ± 0.25 a
FT	0.72 ± 0.43 a	0.82 ± 0.34 a
AB	1.44 ± 0.72 a	1.52 ± 0.50 a
ST	0.32 ± 0.20 a	0.45 ± 0.21 a

Targets	Application rates (L ha)$^{-1}$		
	8.0	12.0	16.0
LAD	0.95 ± 0.44 b	1.64 ± 0.50 ab	2.22 ± 0.78 a
LAB	0.33 ± 0.47 a	0.38 ± 0.35 a	0.25 ± 0.29 a
FT	0.50 ± 0.20 b	0.83 ±0.30 a	1.00 ± 0.47 a
AB	0.92 ± 0.36 b	1.66 ± 0.49 a	1.87 ± 0.53 a
ST	0.18 ± 0.07 b	0.40 ± 0.18 a	0.58 ± 0.15 a

Means followed by different letters in the rows are significantly different from each other according to Tukey's test at $p \leq 0.05$.

Droplet coverage on LAD, FT, apical bud (AB), and ST increased as the application rate was increased. The application rate of 8.0 L ha^{-1} presented lower droplet coverage, and 12 and 16.0 L ha^{-1} were statistically similar, except for the leaf abaxial side (LAB), which did not differ statistically between application rates. However, the intermediate application rate (12.0 L ha^{-1}) presented the highest droplet coverage on LAB.

The application rate of 16.0 L ha^{-1} presented the highest droplet coverage on LAD, being 26.12% and 57.20% higher compared to the rates of 12 and 8.0 L ha^{-1}, respectively.

Similar results were found for the application rate of 16.0 L ha^{-1} in the other targets. The highest application rate was 17.0% and 50.0% higher on FT compared to the rates of 12 and 8.0 L ha^{-1}; 11.22% and 50.80% higher on AB compared to the rates of 12 and 8.0 L ha^{-1}; and 31.03% and 69.00% higher on ST compared to the rates of 12 and 8.0 L ha^{-1}. The differences between 16.0 and 12.0 L ha^{-1} did not reach 50.0%, presenting statistical similarity; in contrast, the application rate of 8.0 L ha^{-1} resulted in differences greater than 50.0% in droplet coverage compared to the other application rates.

3.2. Droplet Density

The application rate factor significantly affected ($p < 0.05$) droplet density on the LAD, AB, and ST (Table 6). All residuals were normally distributed according to the Shapiro–Wilk test. The interaction between the application rate and the spray nozzle was not significant.

Table 6. Analysis of variance (Anova) and normality of residuals for droplet density on different watermelon plant parts.

Droplet Density (Droplet cm^{-2})										
Analysis of variance										
	Leaf adaxial side		**Leaf abaxial side**		**Fruit**		**Apical bud**		**Stem**	
Factor	*p*-value	CV (%)	*p*-value	CV (%)	*p*-value	CV (%)	*p*-value	CV (%)	*p*-value	CV (%)
T	<0.05 *	42.33	0.95 ^{ns}	121.21	0.79 ^{ns}	77.40	<0.05 *	38.29	<0.05 *	55.23
P	0.36 ^{ns}		0.12 ^{ns}		0.50 ^{ns}		0.76 ^{ns}		0.95 ^{ns}	
T × P	0.85 ^{ns}		0.85 ^{ns}		0.65 ^{ns}		0.63 ^{ns}		0.51 ^{ns}	
Normality of residuals										
W=	0.252 ^{ns}		0.808 ^{ns}		0.215 ^{ns}		0.530 ^{ns}		0.043 ^{ns}	

P = spray nozzles; T = application rate; T × P = interaction between spray nozzle and application rate; * = significant at $p < 0.05$; ^{ns} = not significant; CV (%) = coefficient of variation of Anova; W = normality of residuals according to the Shapiro–Wilk test.

The MGA60015 spray nozzle presented the highest mean droplet density on LAB (Table 7), with a difference of 7.02 droplets cm^{-2}, equivalent to 58.80%, 2.43-fold higher compared to the XR110015 nozzle. The XR110015 nozzle resulted in the highest number of droplets on the other targets, However, it presented means close to those of the MGA60015, not differing statistically.

The number of droplets per square centimeter increased on all watermelon plant parts as the application rates increased, except for LAB. The application rates of 16.0 and 12.0 L ha^{-1} were statistically similar for LAD, AB, and ST.

The mean values and the impacts of droplets on water-sensitive papers (Figure 6) showed increases in the mean droplet coverage and density, according to the accessibility of the targets. The highest mean values were found for LAD and AB on the upper part of watermelon plants. The highest application rate (16.0 L ha^{-1}) resulted in a higher droplet density on LAD, with a difference of 15.02 and 25.61 droplet cm^{-2} compared to the application rates of 12.0 and 8.0 L ha^{-1}, respectively. The highest application rate resulted in a 28.74% and 48.07% higher droplet density on AB compared to the application rates of 12.0 and 8.0 L ha^{-1}, respectively.

Watermelon plant parts with more obstacles (FT, LAB, and ST) presented lower mean droplet densities. The application rate of 16.0 L ha^{-1} showed a 17.47% and 21.43% higher droplet density on FT compared to the application rates of 12.0 and 8.0 L ha^{-1}, respectively. The differences between application rates on FT were less than 50% and not significant according to the Anova. The differences between the application rates of 12.0 and 8.0 L ha^{-1} on ST were 1.83 and 6.94 droplet cm^{-2}, respectively. The application rate of 12.0 L ha^{-1} resulted in higher droplet density on LAB, which was 12.04% and 16.12% higher compared to the rates of 8 and 16.0 L ha^{-1}, respectively. The application rates had a similar effect on droplets on water-sensitive papers on LAB (Figure 6), with no difference between application rates, regardless of the spray nozzle type, as indicated by the Anova and Tukey test.

Table 7. Droplet density (droplets cm^{-2}) on watermelon plant parts (leaf adaxial side—LAD; leaf abaxial side—LAB; fruit—FT; apical bud—AB; and stem—ST) using different spray nozzles and application rates. Mean values are followed by standard deviation.

Targets	Spray nozzles		
	MGA60015		XR110015
LAD	33.04 ± 15.96 a		38.91 ± 23.40 a
LAB	11.95 ± 10.15 a		4.92 ± 6.57 a
FT	17.00 ± 8.26 a		21.24 ± 19.47 a
AB	32.53 ± 13.00 a		36.28 ± 21.42 a
ST	8.37 ± 4.01 a		8.49 ± 6.78 a
Targets	Application rates (L ha)$^{-1}$		
	8.0	12.0	16.0
LAD	23.91 ± 10.24 b	34.50 ± 15.46 ab	49.52 ± 23.83 a
LAB	8.18 ± 9.59 a	9.30 ± 9.21 a	7.80 ± 9.70 a
FT	17.26 ± 7.47	18.13 ± 22.75	21.97 ± 11.41
AB	24.71 ±11.18 b	33.91 ±12.51 ab	47.59 ± 20.13 a
ST	4.41 ± 1.70 b	9.52 ± 6.77 ab	11.35 ± 4.39 a

Means followed by different letters in the rows are significantly different from each other according to Tukey's test at $p \leq 0.05$.

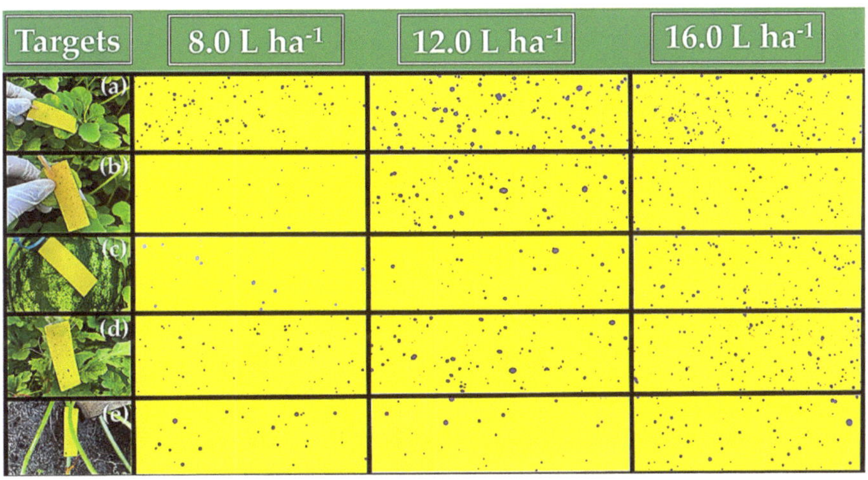

Figure 6. Effects of application rates on droplet coverage and density on the different targets in watermelon plants: leaf adaxial side (**a**), leaf abaxial side (**b**), fruit (**c**), apical bud (**d**), and stem (**e**).

3.3. Droplet Deposition

The application rate was significant for droplet deposition on all targets, except for the leaf abaxial side (Table 8). The spray nozzle factor and the interaction between application rate and spray nozzle were not significant, rejecting the hypothesis of difference between treatments. All residuals were normally distributed according to the Shapiro–Wilk test. The coefficients of variation of the Anova for the targets ranged from 27.96% to 110.97%, indicating a possible effect of the UAV operational conditions connected to external environmental factors but not invalidating the statistical analyses.

Similarly to droplet coverage and density, the mean droplet deposition increased as the application rate increased, with no significant interaction between the factors or statistical difference between spray nozzles, except for the leaf abaxial side. However, application rates of 12.0 and 16.0 L ha^{-1} resulted in a statistically similar droplet deposition, significantly differing from the rate of 8.0 L ha^{-1}.

Table 8. Analysis of variance (Anova) and normality of residuals for droplet deposition on different watermelon plant parts.

Factor	Leaf adaxial side		Leaf abaxial side		Fruit		Apical bud		Stem	
	Droplet deposition (μL cm)$^{-2}$									
	Analysis of variance									
	p-value	CV (%)	*p*-value	CV (%)	*p*-value	CV (%)	*p*-value	CV (%)	*p*-value	CV (%)
T	<0.05 *	43.07	0.58 ^{ns}	110.97	<0.01 **	27.96	<0.01 **	28.92	<0.01 **	49.37
P	0.26 ^{ns}		0.15 ^{ns}		0.21 ^{ns}		0.31 ^{ns}		0.09 ^{ns}	
T × P	0.74 ^{ns}		0.76 ^{ns}		0.37 ^{ns}		0.90 ^{ns}		0.81 ^{ns}	
	Normality of residuals									
W=	0.580 ^{ns}		0.809 ^{ns}		0.353 ^{ns}		0.694 ^{ns}		0.448 ^{ns}	

P = spray nozzles; T = application rate; T × P = interaction between spray nozzle and application rate; * = significant at $p < 0.05$; ** = significant at $p < 0.01$; ^{ns} = not significant; CV (%) = coefficient of variation of Anova; W = normality of residuals according to the Shapiro–Wilk test.

The application rate of 16.0 L ha^{-1} resulted in the highest mean droplet deposition on LAD (0.065 μL cm^{-2}) (Figure 7a), being 23.07% and 53.85% higher compared to the 12.0 and 16.0 L ha^{-1}, respectively. The XR110015 nozzle presented an increase of 0.010 μL cm^{-2} in droplet deposition on LAD compared to the MGA60015 nozzle. The application rate of 12.0 L ha^{-1} was 25.0% and 50.0% higher on LAB compared to 8.0 and 16.0 L ha^{-1}, respectively (Figure 7b). The MGA60015 nozzle resulted in a 2.0-fold increase (0.006 μL cm^{-2}) in droplet deposition on LAB compared to the XR110015 nozzle.

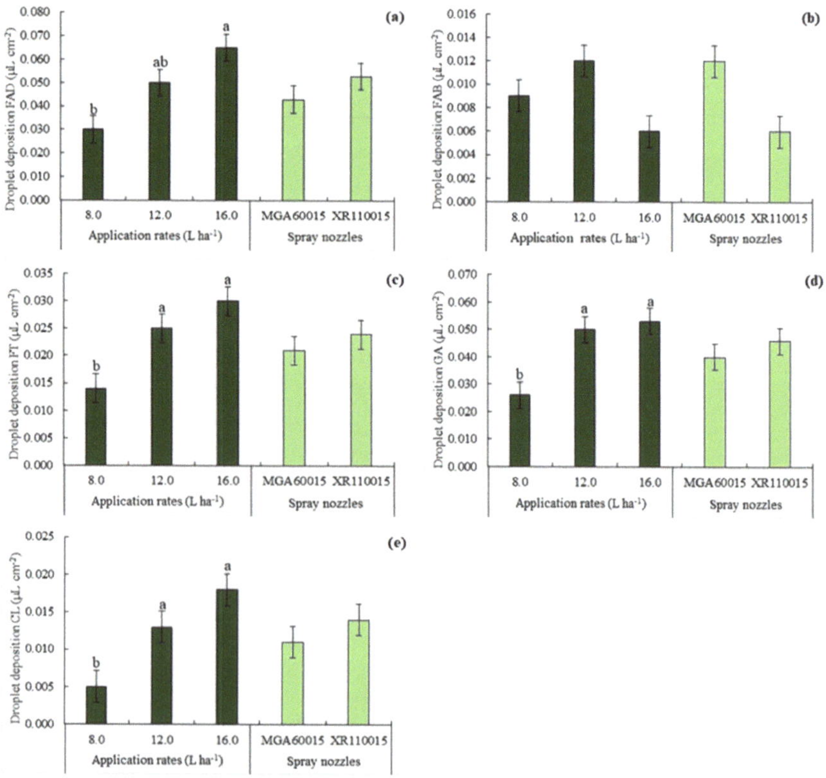

Figure 7. Droplet deposition (μL cm^{-2}) on different targets in watermelon plants: (**a**) leaf adaxial side (LAD), (**b**) leaf abaxial side (LAB), (**c**) fruit (FT), (**d**) apical bud (AB), and (**e**) stem (ST). Bars with different letters are significantly different from each other according to Tukey's test at $p \leq 0.05$.

Droplet deposition on FT (Figure 7c) and AB (Figure 7d) showed similar results to those on LAD, with 16.0 L ha^{-1}, and the XR110015 spray nozzle presented the highest means. The stem is the plant part that experiences great interference by plant parts and is located close to the ground level; however, the application rate of 16.0 L ha^{-1} also presented an increase in droplet deposition, which was 0.005 and 0.0013 µL cm^{-2} higher compared to those of 12.0 and 8.0 L ha^{-1}, respectively (Figure 7e).

3.4. Volume Median Diameter

The application rate and spray nozzle factors were significant for the volume median diameter (VMD) on the leaf adaxial side, but with no significant interaction effect between factors (Table 9). Only the spray nozzle was significant for the fruit. The factors and their interaction had no significant effect on VMD for LAB, AB, or ST.

Table 9. Analysis of variance (Anova) and normality of residuals for volume median diameter on different watermelon plant targets.

Volume median diameter (µm)										
Analysis of variance										
Factor	**Leaf adaxial side**		**Leaf abaxial side**		**Fruit**		**Apical bud**		**Stem**	
	p-value	CV (%)	*p*-value	CV (%)	*p*-value	CV (%)	*p*-value	CV (%)	*p*-value	CV (%)
T	<0.01 **	6.90	0.56 ^{ns}	37.60	0.44 ^{ns}	9.38	0.109 ^{ns}	9.11	0.67 ^{ns}	33.34
P	<0.05 *		0.65 ^{ns}		<0.05 *		0.37 ^{ns}		0.90 ^{ns}	
T × P	0.95 ^{ns}		0.42 ^{ns}		0.79 ^{ns}		0.08 ^{ns}		0.33 ^{ns}	
Normality of residuals										
W=	0.170 ^{ns}		0.401 ^{ns}		0.300 ^{ns}		0.101 ^{ns}		0.015 ^{ns}	

P = spray nozzles; T = application rate; T × P = interaction between spray nozzle and application rate; * = significant at $p < 0.05$; ** = significant at $p < 0.01$; ^{ns} = not significant; CV (%) = coefficient of variation of Anova; W = normality of residuals according to the Shapiro–Wilk test.

The coefficients of variation (Anova) for the targets ranged from 6.90% to 37.60%, indicating the accuracy of the experimental design in obtaining the data. The residuals from the ST data were not normally distributed based on the Shapiro–Wilk test.

The XR110015 spray nozzle resulted in the largest droplet diameters on all watermelon plant parts, ranging from 171.83 to 206.38 µm, and was significant for LAD and ST, as indicated by the Anova. VMD on the watermelon plant parts varied among application rates, with the application rate of 12.0 L ha^{-1} presenting higher means compared to 8.0 and 16.0 L ha^{-1} (Table 10).

There was little variation in VMD among the different targets. The largest interquartile differences (ΔQ) were found for LAB (Q_1 = 146.70 µL; Q_3 = 196.50 µL). The difference in the third quartile (Q_3) between the stem (219.00 µm) and leaf abaxial side (196.50 µm) was 22.50 µL, equivalent to 10.27% (Figure 8).

Table 10. Volume median diameter (µm) on watermelon plant parts (leaf adaxial side—LAD, leaf abaxial side—LAB, fruit—FT, apical bud—AB, and stem—ST) using different spray nozzles and application rates. Mean values are followed by standard deviation.

Targets	Spray nozzles	
	MGA60015	XR110015
LAD	188.43 ± 17.54 b	202.24 ± 20.45 a
LAB	160.21 ± 59.02 a	171.83 ± 69.01 a
FT	186.74 ± 18,24 b	206.38 ± 23.43 a
AB	190.91 ± 21.45 a	197.66 ± 18.65 a
ST	210.08 ± 21.98 a	213.72 ± 20.35 a

Table 10. *Cont.*

Targets	Application rates (L ha)$^{-1}$		
	8.0	12.0	16.0
LAD	181.93 ± 12.73 b	208.88 ± 23.86 a	195.20 ± 12.45 ab
LAB	170.00 ± 78.65 a	180.85 ± 41.48 a	147.22 ± 67.02 a
FT	200.00 ± 27.51 a	200.17 ± 24.81 a	190.00 ± 16.04 a
AB	188.13 ± 25.74 a	206.50 ± 10.35 a	188.23 ± 16.84 a
ST	193.31 ± 24.00 a	212.37 ± 41.03 a	195.71 ± 24.06 a

Means followed by different letters in the rows are significantly different from each other according to Tukey's test at $p \leq 0.05$.

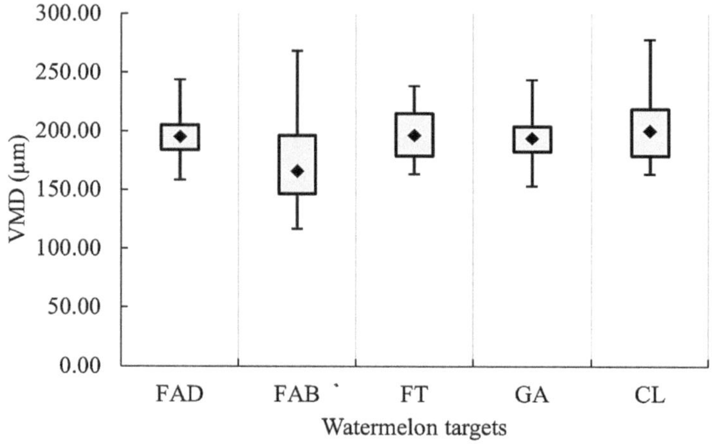

Figure 8. Comparison of volume median diameters (μm) on watermelon plant parts, regardless of application rate and spray nozzle factors. Leaf adaxial side (LAD), leaf abaxial side (LAB), fruit (FT), apical bud (AB), and stem (ST).

The maximum and minimum droplet diameters in the targets were 158.30 and 243.70 μm (LAD), 116.36 and 268.10 μm (LAB), 163.50 and 238.70 μm (FT), 153.20 and 243.70 μm (AB), and 163.50 and 277.70 μm (ST), respectively. Overall, no outliers were found (VMD < 300.00 μm), regardless of the target, i.e., there was no value outside of the interquartile detection range.

4. Discussion

The results indicate that application rates had a significant effect on droplet coverage, density, and deposition on the leaf adaxial side, fruit, apical bud, and stem (ST) of watermelon plants, with increasing means as the application rate was increased. Similar results were found for large [32,33] and small [34] shrub crops and perennial grass [26].

Watermelon plants have a creeping growth habit with no defined canopy layers; thus, droplet distribution on watermelon plants is variable when using UAVs, unlike perennial and annual plants [17,30,35,36]. This explains the absence of significant interaction between the application rate and spray nozzle for droplet coverage, density, deposition, and volume median diameter (VMD), as recently found for papaya plants with the use of different application rates and spray nozzles (MGA60015 and XR110015) [24].

Although the highest application rate (16.0 L ha^{-1}) increased droplet density and deposition, the current investigation evaluated only the efficiency of the application using a dye tracer since the droplets sprayed by UAVs are highly concentrated. Therefore, the highest application rates may not always provide the best control efficacy for the pesticide. In the control of diseases in *Coffea arabica* plants [37] and pests in wheat plants [25], the rates applied using a UAV achieved the same control efficacy as high rates applied using conventional sprayers.

Considering the limitations of battery life and load capacity (8–40 L) [38,39], the UAV operational capacity tends to increase by reducing application rates without losing gains in pest control. The application rate of 12.0 L ha^{-1} did not differ from 16.0 L ha^{-1} for most variables, regardless of the spray nozzle, presenting satisfactory results on the leaf adaxial side, fruit, and apical bud, which experienced no interference from watermelon ramifications and the overlapping of leaves. This indicates the potential for controlling diseases that grow on the leaf adaxial side, such as *Pseudoperonospora cubensis* and *Cercospora citrullina,* and attacks from chewing and sucking insects on fruits with contact insecticides.

The apical bud is one of the main targets covered in watermelon plants, as it hosts sucking insects in its tissues covered by trichomes, including *Frankliniella schultzei,* which is found in the apical bud and inside flowers. The droplet density and deposition resulting from the application rates of 12.0 and 16.0 L ha^{-1} using the XR110015 spray nozzle were promising for the apical bud and stem (where flowers grow). Higher values were found when using the fan nozzle on these plant parts due to the transverse distribution of the spray mixture, symmetrical to the nozzle center, reaching the targets with larger droplets [40]. When using flat jet/fan nozzles and UAVs for perennial crops, droplet distribution was greater in the upper layer and lower in the lower plant layer [41–43].

The leaf abaxial side (LAB) has a greater number of stomata and thinner cuticles; thus, it is one of the main pathogen infection sites. Additionally, the LAB of watermelon plants hosts sucking pests such as *Aphis gossypi* and *Bemisia tabaci.* The higher droplet coverage, density, and deposition on LAB by the MGA60015 nozzle at a rate of 12.0 L ha^{-1} was due to the greater turbulence in droplet distribution. The fine droplet distribution at the cone's outer edge, caused by the turbulence chamber within the nozzle [44] and combined with the downwash effect, facilitates droplet penetration into the lower plant layers [45]. The use of a UAV equipped with conical nozzles for pumpkin crops showed that the operational flight height and speed were significant in increasing deposition and droplet density on LAB [27]. The application rates had no significant effect on the leaf abaxial side using a UAV on watermelon plants, as observed in a laboratory study [46]. Moreover, the angle between the leaf and the airflow is an important factor for droplet deposition on LAB [47].

The VMD found for the XR110015 spray nozzle was higher, as the production of larger droplets is a typical characteristic of flat jet nozzles. The MGA60015 spray nozzle resulted in droplets with smaller diameters, a desirable characteristic of conical nozzles because droplets easily penetrate the canopy of plants with high leaf density [48]. This results in higher values of variables related to the application efficiency for the LAB.

The mean VMD found on the leaf abaxial side, fruits, and stems, which have greater plant barriers (branching and overlapping leaves), indicated that these plant parts should be sprayed with fine droplets to enhance droplet deposition, regardless of the application rate and spray nozzle. However, the risks related to primary and secondary drift and evaporation of droplets under adverse climate conditions increase as the droplet size is reduced [49]. Overall, the VMD values found are consistent with the manufacturers' catalog, which describes fine to medium droplets [50,51].

Although the results denoted the variability of droplet distribution on different watermelon plant parts, further studies should be carried out to confirm the viability of using UAVs in watermelon crops. Nevertheless, several operational parameters (operational flight height and speed, application rates, application range, and spray points) should be tested at different stages of watermelon plants to improve the results related to application efficiency. In addition, the effectiveness of pesticides in pest and disease control and foliar fertilizers should be considered to enhance and leverage the use of UAVs in vegetable crops. Moreover, the importance of evaluating the efficacy of applied products as a parameter to complement assessments of the quality of droplet deposition on target crops has been highlighted [17,37,52–54].

5. Conclusions

The use of unmanned aerial vehicles (UAVs) for spraying on agricultural crops is expanding due to their numerous advantages compared to conventional equipment. However, there are no operational parameters in the technical–scientific literature that elucidate the feasibility of using UAVs for applications to watermelon crops. Nevertheless, the main objective of this study was to evaluate the effect of spray nozzles and application rates on droplet distribution on different targets in watermelon plants using a UAV. The results found after applying the methodology, data collection, and analysis of the results obtained showed the following:

1. Application rates significantly affect droplet distribution on the leaf adaxial and abaxial sides, fruit, apical bud, and stem of watermelon plants, with variable droplet distribution depending on the target.
2. Droplet density and deposition on the leaf adaxial side, fruit, apical bud, and stem were higher when using the XR110015 spray nozzle. Droplet density and deposition on the leaf abaxial side were 58.80% and 50.0% higher, respectively, when using the MGA60015 spray nozzle.
3. Application rates of 12.0 and 16.0 L ha^{-1} can be used as a reference for evaluations of the efficacy of pesticides, as they resulted in statistically similar means regarding application efficiency.
4. The results demonstrated that the use of UAVs for applications to watermelon crops is efficient. However, after understanding the droplet-deposition process through the application efficiency, further studies should be conducted to evaluate the efficacy of this technology for controlling pests and diseases in watermelon crops, with comparisons between applications using UAV and conventional equipment at different phenological stages of watermelon plants.

Supplementary Materials: The following supporting information can be downloaded at: https://www.mdpi.com/article/10.3390/agriculture14081351/s1.

Author Contributions: Conceptualization, E.L.d.V. and L.F.O.R.; methodology, E.L.d.V. and L.F.O.R.; validation, E.L.d.V. and L.F.O.R.; formal analysis, L.F.O.R.; investigation, L.F.O.R.; resources, E.L.d.V. and L.F.O.R.; data curation, L.F.O.R.; writing—original draft preparation, L.F.O.R.; writing—review and editing, E.L.d.V.; visualization, E.L.d.V. and L.F.O.R.; supervision, E.L.d.V.; project administration, E.L.d.V.; funding acquisition, E.L.d.V. All authors have read and agreed to the published version of the manuscript.

Funding: This study was partly financed by the Coordenação de Aperfeiçoamento de Pessoal de Nível Superior—Brazil (CAPES 18/2020) and the Fundação de Amparo a Pesquisa do Espírito Santo (FAPES 140/2021)—Brazil. CAPES/FAPES Cooperation—Graduate Development Program—PDPG.

Data Availability Statement: Data are contained within the article and Supplementary Materials.

Conflicts of Interest: The authors declare no conflicts of interest.

References

1. Tegen, H.; Alemayehu, M.; Alemayehu, G.; Abate, E.; Amare, T. Response of watermelon growth, yield, and quality to plant density and variety in Northwest Ethiopia. *Open Agric.* **2021**, *6*, 655–672. [CrossRef]
2. Bantis, F.; Koukounaras, A. Ascophyllum nodosum and Silicon-Based Biostimulants differentially affect the physiology and growth of watermelon transplants under abiotic stress factors: The case of drought. *Horticulturae* **2022**, *8*, 1177. [CrossRef]
3. FAO. Watermelon Crop Information. 2019. Available online: http://www.fao.org/land-water/databases-and-software/crop-information/watermelon/en/ (accessed on 4 August 2023).
4. da Vitória, E.L.; Alves, D.d.S.; Rossi, M.T.; Favero, R.G.; Fernandes, A.A.; da Silva, M.B.; Freitas, I.L.d.J.; Locatelli, T.; Freitas, S.d.J.; Freitas, S.d.P.; et al. Spray Deposition on Watermelon Crop in Aerial and Ground Application. *J. Agric. Sci.* **2022**, *14*, 172. [CrossRef]
5. Dantas, M.S.; Grangeiro, L.C.; Medeiros, J.F.D.; Cruz, C.A.; Cunha, A. Rendimento e qualidade de melancia cultivada sob proteção de agrotêxtil combinado com mulching plástico. *Rev. Bras. De Eng. Agrícola E Ambient.* **2013**, *17*, 824–829. [CrossRef]
6. Chen, P.; Douzals, J.P.; Lan, Y.; Cotteux, E.; Delpuech, X.; Pouxviel, G.; Zhan, Y. Characteristics of unmanned aerial spraying systems and related spray drift: A review. *Front. Plant Sci.* **2022**, *13*, 870956. [CrossRef]

7. Guo, S.; Li, J.; Yao, W.; Zhan, Y.; Li, Y.; Shi, Y. Distribution characteristics on droplet deposition of wind field vortex formed by multi-rotor UAV. *PLoS ONE* **2019**, *14*, 7. [CrossRef] [PubMed]
8. Meng, Y.; Su, J.; Song, J.; Chen, W.H.; Lan, Y. Experimental evaluation of UAV spraying for peach trees of different shapes: Effects of operational parameters on droplet distribution. *Comput. Electron. Agric.* **2020**, *170*, 105282. [CrossRef]
9. Jiang, Y.; He, X.; Song, J.; Liu, Y.; Wang, C.; Li, T.; Qi, P.; Yu, C.; Chen, F. Comprehensive assessment of intelligent unmanned vehicle techniques in pesticide application: A case study in pear orchard. *Front. Plant Sci.* **2022**, *13*, 959429. [CrossRef] [PubMed]
10. Hu, H.; Kaizu, Y.; Huang, J.; Furuhashi, K.; Zhang, H.; Li, M.; Imou, K. Research on Methods Decreasing Pesticide Waste Based on Plant Protection Unmanned Aerial Vehicles: A Review. *Front. Plant Sci.* **2022**, *13*, 811256. [CrossRef] [PubMed]
11. Xu, Y.; Xue, X.; Sun, Z.; Gu, W.; Cui, L.; Jin, Y.; Lan, Y. Joint path planning and scheduling for vehicle-assisted multiple Unmanned Aerial Systems plant protection operation. *Comput. Electron. Agric.* **2022**, *200*, 107221. [CrossRef]
12. Richardson, B.; Rolando, C.A.; Somchit, C.; Dunker, C.; Strand, T.M.; Kimberley, M.O. Swath pattern analysis from a multi-rotor unmanned aerial vehicle configured for pesticide application. *Pest Manag. Sci.* **2020**, *76*, 1282–1290. [CrossRef] [PubMed]
13. Mahmud, M.S.; He, L.; Heinemann, P.; Choi, D.; Zhu, H. Unmanned aerial vehicle based tree canopy characteristics measurement for precision spray applications. *Smart Agric. Technol.* **2023**, *4*, 100153. [CrossRef]
14. Li, L.; Hu, Z.; Liu, Q.; Yi, T.; Han, P.; Zhang, R.; Pan, L. Effect of flight velocity on droplet deposition and drift of combined pesticides sprayed using an unmanned aerial vehicle sprayer in a peach orchard. *Front. Plant Sci.* **2022**, *13*, 981494. [CrossRef]
15. Arakawa, T.; Kamio, S. Control Efficacy of UAV-Based Ultra-Low-Volume Application of Pesticide in Chestnut Orchards. *Plants* **2023**, *12*, 2597. [CrossRef]
16. Souza, F.G.; Portes, M.F.; Silva, M.V.; Teixeira, M.M.; Furtado Júnior, M.R. Impact of sprayer drone flight height on droplet spectrum in mountainous coffee plantation. *Rev. Bras. Eng. Agrícola Ambient.* **2022**, *26*, 12. [CrossRef]
17. Crause, D.H.; Vitória, E.L.d.; Ribeiro, L.F.O.; Ferreira, F.d.A.; Lan, Y.; Chen, P. Droplet Deposition of Leaf Fertilizers Applied by an Unmanned Aerial Vehicle in Coffea Canephora Plants. *Agronomy* **2023**, *13*, 1506. [CrossRef]
18. Biglia, A.; Grella, M.; Bloise, N.; Comba, L.; Mozzanini, E.; Sopegno, A.; Pittarello, M.; Dicembrini, E.; Alcatrão, L.E.; Guglieri, G.; et al. UAV-spray application in vineyards: Flight modes and spray system adjustment effects on canopy deposit, coverage, and off-target losses. *Sci. Total Environ.* **2022**, *845*, 157292. [CrossRef] [PubMed]
19. Wang, J.; Ma, C.; Chen, P.; Yao, W.; Yan, Y.; Zeng, T.; Lan, Y. Evaluation of aerial spraying application of multi-rotor unmanned aerial vehicle for Areca catechu protection. *Front. Plant Sci.* **2023**, *14*, 1093912. [CrossRef] [PubMed]
20. Chen, P.; Lan, Y.; Huang, X.; Qi, H.; Wang, G.; Wang, J.; Wang, L.; Xiao, H. Droplet deposition and control of planthoppers of different nozzles in two-stage rice with a quadrotor unmanned aerial vehicle. *Agronomy* **2020**, *10*, 2. [CrossRef]
21. Yongjun, Z.; Shenghui, Y.; Chunjiang, Z.; Liping, C.; Lan, Y.; Yu, T. Modelling operation parameters of UAV on spray effects at different growth stages of corns. *Int. J. Agric. Biol. Eng.* **2017**, *10*, 57–66. [CrossRef]
22. Wang, G.; Wang, J.; Chen, P.; Han, X.; Chen, S.; Lan, Y. Droplets deposition and harvest-aid efficacy for UAV application in arid cotton areas in Xinjiang, China. *Int. J. Agric. Biol. Eng.* **2022**, *15*, 9–18. [CrossRef]
23. Kumar, S.; Singh, M.; Singh, S.K.; Bhullar, M.S. Droplet distribution and weed control efficacy of unmanned aerial vehicle sprayer in wheat crop. *J. Agric. Eng.* **2022**, *59*, 126–136. [CrossRef]
24. Ribeiro, L.F.O.; Vitória, E.L.d.; Soprani Júnior, G.G.; Chen, P.; Lan, Y. Impact of Operational Parameters on Droplet Distribution Using an Unmanned Aerial Vehicle in a Papaya Orchard. *Agronomy* **2023**, *13*, 1138. [CrossRef]
25. Wang, G.; Lan, Y.; Qi, H.; Chen, P.; Hewitt, A.; Han, Y. Field evaluation of an unmanned aerial vehicle (UAV) sprayer: Effect of spray volume on deposition and the control of pests and disease in wheat. *Pest Manag. Sci.* **2019**, *75*, 1546–1555. [CrossRef] [PubMed]
26. Zhang, X.Q.; Song, X.P.; Liang, Y.J.; Qin, Z.Q.; Zhang, B.Q.; Wei, J.J.; Li, Y.-R.; Wu, J.M. Effects of spray parameters of drone on the droplet deposition in sugarcane canopy. *Sugar Tech.* **2020**, *22*, 583–588. [CrossRef]
27. Jingxin, X.; Changzheng, H.E.; Limin, Y.; Liang, Z.; Sihui, D.; Ming, L.I.; Wei, Z. Effect of working parameter on droplet deposition in pumpkin top dressing using multi-rotor UAV. *IOP Conf. Ser. Earth Environ. Sci.* **2021**, *792*, 012045. [CrossRef]
28. Alvares, C.A.; Stape, J.L.; Sentelhas, P.C.; Gonçalves, J.D.M.; Sparovek, G. Köppen's climate classification map for Brazil. *Meteorol. Z.* **2013**, *22*, 711–728. [CrossRef] [PubMed]
29. Vitória EL, d.a.; de Oliveira, R.F.; Crause, D.H.; Ribeiro, L.F.O. Effect of flight operative height and genotypes on conilon coffee spraying using an unmanned aerial vehicle. *Coffee Sci.* **2022**, *17*, e172003. [CrossRef]
30. Cunha, J.P.A.R.d.; Silva, M.R.A.d. Spray deposition from a remotely piloted aircraft on the corn crop. *Rev. Ciência Agronômica* **2023**, *54*, e20217862. [CrossRef]
31. *ISO22866*; Equipment for Crop Protection-Methods for Field Measurements of Spray Drift. International Organization for Standardization: Geneva, Switzerland, 2005; pp. 1–17.
32. Guo, S.; Yao, W.; Xu, T.; Ma, H.; Sun, M.; Chen, C.; Lan, Y. Assessing the application of spot spray in Nanguo pear orchards: Effect of nozzle type, spray volume rate and adjuvant. *Pest Manag. Sci.* **2022**, *78*, 3564–3575. [CrossRef]
33. Wang, C.; Liu, Y.; Zhang, Z.; Han, L.; Li, Y.; Zhang, H.; Wongsuk, S.; Li, Y.; Wu, X.; He, X. Spray performance evaluation of a six-rotor unmanned aerial vehicle sprayer for pesticide application using an orchard operation mode in apple orchards. *Pest Manag. Sci.* **2022**, *78*, 2449–2466. [CrossRef] [PubMed]
34. Chen, P.; Lan, Y.B.; Douzals, J.; Ouyang, F.; Wang, J.; Xu, W.C. Droplet distribution of Unmanned Aerial Vehicle under several spray volumes and canopy heights in the cotton canopy. *Int. J. Precis. Agric. Aviat.* **2020**, *3*, 74–79. [CrossRef]

35. Verma, A.; Singh, M.; Parmar, R.P.; Bhullar, K.S. Feasibility study on hexacopter UAV based sprayer for application of environment-friendly biopesticide in guava orchard. *J. Environ. Biol.* **2022**, *43*, 97–104. [CrossRef]
36. Dengeru, Y.; Ramasamy, K.; Allimuthu, S.; Balakrishnan, S.; Kumar, A.P.M.; Kannan, B.; Karuppasami, K.M. Study on Spray Deposition and Drift Characteristics of UAV Agricultural Sprayer for Application of Insecticide in Redgram Crop (*Cajanus cajan* L. Millsp.). *Agronomy* **2022**, *12*, 3196. [CrossRef]
37. Vitória, E.L.d.; Krohling, C.A.; Borges, F.R.P.; Ribeiro, L.F.O.; Ribeiro, M.E.A.; Chen, P.; Lan, Y.; Wang, S.; Moraes, H.M.F.; Furtado Júnior, M.R. Efficiency of Fungicide Application an Using an Unmanned Aerial Vehicle and Pneumatic Sprayer for Control of Hemileia vastatrix and Cercospora coffeicola in Mountain Coffee Crops. *Agronomy* **2023**, *13*, 340. [CrossRef]
38. Lan, Y.B.; Chen, S.D. Current status and trends of plant protection UAV and its spraying technology in China. *Int. J. Precis. Agric. Aviat.* **2018**, *1*, 1–9. [CrossRef]
39. Chen, P.; Xu, W.; Zhan, Y.; Yang, W.; Wang, J.; Lan, Y. Evaluation of Cotton Defoliation Rate and Establishment of Spray Prescription Map Using Remote Sensing Imagery. *Remote Sens.* **2022**, *14*, 17. [CrossRef]
40. Minguela, J.V.; Cunha, J.P.A.R.d. *Manual de Aplicação de Produtos Fitossanitários*; Aprenda Fácil Editora: Viçosa, MG, Brazil, 2010; 588p.
41. Li, X.; Giles, D.K.; Niederholzer, F.J.; Andaloro, J.T.; Lang, E.B.; Watson, L.J. Evaluation of an unmanned aerial vehicle as a new method of pesticide application for almond crop protection. *Pest Manag. Sci.* **2021**, *77*, 527–537. [CrossRef]
42. Irfan, M.; Ahmad, F.; Kahliq, A.; Awais, M.; Hamid, S. Uav Sprayers Use For Mango Orchards Spraying to Control the Environmental Pollution. In *Book of Abstracts*; Mehmet Akif Ersoy University: Burdur, Turkey, 2021; p. 295.
43. Shan, C.; Wu, J.; Song, C.; Chen, S.; Wang, J.; Wang, H.; Wang, G.; Lan, Y. Control Efficacy and Deposition Characteristics of an Unmanned Aerial Spray System Low-Volume Application on Corn Fall Armyworm Spodoptera frugiperda. *Front. Plant Sci.* **2022**, *13*, 900939. [CrossRef]
44. Ruiz, M.C.; Bloise, N.; Guglieri, G.; D'Ambrosio, D. Numerical Analysis and Wind Tunnel Validation of Droplet Distribution in the Wake of an Unmanned Aerial Spraying System in Forward Flight. *Drones* **2022**, *6*, 11. [CrossRef]
45. do Nascimento, V.P.; da Vitória, E.L. Spraying quality using unmanned aerial vehicle in citrus. *Rev. Eng. Na Agric.-Reveng.* **2022**, *30*, 214–221. [CrossRef]
46. Garcerá, C.; Vicent, A.; Chueca, P. Effect of spray volume, application timing and droplet size on spray distribution and control efficacy of different fungicides against circular leaf spot of persimmon caused by plurivorosphaerella nawae. *Crop. Prot.* **2020**, *130*, 105072. [CrossRef]
47. Xu, S.; Wang, X.; Li, C.; Ran, X.; Zhong, Y.; Jin, Y.; Song, J. Effect of airflow angle on abaxial surface deposition in air-assisted spraying. *Front. Plant Sci.* **2023**, *14*, 1211104. [CrossRef]
48. Vitória, E.L.; Leite, J. Espectro De Gotas De Pontas De Pulverização De Jato Cônico Vazio. *Enciclopedia Biosf.* **2014**, *10*, 18.
49. Wang, J.; Lan, Y.; Zhang, H.; Zhang, Y.; Wen, S.; Yao, W.; Deng, J. Drift and deposition of pesticide applied by UAV on pineapple plants under different meteorological conditions. *Int. J. Agric. Biol. Eng.* **2018**, *11*, 5–12. [CrossRef]
50. Ribeiro, L.F.O.; Vitória, E.L.d. Qualidade de pulverização hidropneumática na cultura da macadâmia. *Agrotrópica* **2022**, *34*, 81–88. [CrossRef]
51. Ribeiro, L.F.O.; Ribeiro, M.E.A.; Santos, T.M.; Aiala, M.L.C.; Vitória, E.L.d. Simulated herbicide application between rows of black pepper crops. *Braz. J. Prod. Eng.* **2023**, *9*, 41–55. [CrossRef]
52. Sun, T.; Zhang, S.; Xue, X.; Jiao, Y. Comparison of Droplet Distribution and Control Effect of Wheat Aphids under Different Operation Parameters of the Crop Protection UAV in the Wheat Flowering Stage. *Agronomy* **2022**, *12*, 3175. [CrossRef]
53. Garcia-Ruiz, F.; Campos, J.; Llop-Casamada, J.; Gil, E. Assessment of map based variable rate strategies for copper reduction in hedge vineyards. *Comput. Electron. Agric.* **2023**, *207*, e107753. [CrossRef]
54. Dammer, K.H.; Garz, A.; Hobart, M.; Schirrmann, M. Combined UAV-and tractor-based stripe rust monitoring in winter wheat under field conditions. *Agron. J.* **2022**, *114*, 651–661. [CrossRef]

agriculture

Article

Design and Test of Single-Disc Opener for No-Till Planter Based on Support Cutting

Guangyuan Zhong [1,2], Hongwen Li [1,2,*], Jin He [1], Qingjie Wang [1], Caiyun Lu [1], Chao Wang [1], Zhenwei Tong [1], Dandan Cui [1] and Dong He [1]

1 College of Engineering, China Agricultural University, Beijing 100083, China
2 Key Laboratory of Agricultural Equipment for Conservation Tillage, Ministry of Agricultural and Rural Affairs, Beijing 100083, China
* Correspondence: lhwen@cau.edu.cn; Tel.: +86-010-6273-7300

Abstract: To solve the problem of low straw-cutting efficiency of single-disc openers of no-till planters under conditions of high soil moisture content, a single-disc furrowing and straw-cutting device was designed based on the support-cutting principle. To improve the straw-cutting ability of the disc opener when it operates under high-moisture-content soil conditions and to make sure that the straw that is not cut by the disc coulter can be cut smoothly by the disc opener, the support shovel was designed, and the operation mechanism of the support shovel device was analyzed. The soil moisture content, the support shovel's entry angle, the support shovel's entry gap angle, and the support shovel's tip margin were identified as the factors influencing the device design through the theoretical analysis of the furrowing and straw-cutting device. Through the discrete element method (DEM), a single-factor simulation test was first conducted to analyze how different soil moisture contents affected the device's ability to cut straw, and the results showed that the number of broken bonds was lowest when the soil moisture content was 20 ± 1%, and the time taken for the straw to be wholly cut off was also the longest. Then, a quadratic orthogonal simulation test was conducted to construct a regression model and optimize the parameters at the soil moisture content of 20 ± 1%, and the results revealed that the significant order of each factor's influence on the number of broken bonds is as follows: entry gap angle, entry angle, and shovel tip margin. In addition, the device's overall operation quality was better when the entry angle was 49°, the entry gap angle was 0°, and the shovel tip margin was 10 mm. At this time, the number of broken bonds was predicted to be 506. Finally, the simulation validation test was run, and the number of broken bonds was obtained to be 478, with a relative error of 5.6% from the predicted value. According to the optimal parameters to complete the device trial production and field test, the results show that the average cut-off rate of the device is 71.7%, the stability coefficient of the furrowing depth is 90.87%, and the performance of the furrow opening is excellent, which meets the requirements of a no-tillage seeding operation. This study can provide a reference for the design and improvement of no-tillage seeding machines under conditions of high soil moisture content.

Keywords: single-disc opener; support cutting; soil moisture content; discrete element method (DEM); parameter optimization

Citation: Zhong, G.; Li, H.; He, J.; Wang, Q.; Lu, C.; Wang, C.; Tong, Z.; Cui, D.; He, D. Design and Test of Single-Disc Opener for No-Till Planter Based on Support Cutting. *Agriculture* **2023**, *13*, 1635. https://doi.org/10.3390/agriculture13081635

Academic Editors: Mustafa Ucgul and Chung-Liang Chang

Received: 27 July 2023
Revised: 16 August 2023
Accepted: 17 August 2023
Published: 19 August 2023

1. Introduction

Conservation tillage improves soil organic matter, prevents wind erosion, and conserves moisture [1]. No-till seeding may significantly increase productivity while lowering labor expenses, and it has been extensively used in China [2]. In the two mature Huang-Huai-Hai wheat and maize regions, there is much maize straw during the wheat no-tillage seeding process. In the case of high soil moisture content, the stubble cutter at the front of the no-tillage planter cannot completely cut off the straw, while the traditional single-disc opener has a poor cutting effect, and the straw will block the planter and cause a high seed-drying rate, which affects seed emergence [3]. Therefore, improving the straw-cutting ability of disc openers when operating under high-moisture-content soil conditions and ensuring that the straw not cut by the stubble cutter can be cut off by the disc opener is the key to solving the clogging of wheat no-till planters and improving the quality of seeding.

The traditional single-disc opener utilizes a disc with a slight angle to the direction of movement and a vertical direction to move the soil laterally to form a furrow [4]. Moreover, the disc has a certain ability to cut crop straw [5], and the cutting performance is affected by the different geometry of the disc and its operating parameters [6]. The shape of the disc has a significant impact on the straw-cutting effect [7,8]; at the same working depth, the cutting effect of ripple disc, notched disc, and plain disc decreases in order [9], but the disturbance width of the soil does increase accordingly [6]. Different disc diameters have different straw-cutting capacities [10], but the vertical force required increases with increasing diameter [11]. Insufficient downforce will cause the disc opener to push the incompletely cut straw into the seed furrow, which will cause "hair-pinning", suspend the seed, and hinder its emergence [12,13]. Increasing the downward pressure is more conducive to the cutting of straw [14]. Straw-cutting efficiency is also affected by the operating speed of the opener disc [15], and the efficiency of straw-cutting will be improved with an increase in the operating speed [16]. The disc angle and tilt angle are the most crucial factors in figuring out how well the openers will furrow and cut straw, and the large rake angle can promote the "hair-pinning" of residue in wet or soft soil conditions [17]. At the same time, increasing the disc angle decreases the specific draft, side, and vertical forces, and increasing the tilt angle causes the specific vertical, draft, and side forces to increase [18]. Existing research has focused on the effects of disc geometry and operating parameters on tillage performance but has not yet considered the impact of soil characteristics on the operational performance of single-disc openers.

To address the aforementioned problems, this study designs a furrow opener with a straw-cutting device based on the principle of support cutting to enhance the efficiency of straw that has not been cut off by the stubble cutter to be cut by the single-disc furrow opener under the condition of high soil moisture content, lessen "hair-pinning", and enhance the quality of seeding. The parameters of the device's essential components are established through theoretical analysis, the straw-cutting process is simulated by EDEM, the optimal structural parameter combination is established by the quadratic orthogonal combination test, and the optimized device is put through field tests to confirm its functionality.

2. Materials and Methods

2.1. Structure and Working Principle

2.1.1. Structure

The major components of the device are seen in Figure 1 and include a disc coulter, a scraper, a support shovel, a disc hanging plate, etc. The installation disc angle and tilt angle are both 0°; the scraper is positioned on one side of the disc and can scrape off the soil adhering to the disc as well as assist in the opening of furrows. The support shovel is positioned on the same side as the scraper and passes through the scraper; the disc can cut off the straw under the support provided by the support shovel and open the suitable seed furrow.

2.1.2. Working Principle

During the operation, the front end of the support shovel reaches into the soil with a certain entry angle, and with the forward movement of the machine, the soil is slightly raised by the support shovel. The disc coulter then cuts the soil, which is divided into two parts, completing the initial seed furrow guidance and forming the seed furrow prototype. The soil moves along the disc's sides on both sides, and under the scraper's compression and compacting action forces, soil on one of the disc's sides is pushed away and lifted, widening and stabilizing the narrow seed furrow into a complete seed furrow that satisfies the requirements of agronomy. The scraper also has the function of scraping away the soil adhering to the disc while completing the furrow opening to prevent the furrow-opening disc from adhering to too much soil. Figure 2 depicts the changes in the soil that occur during the single-disc opener's soil-penetration procedure.

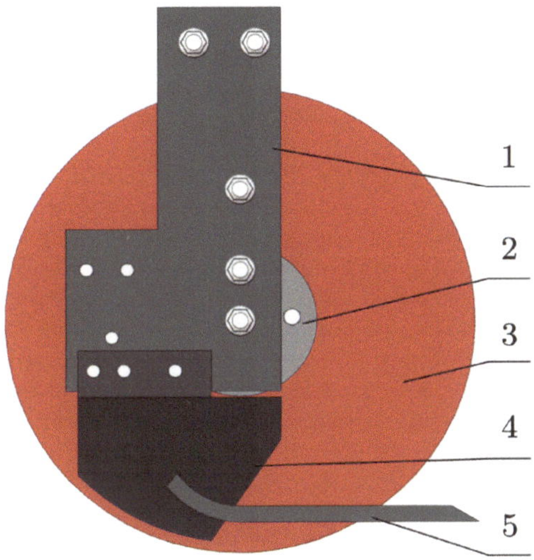

Figure 1. Schematics of opener structure: 1. disc hanging plate; 2. bearing base; 3. disc coulter; 4. scraper; 5. support shovel.

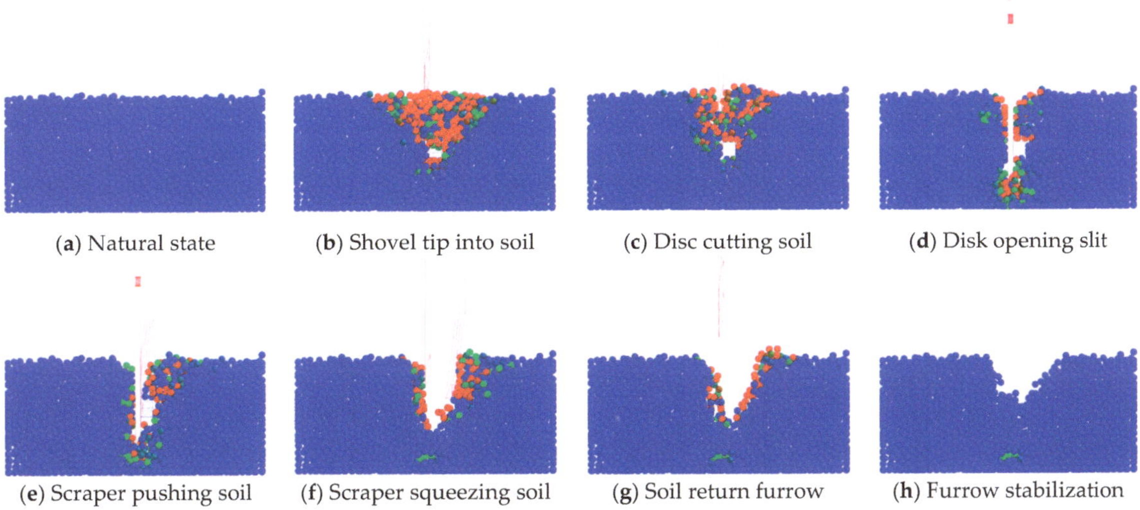

Figure 2. Single-disc opener entering the soil process.

When the opener has not penetrated the soil, the soil is in its natural state (Figure 2a). The support shovel penetrates the soil first; its tip slices the dirt, and the soil on its upper half has a tendency to move higher under its action (Figure 2b). With the advancement of the opener, the soil is warped. The disc cuts the warped soil from the top, and as the disc rotates, the soil is opened up to a slit (Figure 2c,d). The scraper pushes the soil on one side away from the slit and compacts it (Figure 2e,f), and the soil is piled up on one side.

The pile of soil particles begins to flow back into the seed furrow after the opener has left (Figure 2g) until the shape of the seed furrow is stabilized (Figure 2h).

2.2. Design of Key Components

The operating performance of the opener is mainly affected by factors such as the outside environment and its structural design. External factors include the physical and chemical properties of the soil, the operating speed of the machine, and the operating depth of the opener. Structural design factors are the shape of the support shovel, the diameter of the disc, and the shape of the scraper.

2.2.1. Analysis of Straw Forces

The ordinary single-disc coulter balances the disc's cutting force by using the soil's supporting force and the straw's bending resistance. Due to the straw's low stiffness, when the soil is soft, the straw's supporting force fluctuates with its depth, making it difficult to reach the force balance, difficult to cut off [11], and simple to press into the soil by the disc. The seeds are easily aerated by straw when sowing at this time. Increase the support shovel at the bottom so that when the straw penetrates the soil and reaches it, the support force is fixed and may be conducive to straw cutting. If the soil is loose, the disc drives the straw into the soil until it touches the support shovel at a distance of h from the furrow's bottom. Assume that the angle between the support shovel and the soil level is 0°. Use the straw as a mass point M to analyze the instantaneous cutting process. The force of the straw is shown in Figure 3. The support shovel provides an upward supporting force as the disc presses the straw against it, and the only force between the soil and the straw is the horizontal resistance F_w.

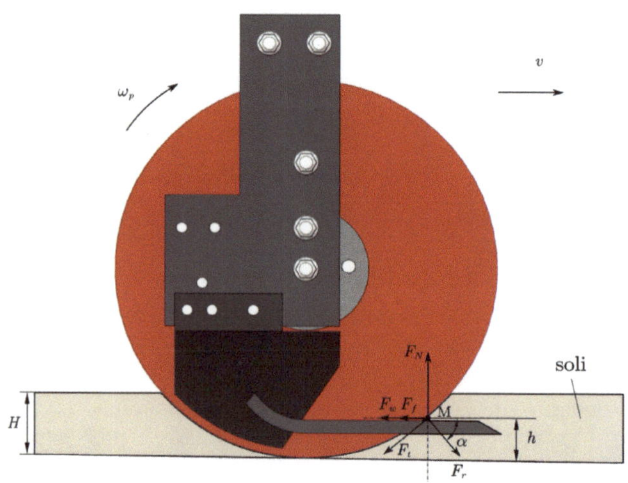

Figure 3. Analysis of straw force during the straw-cutting process: w_p is the rotational speed of the disc, rad/s; v is the forward speed of the operation, m/s; F_N is the support force of the support shovel on the straw, N; F_f is the friction force between the support shovel and the straw, N; F_w is the horizontal force between the soil and the straw, N; F_r is the positive pressure of the disc on the straw, N; F_t is the sliding shear force of the disc for the straw, N; α is the angle between the positive pressure of the disc on the straw and the horizontal plane of the soil, °; H is the depth of furrowing with disc openers, cm; h is the distance from the furrow bottom to the upper surface of the support shovel, cm.

According to the force analysis in Figure 3, the forces are balanced in all directions:

$$\begin{cases} F_r \sin\alpha + F_t \cos\alpha = F_N \\ F_w + F_f + F_t \sin\alpha = F_r \cos\alpha \end{cases} \quad (1)$$

where:

$$\begin{cases} F_t = F_r \tan \varphi_1 \\ F_f = F_N \tan \varphi_2 \end{cases} \tag{2}$$

In the formula, φ_1 is the friction angle between the disc cutter and the straw, °; φ_2 is the friction angle between the support shovel and the straw, °.

Take the diameter of the disc as D. There are

$$\alpha = \arcsin\left(1 - \frac{2h}{D}\right) \tag{3}$$

According to Formula (3), it is clear that when the value of D is determined, the larger h is, the smaller α is, and the straw is easier to push away by the disc at this time; when the value of h is determined, as the diameter D of the disc increases, α also increases, and the better the disc cuts the straw.

The support shovel and the disc are both constructed of 65Mn steel, which means that the values of φ_1 and φ_2 in Equation (2) are equivalent and should be taken as the same value φ. Organizing Equations (1)–(3) yields the following:

$$F_r = \frac{DF_w}{2\left[(1 - \tan^2 \varphi)\sqrt{Dh - h^2} - \tan \varphi (D - 2h)\right]} \tag{4}$$

According to previous research, the friction angle between maize straw and 65Mn steel typically ranges from 23 to 33° [19], and the friction angle between straw and soil is assumed to be 30° in this study. In the Huang-Huai-Hai region, winter wheat is typically planted 3 to 5 cm deep, with fertilizer depths ranging from 7 to 10 cm, vertical spacing distances between fertilizer and seed being greater than or equal to 5 cm [20], disc depth into the soil H set at 10 cm, and distance from the bottom of the furrow to the upper surface of the support shovel h set at 5 cm. According to Kushwaha's study [21], the optimal working diameter of the plane disc is 460 mm, which satisfies the agronomic criteria, and it is established that the disc diameter D is set at 460 mm.

The positive pressure F_r of the disc on the straw is a prerequisite to ensuring that the disc can cut off the straw. According to Equation (4), it can be seen that when the diameter of the disc, the friction angle between the straw and the steel plate, and the distance from the bottom of the groove to the upper surface of the support shovel are determined, the soil force F_w is the main factor affecting the cutting of the straw.

2.2.2. Force Analysis of the Support Shovel

The support shovel's tip first enters the soil, and as it advances, the soil will produce tillage resistance as the tip cuts through it. As the operating speed changes, the soil's shear strength will also change because the tip of the support shovel will cause the soil to be squeezed and deformed until it is crushed, which results in a dynamic change in the resistance. The support shovel moves through the soil at a uniform speed, assuming that the soil is homogeneous and isotropic. At this point, the tillage resistance is mostly seen in the soil's cohesion and pressure, as well as its friction with and adhesion to the support shovel [22]. Figure 4 depicts the force analysis of the supporting shovel in this condition.

The cohesion of the soil itself, the sliding friction force of the soil on the upper and lower surfaces, and the adhesion force of the soil to them must all be overcome throughout the advancing process of the support shovel [23]. So, the amount of tillage resistance P during cultivation can be expressed as follows:

$$P = (S + f_1 + T_1)\cos\beta + N_1 \sin\beta + (f_2 + T_2 + f_3 + T_3)\cos\varepsilon + (N_3 - N_2)\sin\varepsilon \tag{5}$$

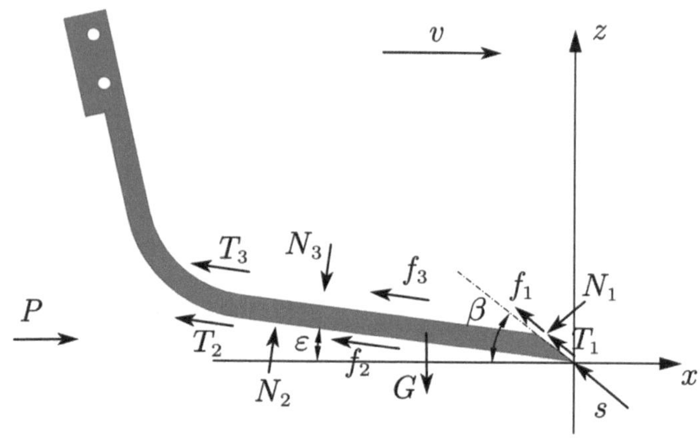

Figure 4. Force analysis of support shovel: P is the tillage resistance of the supporting shovel during tillage, N; S the soil cohesion, N; f_1, f_2, and f_3 are the soil's sliding friction forces on the edge, the shovel body's upper surface, and the bottom, respectively, N; N_1, N_2, and N_3 are the soil's normal pressures on the edge, the shovel body's upper surface, and the bottom, respectively, N; T_1, T_2, and T_3 are the soil's adhesion on the edge, the shovel body's upper surface, and the bottom, respectively, N; G is the gravity of the supporting shovel, N; β is the entry angle, °; ε is the entry gap angle, °.

Among them, the cohesion of the soil S is the force between soil particles that are bonded together as a result of molecular attraction; the strength of these forces varies depending on the soil's texture, structure, exchangeable cation composition, content of organic matter, content of soil water, etc. [24]. Better cohesiveness results in better resistance to tillage. The soil type, the physical characteristics of the soil, the material to which the soil adheres to the support plate, and the contact area are the main factors influencing the magnitude of the adhesion forces T_1, T_2, and T_3, with the soil moisture content having the largest impact on the physical characteristics of the soil [25]. The entry gap angle ε is related to the installation design of the support shovel, and the right entry gap angle can efficiently reduce soil entry resistance, facilitate soil cutting, and improve operational efficiency. The entry gap angle range of the common sliding opener is typically 0° to 12° [26], and it is important to consider both the function of the straw cutting and the design of the support shovel in order to determine the range of the entry gap angle of 0° to 5°. The entry angle β is related to the shape design of the shovel tip; the greater the entry angle, the more soil resistance the opener will encounter; on the other hand, the decrease will result in a loss of support shovel strength and shorten the shovel's useful life [27]. The entry angle is designed in reference to the entry angle of the common sliding opener, and the entry angle range of the common sliding opener is generally 25° to 55°; therefore, the entry angle takes the value range of 25° to 55° [28], and this paper is optimized by the discrete element method simulation of this parameter. The normal pressures N_1, N_2, and N_3 on the contact surfaces of the soil and the support shovel result in sliding friction forces f_1, f_2, and f_3, which are correlated with the friction angle between the soil and the furrow opener. When determining the friction angle, the sliding friction increases in proportion to the normal pressure on the contact surface, which, in turn, raises the resistance to plowing. Additionally, N_1, N_2, and N_3 are all connected to the combined impact of the soil's size, shape, and support shovel.

It can be shown that the soil moisture content is a significant element impacting the physical properties of the soil by analyzing the force during the work of the supporting shovel in the soil. The force of the soil on the straw is the main factor affecting the cutting of the straw, and the force of the soil on the straw is greatly affected by the physical properties of the soil, which is the conclusion reached after analyzing the force of the straw in the process of cutting the straw in the disc, so the soil moisture content also influences the process of cutting straw. The force on the support shovel is mostly determined by the entry angle and entry gap angle, which also affect how much resistance the support shovel experiences. As a result, variables including soil moisture content, entry angle, and entry gap angle must be taken into account.

2.2.3. Design of the Support Shovel Structure

Figure 5 illustrates the structure of the support shovel. Decide on the main body's dimensions, which shall be 14 mm for the width b and 16 mm for the thickness c, to ensure that the support shovel has adequate strength. When cutting a straw with a disc and a support shovel, the length l should guarantee that the straw can be supported properly. At this time, $l > l_1 + d/2$, where the diameter of maize straw d is gradually increasing from the top down, measures the diameter of the straw's bottom, which ranges from 21 to 30 mm. Take the bottom of the diameter of the straw at the time of the largest value of 30 mm. The length of l_1 can be calculated by Formula (6), combined with Figure 5a substituting the relevant data that can be obtained l_1 = 181.1 mm; at this time, l > 196.1 mm. To prevent the straw from sliding out along the shovel tip, a margin of at least 10mm should be set. The design margin should be smaller than 50 mm since l is too lengthy to support the shovel's strength, and the length of l after rounding is between 206 and 246 mm. According to the support shovel arc section arc angle α_2 and the support shovel arc section arc radius r_1, the support shovel arc section in horizontal plane projection length l_2 can be calculated. Set the support shovel arc section arc angle α_2 for 70° and the support shovel arc section arc radius r_1 for 74 mm.

$$l_1 = \sqrt{\left(\frac{D+d}{2}\right)^2 - \left(\frac{D-d-2h}{2}\right)} \tag{6}$$

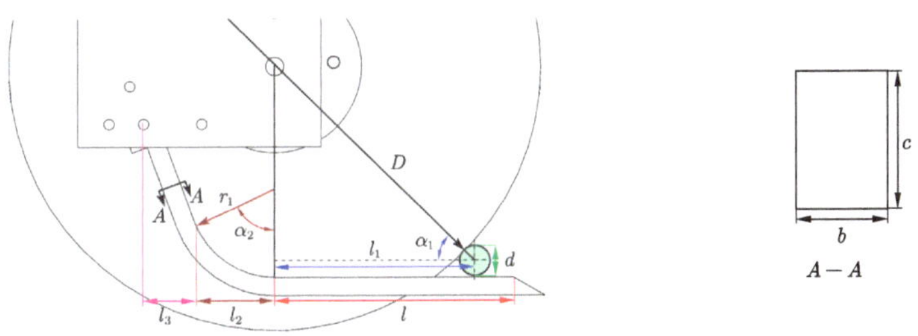

(**a**) Main view (**b**) Section view

Figure 5. Sketch of support shovel structure: l is the distance from the apex of the cutting edge of the support shovel to the vertical position of the center point of the disc, mm; l_1 is the distance from the center of the straw to the vertical position of the center point of the disc, mm; l_2 is the length of the circular arc section of the support shovel projected on the horizontal plane, mm; l_3 is the length of the tilted section of the support shovel projected on the horizontal plane, mm; d is the diameter of the straw, mm; r_1 is the radius of the arc of the circular arc section of the support shovel, mm; α_1 is the angle of the line connecting the center of the disc to the center of the straw and the plumb line, °; α_2 is the angle of the arc of the circular arc section of the support shovel, °.

Through the design of the support shovel structure, it is possible to draw the following conclusions: the appropriate shovel tip margin can prevent the straw from slipping out of the support shovel during the process of cutting straw with the support cutting disc, which is the key to making the support shovel play a supporting role and determines the effectiveness of the support shovel in cutting straw.

2.2.4. Design of Scraper

The disc cuts out slits in the soil, but because they are not wide enough for seeds to fall through, it is required to install auxiliary devices to make them wider. On the disc's one side, soil can be compressed by the scraper to create the seed furrow. The scraper adopts an inclined structure with a narrow bottom and wide top, as shown in Figure 6. The lower part of the narrower is for the soil back to the gap reserved for the convenience of wet soil back to the seed furrow, so the lower periphery of the scraper is designed as a circular arc, the arc center and disc center coincide, in order to avoid the bottom of the furrow is too wide, the periphery of the scraper radius of the arc should be less than the radius of the disc, take r_2 = 220 mm. Winter wheat is typically planted 3 to 5 cm deep in the Huang-Huai-Hai region, with a furrow width of 40 mm, a disc thickness of 6 mm, a lower scraper e_1 scraper width of 34 mm, and an upper scraper e_2 scraper width of 40 mm. The scraper's length is too short, putting too much pressure on one side of the soil while also making it difficult to direct seeds. Its length is also too lengthy, which degrades the quality of the seed furrow that leads back to the soil and is recognized as l_4 = 199 mm.

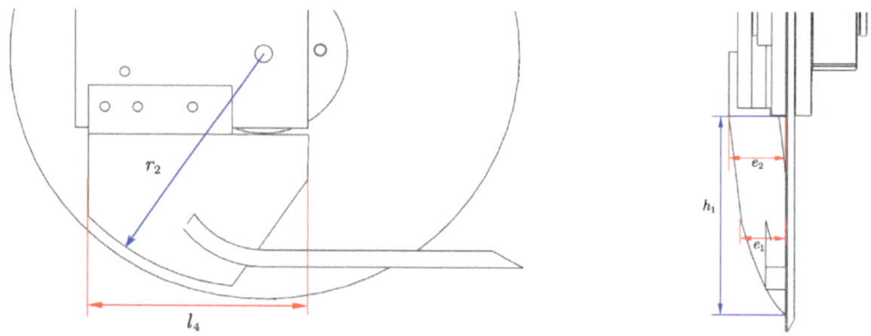

(**a**) Main view (**b**) Side view

Figure 6. Sketch of the scraper: l_4 is the length of the scraper, mm; r_2 is the radius of the peripheral arc of the scraper, mm; e_1 is the width of the lower part of the scraper, mm; e_2 is the width of the lower part of the scraper, mm; h_1 is the height of the scraper, mm.

Through the analysis above, the soil force on the straw is the primary element determining straw cutting, and the soil force on the straw is significantly influenced by the physical properties of the soil, of which the soil water content is a significant influence. The resistance of the support shovel in the movement and the support shovel in the design process are affected by the combined effects of soil moisture content, soil entry angle, soil clearance angle, and shovel tip margin.

2.3. *Discrete Element Simulation of Furrow Openers*

Due to the complexity of the motion involved in cutting straw with a disc in the soil, discrete element method simulation is a better method for simulating the interaction between the working parts and the soil as well as the straw [29]. The support-cutting "device–soil–straw" interaction model was developed using the discrete element simulation program EDEM in conjunction with the aforementioned analysis to simulate the working conditions of the support cutting device, identify the best parameter combinations for the support cutting device, and lay the groundwork for the ensuing field test.

2.3.1. Model of Support Cutting

Simulation test on the disc and support shovel. Set the soil simulation model's dimensions to 1500 mm in length, 360 mm in breadth, and 200 mm in height to ensure the operating range of the working parts. Creating a 3D geometric model in SolidWorks. This model was then saved in .step format and imported into the EDEM software. The device material was 65 Mn steel with a density of 7800 kg·m^{-3}, a shear modulus of 7.0 × 10^{10} Pa, and a Poisson's ratio of 0.3. The machine's forward speed is set to 0.83 m/s. The slip rate is low and negligible when the rotating speed of the disc is small [30]. The simulation can employ the active rotation of the disc in place of the passive rotation of the disc, and the disc is set to rotate periodically at 3.62 rad/s.

2.3.2. Model of Straw

The simulation test involves the disc cutting and crushing of straw. The maize straw is equivalent to an isotropic material, and the Hertz–Mindlin with bonding contact model in EDEM software is utilized to better represent the crushing state of straw. The parameters of maize straw in the literature [31,32] were used to set up the straw model. These parameters include the Poisson's ratio of straw, which is 0.40; its density, which is 470 kg·m^{-3}; its shear modulus, which is 1.7×10^6 pa; its contact model, which is the Hertz–Mindlin with bonding model, which sets the radius of the particles at 2 mm and the radius of the bonded disc at 3 mm; its normal stiffness per unit area, which is 9.6×10^6 N·m^{-3}; its shear stiffness per unit area, which is 6.8×10^6 N·m^{-3}; its critical normal stress, which is 8.72×10^6 Pa; and its critical shear stress, which is 7.5×10^6 Pa. The diameter of straw is taken to be 25 mm, the length is taken to be 180 mm, and the coefficient of restitution of the straw–straw contact is 0.49, the coefficient of static friction is 0.14, and the coefficient of rolling friction is 0.08; the coefficient of restitution of the contact between straw and 65 Mn steel is 0.66, the coefficient of static friction is 0.23, and the coefficient of rolling friction is 0.12, and the straw model is shown in Figure 7.

(**a**) Main view (**b**) Side view

Figure 7. Corn straw model.

2.3.3. Model of Soil

Considering that soil cohesion and adhesion are different in soil conditions with different water content, the soil is modeled with soil parameters of 12 ± 1%, 16 ± 1%, and 20 ± 1% water content, respectively, with reference to the literature [33]. The spherical particles with a radius of 4 mm were chosen to construct the soil model, and the particle contact model was the Hertz–Mindlin with bonding model. The model can be used to measure the soil's water content using the bonding radius and by modifying the bonding bonds' parameters to simulate the various soil properties [34], soil-related parameter references [31,33–40], and the precise simulation parameter settings, as shown in Table 1.

Table 1. Basic parameters of the soil model.

Parameters	Value		
Moisture content/%	12 ± 1	16 ± 1	20 ± 1
Density/(kg·m^{-3})	2050	2090	2150
Poisson's ratio	0.35	0.38	0.41
Shear modulus/Pa	0.85×10^6	1.05×10^6	1.24×10^6
Coefficient of restitution (soil–soil)	0.15	0.13	0.1
Coefficient of static friction (soil–soil)	0.532	0.364	0.268
Coefficient of rolling friction (soil–soil)	0.25	0.22	0.2
Normal stiffness per unit area/(N·m^{-3})	1.3×10^7	1.9×10^7	2.5×10^7
Shear stiffness per unit area/(N·m^{-3})	9×10^6	1.4×10^7	1.9×10^7
Critical normal stress/Pa	50,000	55,000	62,000
Critical shear stress/Pa	25,000	29,000	35,000
Coefficient of restitution (soil–straw)	0.6	0.5	0.4
Coefficient of static friction (soil–straw)	0.573	0.539	0.483
Coefficient of rolling friction (soil–straw)	0.21	0.18	0.16
Coefficient of restitution (soil–steel)	0.18	0.15	0.12
Coefficient of static friction (soil–steel)	0.351	0.571	0.65
Coefficient of rolling friction (soil–steel)	0.05	0.05	0.05
Bonded disk radius	4.34	4.47	4.62

2.3.4. Overall Model

Create a bed of particles to quickly generate the soil bin. In this procedure, maize straw is manufactured in the form of the particle factory API (Application Programming Interface), and the particle bonding time is set at 3.41 s. The soil layer is 200 mm thick in total, and the straws can be arranged in two different ways. The first one involves placing straw on the soil's surface to simulate the effect of straw being sliced by discs there. The second one involves placing straws in the middle and upper parts of the soil. The soil particles are divided into two layers, with the bottom layer being 150 mm tall, and the straw was laid on the surface of this layer, which was then covered with a 50 mm thick layer of soil to simulate the state of the straw when it is not cut off but is pressed into the soil and the support shovel just supports the straw. Four straws were placed in the soil bin, spaced 300 mm apart to reduce computational complexity, and the discrete element model is depicted in Figure 8.

(**a**) Straw covers the soil surface (**b**) Straw in the upper-middle soil

Figure 8. Overall discrete element model.

2.3.5. Simulation Test Scheme

According to the previous analysis, the test factors were determined as follows: soil moisture content of 12 ± 1%, 16 ± 1%, and 20 ± 1%; entry angle of 25–55°; entry gap angle of 0–5°; and shovel tip margin of 10–50 mm. Among them, soil moisture content is an external factor, so a single-factor test is used to analyze its influence on the device in the process of cutting straw. The entry angle, the entry gap angle, and the shovel tip margin are the structural parameters in the design process of the device, so the quadratic orthogonal test is used to optimize the design of these factors. First, a single-factor simulation test using straw that was covered in soil was conducted to ascertain the impact of soil conditions with various moisture contents on the device for cutting straw. The soil moisture contents used were 12 ± 1%, 16 ± 1%, and 20 ± 1%, with entry angles of 25°, entry clearance angles of 0°, and shovel tip margins of 206 mm. Then, using the Design-Expert software for the entry angle, the entry gap angle, and the shovel tip margin of the multifactor simulation optimization test to determine the best parameter combinations for the high soil moisture content case, that is, simulating the soil moisture content of 20 ± 1%, the state of the straw after being pressed into the soil, with the entry angle of 25 to 55°, the entry gap angle of 0 to 5°, and the shovel tip margin of 10 to 50 mm.

2.3.6. Data Collection and Processing

The Number of Broken Bonds

The bond of the straw model will break during the simulation due to the action of shear stress. From the contact between the disc and the first straw to the last straw being completely cut off, the number of broken bonds with simulation time is a step-like regular change calculation of the length of each step change can be obtained by calculating the average time for the disc to cut off a straw. Figure 9 illustrates the bonds break regularity at 12% soil moisture content. The red dashed line denotes the moment when the disc first came into contact with the first straw, and the blue dashed line denotes the moment when the disc severed the first straw.

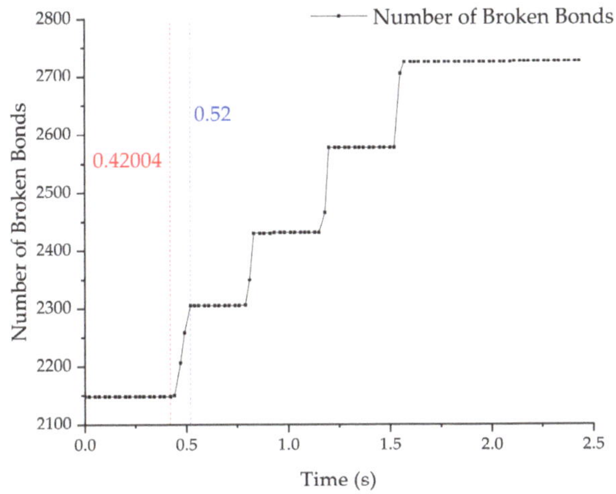

Figure 9. Number and time of broken bonds.

Tillage Component Resistance

After the simulation, the horizontal and vertical resistance of each component can be exported through the post-processing module of the software.

2.4. Field Experiment

2.4.1. Field Test Conditions and Equipment

On 15–18 November 2022, the test was conducted in Xinglongtun, Jiangshan Town, Laixi City, Qingdao City (36°36′17″ N, 120°36′50″ E). The test field was located in the wheat–maize double-ripening area of the Huang-Huai-Hai region, and the conservation-tillage model was practiced. The previous crop was maize at the time of the test, and the ground was covered in straw and stubble with an average moisture content of 35.21% for straw and roughly 17.82% for soil.

According to the optimal parameters obtained from the regression model, it is possible to complete the trial production of a single-disc furrowing and straw-cutting device. It is connected to the tractor by a three-point suspension during operation and is powered by a LUZHONG-1004A tractor, as shown in Figure 10a.

(**a**) Test equipment

(**b**) Test conditions

Figure 10. Field experiment.

2.4.2. Test Indicators and Methods

Straw Cut-Off Rate

Twenty straws, each measuring approximately 55 cm in length, were distributed uniformly throughout the planting belt after the ground was cleared of any remaining straw (Figure 10b). The machine drives forward for a while before passing uniformly across the straw-laid planting belt under the conditions of 3 km/h forward speed and 100 mm furrow depth. After the operation, the straw was considered cut when it broke into two parts, and the rest was considered uncut. The total number of cut-off straws was counted, the cut-off rate of straws was calculated, the test was repeated three times, the average value was taken, and the cut-off rate was calculated by the formula:

$$\eta = \frac{Q_d}{Q_t} \times 100\% \tag{7}$$

where η is the straw-cutting rate, %; Q_d is the number of cut-off straws; Q_t is the total number of straws.

Stability Coefficient of Furrowing Depth

The test verified that the groove profile was opened by the furrow opener, as shown in Figure 11a (H_1 is the depth of the furrow opening). The measurement of furrow depth is shown in Figure 11b. When measuring the depth of the furrow, the method of taking the average value of segment measurement in the literature [41] was adopted, five points were randomly chosen to measure the furrow depth in the operating interval with the presence of straw, and the average value was obtained. The test was repeated three times, and the stability coefficient of the furrow depth was obtained by calculating the average and standard deviation of the furrow depth.

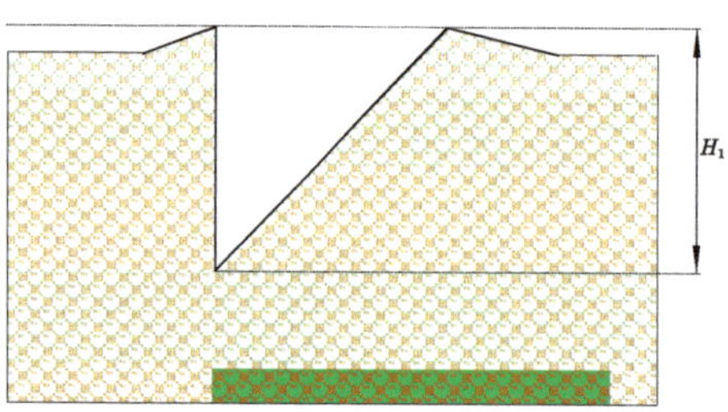

(**a**) Groove profile

(**b**) Furrow depth measurement

Figure 11. Measurement of furrowing depth.

3. Results and Discussion

3.1. Single-Factor Simulation Test

3.1.1. Influence of Soil Moisture Content on the Number of Broken Bonds in Straw

The results of the single-factor simulation test are shown in Table 2. The largest number of bonds broken was 574 when the soil moisture content was 12 ± 1%, but the average time for a straw to be entirely cut was the shortest, 0.067 s. This shows that the soil was hard and conducive for cutting straw when the soil moisture content was low. When the soil moisture content is 16 ± 1%, the number of broken bonds is 512, and the average time for a straw to be completely cut is 0.082 s. The least number of broken bonds occurs when the soil moisture level is 20 ± 1%. At this point, the average time required to entirely cut a stalk is 0.085 s, which is the longest time. The amount of time it took to completely cut off the straw increased as the soil's moisture content increased, indicating that as the soil's moisture content increased, the friction between soil particles and the straw decreased, making it more likely for the straw to slip on the soil's surface or go deeper into the soil. As a result, the disc had to move farther before the straw was completely cut off.

Table 2. Single-factor simulation test results.

Moisture Content/%	The Number of Broken Bonds	The Average Cut-Off Time/s
12 ± 1	578	0.067
16 ± 1	512	0.082
20 ± 1	505	0.085

3.1.2. Influence of Soil Moisture Content on the Device's Operational Resistance

The device is subjected to horizontal resistance in the direction of advance during the simulation process, as well as vertical resistance from the vertical, horizontal plane downward. The disc and support shovel were the analysis's objectives, and the resistance of the device was analyzed for various soil moisture contents.

The simulation lasts 2.43 seconds. The disc is assumed to finish contacting the first straw and entirely cut the last straw between 0.41 and 1.55 s, and this stage falls under the category of the disc cutting straw. Figure 12 shows the change curves of the resistance values.

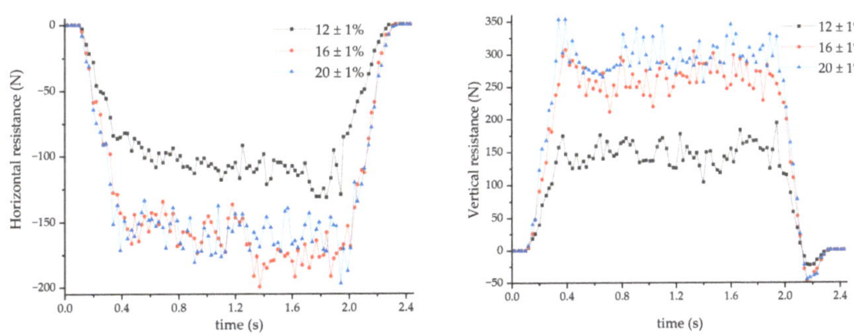

(a) Horizontal resistance (b) Vertical resistance

Figure 12. Variation in resistance values to discs under different soil moisture content conditions.

Figure 12a shows that as the disc is cutting straws, its horizontal resistance is negative, suggesting that the force acting in this direction is the opposite of the advancing direction. The horizontal resistance of the disc is low and steady, with a narrow fluctuation range when the moisture content is low, and the fluctuation range widens as the moisture content increases. The soil cohesion is less affected by the water content, and the disc's horizontal resistance is less sensitive to changes in water content when the water content is within a specified range.

Figure 12b shows that the vertical resistance to the disc during the stage of cutting straw by them was all positive, demonstrating that the disc was supported upward by the soil during its operation. When the water content of the soil is low, the overall vertical resistance of the disc is low. However, as the water content rises, the vertical resistance significantly increases. This is likely because the viscous soil creates a buildup after the support shovel picks up the soil, which exerts more force on the disc.

The device was fully in the simulated soil bin from 0.6 to 1.83 s of the simulation period. The resistance to the support shovel during this time was analyzed, and Figure 13 shows the variation curve of the resistance value with the two green dashed lines denoting 0.6 and 1.83 s, respectively.

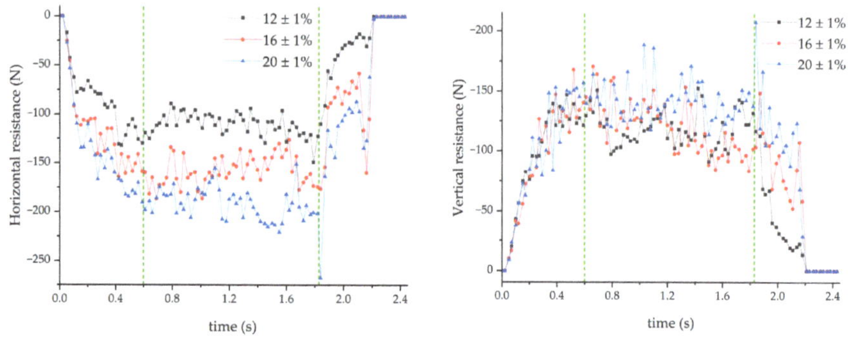

(**a**) Horizontal resistance (**b**) Vertical resistance

Figure 13. Variation in resistance values to support shovels under different soil moisture content conditions.

Figure 13a shows that the horizontal resistance to the support shovels is uniformly negative, showing that the force operates in the opposite direction to that of forward speed. When the soil moisture content is 12 ± 1%, the support shovel's horizontal resistance ranges from −88.94 to −149.04 N, with 149.04 N as the maximum value; when the soil moisture content is 16 ± 1%, the horizontal resistance ranges from −126.08 to −186.25 N, with 186.25N as the maximum value; and when the soil moisture content is 20 ± 1%, the horizontal resistance ranges from −150 to −220.66 N, with 220.66 N as the maximum value. The horizontal resistance and fluctuation range increase as the soil moisture content rises. This is because the soil's adhesion to the support shovel changes as the moisture content of the soil fluctuates. The soil's adhesion to the support shovel also increases as the moisture content increases, and as the soil and the support shovel generate relative sliding, tangential friction along the direction of movement rises as well, increasing the horizontal resistance of the support shovel.

Figure 13b shows that the vertical resistance of the support shovel is all negative, indicating that the support shovel is subjected to downward vertical resistance. When the soil moisture content is 12 ± 1%, the vertical resistance of the support shovel ranges from −91.21 to −152.68 N, with the maximum value of 152.68N; when the soil moisture content is 16 ± 1%, the vertical resistance ranges from −83.57 to −170.69 N, with the maximum value of 170.69N; when the soil moisture content is 20 ± 1%, the vertical resistance ranges from −113.94 to −188.94 N, with the maximum value of 188.94 N. The support shovel's vertical resistance is subjected to a wide range of fluctuations during the simulation process, but when the soil moisture content is 16 ± 1% and 20 ± 1%, the maximum resistance value is produced by a sudden change. The reason may be due to the increase in soil moisture content, so that the internal structure of the soil changes, the soil support for the straw may not be sufficient, the disc cannot quickly cut off the straw, the straw is pressed down and pushed forward, and when the soil is piled up to a certain extent, the straw that is between the disc and the soil layer on the upper part of the support shovel is cut off, resulting in a sudden change in resistance in the vertical direction.

3.2. Quadratic Orthogonal Simulation Test

3.2.1. Test Design

To find the optimal parameter, Design-Expert software was used to carry out a Box–Behnken test design for a total of 17 groups of experiments. The factor-level coding is shown in Table 3, and the experimental scheme and results are shown in Table 4, where X_1, X_2, and X_3 are the factor coding values, and the number of broken bonds Y is the experimental index.

Table 3. Coding table of the test factor level.

Levels	Entry Angle β/(°)	Entry Gap Angle ε/(°)	Shovel Tip Margin/mm
−1	25	0	10
0	40	2.5	30
1	55	5	50

Table 4. Experiment scheme and results.

Test Serial Number No.	Factors and Levels			Response Index
	X_1	X_2	X_3	The Number of Broken Bonds/Y
1	−1	−1	0	482
2	1	−1	0	505
3	−1	1	0	339
4	1	1	0	355
5	−1	0	−1	384
6	1	0	−1	412
7	−1	0	1	282
8	1	0	1	398
9	0	−1	−1	492
10	0	1	−1	367
11	0	−1	1	485
12	0	1	1	309
13	0	0	0	323
14	0	0	0	335
15	0	0	0	313
16	0	0	0	319
17	0	0	0	308

3.2.2. Analysis of Experimental Results and Establishment of Regression Model

The results of the simulation tests were analyzed by multiple fitting and regression analysis using Design-Expert data-processing software to obtain a regression model for the number of broken bonds Y. Table 5 shows the ANOVA results of the model. The regression model is significant, where X_1, X_2, and X_3 in the linear term and X_2^2 in the quadratic term were extremely significant ($p < 0.01$); X_1^2, X_3^2 in the quadratic term, and X_1X_3 in the interaction term were significant ($0.05 < p < 0.1$); and the rest of the terms were not significant. The three selected test factors have a quadratic relationship with the number of bond breaks, and the primary and secondary order of the degree of influence of each test factor on the test index Y is as follows: the entry gap angle, the entry angle, and the shovel tip margin. The non-significant terms were removed, and the fitting process was repeated to determine the regression mathematical model equation for the number of broken bonds Y:

$$Y = 319.6 + 22.88X_1 - 74.25X_2 - 22.63X_3 + 20X_1X_3 + 28.2X_1^2 + 72.45X_2^2 + 21.2X_3^2 \quad (8)$$

To further analyze the interaction effects of entry angle, entry gap angle, and shovel tip margin on the number of bond breaks Y, the response surfaces of the influencing factors on the test indexes were established, as shown in Figure 14. When the entry angle is between 25° and 39°, as shown in Figure 14, the number of broken bonds tends to go lower as the shovel tip margin goes up. This is because the entry angle is smaller, the soil is disturbed less, the contact time between the straw and the disc is short, and the straw can be quickly severed. The entry angle is between 39° and 55°, and the number of broken bonds with the increase in the shovel tip margin was first reduced and then increased. This change is because the entry angle is small, the shovel tip margin on the number of broken bonds is greater than the entry angle, and the number of broken bonds is a reduction trend. Increased entry angle causes the support shovel to disturb the soil more, but it has less of an impact on the number of broken bonds than increased shovel tip margin, which causes more soil to warp ahead of time and advances the point at which the straw first makes contact with the disc and the support shovel. As the opener moves forward, the straw slides along the support shovel, increasing the time of the cutting of the straw, and the cutting surface of the straw produces tearing, which increases the number of broken bonds.

Table 5. Analysis of variance for the number of broken bonds.

Source	Sum of Square	Free Degree	Mean Square	*F* Value	*p* Value
Model	84,452.49	9	9383.61	30.72	<0.0001 **
X_1	4186.13	1	4186.13	13.70	0.0076 **
X_2	44,104.50	1	44,104.50	144.37	<0.0001 **
X_3	4095.13	1	4095.13	13.40	0.0081 **
X_1X_2	12.25	1	12.25	0.0401	0.8470
X_1X_3	1936.00	1	1936.00	6.34	0.0400 *
X_2X_3	650.25	1	650.25	2.13	0.1879
X_1^2	3348.38	1	3348.38	10.96	0.0129 *
X_2^2	22,101.06	1	22,101.06	72.35	<0.0001 **
X_3^2	1892.38	1	1892.38	6.19	0.0417 *
Residual	2138.45	7	305.49		
Lack of Fit	1711.25	3	570.42	5.34	0.0696
Pure Error	427.20	4	106.80		
Cor Total	86,590.94	16			

* means that the impact is significant ($0.01 \leq p < 0.05$); ** means that the impact is extremely significant ($p < 0.01$).

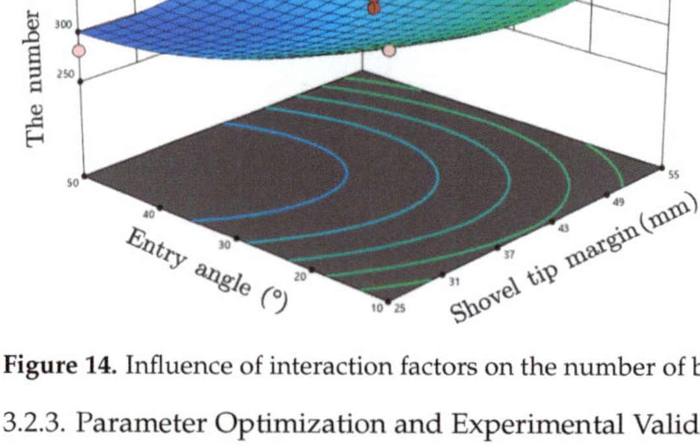

Figure 14. Influence of interaction factors on the number of broken bonds.

3.2.3. Parameter Optimization and Experimental Validation

A multi-objective optimization solution was performed on the entry angle, entry gap angle, and shovel tip margin based on the results of the response surface analysis and the actual operational requirements, using the numerical optimization function of the

Design-Expert software, with the aim of increasing the number of broken bonds. Then, specify the objective of the number of broken bonds Y to be maximized in the numerical module's criterion section. The constraint conditions of the objective function and each parameter variable are shown in Equation (9).

$$\begin{cases} \max Y(\beta, \varepsilon, l) \\ s.t \begin{cases} 25 \leqslant \beta \leqslant 55 \\ 0 \leqslant \varepsilon \leqslant 5 \\ 10 \leqslant l \leqslant 50 \end{cases} \end{cases} \quad (9)$$

The objective function is solved to obtain a variety of sets of optimum parameter combinations. At the entry angle of 49.44°, the entry gap angle of 0.05°, and the shovel tip margin of 10.10 mm, the number of straw bonds broken is 506. Round the optimal combination of parameters, set the entry angle at 49°, the entry gap angle at 0°, and the shovel tip margin at 10 mm, and run a simulation verification to obtain the number of broken bonds at 478. The number of broken bonds and optimization prediction of the results were obtained with a certain degree of difference. The difference between the two is 5.6%, which is small and meets the requirements of the operation.

3.3. Field Tests

The single-disc furrowing and straw-cutting device operates in the presence of straws, can cut off the surface straw, and produces a stable depth of the seed furrow. The results of the field test are provided in Tables 6 and 7, and the opener effect is displayed in Figure 15. As shown in Table 6, the average cut-off rate of the device was 71.7%. From Table 7, the stability coefficient of furrow depth for the device is 90.87%.

Table 6. Cut-off rate.

NO.	Total Number of Straws	Number of Straw Cut Off	Cut-Off Rate/%	Average Cut-Off Rate/%
1	20	15	75	
2	20	16	80	71.7%
3	20	12	60	

Table 7. Stability coefficient of furrowing depth.

NO.	H/mm	Average Value/mm	Standard Deviation	Coefficient of Variation/%	Stability Coefficient/%
1	10.44				
2	8.74	9.49	0.866	9.13	90.87
3	9.3				

Figure 15. Effect of furrowing.

In the test process, if the straw is pulled by the disc out of the planting belt and the straw cannot be completely cut off by the disc, the cutting efficiency is reduced. The forward speed of the machine is also a significant factor affecting the efficiency of the disc opener in cutting the straw [9], and the test time when the forward speed of the machine is low also reduces the efficiency of the cutting of the straw. During the operation, it was difficult to keep the furrowing depth of the device stable. Due to the poor profiling ability of the device itself during the test, coupled with the influence of straw on the surface of the ground, it was difficult to improve the stability coefficient of the furrowing depth.

4. Conclusions

(1) A single-disc opener was created based on the principle of support cutting to address the issue of high soil moisture content, large amounts of straw, and the stubble-cutting device's inability to completely cut off the straw in the no-tillage sowing operation of wheat in the biannual ripening area of maize and wheat in the Huang-Huai-Hai region. The cutting mechanism of the device was revealed through the analysis of the force of the straw and the analysis of the device's performance.

(2) The "device-straw-soil" discrete element model was created, and a single-factor simulation test was conducted to examine the effects of soil conditions with different moisture contents on the number of broken bonds and the changes in resistance of the device to cut off the straw. A quadratic orthogonal test was conducted to establish a regression model for the number of broken bonds, and the following parameter combinations were found to be the most favorable: entry angle 49°, entry gap angle 0°, and shovel tip margin 10 mm.

(3) Results from field tests indicate that the average cut-off rate of the device on maize straw is 71.7%, and the stability coefficient of furrow depth is 90.87%, which shows that the machine has suitable operating performance and produces a stable depth of seed furrow.

Author Contributions: Conceptualization, G.Z. and H.L.; methodology, G.Z. and H.L.; software, Z.T. and D.H.; writing—original draft preparation, G.Z. and D.C.; writing—review and editing, J.H., Q.W., C.L. and C.W.; funding acquisition, H.L. All authors have read and agreed to the published version of the manuscript.

Funding: This research was funded by the China Agriculture Research System of MOF and MARA: CARS-03 and the 2115 Talent Development Program of China Agricultural University and Chinese Universities Scientific Fund: 2021TC105.

Institutional Review Board Statement: Not applicable.

Data Availability Statement: Not applicable.

Conflicts of Interest: The authors declare no conflict of interest.

References

1. Kassam, A.; Friedrich, T.; Derpsch, R. Global Spread of Conservation Agriculture. *Int. J. Environ. Stud.* **2019**, *76*, 29–51. [CrossRef]
2. He, J.; Li, H.; Chen, H.; Lu, C.; Wang, Q. Research Progress of Conservation Tillage Technology and Machine. *Trans. Chin. Soc. Agric. Mach.* **2018**, *49*, 1–19. [CrossRef]
3. Zhuang, J.; Jia, H.; Ma, Y.; Li, Y.; Di, Y. Design and Experiment of Sliding-knife-type Disc Opener. *Trans. Chin. Soc. Agric. Mach.* **2013**, *44*, 83–88. [CrossRef]
4. *ASABE Standard S477.1*; Terminology for Soil-Engaging Components for Conservation-Tillage Planters, Drills and Seeders. ASABE: St. Joseph, MI, USA, 2013.
5. Baker, C.J.; Saxton, K.E.; Ritchie, W.R.; Chamen, W.C.T.; Reicosky, D.C.; Ribeiro, F.; Justice, S.E.; Hobbs, P.R. *No-Tillage Seeding in Conservation Agriculture*, 2nd ed.; FAO: Rome, Italy, 2006.
6. Xu, G.; Xie, Y.; Peng, S.; Liang, L.; Ding, Q.S. Performance Evaluation of Vertical Discs and Disc Coulters for Conservation Tillage in an Intensive Rice–Wheat Rotation System. *Agronomy* **2023**, *13*, 1336. [CrossRef]
7. Torotwa, I.; Ding, Q.; Makange, N.R.; Liang, L.; He, R. Performance Evaluation of a Biomimetically Designed Disc for Dense-Straw Mulched Conservation Tillage. *Soil Tillage Res.* **2021**, *212*, 105068. [CrossRef]

8. Kumar, N.; Sawant, C.P.; Sharma, R.K.; Chhokar, R.S.; Tiwari, P.S.; Singh, D.; Roul, A.K.; Tripathi, S.C.; Gill, S.C.; Singh, G.P. Combined Effect of Disc Coulters and Operational Speeds on Soil Disturbance and Crop Residue Cutting under No-Tillage System in Soil Bin. *J. Sci. Ind. Res.* **2021**, *80*, 739–749. [CrossRef]
9. Zeng, Z.; Thoms, D.; Chen, Y.; Ma, X. Comparison of Soil and Corn Residue Cutting Performance of Different Discs Used for Vertical Tillage. *Sci. Rep.* **2021**, *11*, 2537. [CrossRef]
10. Ahmad, F.; Weimin, D.; Qishuo, D.; Hussain, M.; Jabran, K. Forces and Straw Cutting Performance of Double Disc Furrow Opener in No-Till Paddy Soil. *PLoS ONE* **2015**, *10*, e0119648. [CrossRef]
11. Magalhães, P.S.G.; Bianchini, A.; Braunbeck, O.A. Simulated and Experimental Analyses of a Toothed Rolling Coulter for Cutting Crop Residues. *Biosyst. Eng.* **2007**, *96*, 193–200. [CrossRef]
12. Bianchini, A.; Magalhães, P.S.G. Evaluation of Coulters for Cutting Sugar Cane Residue in a Soil Bin. *Biosyst. Eng.* **2008**, *100*, 370–375. [CrossRef]
13. Doan, V.; Chen, Y.; Irvine, B. Effect of Oat Stubble Height on the Performance of No-till Seeder Openers. *Can. Biosyst. Eng./Le Genie des biosystems au Canada* **2005**, *47*, 2.37–2.44.
14. Lu, C. Study on Anti-Blocking Technology and Device of Rotary Cuttingwith Slide Plate Pressing Straw for No-till Planter. Ph.D. Thesis, China Agricultural University, Beijing, China, 2014.
15. Sarauskis, E.; Masilionyte, L.; Romaneckas, K.; Kriauciuniene, Z.; Jasinskas, A. The Effect of the Disc Coulters Forms and Speed Ratios on Cutting of Crop Residues in No-Tillage System. *Bulg. J. Agric. Sci.* **2013**, *19*, 620–624.
16. Ahmad, F.; Weimin, D.; Qishou, D.; Rehim, A.; Jabran, K. Comparative Performance of Various Disc-Type Furrow Openers in No-Till Paddy Field Conditions. *Sustainability* **2017**, *9*, 1143. [CrossRef]
17. Murray, J.R.; Tullberg, J.N.; Basnet, B.B. *Planters and Their Components: Types, Attributes, Functional Requirements, Classification and Description (ACIAR Monograph No. 121)*; ACIAR: Canberra, Australia, 2006.
18. Malasli, M.Z.; Celik, A. Disc Angle and Tilt Angle Effects on Forces Acting on a Single-Disc Type No-till Seeder Opener. *Soil Tillage Res.* **2019**, *194*, 104304. [CrossRef]
19. Wang, Q.; Li, H.; He, J.; Lu, C.; Su, Y. Design and experiment on twist type ridge-clear device. *Trans. CSAE* **2010**, *26*, 109–113.
20. Ministry of Agriculture and Rural Affairs of the People's Republic of China. *Technical Specification for Mechanized Production of Wheat*; China Agriculture Press: Beijing, China, 2021. Available online: https://www.sdtdata.com/fx/fmoa/tsLibCard/183938.html (accessed on 23 July 2023).
21. Kushwaha, R.L.; Vaishnav, A.S.; Zoerb, G.C. Soil Bin Evaluation of Disc Coulters Under No-Till Crop Residue Conditions. *Trans. ASAE* **1986**, *29*, 40–44. [CrossRef]
22. Zhang, X.; You, Y.; Wang, D.; Wang, Z.; Liao, Y.; Lv, J. Design and Experiment of Soil-breaking and Root-cutting Cutter Based on Discrete Element Method. *Trans. Chin. Soc. Agric. Mach.* **2022**, *53*, 176–187. [CrossRef]
23. You, Y.; He, C.; Wang, D.; Wang, G. Interaction Relationship between Soil and Very Narrow Tine during Penetration Process. *Trans. Chin. Soc. Agric. Mach.* **2017**, *48*, 50–58. [CrossRef]
24. Lv, Y.; Li, B. *Soil Science*; China Agriculture Press: Beijing, China, 2006; pp. 73–75.
25. Wang, L.; Liao, J.; Hu, H.; Liu, L.; Bai, X.; Chen, C. Research status and prospect of adhesion reduction and desorption technology for agricultural machinery parts touching soil. *J. Chin. Agric. Mech.* **2021**, *42*, 214–221. [CrossRef]
26. Zhang, X.; Li, H.; Du, R.; Ma, S.; He, J.; Wang, Q.; Chen, W.; Zheng, Z. Effects of key design parameters of tine furrow opener on soil seedbed properties. *Int. J. Agric. Biol. Eng.* **2016**, *9*, 67–80. [CrossRef]
27. Chinese Academy of Agricultural Mechanization Sciences. *Agricultural Machinery Design Manual*; China Agricultural Science and Technology Press: Beijing, China, 2007.
28. Gou, W.; Ma, R.; Yang, W.; Fan, G.; Lei, X.; Hui, K.; Yang, H. Design of opener on no-till wheat seeder. *Trans. CSAE* **2012**, *28*, 21–25.
29. Ucgul, M.; Fielke, J.M.; Saunders, C. Three-Dimensional Discrete Element Modelling (DEM) of Tillage: Accounting for Soil Cohesion and Adhesion. *Biosyst. Eng.* **2015**, *129*, 298–306. [CrossRef]
30. Zhao, S.; Wang, J.; Yang, C.; Chen, J.; Yang, Y. Design and Experiment of Stubble Chopper under Conservation Tillage. *Trans. Chin. Soc. Agric. Mach.* **2019**, *50*, 57–68. [CrossRef]
31. Zhu, H.; Qian, C.; Bai, L.; Zhao, H.; Ma, S.; Zhang, X.; Li, H. Design and experiments of active anti-blocking device with forward-reverse rotation. *Trans. CSAE* **2022**, *38*, 1–11. [CrossRef]
32. Zhang, F.; Song, X.; Zhang, X.; Zhang, F.; Wei, W.; Dai, F. Simulation and experiment on mechanical characteristics of kneading and crushing process of corn straw. *Trans. CSAE* **2019**, *35*, 58–65. [CrossRef]
33. Zhao, S.; Liu, H.; Tan, H.; Cao, X.; Zhang, X.; Yang, Y. Design and performance experiment of opener based on bionic sailfish head curve. *Trans. CSAE* **2017**, *33*, 32–39. [CrossRef]
34. Pan, S.Q.; Cao, Z.F.; Jiang, S.S.; Yu, J.Q. Experimental Research on the Core Ploughshare Furrow Opener Based on the Discrete Element Method. In Proceedings of the 2014 International Conference on Mechanics and Civil Engineering (ICMCE 2014), Wuhan, China, 13–14 December 2014. [CrossRef]
35. Zhao, S.; Gu, Z.; Yuan, Y.; Lv, J. Bionic Design and Experiment of Potato Curved Surface Sowing Furrow Opener. *Trans. Chin. Soc. Agric. Mach.* **2021**, *52*, 32–42+64. [CrossRef]
36. Zhao, S.; Liu, H.; Yang, C.; Yang, L.; Gao, L.; Yang, Y. Design and Discrete Element Simulation of Interactive Layered Subsoiler with Maize Straw Returned to Filed. *Trans. Chin. Soc. Agric. Mach.* **2021**, *52*, 75–87. [CrossRef]

37. Tian, X.; Cong, X.; Qi, J.; Guo, H.; Li, M.; Fan, X. Parameter Calibration of Discrete Element Model for Corn Straw-Soil Mixture in Black Soil Areas. *Trans. Chin. Soc. Agric. Mach.* **2021**, *52*, 100–108+242. [CrossRef]
38. Sun, J. Structural and Mechanical Characteristics of Corn Stubble and Its Tribological Properties against Soil. Master's Thesis, Jilin Agricultural University, Changchun, China, 2011.
39. Li, J.; Tong, J.; Hu, B.; Wang, H.; Mao, C.; Ma, Y. Calibration of parameters of interaction between clayey black soil with different moisture content and soil-engaging component in northeast China. *Trans. CSAE* **2019**, *35*, 130–140. [CrossRef]
40. Zhao, S.; Liu, H.; Hou, L.; Zhang, X.; Yuan, Y.; Yang, Y. Development of deep fertilizing no-tillage segmented maize sowing opener using discrete element method. *Trans. CSAE* **2021**, *37*, 1–10. [CrossRef]
41. Jia, H.; Meng, F.; Liu, L.; Shi, S.; Zhao, J.; Zhuang, J. Biomimetic Design and Experiment of Core-share Furrow Opener. *Trans. Chin. Soc. Agric. Mach.* **2020**, *51*, 44–49+77. [CrossRef]

agriculture

MDPI

Article

Discrete Element Method Simulation and Field Evaluation of a Vibrating Root-Tuber Shovel in Cohesive and Frictional Soils

Emmanuel Awuah [1], Kojo Atta Aikins [2], Diogenes L. Antille [3,4], Jun Zhou [1,*], Bertrand Vigninou Gbenontin [1], Peter Mecha [1] and Zian Liang [1]

1 College of Engineering, Nanjing Agricultural University, No. 40 Dianjiangtai, Pukou District, Nanjing 210031, China; eawuah@njau.edu.cn (E.A.); bertrandfirst@gmail.com (B.V.G.); mechapeter24@gmail.com (P.M.); lza1997@stu.njau.edu.cn (Z.L.)
2 Department of Agricultural and Biosystems Engineering, Kwame Nkrumah University of Science and Technology, Kumasi AK-385-1973, Ghana; kaaikins.coe@knust.edu.gh
3 CSIRO Agriculture and Food, Black Mountain Science and Innovation Precinct, Canberra, ACT 2601, Australia; dio.antille@csiro.au
4 Centre for Agricultural Engineering, University of Southern Queensland, Toowoomba, QLD 4350, Australia
* Correspondence: zhoujun@njau.edu.cn; Tel.: +86-13914467155

Abstract: Soil-cutting forces are key indicators of root-tuber harvesters and other soil-engaging tools' performance. To improve operational efficiency, minimise soil disturbance, and reduce fuel consumption, the draught and vertical forces involved in root and tuber crop harvesting must be minimised. Two field experiments assessed the harvester's performance at a depth of 200 mm, varying frequencies, and travel speeds on clay and sandy loam soils. Discrete element models (DEM) were developed and subsequently used to replicate the field experiments and evaluate S-shaped and fork-shaped shovels. Linear regression and ANOVA ($p < 0.05$) were used to analyse the data. Draught force concurrently increased with speed in both soil textures but decreased with vibration frequency. The draught force decreased by approximately 41% in clay soil and 21% in sandy loam soil when the harvester was operated between 5 Hz and 14.5 Hz and between 10 Hz and 12.5 Hz, respectively. DEM simulations had relative errors of 4% (clay) and 4.7% (sandy loam) for draught force and drawbar power compared to experimental data. The S-shaped shovel was more efficient at crushing and translocating soil–crop mass to the rear of the harvester than the fork-shaped shovel. These DEM soil–crop models are reliable for evaluating other root-tuber harvesting tools.

Keywords: clay; frequency; sandy loam; soil reaction forces; soil–crop model; Jerusalem artichoke

Citation: Awuah, E.; Aikins, K.A.; Antille, D.L.; Zhou, J.; Gbenontin, B.V.; Mecha, P.; Liang, Z. Discrete Element Method Simulation and Field Evaluation of a Vibrating Root-Tuber Shovel in Cohesive and Frictional Soils. *Agriculture* **2023**, *13*, 1525. https://doi.org/10.3390/agriculture13081525

Academic Editor: Jin He

Received: 6 July 2023
Revised: 24 July 2023
Accepted: 24 July 2023
Published: 31 July 2023

1. Introduction

Harvesting Jerusalem artichokes (*Helianthus tuberosus* L.) is an integral part of the production and marketing. This step influences the cost-effectiveness of the crop and can result in considerable economic losses if not performed accurately [1]. The harvesting operation generally consists of digging up the crop, separating the tubers from the soil, and picking, and is performed either manually or with mechanical harvesters. Unlike potato harvesting, the mechanical harvesting of Jerusalem artichokes is not advanced. Therefore, the development, performance evaluation, and optimisation of a specialised Jerusalem artichoke mechanical harvester are crucial.

The energy requirements or soil-cutting forces (draught, vertical and lateral forces) for digging tubers are directly linked to the soil-cutting tool (shovel) design and are essential performance indicators [2,3]. Due to the increasing size of harvesters, it has become even more crucial to keep the draught and vertical forces required to dig root and tuber crops minimal. This enables the use of smaller tractors with low overall drawbar power requirements to pull root and tuber harvesters [4,5].

Many researchers have proposed the use of vibrating blades for harvesting root and tuber crops [4–7]. The goal is to reduce power consumption while improving tuber separation efficiency, energy utilisation efficiency, scouring of soil-working components (reducing adhesion), and reducing soil compaction [8,9]. Limited work has been done on the adoption of oscillating blades in root-tuber harvesting machines. Meanwhile, the oscillation parameters must be selected to obtain the desired draught reduction and optimum soil breakup without a prohibitive increase in power expenditure.

An in-depth understanding of how soil interacts with tools is crucial for conceptualising and designing energy-efficient and adaptable implements for root and tuber harvesting [4,10]. However, understanding soil–tool interaction phenomena is challenging due to the anisotropic behaviour of soils and the transient stochastic loads they encounter during loosening [11]. The critical-state soil mechanics of agricultural tools are still being investigated. Some approaches have been used to study soil–tool interactions and soil failure, including empirical studies, analytical modelling, and numerical simulations using computational fluid dynamics (CFD), finite element method (FEM), and the discrete element method (DEM) [12–14].

Numerical computations based on discrete element methods (DEM) utilise the laws of motion and mechanical interaction properties of elements or particles within complex discrete systems such as grains and soils [15]. Recent developments in discrete element theories and related software have made it easier for scholars to apply DEM to design and optimise agricultural machinery [16–19]. In addition to soil–tool interaction, crop–machine interaction is also included in such advancements. For instance, a dry direct-seeding rice precision planter with film mulching was developed by Li et al. [20] using DEM. Their study determined optimum working parameters based on DEM simulation results under different soil conditions. Wan et al. [7] studied the effect of oscillations on shovel–rod components for liquorice harvesting using the discrete element method. The results showed that each 1 mm increase in amplitude decreased draught force by 463.35 N and increased total torque and specific energy consumption by 35.03 N m and 4.3 kJ m^{-3}, respectively. However, a 1 Hz increase in vibration frequency increased specific energy consumption by 3.12 kJ m^{-3}, whereas draught force and total torque decreased by 375.75 N and 28.44 Nm, respectively. Li et al. [21] investigated the mechanism for soil separation and its effectiveness in removing soil by combining DEM with the multi-body dynamics (MBD) method. However, the energy requirements for digging the tuber from the soil were not analysed. The model was validated through field experiments, and the results exhibited a relative error of 3.81%. Wanru et al. [22] utilised the Hertz–Mindlin approach along with a flexible bonding contact model to simulate and establish the process of harvesting tiller taro using DEM.

A key step in ensuring the accuracy of a DEM model is the determination and use of accurate simulation input parameters, including intrinsic and contact parameters [6,23,24]. The intrinsic properties of a material, such as the Poisson's ratio, Young's modulus, and bulk density, can be determined experimentally in the laboratory, or the values can be obtained from the literature [25]. As a result of differences in geometry and surface roughness between simulated and actual particles, it is necessary to simulate and use DEM to calibrate the contact parameters (primarily friction coefficients and coefficient of restitution) [26,27]. A combination of physical experiments and simulation tests is frequently used to calibrate contact parameters. For instance, mung beans were simulated using the Hertz–Mindlin contact model with bonding by Zhang et al. [28]. They performed physical and DEM simulation experiments between mung bean seeds and two machine parts.

Although some field and laboratory studies have been conducted to study the draught and power requirements of vibrating soil tools, there are only a few investigations in which the problem was numerically approached. Additionally, there is no indication that these investigations are related to the specific situation of a mixed medium such as soil with tubers or roots embedded within it. Despite the importance of understanding cutting-tool interactions, little progress has been made in this frontier due to the complexity of tool–medium interactions. In addition, there has been little progress in modelling the interaction of full-scale implements and soil. Thus, the objectives of this study were to (i) investigate the performance of a Jerusalem artichoke harvester vibrating digging shovel (full scale) on two soil textures (clay and sandy loam soils), (ii) develop DEM soil models including the artichoke crop, (iii) validate the developed DEM models by comparing the field draught force and drawbar power measurement with the predicted DEM values, and (iv) assess the shovel's ability to dig and translocate the soil–crop mixture to the rear. The soil–crop DEM model developed will help evaluate root-tuber harvesting machines considering the entire working width of the digging tool.

2. Materials and Methods

2.1. Description of the Jerusalem Artichoke Harvester

The harvester is a 4U-1600A Jerusalem artichoke harvester, which is a semi-trailed harvester (Figure 1). Generally, the harvester has five main working units: depth adjustment, shovel and vibration mechanism, soil–tuber conveyor, cleaning cylinder, and tuber conveyor. The effective working width of the harvester is 1600 mm. The machine has a hydraulic system that powers the working units (Supplementary Figure S1). The harvester has a programmable logic control (PLC) system to control the hydraulic system and collect data such as draught force, vibration frequency, and forward speed (Supplementary Figure S2). Typically, the harvester can dig up to a depth of 300 mm. However, with minimal adjustment, it can exceed this limit.

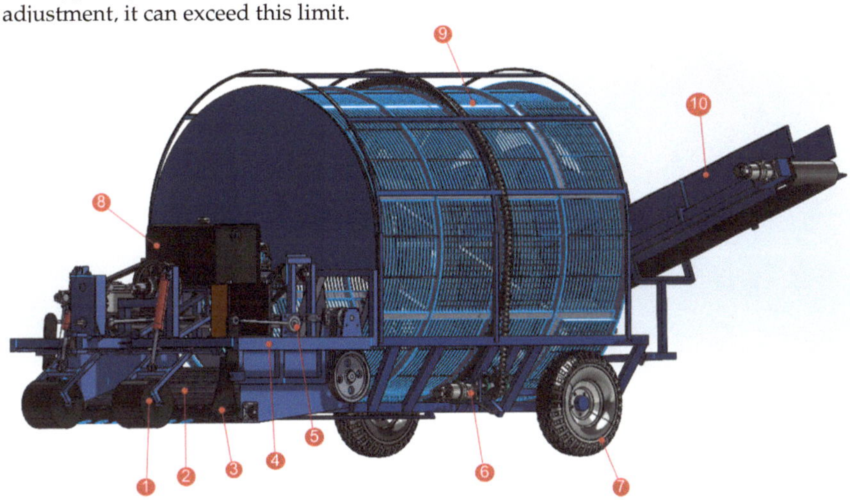

Figure 1. 4U-1600A Jerusalem artichoke harvester: (1) diablo rollers (depth adjustment), (2) soil–tuber conveyor, (3) shovel, (4) harvester body, (5) vibration mechanism, (6) hydraulic motor, (7) wheel, (8) hydraulic fluid tank, (9) cleaning cylinder, and (10) tuber conveyor.

2.2. Determination of Mechanical and Physical Properties of Jerusalem Artichoke Crop

Fresh Jerusalem artichoke tubers (Figure 2a) were obtained, and their average density was determined to be 1184 kg m^{-3} using the water displacement method (Figure 2b). Standard compression test specimens with a diameter of 10 mm and a length of 20 mm were prepared. A single-factor uniaxial compression (Figure 2c) was conducted using a TMS-Pro texture analyser (FTC, Washington, DC, USA). The shear modulus, elastic modulus, and Poisson's ratio of the Jerusalem artichoke tubers were calculated as 4.23×10^6 Pa, 1.19×10^7 Pa, and 0.408, respectively, using Equations (1)–(5). The average diameter of the Jerusalem artichoke roots was determined to be 5.5 mm. The friction coefficients (static and rolling) and the coefficient of restitution of the Jerusalem artichoke tubers and roots were determined using the inclined plane method (Figure 2d). A three-point bending test was carried out on the roots (Figure 2e). From the test, the elastic modulus of the roots was determined to be 1.32×10^7 Pa. The approaches used in determining the mechanical and physical properties of the tubers and roots were utilised for the stem. A summary of the mechanical and physical properties is presented in Supplementary Table S1.

$$\sigma = \frac{p}{A} \quad (1)$$

$$\varepsilon = \frac{\Delta L}{L} \quad (2)$$

$$E = \frac{\sigma}{\varepsilon} \quad (3)$$

$$v = -\frac{\varepsilon_{lateral}}{\varepsilon_{axial}} \quad (4)$$

$$G_* = \frac{E}{(2 \times (1 + v))} \quad (5)$$

where A is the cross-sectional area (mm^2), p is the load applied (N), σ is the axial stress (N m^{-2}), ε is the axial strain, ΔL is the change in specimen length (mm), L is the original specimen length (mm), E is the elastic modulus (Pa), v is the Poisson's ratio, and G is the shear modulus (Pa).

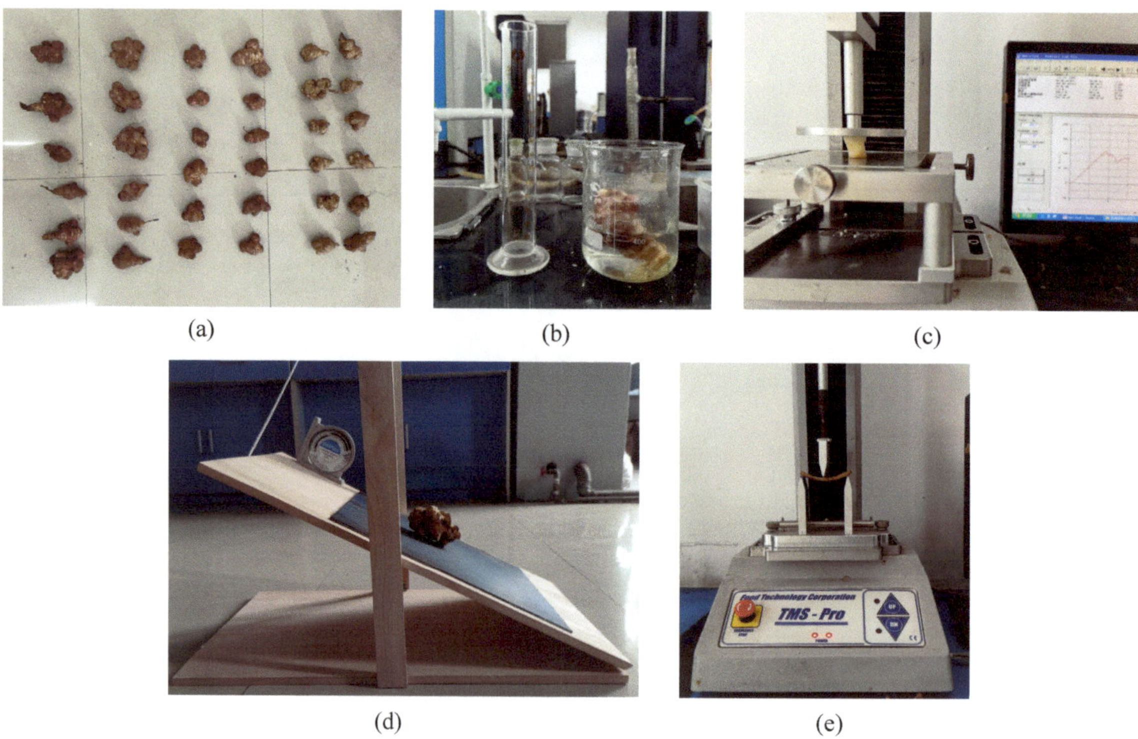

Figure 2. Mechanical and physical properties experiment for Jerusalem artichoke crop: (**a**) sampled tubers from artichoke field (different sizes and shapes), (**b**) tuber density determination, (**c**) uniaxial compression test, (**d**) inclined plane method for friction coefficients, and (**e**) root bending test.

2.3. Determination of Soil Mechanical and Physical Properties

Soil moisture content and cone penetration resistance were measured randomly in the experimental fields before the field test. The moisture content was determined using an on-site soil moisture meter (volumetric water content), whereas the cone penetration resistance was measured using a standard American Society of Agricultural and Biological Engineers (ASABE) cone penetrometer (Supplementary Figure S3). The clay soil had a mean moisture content of 17.14% and a cone index of 596.87–3744.86 kPa at 50–250 mm depth. On the other hand, the sandy loam soil had a moisture content of less than 25% and a cone penetration resistance of 800–1500 kPa at a depth of 150–250 mm.

2.4. Discrete Element Method Simulation Setup and Analysis

A DELL Precision 7920 Tower with Intel® Xeon® CPU 4214R @ 2.4 GHz, 12 cores (24 threads), and 32GB RAM computer running EDEM® 2020 bulk material simulation software was used to perform the DEM simulations. The clay soil was modelled without the Jerusalem artichoke crop. However, the crop was incorporated into the sandy loam model to mimic field conditions. Therefore, the measured mechanical and physical properties determined in Section 2.2 were utilised to set up the DEM soil–crop model described in detail in Section 2.4.1. The two virtual soil bins were both 2500 mm × 2500 mm × 400 mm (width, breadth, and depth) in size. Furthermore, the DEM soil models were fitted with a random particle size distribution (minimum: 0.5, maximum: 1.5 radius scales).

A hysteretic spring contact model (HSCM) with a linear cohesion model (LCM) was used to establish the clay soil DEM model. The Hertz–Mindlin (HM) with Johnson–Kendall–Roberts (JKR) base contact model and parallel bond (PB, version 2) were utilised to set up the sandy loam soil [29]. The JKR model was added to account for the cohesion between particles, whereas PB was used to create the crop model. The soil–crop model was calibrated using the static angle of repose method, similar to what was used by Awuah et al. [6]. The calibration was carried out by using an experimental design (I-optimal design, Supplementary Table S3) based on response surface methodology (RSM). Figure 3 shows the static angle of the repose experiment and the DEM results.

(a) (b)

Figure 3. Static angle of repose for soil–crop mass calibration: (**a**) experiment (the red line is a laser pointer which serves as a visual reference for the angle being measured) and (**b**) DEM simulation.

Linear regression and ANOVA were performed to analyse the static angle of repose results (Supplementary Table S4). The predicted R^2 of 0.82 was in reasonable agreement with the adjusted R^2 of 0.93. A mean static angle of repose of 30.06° was obtained with a standard deviation (Std.Dev). of 1.47 and coefficient of variation (CoV) of 4.91%. The regression model was assessed for adequacy by using the diagnostic plots shown in Figure 4. The diagnostic plots did not show any outliers in the data, implying that the data fit the designed model. Numerical optimisation was performed to obtain optimal input parameters by targeting the measured static angle of repose (29.85°). The optimal values are shown in Figure 5, whereas Table 1 lists the input parameters used for the DEM soil–crop mixture model simulation. The procedure used to determine the optimal input parameters for the soil–crop mixture was employed to obtain the optimal input parameters for the clay–soil DEM model (see Supplementary Tables S5 and S6, and Supplementary Figure S4). The DEM input parameters used for the clay soil are shown in Table 2.

Table 1. DEM input parameters for the soil–crop mixture model (sandy loam soil).

Parameter and Unit	Value	Remarks
Poison's ratio: soil	0.3	Selected
Poison's ratio: steel	0.3	[30]
Poisson's ratio: root	0.38	Measured
Poison's ratio: tuber	0.48	Measured
Poison's ratio: stem	0.35	Measured
Particles' solid density (kg m^{-3})	2600	[6]
Density of steel (kg m^{-3})	7865	[30]
Density of root (kg m^{-3})	1132	Measured
Density of tuber (kg m^{-3})	1184.4	Measured
Density of stem (kg m^{-3})	250.75	Measured
Shear modulus (Pa): soil	1.7×10^7	[6]
Shear modulus (Pa): steel	7.9×10^{10}	[30]
Shear modulus (Pa): root	4.78×10^6	Measured
Shear modulus (Pa): tuber	4.23×10^6	Measured
Shear modulus (Pa): stem	2.72×10^8	Measured
Coefficient of restitution: soil–soil	0.6	[6]
Coefficient of restitution: soil–steel	0.6	[6]
Coefficient of restitution: soil–root	0.439	Calibrated
Coefficient of restitution: soil–tuber	0.514	Calibrated
Coefficient of restitution: soil–stem	0.554	Calibrated
Coefficient of restitution: root–steel	0.32	Measured
Coefficient of restitution: tuber–steel	0.62	Measured
Coefficient of restitution: stem–steel	0.53	Measured
Coefficient of static friction: soil–soil	0.45	[6]
Coefficient of static friction: root–soil	0.195	Calibrated
Coefficient of static friction: tuber–soil	0.212	Calibrated
Coefficient of static friction: stem–soil	0.166	Calibrated
Coefficient of static friction: soil–steel	0.45	[6]
Coefficient of static friction: root–steel	0.511	Measured
Coefficient of static friction: tuber–steel	0.446	Measured
Coefficient of static friction: stem–steel	0.5	Measured
Coefficient of rolling friction: soil–soil	0.18	[6]
Coefficient of rolling friction: root–soil	0.015	Calibrated
Coefficient of rolling friction: tuber–soil	0.175	Calibrated
Coefficient of rolling friction: stem–soil	0.069	Calibrated
Coefficient of rolling friction: root–steel	0.21	Measured
Coefficient of rolling friction: tuber–steel	0.32	Measured
Coefficient of rolling friction: stem–steel	0.05	Measured
Normal stiffness per unit area (N m^{-3})	1×10^9	Selected
Shear stiffness per unit area (N m^{-3})	2.5×10^7	Calibrated
Normal strength (Pa)	1.3×10^6	Calibrated
Shear strength (Pa)	1.15×10^6	Calibrated
Bonded disk scale	1	Selected
JKR surface energy (J m^{-2})	10	Selected

Table 2. DEM input parameters used for the soil–tool interaction simulation (clay soil).

Parameter and Unit	Value	Remarks
Poison's ratio: soil	0.3	Selected
Poison's ratio: steel	0.3	[30]
Particles' solid density (kg m^{-3})	2600	Selected
Density of steel (kg m^{-3})	7865	[30]
Shear modulus (Pa): soil	2×10^7	Calibrated
Shear modulus (Pa): steel	7.9×10^{10}	[30]
Yield strength (Pa): soil (single sphere, dual sphere, and triple sphere)	$2.21 \times 10^6, 2.56 \times 10^6$, and 2.37×10^6	Default value in EDEM® 2020

Table 2. *Cont.*

Parameter and Unit	Value	Remarks
Yield strength (Pa): steel	1×10^9	Default value in EDEM® 2020
Coefficient of restitution: soil–soil	0.467	Calibrated
Coefficient of restitution: soil–steel	0.05	Selected
Coefficient of static friction: soil–soil	0.388	Calibrated
Coefficient of static friction: soil–steel	0.45	Selected
Coefficient of rolling friction: soil–soil	0.192	Calibrated
Coefficient of rolling friction: soil–steel	0.15	Selected
Damping factor	0.5	Default value in EDEM® 2020
Stiffness factor	0.85	Default value in EDEM® 2020
Cohesive energy density (J m^{-3})	20,965.7	Calibrated

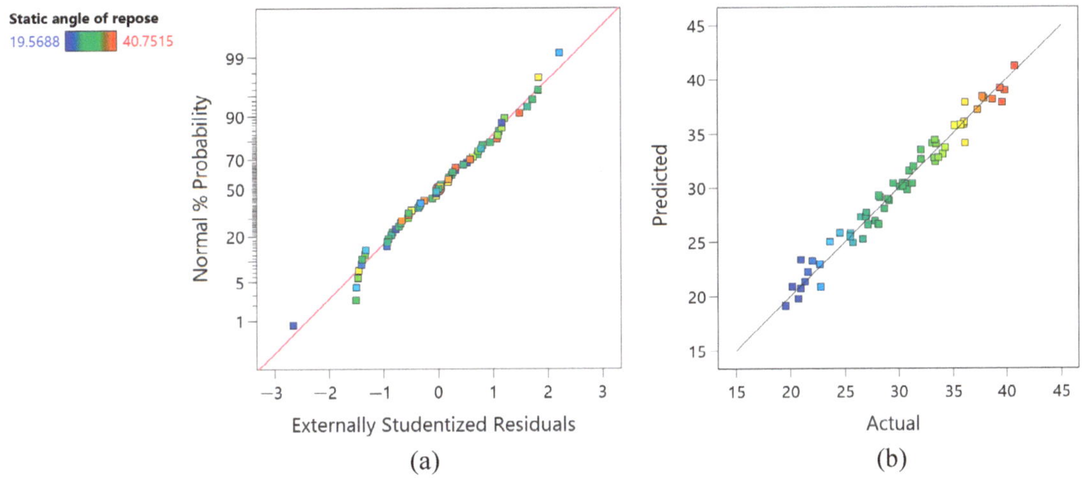

Figure 4. Diagnostic plots: (**a**) normal plot of residuals and (**b**) predicted versus actual plot.

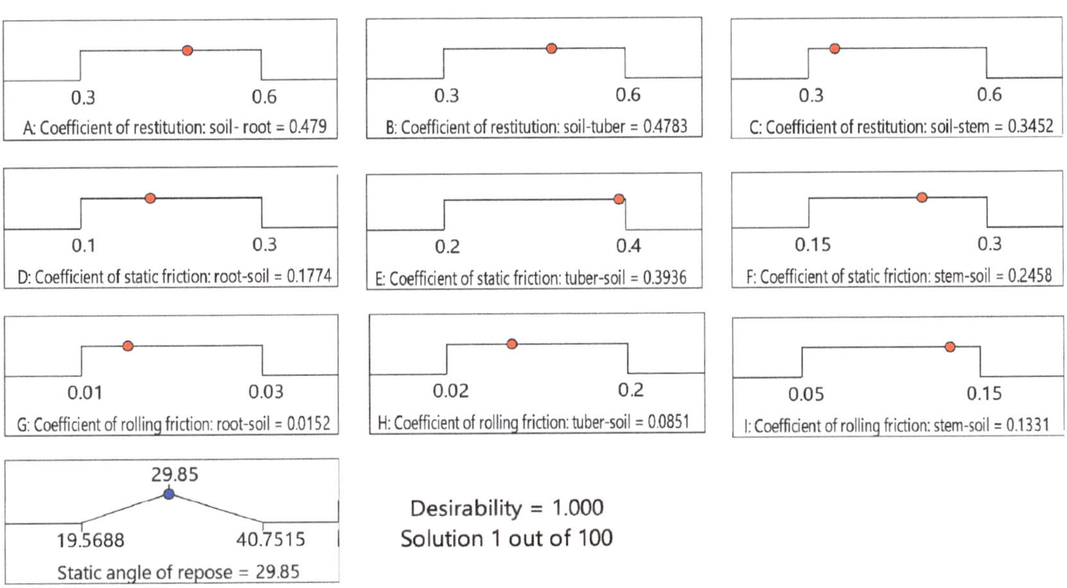

Figure 5. Optimal DEM input parameters from the static angle of repose simulation.

2.4.1. Discrete Element Method Soil–Crop Modelling

Generally, Jerusalem artichoke tubers have irregular shapes compared to those of potatoes. Thus, modelling such forms is challenging. However, a relatively less complicated tuber was used to create the DEM model (Figure 6a). The Jerusalem artichoke DEM crop model was first created using SolidWorks from a scanned image of the tuber (Figure 6b). The 3D model was then imported into EDEM as a template and meshed (Figure 6c). The complete crop model was then developed using the meta-particle feature in EDEM [29] and the PB contact model (Figure 6d). Meta particle allows flexible materials such as roots to be modelled. The virtual soil bin was created with the Jerusalem artichoke crop models embedded at a tuber depth of 150 mm and a spacing of 829 × 425 mm (Figure 7). The particles were compressed to the desired bulk density (1536 kg m^{-3}) and depth to prevent the soil bin from becoming excessively loose. The particle size distribution employed created an interlocking effect, which also aided with realistic particle behaviour. The fork-shaped and the S-shaped shovels were then imported to evaluate their ability to dig and translocate the soil–crop mass to the rear of the digging tool, a feature desirable for a tuber-harvesting machine.

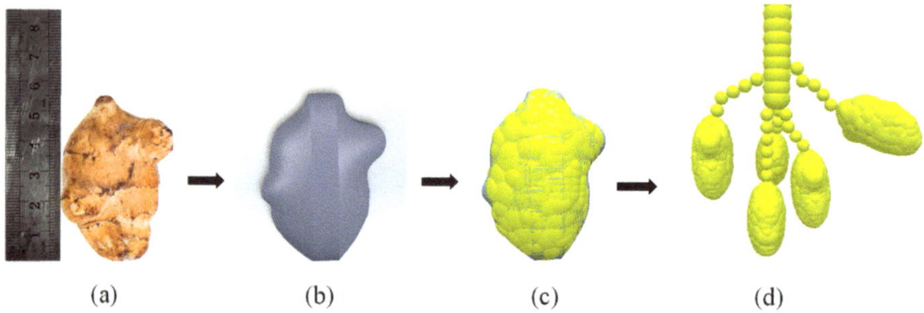

Figure 6. DEM crop–modelling technique: (**a**) Fresh Jerusalem artichoke tuber, (**b**) 3D computer-aided design model, (**c**) DEM tuber particle, and (**d**) DEM crop model.

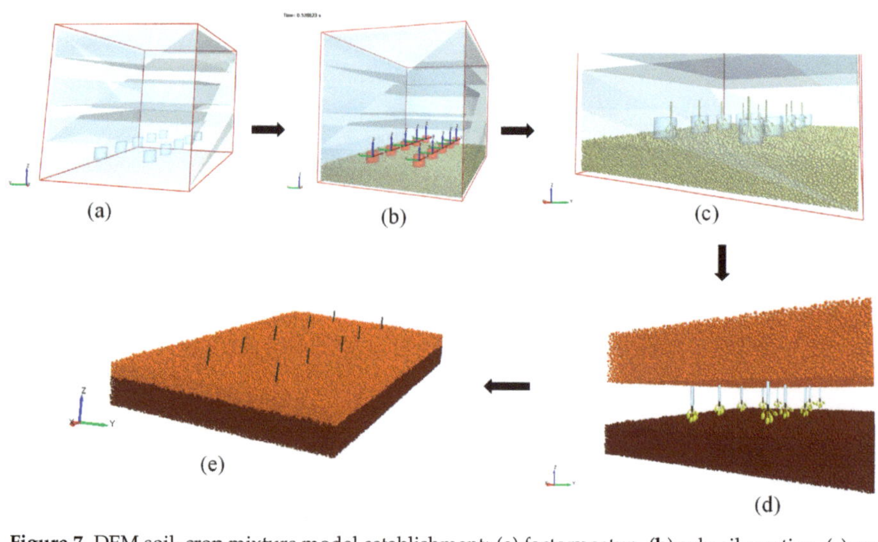

Figure 7. DEM soil–crop mixture model establishment: (**a**) factory setup, (**b**) subsoil creation, (**c**) crop creation, (**d**) topsoil creation, and (**e**) soil–crop model.

2.5. Experimental Design and Analysis

Field performance evaluation was conducted on two soil types (clay and sandy loam soils) to assess the shovel's ability to work in diverse soil conditions. First, the harvester was evaluated in a clay-textured soil experimental field without Jerusalem artichoke crop on 8 and 9 September 2022, in Yu Cheng City, Shandong Province. The second evaluation was performed on 14 November 2022, in a Jerusalem artichoke plantation field with sandy loam soil in Wanggang Xinzha, Yancheng City, Jiangsu Province. Figure 8 shows the tractor–harvester setup used in the two fields.

Figure 8. Tractor–harvester setup used in this study.

Single-factor tests were used to evaluate the harvesters. A mixture of 5, 9.75, 12.5, and 14.5 Hz frequencies and 1, 2, and 4 km h^{-1} forward speeds were employed in the clay soil field at a digging depth of 200 mm (Figure 9a). However, the sandy loam soil field was evaluated using 10 Hz and 12.5 Hz frequencies plus 1 and 2 km h^{-1} forward speeds at a digging depth of 200 mm. These parameters were selected based on the optimal parameters of previous studies [6] and the available tractor power. The digging depth of 200 mm was used for the Jerusalem artichoke field because the tubers had developed to an average maximum depth of 150 mm (Figure 9b).

Figure 9. The depth of operation for the two field experiments: (**a**) clay soil field and (**b**) sandy loam field.

The fork-shaped shovel was fitted as the digging shovel for this field experiment. However, the S-shaped shovel was combined with the fork-shaped shovel to assess the tools' ability to dig, crush, and transport the soil/soil–crop mass to the other parts of the harvester using DEM. The field experiments were replicated using the virtual DEM soil bin created. Figure 10 shows the DEM simulation setup and the digging shovels used for this study. The shovel's vibratory motion involves rotation around its pin and striking with a translational motion, causing it to penetrate, lift, and slightly retract from the ground before returning to its original position. The velocity profile feature in EDEM simulation software was utilised to assess the movement of the soil particles and the crop during harvesting. In contrast, the bulk density sensor feature was used to evaluate the tool's ability to crush the soil. General linear regression was used to analyse the effect of the parameters on the draught force and drawbar power. Linear regression and analysis of variance (ANOVA) were performed to check the adequacy of the models.

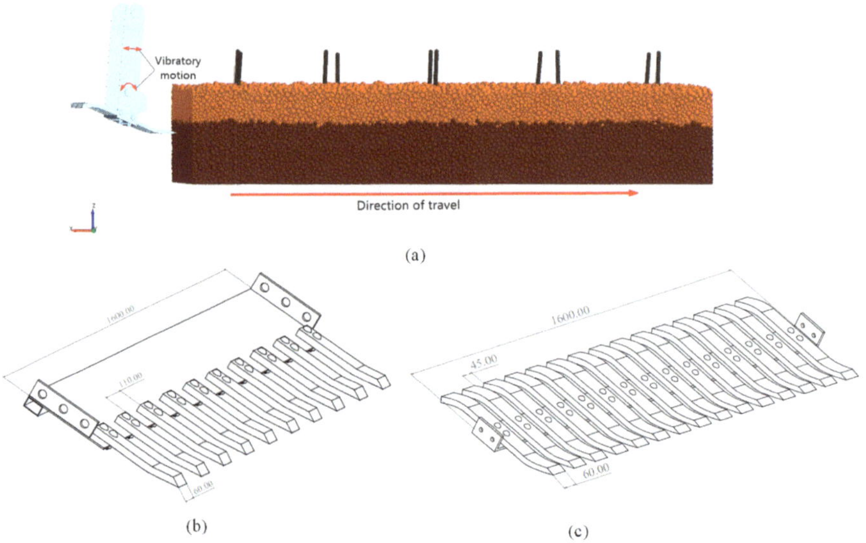

Figure 10. (**a**) DEM soil–shovel simulation setup, (**b**) fork-shaped shovel, and (**c**) S-shaped shovel.

3. Results and Discussion

This section presents the field experiment results and the discrete element simulations. Tables 3 and 4 show a good correlation between the adjusted R^2 and the predicted R^2 (i.e., the difference was less than 0.2) for all the response variables, suggesting that the design model was accurate. Additionally, from Table 5, all the main model treatments and the interaction of the terms were statistically significant at $p < 0.05$, implying that the treatments affected the measured variables.

Table 3. Regression model fit summary for responses (clay field experiment and DEM simulation).

Source	Sequential p-Value	Adjusted R^2	Predicted R^2	Remark
Experiment draught force				
Linear	<0.0001 *	0.9038	0.8475	
2FI	0.0798 **	0.9280	0.7966	
Quadratic	0.0512 **	0.9643	0.9324	Suggested
Cubic	0.0340 *	0.9947	0.9664	Aliased
DEM draught force				
Linear	<0.0001 *	0.9017	0.8430	
2FI	0.0908 **	0.9243	0.7748	
Quadratic	0.0394 *	0.9657	0.9120	Suggested
Cubic	0.0240 *	0.9959	0.9538	Aliased
Experiment drawbar power				
Linear	<0.0001 *	0.9473	0.9011	
2FI	0.0006 *	0.9874	0.9692	Suggested
Quadratic	0.0963 **	0.9923	0.9734	
Cubic	0.0085 **	0.9995	0.9977	Aliased
DEM drawbar power				
Linear	<0.0001 *	0.9450	0.8974	
2FI	0.0010 *	0.9850	0.9623	Suggested
Quadratic	0.0777 **	0.9915	0.9693	
Cubic	0.0073 *	0.9995	0.9968	Aliased

*: Statistically significant ($p < 0.05$); **: statistically not significant ($p > 0.05$).

Table 4. Regression model fit summary for the sandy loam soil (field and DEM simulation).

Source	Experiment Draught Force	DEM Draught Force
Std. Dev.	368.33	362.5
Mean	11,963.1	11,411.15
CoV%	3.08	3.18
R^2	0.9952	0.9951
Adjusted R^2	0.9857	0.9852
Predicted R^2	0.9235	0.9212
Adequacy precision	22.8498	22.4509

3.1. Effect of Speed and Frequency on Draught Force

Figure 11 shows the effect of vibration frequency and forward speed on draught force for the clay soil field. The draught force increased as speed increased for the field experiment and DEM simulation. For example, at 5 Hz frequency, the draught increased from 16,542.17 to 26,019.39 N when forward speed increased from 1 to 4 km h^{-1}. Soehne [31] found that draught force was a function of soil acceleration and, consequently, proportional to the square of speed. McLaughlin and Campbell [32] also observed a similar outcome. They noted that accelerating the movement of soil particles increased frictional forces on tines.

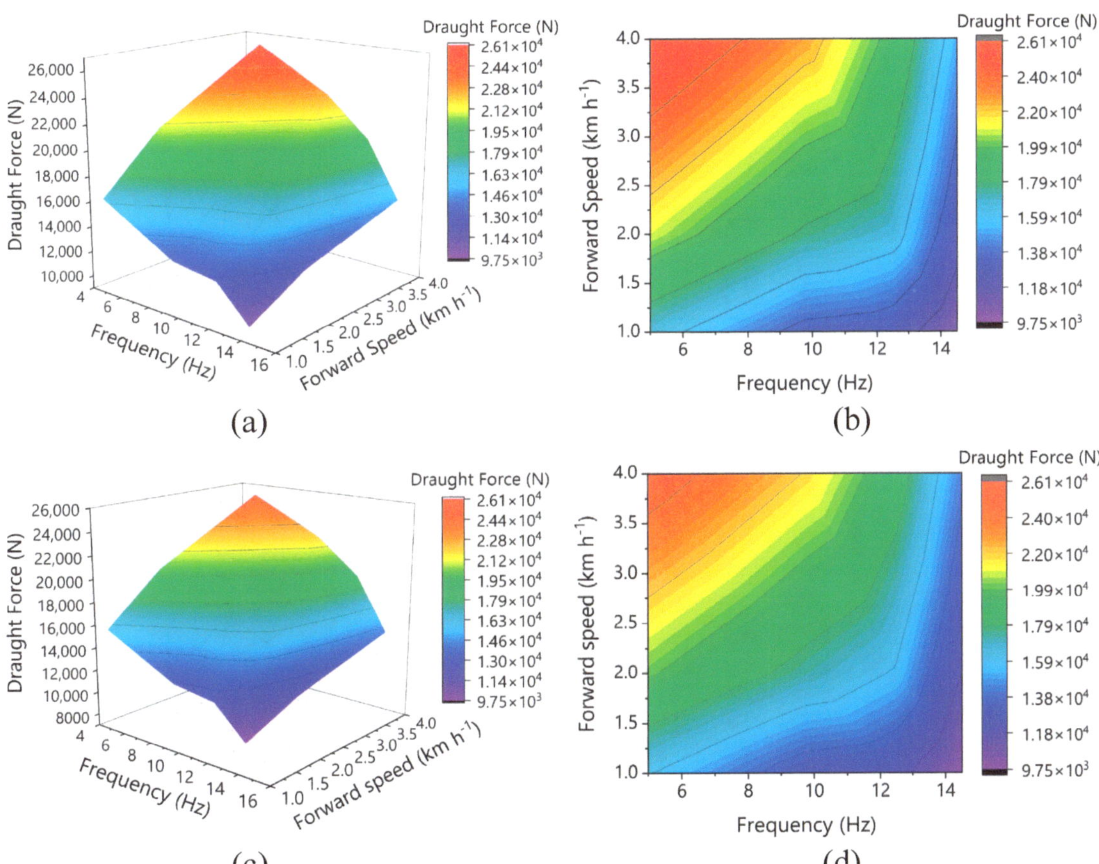

Figure 11. Effect of forward speed and vibration frequency on draught force: (**a**,**b**) clay field experiment 3D and 2D contour plots, and (**c**,**d**) DEM simulation 3D and 2D contour plots.

Table 5. ANOVA summary result for the clay soil (field and DEM simulation).

Experiment Draught Force (Quadratic)			DEM Draught Force (Quadratic)			Experiment Drawbar Power (2FI)			DEM Drawbar Power (2FI)		
Source	F-Value	*p*-Value	Source	F-Value	*p*-Value	Source	F-Value	*p*-Value	Source	F-Value	*p*-Value
Model	60.51	<0.0001 *	Model	62.89	<0.0001 *	Model	288.85	<0.0001 *	Model	242.23	<0.0001 *
A	150.65	<0.0001 *	A	159.85	<0.0001 *	A	66.6	<0.0001 *	A	57.39	0.0003 *
B	157.21	<0.0001 *	B	159.02	<0.0001 *	B	810.21	<0.0001 *	B	678.05	<0.0001 *
AB	8.13	0.0291 *	AB	8.15	0.029 *	AB	29.72	0.0006 *	AB	25.06	0.0010 *
A^2	5.33	0.0603 **	A^2	7.54	0.0335 *						
B^2	4.83	0.0.0704 **	B^2	4.09	0.0895 **						
Std.Dev.	876.40		Std.Dev.	836.44		Std.Dev.	1.01		Std.Dev.	1.06	
Mean	16,990.08		Mean	16,321.74		Mean	12.07		Mean	11.59	
CoV%	5.16		CoV%	5.12		CoV%	8.37		CoV%	9.17	

A: forward speed; B: frequency; Std.Dev.: standard deviation; CoV: coefficient of variation; 2FI: two-factor interaction; *: statistically significant ($p < 0.05$); **: statistically not significant ($p > 0.05$).

Different studies have reported different results about draught force and forward speed [33–35]. Several factors have contributed to this difference, including field conditions and the type of tillage tool used in the studies. A similar trend was observed when the harvester was evaluated in sandy loam soil, as shown in Figure 12. However, the sandy loam soil's draught forces (i.e., measured and predicted DEM values) were relatively smaller than those of the clay soil. The reason may be due to the high cohesion in clay soil.

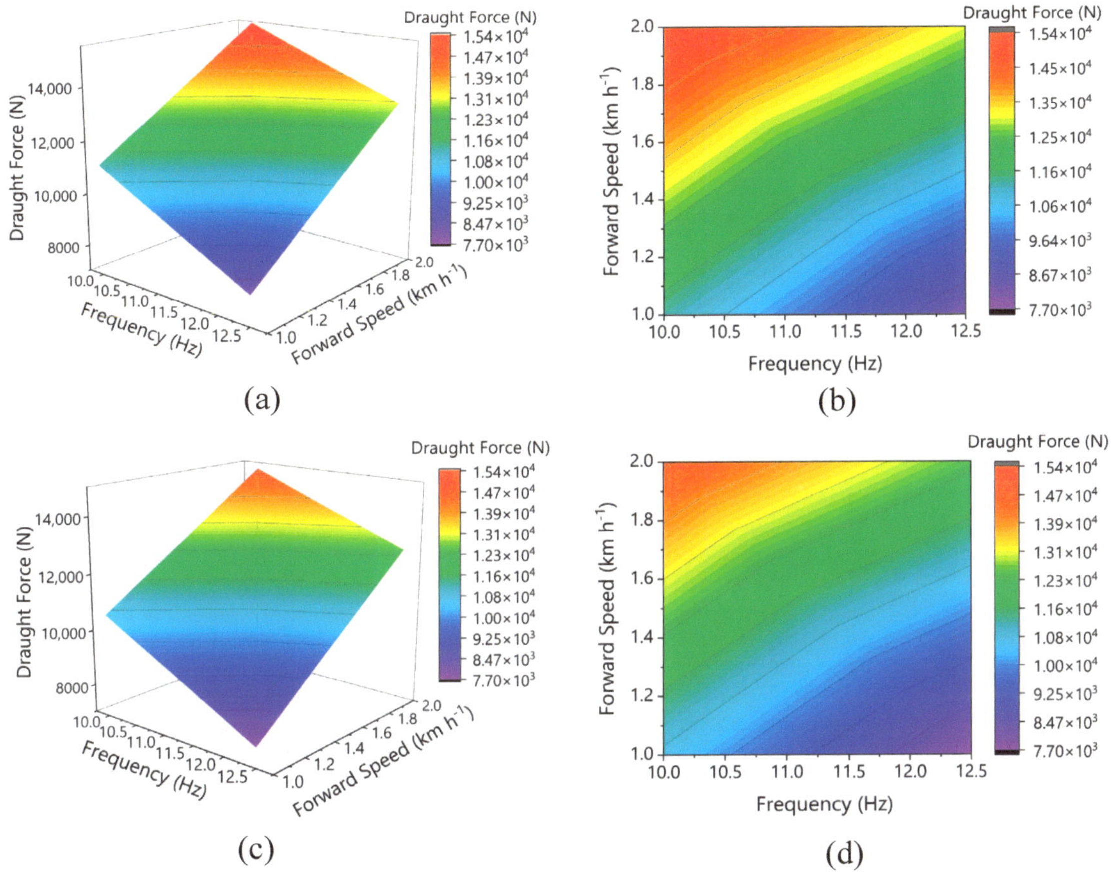

Figure 12. Effects of forward speed and vibration frequency on draught force: (**a**,**b**) sandy loam field experiment 3D and 2D contour plots, and (**c**,**d**) DEM simulation 3D and 2D contour plots.

In contrast to forward speed, draught force decreased with increasing vibration frequency. For instance, the draught force decreased from 21,176.69 to 12,604.87 N (40.48% reduction) when the harvester was operated at 5–14.5 Hz in clay soil (Figure 13a). Again, a reduction of 20.48% (13,327.93–10,598.26 N) was obtained when the harvester was operated in sandy loam soil at 10–12.5 Hz (Figure 13b). This reduction can be attributed to the reduced soil–steel friction angle and soil friction coefficient on the steel due to vibrations [36]. Vasilenko et al. [36] observed in a laboratory that the average value of the friction angle of soil on steel, $\varphi = 31.4°$, was reduced to $\varphi = 26.5°$ when it was subjected to a vibration frequency of 22–24 Hz. Conversely, when the metal sheet vibrated at 22–24 Hz, its friction coefficient was reduced from 0.61 to 0.5. They later confirmed this phenomenon under field experimental conditions using a plough. Their results showed that draught force was reduced by 14.5% when the plough was operated at a depth of 300 mm, a forward speed of 7.74 km h^{-1}, a 5 mm amplitude, and an 8–10 Hz frequency. Other researchers have reported similar reduction trends [8,37,38].

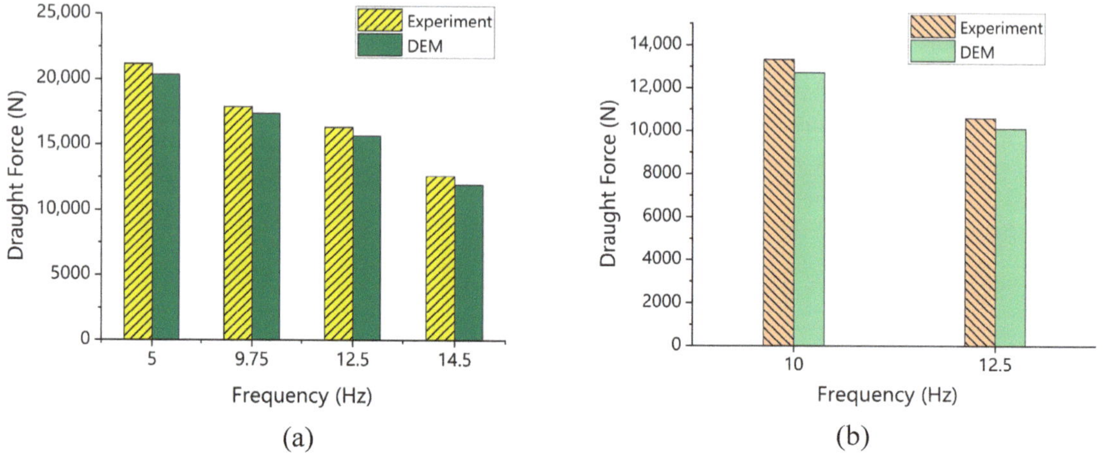

Figure 13. Effect of vibration frequency on draught force for the experiment and DEM simulation: (**a**) clay soil and (**b**) sandy loam.

3.2. *Effect of Speed and Frequency on Drawbar Power*

Figure 14 shows the relationship between drawbar power, forward speed, and vibration frequency for the field evaluation and DEM simulation. Drawbar power increased as forward speed increased but decreased with increasing vibration frequency. The observed trend is reasonable since drawbar power is influenced by draught force. Therefore, the phenomenon used to explain the draught force trend holds. The drawbar power recorded for the experimental fields with clay and sandy loam soils were minimum at vibration frequencies of 12.5 Hz and 14.5 Hz, with values of 2.26 kW and 2.85 kW, respectively (Tables 6 and 7). The results suggest that a relatively smaller tractor should pull the harvester when harvesting in sandy loam soil. To reduce drawbar power during harvesting operation, it is recommended to set the frequency and speed to between 12–14.5 Hz and 1–2 km h^{-1}, respectively.

Table 6. Comparison of clay soil field and DEM simulation result for draught force and drawbar power.

Frequency (Hz)	Forward Speed (km h^{-1})	Experiment Draught Force (N)	DEM Draught Force (N)	Experiment Drawbar Power (kW)	DEM Drawbar Power (kW)	RE (%)
5	1	16,542.170	15,926.626	4.595	4.424	3.721
5	2	20,951.889	20,105.592	11.640	11.170	4.039
5	4	26,036.022	25,019.387	28.929	27.799	3.905
9.75	1	13,288.189	12,867.725	3.691	3.574	3.164
9.75	2	17,684.548	17,270.654	9.825	9.595	2.340
9.75	4	22,677.284	21,915.671	25.197	24.351	3.358
12.5	1	12,819.620	12,362.924	3.561	3.434	3.562
12.5	2	16,586.874	15,699.174	9.215	8.722	5.352
12.5	4	19,479.821	18,863.692	21.644	20.960	3.163
14.5	1	10,245.086	9795.835	2.846	2.721	4.385
14.5	2	12,573.063	11,794.365	6.985	6.552	6.193
14.5	4	14,996.467	14,239.182	16.663	15.821	5.050

RE: relative error.

Table 7. Comparison of sandy loam soil field and DEM simulation result for draught force and drawbar power.

Frequency (Hz)	Forward Speed (km h^{-1})	Experiment Draught Force (N)	DEM Draught Force (N)	Experiment Drawbar Power (kW)	DEM Drawbar Power (kW)	RE (%)
10	1	11,232.587	10,669.730	3.120	2.964	5.011
10	2	15,423.266	14,753.971	8.568	8.197	4.340
12.5	1	8134.596	7705.821	2.260	2.141	5.271
12.5	2	13,061.930	12,515.067	7.257	6.953	4.187

RE: relative error.

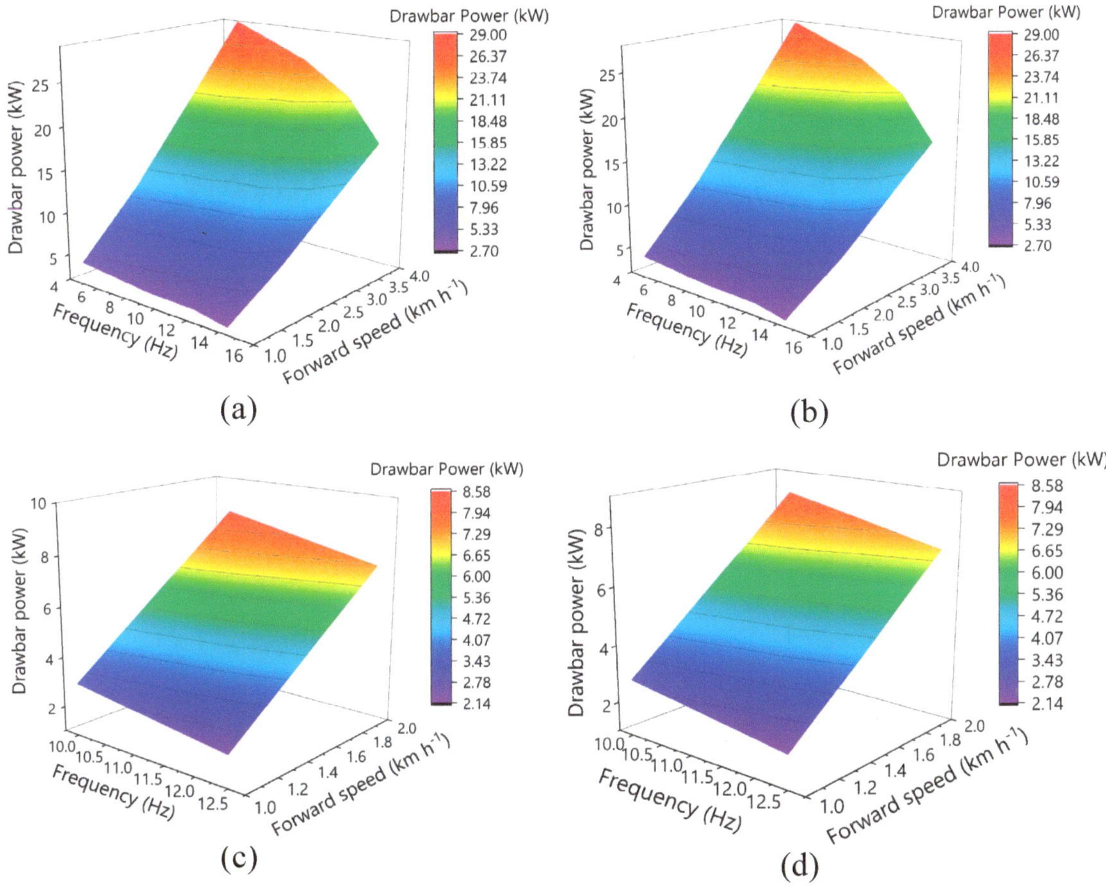

Figure 14. Comparison of field data and DEM for drawbar power: (**a**,**b**) clay soil field and DEM results, respectively; (**c**,**d**) sandy loam soil field and DEM results, respectively.

3.3. Discrete Element Method Model Validation

Tables 6 and 7 compare the two soil textures' field experiment data and DEM simulation results for draught force and drawbar power. It is evident from the results that DEM closely predicted the draught force for the two scenarios (i.e., in clay and sandy loam soils). Relative error (RE) was used to assess the prediction accuracy shown in Tables 6 and 7 for the clay and sandy loam soil, respectively. The average RE for the clay soil was approximately 4%, whereas the RE for the sandy loam soil was 4.7%.

3.4. Particle Displacement Analysis

The fork-shaped and S-shaped shovels were tested to see whether they could dig and transfer the soil–crop mass to the rear of the digging tool, a vital feature for tuber harvesting. From Figure 15, the S-shaped shovel was better than the fork-shaped shovel regarding soil translocation when the tool was run at 14.5 Hz, 4 km h^{-1} forward speed, and 200 mm depth in clay soil. In addition, at those same operating parameters, the S-shaped shovel produced a smaller soil bulk density (706.35 kg m^{-3}) than the fork-shaped shovel (864.53 kg m^{-3}) after the tool passed (Figure 16).

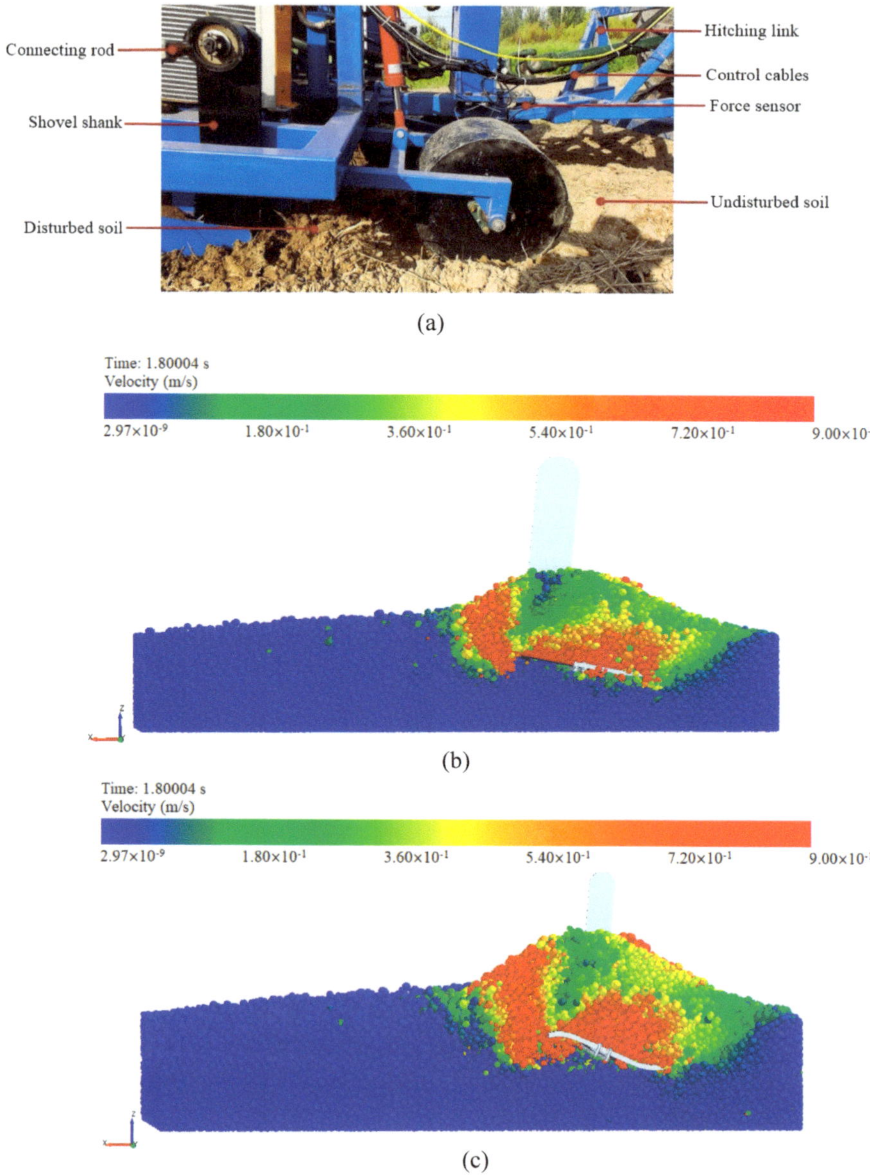

Figure 15. Soil translocation properties of the shovels evaluated in clay soil: (**a**) field experiment for fork shovel, (**b**) DEM result for fork-shaped shovel, and (**c**) DEM result for S-shaped shovel.

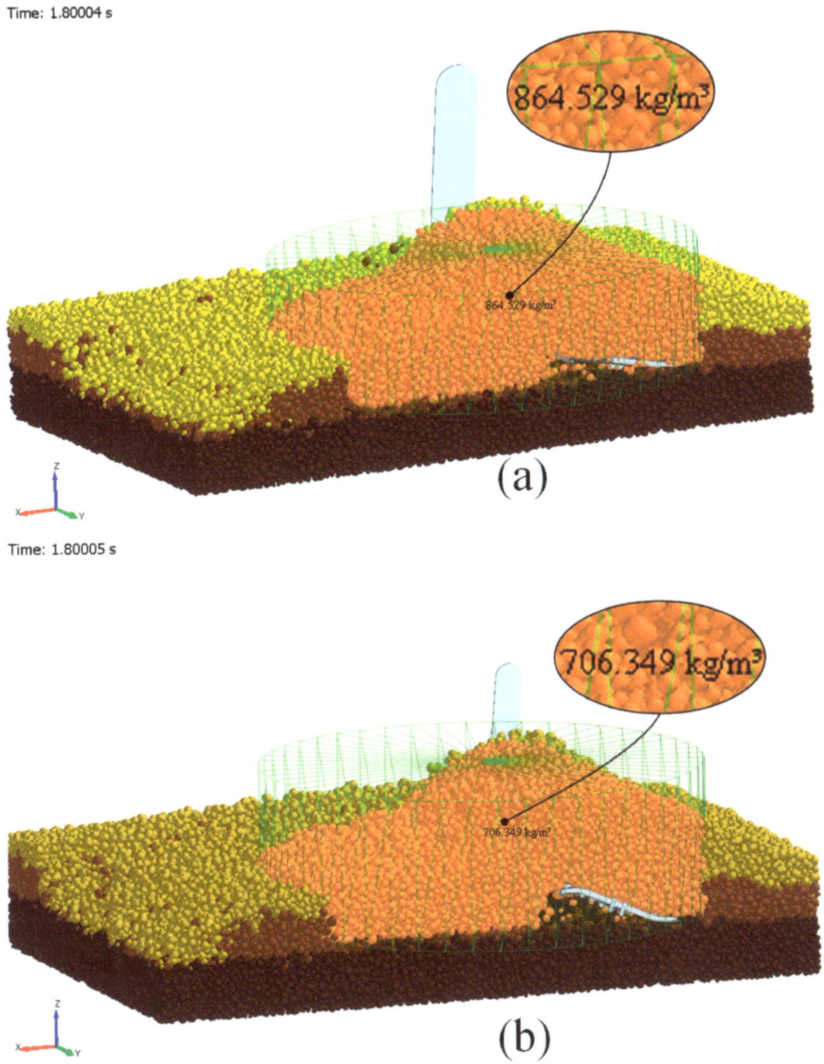

Figure 16. DEM predicted soil bulk density for clay soil evaluation: (**a**) fork-shaped shovel and (**b**) S-shaped shovel.

The lower bulk density suggests that the shovel can pulverise the soil fairly well, thereby increasing porosity. Porosity, the percent by volume of a soil sample not occupied by solids, is directly related to bulk and particle densities. If particle density remains constant as bulk density increases, the porosity decreases [39]. Furthermore, the results indicate that the S-shaped shovel performed better in cohesive soils than the fork-shaped shovel.

Figures 17 and 18 show the soil–crop mass translocation and DEM bulk density estimation, respectively, when the tools were run at 12 Hz frequency, 2 km h^{-1}, and 200 mm depth in sandy loam soil. Once again, the results show that the S-shaped shovel was able to translocate the soil–crop mass to the rear better than the fork-shaped shovel. The predicted bulk density for the S-shaped shovel (889.13 kg m^{-3}) was also smaller than that of the fork-shaped shovel (993.91 kg m^{-3}).

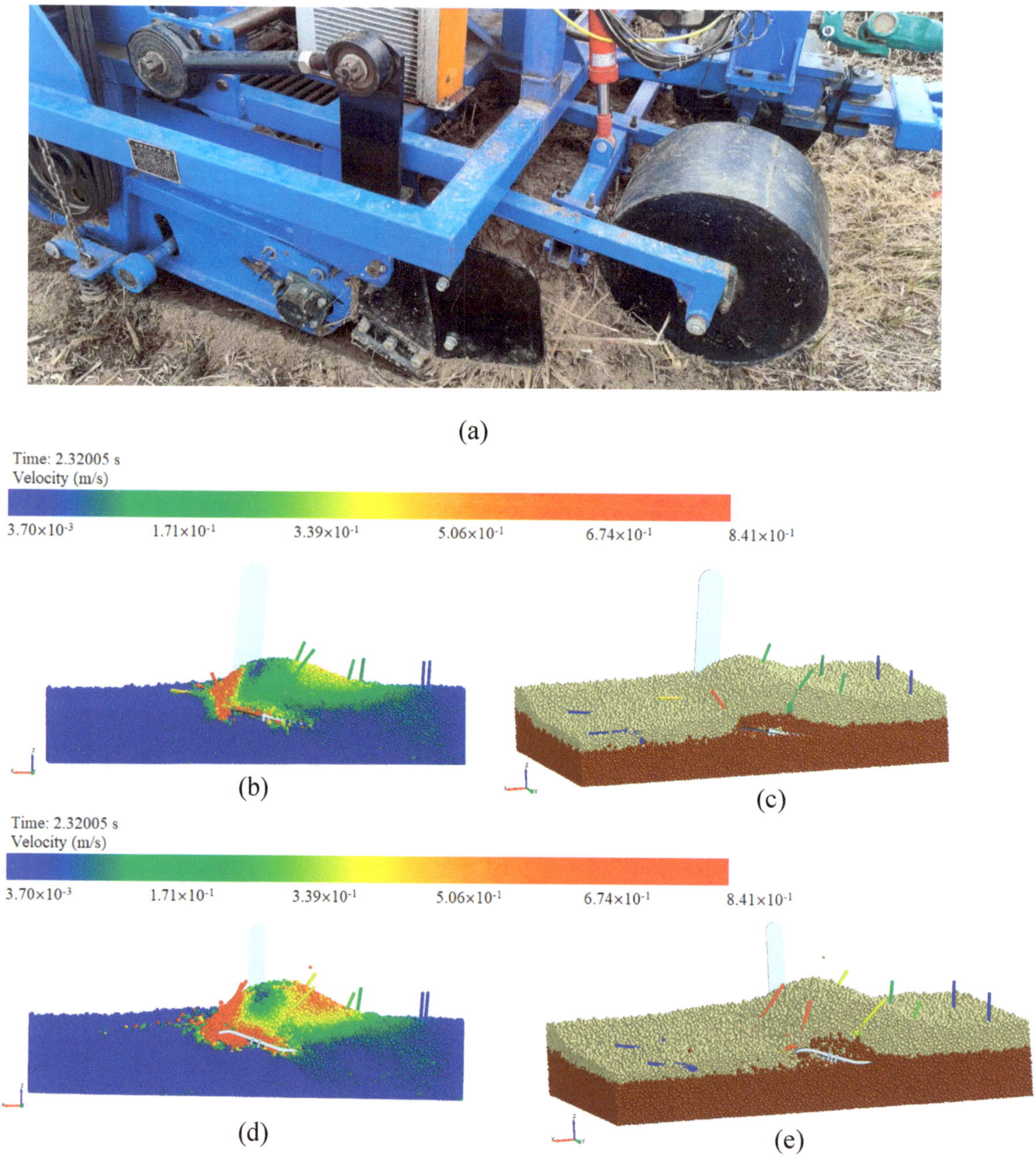

Figure 17. Soil–crop mass translocation properties of the tools evaluated in sandy loam soil: (**a**) field experiment result, (**b**,**c**) DEM result for the fork-shaped shovel, and (**d**,**e**) DEM result for the S-shaped shovel.

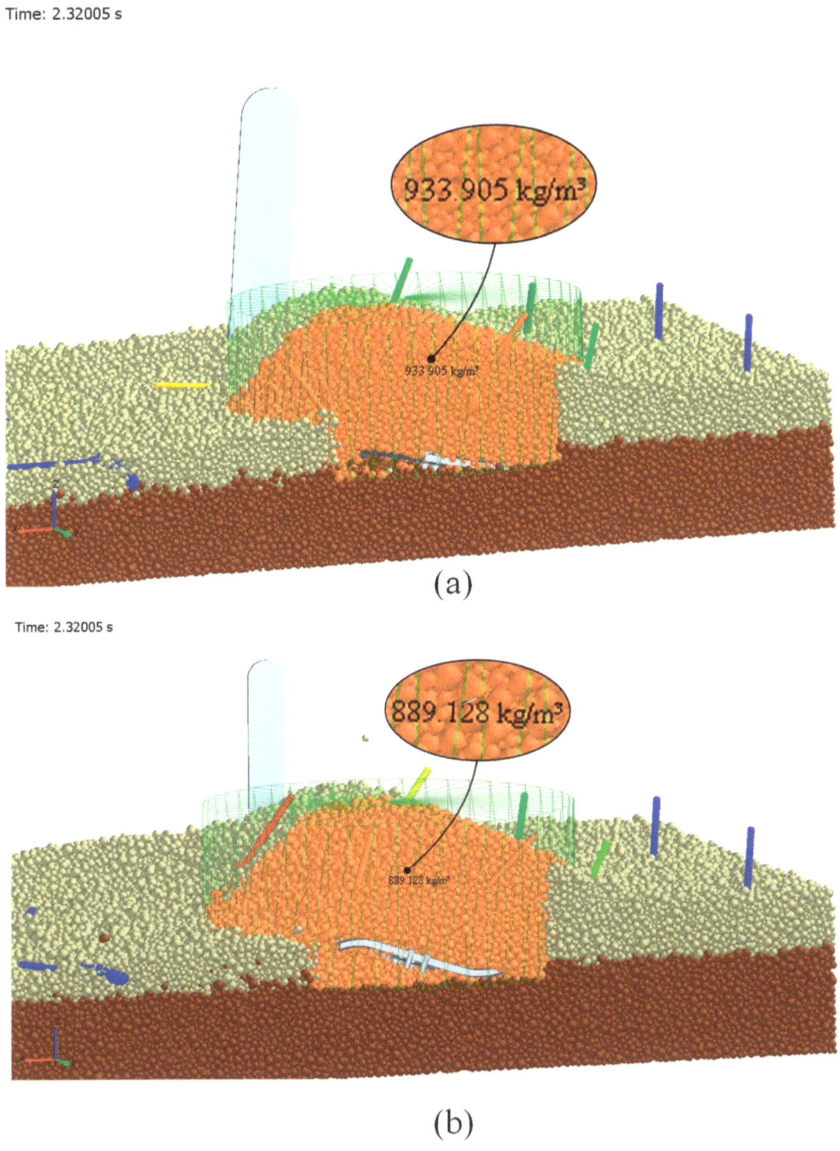

Figure 18. DEM-predicted bulk density for sandy loam soil evaluation: (**a**) fork-shaped shovel and (**b**) S-shaped shovel.

4. Conclusions

In this study, the field performance of a vibrating digging shovel was evaluated. DEM models were successfully developed for two soil textures (clay and sandy loam). The developed DEM models were used to simulate the soil–tool and soil–crop interactions using a hysteretic spring, Hertz–Mindlin, and parallel bond contact models. Full-scale shovels were evaluated to mimic the actual harvesting scenarios. Reasonable draught force and drawbar power predictions were made with DEM with mean relative error values of 4% and 4.7% for the clay and sandy loam soils, respectively, compared to experimental data. With increasing forward speed (1–4 km h^{-1}), draught requirements on the tine increased by 46.38%. It was also found that increasing vibration frequency from 5 to 14.5 Hz decreased both draught force and drawbar power by 42.4%. The S-shaped shovel could crush and translocate soil–crop mass to the rear better than the fork–shaped shovel. This suggests that the S-shaped shovel works well in different soil conditions ranging from frictional to cohesive soils. The findings of this study also show that draught force and drawbar power were generally higher for the clay soil compared to the sandy loam soil evaluated. The methodology used to develop the soil–crop model can be applied to other root and tuber crops, facilitating the virtual evaluation of digging tools or entire harvesters.

Supplementary Materials: The following supporting information can be downloaded at: https://www.mdpi.com/article/10.3390/agriculture13081525/s1, Supplementary Figure S1. Hydraulic system circuit diagram; Supplementary Figure S2. Control and measuring equipment: (a) PTO input, (b) hydraulic cylinder, (c) user interface module, (d) information acquisition module, (e) hydraulic pump, (f) hydraulic valve module, (g) PLC unit, and (h) hydraulic motor; Supplementary Figure S3. Field soil moisture meter (left) and soil cone penetrometer (right); Supplementary Figure S4. DEM input parameter calibration for the clay soil: (a) static angle of repose test, (b) static angle of repose measurement using EDEMPy script, (c) normal plot of residuals, (d) predicted versus actual diagnostic plot, (e) static angle of repose 3D surface plot for rolling friction and static friction, (f) static angle of repose 3D surface plot for static friction and restitution, (g) numerical optimisation result; Supplementary Table S1. Mechanical and physical properties of Jerusalem artichoke crop; Supplementary Table S2. Static angle of repose factors and level; Supplementary Table S3. I-optimal experimental design and data for static angle of repose simulation; Supplementary Table S4. ANOVA results for static angle of repose simulation (sandy loam soil); Supplementary Table S5. Box-Behnken experimental design and result for the angle of repose simulation (Clay soil). Supplementary Table S6. ANOVA results for static angle of repose simulation (clay soil).

Author Contributions: Conceptualisation, E.A.; methodology, E.A., K.A.A. and D.L.A.; software, E.A., K.A.A. and B.V.G.; validation, E.A., D.L.A., K.A.A. and J.Z.; formal analysis, B.V.G., Z.L. and P.M.; investigation, P.M. and Z.L.; data curation, E.A.; writing—original draft preparation, E.A.; writing—review and editing, E.A., K.A.A., D.L.A., J.Z., B.V.G. and P.M.; visualisation, E.A., D.L.A. and K.A.A.; supervision, J.Z.; project administration, J.Z.; funding acquisition, J.Z. All authors have read and agreed to the published version of the manuscript.

Funding: This work was supported by the China National Key Research and Development Plan Project (grant number: 2019YFD0900701).

Institutional Review Board Statement: Not applicable.

Data Availability Statement: Not applicable.

Acknowledgments: The authors are grateful to Yatai Machinery Co., Ltd. (Yu Cheng City, Shandong Province, China), and Nanjing Agricultural University (College of Engineering) for the technical assistance and operational support to conduct this research.

Conflicts of Interest: The authors declare no conflict of interest.

Nomenclature

A	Cross-sectional area (mm^2)
ANOVA	Analysis of variance
ASABE	American Society of Agricultural and Biological Engineers
CoV	Coefficient of variation
CFD	Computational fluid dynamics
DEM	Discrete element method
E	Elastic modulus (Pa)
FEM	Finite element method
G	Shear modulus (Pa)
HSCM	Hysteretic spring contact model
JKR	Johnson–Kendall–Roberts
L	Original specimen length (mm)

LCM	Linear cohesion model
P	Load applied (N)
PB	Parallel bond
RE	Relative error
RSM	Response surface methodology
Std.Dev.	Standard deviation
2FI	Two-factor interaction
v	Poisson's ratio
ΔL	Change in specimen length (mm)
σ	Axial stress (N m^{-2})
ε	Axial strain

References

1. Milne, F. The Canon of Potato Science: 34. Potato Harvesting. *Potato Res.* **2007**, *50*, 347–349. [CrossRef]
2. Collins, B. Effect of soil characteristics, seeding depth, operating speed, and opener design on draft force during direct seeding. *Soil Tillage Res.* **1996**, *39*, 199–211. [CrossRef]
3. McKyes, E. *Soil Cutting and Tillage*; Elsevier: Amsterdam, The Netherlands, 1985; ISBN 044460104X.
4. Fu, W.; Chen, H.; Kan, Z. Optimizing parameters on vibration break shovel of radish harvester. *Trans. Chin. Soc. Agric. Eng.* **2011**, *27*, 46–50.
5. Kang, W.S.; Wen, X.Z. Developing a small commercial vibrating potato digger (I)-assessment of kinematic design parameters. *Appl. Eng. Agric.* **2005**, *21*, 807–811. [CrossRef]
6. Awuah, E.; Zhou, J.; Liang, Z.; Aikins, K.A.; Gbenontin, B.V.; Mecha, P.; Makange, N.R. Parametric analysis and numerical optimisation of Jerusalem artichoke vibrating digging shovel using discrete element method. *Soil Tillage Res.* **2022**, *219*, 105344. [CrossRef]
7. Wan, L.; Li, Y.; Zhang, C.; Ma, X.; Song, J.; Dong, X.; Wang, J. Performance Evaluation of Liquorice Harvester with Novel Oscillating Shovel-Rod Components Using the Discrete Element Method. *Agriculture* **2022**, *12*, 2015. [CrossRef]
8. Shahgoli, G.; Fielke, J.; Desbiolles, J.; Saunders, C. Optimising oscillation frequency in oscillatory tillage. *Soil Tillage Res.* **2010**, *106*, 202–210. [CrossRef]
9. Masienko, I.; Vasilenko, A.; Eranova, L. Theoretical study of the forced oscillation effect on subsoil tillage. *E3S Web Conf.* **2020**, *193*, 1028. [CrossRef]
10. Hua, Z.; Jianmin, W.; Wei, S. The design and experiment of 4UM-640 vibration potato digger. *Agric. Res. Arid. Area* **2014**, *32*, 264–268. [CrossRef]
11. Liu, G.; Xia, J.; Zheng, K.; Cheng, J.; Wang, K.; Zeng, R.; Wang, H.; Liu, Z. Effects of vibration parameters on the interfacial adhesion system between soil and metal surface. *Soil Tillage Res.* **2022**, *218*, 105294. [CrossRef]
12. Ucgul, M.; Saunders, C.; Fielke, J.M. Comparison of the discrete element and finite element methods to model the interaction of soil and tool cutting edge. *Biosyst. Eng.* **2018**, *169*, 199–208. [CrossRef]
13. Karmakar, S.; Kushwaha, R.L. Simulation of soil deformation around a tillage tool using computational fluid Dynamics. *Trans. ASAE* **2005**, *48*, 923–932. [CrossRef]
14. Aikins, K.A.; Jensen, T.A.; Antille, D.L.; Barr, J.B.; Ucgul, M.; Desbiolles, J.M. Evaluation of bentleg and straight narrow point openers in cohesive soil. *Soil Tillage Res.* **2021**, *211*, 105004. [CrossRef]
15. Aikins, K.A.; Ucgul, M.; Barr, J.B.; Awuah, E.; Antille, D.L.; Jensen, T.A.; Desbiolles, J.M.A. Review of Discrete Element Method Simulations of Soil Tillage and Furrow Opening. *Agriculture* **2023**, *13*, 541. [CrossRef]
16. Zhao, H.; Huang, Y.; Liu, Z.; Liu, W.; Zheng, Z. Applications of Discrete Element Method in the Research of Agricultural Machinery: A Review. *Agriculture* **2021**, *11*, 425. [CrossRef]
17. Horabik, J.; Molenda, M. Parameters and contact models for DEM simulations of agricultural granular materials: A review. *Biosyst. Eng.* **2016**, *147*, 206–225. [CrossRef]
18. Horabik, J.; Wiącek, J.; Parafiniuk, P.; Bańda, M.; Kobyłka, R.; Stasiak, M.; Molenda, M. Calibration of discrete-element-method model parameters of bulk wheat for storage. *Biosyst. Eng.* **2020**, *200*, 298–314. [CrossRef]
19. Ucgul, M.; Saunders, C. Simulation of tillage forces and furrow profile during soil-mouldboard plough interaction using discrete element modelling. *Biosyst. Eng.* **2020**, *190*, 58–70. [CrossRef]
20. Li, H.; Zeng, S.; Luo, X.; Fang, L.; Liang, Z.; Yang, W. Design, DEM Simulation, and Field Experiments of a Novel Precision Seeder for Dry Direct-Seeded Rice with Film Mulching. *Agriculture* **2021**, *11*, 378. [CrossRef]
21. Li, Y.; Hu, Z.; Gu, F.; Wang, B.; Fan, J.; Yang, H.; Wu, F. DEM-MBD Coupling Simulation and Analysis of the Working Process of Soil and Tuber Separation of a Potato Combine Harvester. *Agronomy* **2022**, *12*, 1734. [CrossRef]
22. Wanru, L.; Guozhong, Z.; Yong, Z.; Haopeng, L.; Nanrui, T.; Qixin, K.; Zhuangzhuang, Z. Establishment of discrete element flexible model of the tiller taro plant and clamping and pulling experiment. *Front. Plant Sci.* **2022**, *13*, 1019017. [CrossRef]
23. Aikins, K.A.; Ucgul, M.; Barr, J.B.; Jensen, T.A.; Antille, D.L.; Desbiolles, J.M. Determination of discrete element model parameters for a cohesive soil and validation through narrow point opener performance analysis. *Soil Tillage Res.* **2021**, *213*, 105123. [CrossRef]

24. Syed, Z.; Tekeste, M.; White, D. A coupled sliding and rolling friction model for DEM calibration. *J. Terramech.* **2017**, *72*, 9–20. [CrossRef]
25. Chen, Z.; Wassgren, C.; Veikle, E.; Ambrose, K. Determination of material and interaction properties of maize and wheat kernels for DEM simulation. *Biosyst. Eng.* **2020**, *195*, 208–226. [CrossRef]
26. Coetzee, C.J. Review: Calibration of the discrete element method. *Powder Technol.* **2017**, *310*, 104–142. [CrossRef]
27. Aikins, K.A.; Antille, D.L.; Ucgul, M.; Barr, J.B.; Jensen, T.A.; Desbiolles, J.M. Analysis of effects of operating speed and depth on bentleg opener performance in cohesive soil using the discrete element method. *Comput. Electron. Agric.* **2021**, *187*, 106236. [CrossRef]
28. Zhang, S.; Zhang, R.; Chen, T.; Fu, J.; Yuan, H. Calibration of simulation parameters of mung bean seeds using discrete element method and verification of seed-metering test. *Trans. Chin. Soc. Agric. Mach.* **2022**, *53*, 71–79.
29. EDEM. *EDEM 2020 Documentation*; DEM Solutions: Edinburgh, UK, 2020.
30. Budynas, R.G.; Nisbett, J.K.; Shigley, J.E. *Shigley's Mechanical Engineering Design*, 11th ed.; McGraw-Hill Education: New York, NY, USA, 2020.
31. Soehne, W. *Some Basic Considerations of Soil Mechanics as Applied to Agricultural Engineering*; National Institute of Agricultural Engineering: Silsoe, Bedford, UK, 1958.
32. McLaughlin, N.B.; Campbell, A.J. Draft-speed-depth relationships for four liquid manure injectors in a fine sandy loam soil. *Can. Biosyst. Eng.* **2004**, *46*, 1–25.
33. Tekeste, M.Z.; Balvanz, L.R.; Hatfield, J.L.; Ghorbani, S. Discrete element modeling of cultivator sweep-to-soil interaction: Worn and hardened edges effects on soil-tool forces and soil flow. *J. Terramech.* **2019**, *82*, 1–11. [CrossRef]
34. Kim, Y.-S.; Siddique, M.A.A.; Kim, W.-S.; Kim, Y.-J.; Lee, S.-D.; Lee, D.-K.; Hwang, S.-J.; Nam, J.-S.; Park, S.-U.; Lim, R.-G. DEM simulation for draft force prediction of moldboard plow according to the tillage depth in cohesive soil. *Comput. Electron. Agric.* **2021**, *189*, 106368. [CrossRef]
35. Mak, J.; Chen, Y. Simulation of Draft Forces of a Sweep in a Loamy Sand Soil Using the Discrete Element Method. *Can. Biosyst. Eng.* **2014**, *56*, 2.1–2.7. [CrossRef]
36. Vasilenko, V.V.; Vasilenko, S.V.; Achkasova, N.N. Effect of vibration on resistance force of plough. In *International Scientific and Practical Conference "Agro-SMART-Smart Solutions for Agriculture" (Agro-SMART 2018)*; Atlantis Press: Paris, France, 2018; pp. 779–783. ISBN 9462526303.
37. Sun, W.; Wu, J.M.; Shi, L.R.; Zhang, H.; Li, T.; Kao, J. Drag Reduction Performance Parameter Optimization of Vibration Digging Shovel. In *Advanced Materials Research*; Peilong, X., Hongzong, S., Yiqian, W., Pin, W., Eds.; Scientific.Net: Baech, Switzerland, 2014; pp. 777–780.
38. Gowripathi Rao, N.R.N.V.; Chaudhary, H.; Sharma, A.K. Design and development of vibratory cultivator using optimization algorithms. *SN Appl. Sci.* **2019**, *1*, 1287. [CrossRef]
39. Coetzee, C.J. Calibration of the discrete element method and the effect of particle shape. *Powder Technol.* **2016**, *297*, 50–70. [CrossRef]

agriculture

Article

Propellers Spin Rate Effect of a Spraying Drone on Quality of Liquid Deposition in a Crown of Young Spruce

Aleksandra Pachuta [1], Bogusława Berner [1], Jerzy Chojnacki [1,*], Gerhard Moitzi [2], Jiří Dvořák [3], Anna Keutgen [4], Jan Najser [5], Jan Kielar [5], Tomáš Najser [5] and Marcel Mikeska [5]

1 Mechanical Faculty, Koszalin University of Technology, Racławicka Str. 15-17, 75-620 Koszalin, Poland; aleksandra.pachuta@s.tu.koszalin.pl (A.P.); bogusiaberner.b@gmail.com (B.B.)
2 Experimental Farm Groß-Enzersdorf, Department of Crop Sciences, University of Natural Resources and Life Sciences, Vienna (BOKU), Schloßhofer Straße 31, 2301 Groß-Enzersdorf, Austria; gerhard.moitzi@boku.ac.at
3 Department of Forestry Technologies and Construction, Faculty of Forestry and Wood Sciences, Czech University of Life Sciences, Kamýcká 129, 6 Suchdol, 165 21 Prague, Czech Republic; dvorakj@fld.czu.cz
4 Institute of Vegetables and Ornamentals, Department of Crop Sciences, University of Natural Resources and Life Sciences, Vienna (BOKU), Gregor-Mendel-Straße 33, 1180 Vienna, Austria; anna.keutgen@boku.ac.at
5 ENET Centre, CEET, VSB-Technical University of Ostrava, 17. listopadu 2172/15, 708 00 Ostrava, Czech Republic; jan.najser@vsb.cz (J.N.); jan.kielar@vsb.cz (J.K.); tomas.najser@vsb.cz (T.N.); marcel.mikeska@vsb.cz (M.M.)
* Correspondence: jerzy.chojnacki@tu.koszalin.pl; Tel.:+48-943478359

Abstract: The study aimed to assess the quality of spraying of ornamental conifer using a multi-rotor drone. We examined how the speed of drone movement and the propellers' spin speed can affect the deposition quality of the sprayed liquid in the crown of blue spruce (*Picea pungens* Engelm.). Due to the avoidance in the future of droplet drift by air movements, an air injector atomiser for liquid spraying was used, and a low altitude of 0.6 m of the drone flying above the tree was used in the study. The drone moved at two speeds: 0.57 m·s^{-1} and 0.94 m·s^{-1}. The propellers' spin speeds were adjusted based on the drone's weight with the spray liquid tank filled and empty. The propellers' zero-spin rate was also included to compare the drone to a field sprayer. The tests were conducted in a laboratory setting. Volume and uniformity of liquid volume settled on the levels of samplers positioned on a tripod within the tree canopy were assessed. The samplers were placed in two zones of the tree: near the tree trunk and at a distance of 0.21 m from the trunk. Airstream speed generated by drone propellers was also evaluated inside the tree. The findings indicated that the rotations of propellers and air speed significantly influenced the quality of liquid deposition on samplers located away from the trunk. The results also showed that using a drone instead of a field sprayer could benefit the quality of the spray application. The weight of the multi-rotor drone, determined by the spray liquid tank's filling level, can significantly impact the quality of spray deposition in the tree. Based on the investigations, it can be recommended that low-altitude spraying drones be adopted for studies and future strategies in precision agriculture using autonomous inspection-spraying drones.

Keywords: unmanned aerial vehicle; UAV; low altitude; UASS; spraying; drone weight; airstream; conifer tree; tree crown zones

Citation: Pachuta, A.; Berner, B.; Chojnacki, J.; Moitzi, G.; Dvořák, J.; Keutgen, A.; Najser, J.; Kielar, J.; Najser, T.; Mikeska, M. Propellers Spin Rate Effect of a Spraying Drone on Quality of Liquid Deposition in a Crown of Young Spruce. *Agriculture* **2023**, *13*, 1584. https://doi.org/10.3390/agriculture13081584

Academic Editors: Chung-Liang Chang and Mustafa Ucgul

Received: 29 June 2023
Revised: 1 August 2023
Accepted: 6 August 2023
Published: 9 August 2023

1. Introduction

Protection against pests of ornamental tree crops is a crucial aspect of agriculture, forest nursery production, and forestry. Traditionally, crop protection treatments for young ornamental trees and forest nurseries are applied using a backpack or tractor-mounted field sprayers. However, this equipment is not always suitable for these tasks [1]. For trees in forests, pest spraying is mainly carried out using manned aircraft such as planes and helicopters [2,3]. Although treatments performed with aircraft are effective, this method is expensive due to the high cost of aviation equipment and infrastructure, as well as the cost associated with the spraying process itself. The deposition quality of the liquid droplets on trees is strongly influenced by the technical conditions of the flights, such as the aircraft type and speed, the liquid dose, the droplet size, the spray release height above the trees, as well as by the weather conditions, especially wind speed and air humidity. These factors can also cause droplet drift and evaporation [4–6].

Recently, unmanned aerial vehicles (UAVs), commonly known as drones, have emerged as a potential solution for protecting crops against pests [7–10]. These drones can be divided into two groups: inspection drones and work drones. Inspection drones are used to raid agricultural crops and forests to assess their development and health and even monitor the weather to evaluate pest possibility [11–15]. They are equipped with various types of cameras and scanners. The most popular are thermal, multispectral, or hyperspectral cameras, which allow selected spaces to be photographed in a specific wavelength of light, including colours invisible to the human eye. Working drones are used to carry out work from the air. In agriculture, they can be applied, besides spraying pesticides, as flying machines for sowing seeds, spreading and spraying chemical and organic fertilizers, and crop planting [7,9,15–17]. Due to the tremendous potential of drones for crop assessment and programming pesticide application rates based on inspection data, drones can be used to apply chemicals with variable rate application (VRA) [18]. Research is also being conducted into the intelligent use of drones as autonomous robots in precision agriculture, not only for crop protection but to perform other fieldwork [19,20]. Research shows that the future of drone use in precision agriculture also involves using artificial intelligence (AI) in evaluating the information obtained and steering not only individual unmanned aircraft in agricultural operations but a swarm of working drones on the field [20–22].

By construction, drones are divided into fixed-wing and rotor drones. Of the two types of drones, rotor drones, particularly multi-rotors with an electric power source, are most commonly used for crop protection.

Attempts have been made to compare the effects of spraying plants with drones to the effects of using ground-based sprayers, but this is not easy due to differences in the construction and conditions of use of the two types of equipment. Hence, conclusions are inconclusive [23]. A comparison of the penetration in the cotton plant canopy of liquid sprayed with a boom sprayer positioned above the plants at the height of 0.5 m with the penetration of liquid sprayed with a drone at heights of 1.5 and 2.0 m provided the result that liquid sprayed with a drone penetrates the plant canopy less than liquid sprayed with a sprayer [24].

Studies are being carried out on the effects of using unmanned aerial spraying systems (UASSs) to control pests in agricultural fields and plantations of trees in orchards [7,25–29]. It has been observed that the effectiveness of spray penetration, especially of the lower parts of the tree crown, is influenced by such factors as the flight height and speed of the drone above the tree crown and the shape of the tree crown.

There are also attempts to use drones for spraying trees in forests [30,31]. Compared with classic aerial treatments using human-manned aviation, the advantages of using drones in protecting forest crops have many advantages. The most important of these is the possibility of spraying at lower altitudes, smaller areas of plantations, and even individual trees, as well as lower treatment costs. Drones reduce spraying time and can perform treatments on trees of varying heights while maintaining precise control over their flight height. UAVs are also suitable for applying forest protection products to objects at different altitudes, such as mountainous areas. Drones also offer precise removal of parasites inside tree crowns [32]. They are well-suited for spraying small, limited areas and even single trees. During aerial spraying with drones, the rotors' propellers generate airflow. This flow and external natural air movements can alter droplet trajectories and affect the deposition of liquid droplets within the tree crown and their drift [33,34]. The severe disadvantage of aerial spraying is the drift of droplets sprayed with UASS and containing very often significantly concentrated pesticides. This drift is often caused by the influence of the local climate and gusts of air in open spaces, occurring even without wind. The drift of liquid sprayed from drones is much greater than when spraying with ground-based equipment and can lead to environmental contamination [35]. Avoiding the drift of sprayed liquid is a big challenge for drone spraying [9,33,35]. The European Union has strict regulations regarding aerial plant spraying due to the potential drift of sprayed pesticides and environmental contamination. Spraying from aerial vehicles, human-manned or UAV, is permitted only in the absence of any other option. This is outlined by the European Parliament and Council 2009/128/EC of 20 October 2009 [36].

Farmers using drones for crop protection expect, in addition to the typical gains resulting from not being able to carry out treatments with ground-based sprayers, additional profit resulting from a reduction in the cost of crop protection by UASS to the costs of using sprayers. In order to increase efficiency, attempts are being made to widen the swath using very high flight altitudes above the plants, even reaching up to 10 m, high drone travel speeds of about 6.0 $m \cdot s^{-1}$, and using very condensed pesticide solutions in the spray liquid and also by spraying droplets very finely [8,34,35]. Although most drone spraying treatments experiments have been carried out at heights of 1.2 to 2.5 m above trees mainly fruit trees and hazel [11,27,28], even at these spraying heights the long fall of fine droplets can lead to drift and damage the environment. There are no regulations in the European Union, and probably in many countries around the world, concerning the parameters and environmental conditions for safe UASS spraying. On the other hand, there needs to be more information and research on the cost-effectiveness of drone spraying concerning effectiveness and environmental safety.

The flight height of multi-rotor UASS when spraying plants has the most significant impact on reducing the drift of the spray. Lowering the flight height of spraying drones is recommended as a droplet drift prevention strategy [9]. In addition to the short path of the droplets from the nozzle to the plants, the strong airflow generated by the drone's rotors being at a short distance between the rotors and the plants can further accelerate their arrival at the target and improve penetration into crop canopy [35,37]. The low flight altitude of spraying drones also means that precise spraying of individual plants, or even parts of plants, can be carried out. Low drone flight altitudes can also be recommended for inspection drones as they allow very accurate information on pests not only at the scale of the crop area but also at the scale of individual plants and trees and even at the scale of their leaves [19–21]. The low flight altitude also favours the effectiveness of autonomous drones with the ability to identify pests and immediately pesticide spray their location on the plant.

The weight of the multi-rotor drones used for crop spraying changes during the spraying process because the tank of liquid is emptied. At this time, the drone maintains a constant altitude. The drone's rotors must generate enough strong thrust from the airflow to overcome the drone's total weight and lift it above the trees when the tank is full and when the tank is nearly empty. This situation means that, at the same treatment altitude, a drone with a full tank generates greater thrust force from its rotors compared with the thrust force produced when the tank is nearly empty or empty. The thrust of a multi-rotor drone depends on the rotational speed of its rotors [38]. As the thrust force and rotational speed of the rotors decrease during the spraying time, the airflow speed affecting the droplet stream also decreases. Many studies investigating the use of drones for tree spraying overlook the disparities between a drone's weight with a full or empty liquid tank and the outcomes. These studies predominantly focus on factors such as tree type, droplet size, drone flight parameters, and the biological effects of the treatment. In the case of multi-rotor unmanned aerial spraying systems operating at very low altitudes, such an influence can exist, and varying airflow parameters during treatment can affect the quality of liquid deposition.

Forest and ornamental trees are susceptible to many pests, including insects, nematodes, fungi, bacteria, and viruses. The effectiveness of tree protection depends not only on the specific pesticide used to control a particular pest but also on its even deposition, along with the spray of the entire tree crown, or at least its areas where pests mainly feed. Studies show that the shape of the tree crown can affect the volume of spray penetrating its various parts [39]. Studies on the deposition of fluid sprayed from drones onto plants are conducted not only in open spaces, forests, and fields but also in laboratory test beds. Laboratory tests enable the simulation of natural conditions with precise control and measurement of technical parameters during the treatments performed [40,41]. They also allow for detailed observation of the physical phenomena accompanying the spraying of plants using drones.

The research aim was determined because of the intention to reduce droplet air drift when spraying drones. For this reason, the research focused on the low height of the drone's movement above the object being sprayed. It was assumed that the drone's height should be comparable to the distance of the nozzles from the plants in ground sprayers.

The main aim of the research was to evaluate the influence of the drone's rotors speed and speed of drone movement on the quality of spray deposition in the tree crown in conditions of low-altitude drone spraying. The studies were also intended to answer the question of whether the rotational speed of the propellers generating the airflow from the drone when the tank is full and when the tank is empty can significantly influence the quality of liquid deposition in the tree crown.

The research also aimed to assess the influence of the airstream generated by the drone's propeller rotation and the degree of foliage in the tree crown zones on the quality of liquid deposition on the tree.

In addition, the study aimed to compare the droplet deposition results on the tree obtained from the spraying drone with the results from the same treatment that can be performed with a conventional field sprayer without airflow.

2. Materials and Methods

The studies were primarily focused on using drones to spray crops of young ornamental and forest trees. Therefore, a young tree of blue spruce (*Picea pungens* Engelm.) was selected for the research. This spruce species is commonly cultivated for gardens, parks, and Christmas tree plantations, as well as for industrial purposes in forest plantations. Blue spruce trees are susceptible to pests, particularly parasitic fungi, and small insects [42–44]. Protective treatments are typically required for spruce trees grown in plantations, with ground sprayers and, mainly, field sprayers commonly employed for these treatments.

2.1. Test Stands

For the experiment, a laboratory stand was prepared, which ensured precise and repeatable measurements of the drone's movement speed, altitude above the tree, and propeller spin speeds. On the stand, the drone was affixed to a rolling cart on rails supported by fixed structures at opposite ends. The cart was then pulled by an electric motor and a rope connected to the cart. By adjusting the frequency of the current supplied to the electric motor, the researchers can control the drone's movement speed. Studies carried out in a stable laboratory environment allowed for a more accurate analysis of the results compared with field tests. Additionally, this stationary stand prevented the impact of air gusts, which can affect the air stream's properties and the trajectory and shape of the sprayed liquid and hinder research [33,35,45].

For the studies, a hexacopter drone with 15 $\times$ 5.2″ propellers and 500 W, 400 kV brushless electric motors powered by a LiPo 16,000 mAh, 22.2 V battery was used. The drone's propellers were controlled using a transmitter station. An optical digital tachometer (UT-372), produced by Uni-Trend Technology Co. Ltd. (Dongguan, China), connected to a computer was employed to measure the speed of the propellers' spin.

Manufacturers of unmanned aerial spraying systems typically equip drones with fine droplet nozzles for better crop coverage. The fine droplets produced by pressurised and rotary atomisers are highly susceptible to drift by air movement. Bearing in mind that the conditions of the study assumed not to allow droplets to drift by gusts of air for testing, the fine droplet atomiser was abandoned, and the air-injector compact nozzle (IDK 90-015) manufactured by Lechler GmbH (Metzingen, Germany), was mounted on the drone. For delivering pressurized liquid to a spray nozzle, a small ground sprayer was connected to it by a flexible pipe. The liquid pressure in the system remained constant at 0.20 MPa. According to the manufacturer's specifications, at this pressure, the nozzle produced average droplet diameters, measured according to the ISO 25358 classification standard in the coarse droplet range [46]. A nozzle with a spray angle of 90° was used to accurately match the shape of the droplet jet to the shape of the crown of the tree sprayed so that the spray jet covered the entire volume of the crown, assuming that this was the most favourable way to spray similar trees from above. Figure 1 illustrates the positioning of the spray nozzle relative to the drone's frame, rotors, and tree, as well as the drone's altitude to the spruce being sprayed.

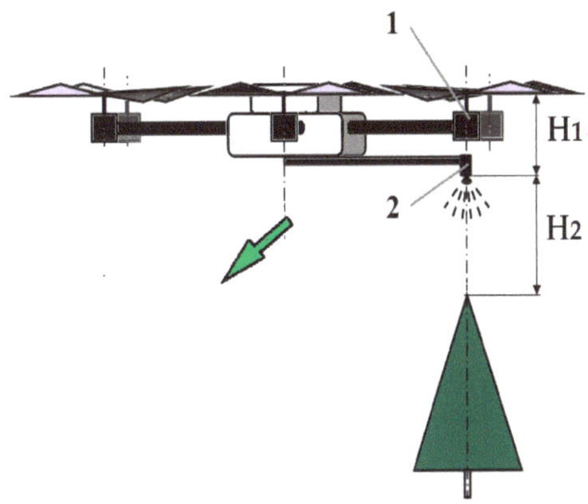

Figure 1. Position of the drone relative to the sprayed tree: 1—drone's front arm; 2—spray nozzle, the arrow indicates the direction of the drone's movement.

The hexacopter had a specific configuration with a single arm positioned in the front and rear in the direction of movement and two arms mounted on each side of it. The liquid spray nozzle (2) was situated under the first front arm (1), located on one side of the multi-rotor drone. The spray nozzle's axis of symmetry aligned with the rotor's axis directly beneath it. The nozzle outlet was located below the lower level of the rotor propellers at a distance of H_1 = 0.15 m.

In order to ensure accurate positioning, the spruce tree was placed in a pot, allowing precise alignment with the drone. The top and centre of the tree canopy were positioned along the nozzle's symmetry axis. The tree's height above the soil level in the pot was 0.90 m. Bearing in mind that there is a need to limit the drift of drops, the lowest possible height of the nozzle over the tree was selected [9]. The nozzle was positioned at the height of H_2 = 0.60 m above the tree's crown. This altitude was determined based on previous tests and literature review, considering it also the lowest feasible altitude for flying the drone over spruce trees without risking a collision. This decision was suggested by the lowest flight altitude of the drone over the citrus tree described in the paper [39].

In order to evaluate the deposition of sprayed liquid in trees, samplers are used. These can be attached to branches or indirectly positioned on metal stands, artificial trees, or nets that simulate canopy space [5,39,47]. A tripod with samplers attached to its arms was employed to assess liquid deposition in this research. The stand consisted of a vertical metal rod connected to three sets of horizontally placed metal rods at three different heights. Each set comprised four rods intersecting at right angles. The vertical rod was attached to the tree trunk. The levels of sampler bars were marked with the letters "A", "B", and "C" from the top. In addition, in order to be able to assess the influence of the crown of the tree on the changes in the velocity of the airflow inside the spruce, a second tripod, identical to the tripod placed in the crown of the tree, was constructed to take airstream measurements without the tree. Figure 2 illustrates the tree with the tripod mounted and the tripod itself.

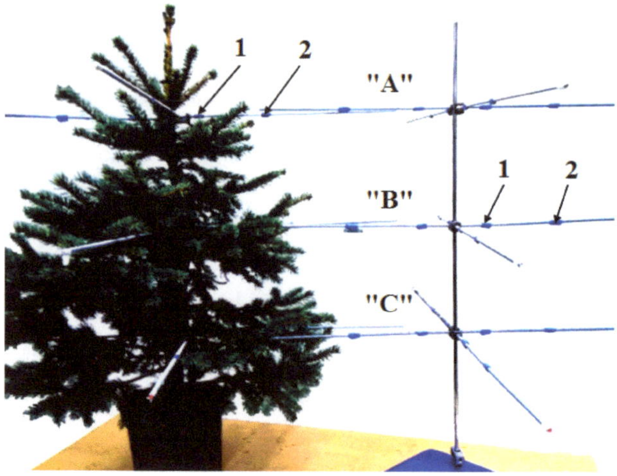

Figure 2. The tripod mounted inside the tree crown and the tripod without the tree: 1—places to mount "trunk" samplers; 2—places to mount "branch" samplers.

The distance between the first level of rods (A) and the top of the tree was 0.18 m. The gap between each level (A, B, and C) was 0.24 m. Placing the samplers on the rods ensured precise positioning relative to the drone and nozzle during the tests.

For measuring liquid deposition inside of the tree crown, plastic labels (0.02 m × 0.04 m) were used as samplers. They were glued to holders attached at the marked positions on the horizontal bars of the tripod. Each rod at every level had two types of samplers: "trunk" samplers placed 0.06 m from the trunk's centre and "branch" samplers positioned at a distance of 0.21 m, also measured from the trunk's centre. As there were four rods at each tripod level, so there were a total of four "trunk" samplers and four "branch" samplers (Figure 2). The "trunk" samplers were used to estimate of the amount of liquid penetrating the branches in the central part of the tree canopy near the trunk. The "branch" samplers assessed the liquid reaching the branches at a distance of approximately two-thirds of the average branch length, counting from the trunk's centre (the vertical zone of the tree, away from the trunk).

Two speeds of moving drone, 0.57 $m \cdot s^{-1}$ and 0.94 $m \cdot s^{-1}$, were taken to study the influence of the drone's linear speed over the tree on the volume of liquid reaching different parts of the spruce canopy. The speeds adopted were due to the design and capabilities of the stand.

The study also assessed how the air stream generated by the drone's rotor thrust, depending on the drone's weight, affected the liquid distribution inside the tree canopy. Since the drone was fixed to the cart was unable to generate an air stream, depending on the water's weight inside the tank, in order to calculate propellers' rotation speeds, the weight of the commercial spraying drone M4E (TTAviation) [48], similar in technical parameters to the drone used in the research, was taken. The calculations utilized a mathematical Formula (1) derived from a previous study of the same drone model [49].

$$F = 9 \cdot 10^{-7} \cdot n^{2.137}, \tag{1}$$

where F is the thrust force, and N and n is the rotors speed, rpm.

For the desired thrust force, F_1 = 72.2 N, when the tank was empty, calculated rotations were n_1 = 5000 rpm. For F_2 = 120.4 N, when the tank was full, rotations were n_2 = 6350 rpm. The ratio of the liquid weight in the tank to the total weight of the drone was determined to be 39.8%. Additionally, to assess the settling of spray in the spruce tree canopy without the influence of airflow from the drone's rotors, the study was performed with the rotors' speed n_0 = 0.0 rpm. This value of the propellers speed also simulated tree spraying with an above-ground sprayer.

All measurements for the designated three drone propeller speeds, the adopted two drone travel speeds, and a tripod placed inside the tree and a tripod without a tree were repeated three times. Only one tree placed under the drone was used for the tests.

Previous studies [34,41,50,51] suggest that the air stream generated by the UAV plays a crucial role in the delivery and deposition of droplets during plant spraying. In order to analyse the influence of the air stream on liquid deposition inside the spruce tree, changes in air velocity within the tree's canopy caused by its attenuation due to branches and leaves were measured. Testo 405i anemometers were used to make the measurements. During the measurement, the drone remained stationary above the tree, with the spraying nozzle axis of symmetry aligned with the centre of the tree trunk. The anemometers were installed on a beam at a tripod at the heights corresponding to each level of the samplers: "A", "B", and "C." The method of fixing the anemometers to enable them to be adjusted is shown in Figure 3.

Figure 3. Example of the method of mounting anemometers on the tripod for placement in the tree crown.

The measurement probes of the anemometers were positioned precisely at the mounting points of the liquid samplers but only on rods positioned transverse to the direction of the drone's movement. Once the position of the anemometer probes was established, the rods were removed from the tripod so as not to suppress airflow. The air velocity measurements were conducted on the same test stand with and without the presence of the spruce tree. During the measurements without the tree, the anemometers were in the same place where they were placed during the measurements with the tree because a second tripod was used to position their probes.

2.2. Quality of Liquid Deposition

The quality of liquid deposition sprayed from the drone was evaluated based on the volume of liquid deposited on each level of the tripod and the uniformity of liquid deposition at all levels, separately for "trunk" and "branch" samplers. Distilled water, stained with aqueous nigrosin dye at a concentration of 0.5%, was used as the liquid in the study. As the drone moved over the tree, liquid droplets were deposited on the samplers. After the liquid dried, samplers from each location and level were collected and placed in separate containers. Each measurement involving the liquid spray on the tree was repeated three times under the same technical parameters. The dye on the samplers collected in the containers was washed off with 5.0 mL of distilled water. The dyed liquid was analysed using a photo colorimeter to determine the dye concentration in parts per million (ppm). Since the volume of liquid used to wash off the dye remained constant in all containers, and the amount of dye washed off the samplers was proportional to the volume of spray liquid deposited on them, the dye concentration measured by the photo colorimeter was converted to the volume of spray liquid in μL deposited on one cm^2 of the sampler.

The uniformity of liquid deposition at tree levels was assessed using the coefficient CV, calculated from the results of liquid volume deposited on the samples at the tripod levels, according to the following Equation (2):

$$CV = \frac{1}{v_m}\sqrt{\frac{\Sigma(v_i - v_m)^2}{3}} \quad (2)$$

where CV is the coefficient of uniformity of liquid deposition at three levels; v_m is an average volume of liquid deposited on samplers, $\mu L \cdot cm^{-2}$; and v_i is the volume of liquid deposited on the *i*-th sampler, $\mu L \cdot cm^{-2}$.

The CV value was calculated using the data obtained from each repetition of the liquid deposition measurement on the samplers.

2.3. Spruce Tree Crown Evaluation

The leaf area index (LAI) was adopted to characterize the density of branches with needles at a particular level of the spruce tree crown. This index was used to examine the relationship between tree structure and airflow velocity reduction at different levels [52]. Formula (3) was employed to calculate the leaf area index:

$$LAI_i = \frac{F_{li}}{F_{si}} \quad (3)$$

where LAI_i is the leaf area index value for the selected tree level—i, F_{li} is an area of needle leaves on branches above the selected level, and F_{si} is the field area under the branches for the selected level.

The area of needle leaves—F_{li}, was calculated by adding up the lengths of branches within the space above the selected level measured, next the average number of needles per unit of branch length was determined, along with the average area of one needle. These factors were then multiplied together. To calculate the field area under the branches for the selected level—F_{si}, for a specific level the distance from the centre of the trunk to the end of the longest branch within the crown of the tree, placed inside or above that level, was measured. The area of a circle with a radius equal to this distance was then calculated to determine the field area under the branches.

The leaf area index for the upper part of the tree (LAI_A) was determined up to level "A" by dividing the calculated leaf area above this level (F_{lA}) by the circle area under this part of the tree (F_{sA}), which was determined using the length of the longest branch at this level as the radius. The leaf area index for the middle part of the crown (LAI_B) at level "B" was determined by dividing the combined leaf area from levels "A" and "B" (F_{lB}) by the corresponding field area under this part of the tree (F_{sB}). For level "C," the leaf area index (LAI_C) was calculated by dividing the summed leaf area above level "C" (F_{lC}) by the area of the circle determined using the longest branch of the tree as the radius (F_{sC}).

2.4. Statistical Methods

Mean values, calculated from the measurement results, were presented in a bar chart, with the measurement error range as ±standard deviation. After analysing the data for normality of distribution (Shapiro–Wilk test), analyses of variance were performed to assess the significance of the input factors on the results obtained. Statistica 13.3 software (StatSoft) was used to perform both analyses. A value of 0.05 was taken as a limit for the calculated significance p-factor.

3. Results

3.1. Effect of Propellers Rotations on Liquid Deposition

After each tree spraying, the samplers, after drying, were disconnected from the tripod rods and subjected to further processing to assess the capacity of the deposited liquid. The average results and their corresponding standard deviations are presented in Figure 4. Separate graphs were created for the liquid deposited on the "trunk" samplers and the liquid deposited on the "branch" samplers, taking into account the varying spin rates of the propellers. Figure 4a shows the results of liquid deposition on the tripod levels obtained at the speed of 0.57 m·s^{-1}, and Figure 4b at the speed of 0.94 m·s^{-1}.

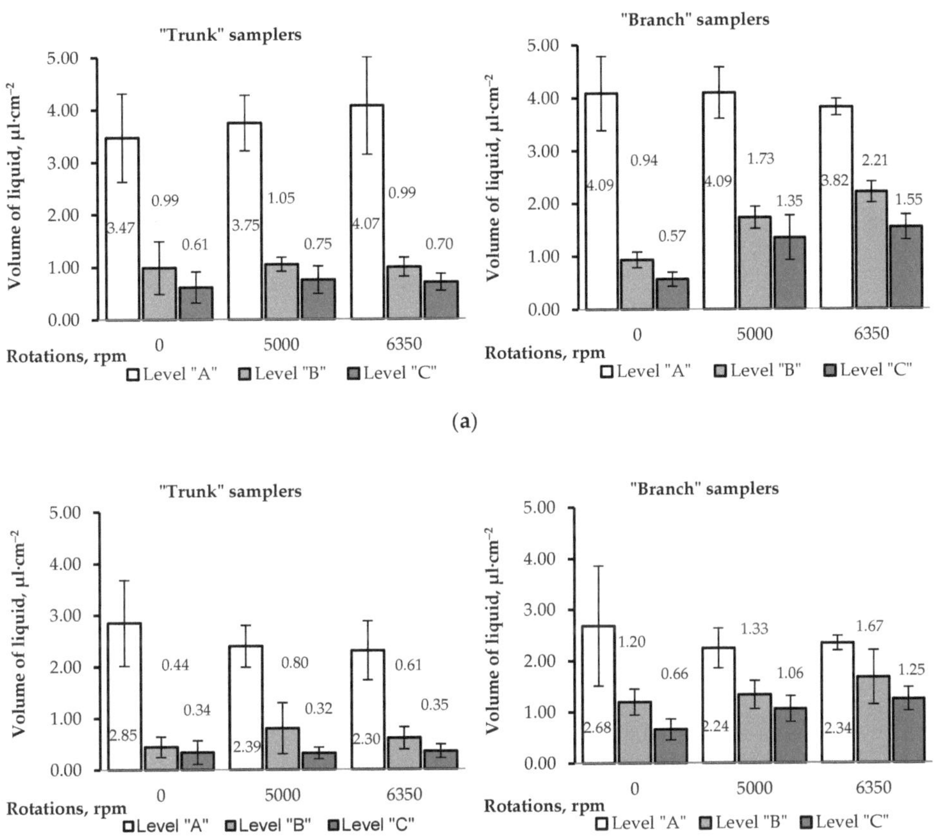

Figure 4. Influence of the propellers speeds and drone velocity on the volume of spray deposited on the samplers at (**a**) 0.57 m·s^{-1}, (**b**) 0.94 m·s^{-1}.

The findings presented in Figure 4a,b indicate that both the speed of the drone's movement over the tree and the speed of its propellers impact the volume of liquid deposited on the samplers. This observation holds for both drone movement speeds of 0.57 m·s^{-1} and 0.94 m·s^{-1}. Variance analysis was performed to evaluate the influence of the drone's movement speed and the presence or absence of rotation of propellers on the liquid volume deposited on the samplers. Similarly, the significance of the propellers' spin speeds, considering the drone's weight with an empty or full tank on the liquid volume at different levels of the tree canopy was examined.

The analysis revealed that the absence or presence of the rotation of propellers at 0.0 rpm, 5000 rpm, and 6350 rpm significantly affected the volume of deposited liquid, but only for the "branch" samplers (p = 0.010456). Additionally, the effect of rotation speeds due to the drone's weight, with an empty or full tank on liquid deposition, was found to be significant only for the "branch" samplers at the middle level ("B") when the drone was moving at a speed of 0.57 m·s^{-1} (p = 0.043965). At the other sampler levels, both "trunk" and "branch," and at both drone travel speeds, the effect of propellers' rotations speeds resulting from tank filling did not exhibit significance.

Variance analysis conducted on "trunk" samplers revealed a significant relationship between drone travel speed and the volume of liquid collected by these samplers (p = 0.000008). Likewise, a significant impact of drone travel speed on the volume of liquid deposited on the "branch" samplers was observed (p = 0.000002).

Additionally, the total volume of liquid deposited on 1 cm^2 of the sampler surface was determined when the individual volumes from repeated measurements were summed across the same drone travel speeds, rotations of propellers, and sampler types. The results of this summation are presented in Table 1.

Table 1. Influence of parameters for performing tree spraying on the total volume of liquid deposited on the samplers.

Conditions		Total Volume of Settled Liquid per Unit Area of Samplers, µL·cm^{-2}		
V, m·s^{-1}	**Rotations of Propellers, rpm**	**0**	**5000**	**6350**
0.57	"Trunk" sampl.	5.07 ± 1.39	5.55 ± 0.44	5.77 ± 0.83
	"Branch" sampl.	5.59 ± 0.84	7.17 ± 0.79	7.58 ± 0.34
0.94	"Trunk" sampl.	3.63 ± 0.99	3.51 ± 0.42	3.26 ± 0.31
	"Branch" sampl.	4.54 ± 0.97	4.63 ± 024	5.26 ± 0.69

In order to compare the effect of the relationship between the drone's movement speeds and the volume of liquid deposited on each level of the samplers, the ratios of the volume of liquid deposited on the samplers at a drone movement speed of 0.57 m·s^{-1} to the volume of liquid deposited on the samplers at a speed of 0.94 m·s^{-1} were calculated. For this purpose, the corresponding results of the repetitions from the measurements for the same tripod levels and speed of drone propellers were divided by each other. And then, the mean values and standard deviations were determined. The ratios of the volume of liquid deposited at all tripod levels (total relations) for the same drone propellers spin rate were also calculated in a similar way. The results are shown in Table 2. The calculated ratio of higher drone speed to lower speed is 1.65.

Table 2. The ratio between the volume of liquid deposited on the samplers at the drone's movement speed of 0.57 m·s^{-1} to the volume of liquid deposited on the samplers at 0.94 m·s^{-1}.

	"Trunk" Samplers			"Branch" Samplers		
Rotations of Propellers, rpm	**0**	**5000**	**6350**	**0**	**5000**	**6350**
Level "A"	1.25 ± 0.33	1.58 ± 0.19	1.82 ± 0.55	1.71 ± 0.77	1.84 ± 0.12	1.64 ± 0.07
Level "B"	2.54 ± 1.86	1.58 ± 0.73	1.76 ± 0.61	0.80 ± 0.13	1.34 ± 0.46	1.46 ± 0.68
Level "C"	2.65 ± 2.47	2.80 ± 1.90	2.36 ± 1.52	0.92 ± 0.32	1.27 ± 0.11	1.28 ± 0.34
Total relations	1.48 ± 0.65	1.59 ± 0.22	1.77 ± 0.21	1.26 ± 0.26	1.55 ± 0.09	1.46 ± 0.27

3.2. The Influence of the Speed of Propellers on the Uniformity of Liquid Deposition

The coefficient of uniformity for liquid deposition at tree levels was calculated using Formula (2), separately for the "trunk" samplers and the "branch" samplers, considering the speed of drone propellers' spin set to "zero" and the rotations resulting from two different weights of the multi-rotor drone. CV values were determined separately for the displacement speed of 0.57 m·s^{-1} and 0.94 m·s^{-1}. The results are presented in Figure 5.

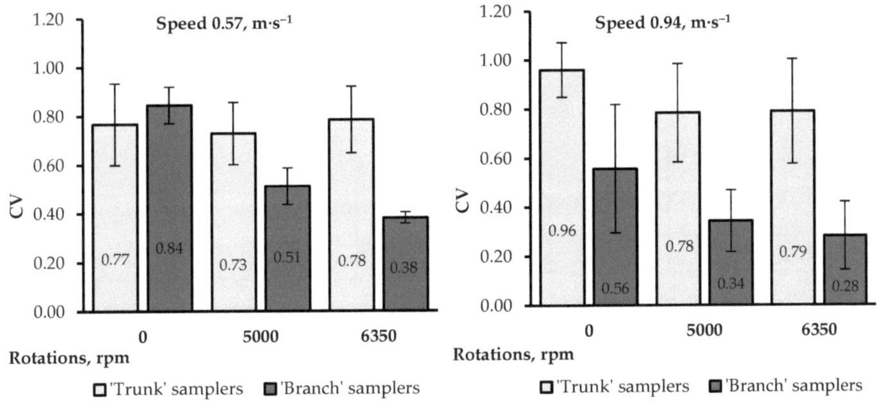

Figure 5. Effect of the propeller speeds on the uniformity of liquid deposition in the spruce crown.

The variance analysis conducted on the results of liquid distribution uniformity at the sampler levels revealed the significant influence of both speed of the propellers and drone movement speed on improving the CV index, but only for the "branch" samplers. In terms of propellers rotation, the significance of its effect on the coefficient of uniformity of deposited liquid was 0.00164. Regarding the effect of the drone's displacement speed, the significance level was 0.014855. Further analysis of the propeller speeds, which were solely influenced by the drone's weight, showed a significant effect on improving the uniformity of liquid deposition for the "branch" samplers at a speed of 0.57 m·s^{-1} (p = 0.046004). No significant effect was observed at a speed of 0.94 m·s^{-1}.

3.3. Leaf Area Index—LAI

Table 3 presents the calculated leaf area index using Formula (3). In the table were also presented the total area of the leaves and the measured sizes of the longest branches for the selected level, which were used as radii for the field area under the branches for the selected level calculation.

Table 3. LAI values for different levels of the tree crown.

Total Area of the Leaves, cm^2		Radius of the Area Circle, cm	LAI Value	
F_{IA},	803.84	16.0	LAI_A	0.658
F_{IB}	3957.2	35.5	LAI_B	2.071
F_{IC}	5539.0	42.0	LAI_C	4.490

3.4. Airstream Velocity in the Crown of the Tree

The results of the velocities of the airstream measurements are presented in Figure 6a,b. Because the velocity of the airstream was measured separately on the positions of the "trunk" samplers and separately at the "branch" samplers at levels "A", "B", and "C," the charts of passing it through the tree crown and without tree were shown side by side.

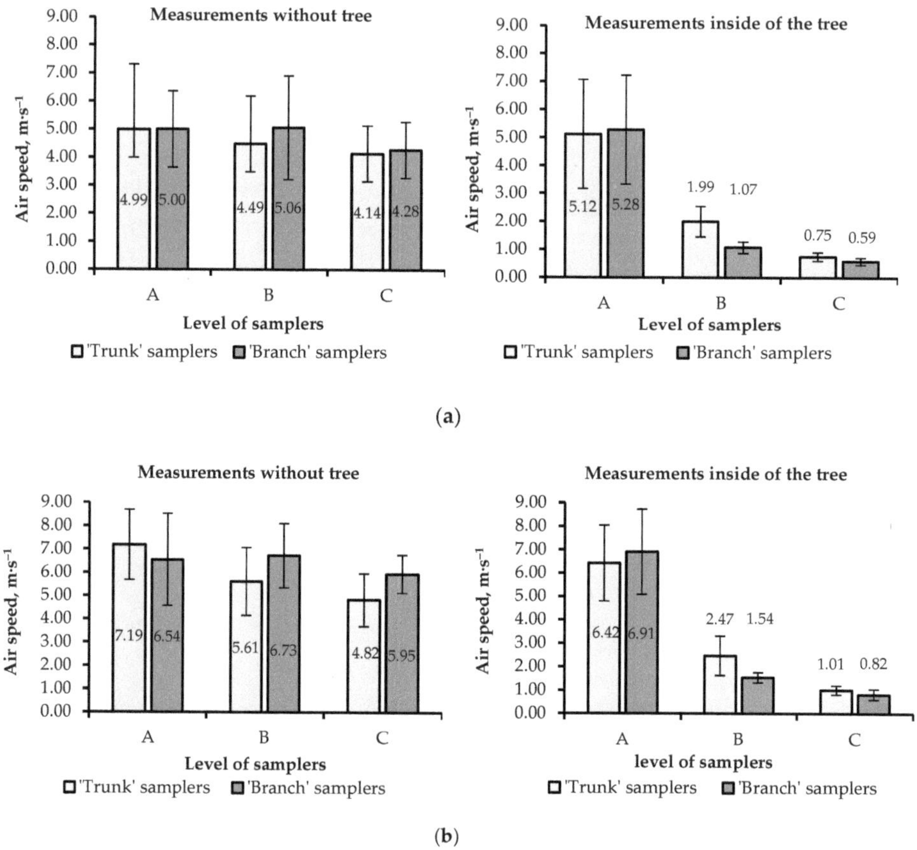

Figure 6. Comparison of changes in airflow speed down the tripod, caused by the foliage of the spruce tree crown at rotations of propellers 5000 rpm (**a**) and 6350 rpm (**b**).

4. Discussion

The analysis of the results concerning the impact of the rotational speed of drone propellers on airflow velocity through the crown of the tree and the quality of the liquid deposition at the "trunk" sampler levels revealed that although the rotations of drone's propellers, at 5000 and 6350 rpm (as shown in Figure 6a,b), generated airflow within the tree crown where the "trunk" samplers were positioned, no evidence was found indicating that this airflow increased the capacity of the liquid deposited on these samplers. The volumes of deposited liquid and the uniformity of deposition were comparable across all speeds of propellers, ranging from 0.0 to 6350 rpm (as shown in Figure 4a,b and Figure 5). This phenomenon was likely attributed to the high density of branches and coniferous leaves in that part of the tree crown and the coarse liquid droplets generated by the IDK 90-015 nozzle. These droplets were able to penetrate this part of the tree crown similarly, regardless of the presence or absence of airflow. The branches near the trunk are stiff and do not deflect under the influence of the air stream. Hence, the speed of the propellers and the resulting airflow had no effect on the volume of the deposited liquid in this zone of the tree crown.

In contrast, at the "branch" sampler levels, the effect of the drone's speed of propellers on increased liquid deposition volume was observed on "B" and "C" tripod levels at both drone movement speeds (as depicted in Figure 4a,b). The total volume of liquid deposited on all levels of this kind of sampler also increased (Table 1). This increase occurred despite a higher leaf area index (as presented in Table 3) and lower airflow velocity through the crown compared with the "trunk" samplers (as shown in Figure 6a,b). The uniformity of liquid deposition on the "branch" sampler levels (Figure 5) improved. This phenomenon can be attributed to the narrowing of the droplet stream angle under the influence of airflow generated by the drone's propellers rotations [40,53]. Consequently, there was an increase in liquid volume within the narrower stream width. The "branch" samplers were positioned in the region of the droplet stream where this volume increase occurred. Not only did the amount of liquid deposited on these samplers increase, but it also improved the uniformity of liquid deposition at tree levels (CV). The airflow generated by the propellers further facilitated the liquid deposition at the lower levels of the tripod on the "branch" samplers.

Analysing both the improvement in the results of liquid deposition on the "branch" samplers and the improvement in the uniformity of liquid deposition on these samplers under the influence of airflow generated by the rotation of the drone rotors in comparison with the situation when the rotors were not rotating (field sprayer), it can be concluded that when comparing the quality of spraying young spruce trees with a drone and a field sprayer, spraying with a drone provides better results of liquid deposition. The possibility of carrying out tests on a laboratory test site made it possible to compare the design of the equipment and the comparability of the conditions and parameters of the treatment, which was not possible previously [23,25].

The analysis of the drone's speed of propellers, which varied with the weight of the drone (full tank vs. empty tank), showed the potential impact of the weight of the liquid in the drone's tank on the quality of spray applied to the plants. A significant effect was observed at level "B" when the speed was 0.57 $m \cdot s^{-1}$. The higher propellers rotations resulting from the increased weight of the drone also significantly improved the uniformity of liquid deposition on the "branch" samplers.

The analysis of variance confirmed the significance of the drone's movement speed on the volume of liquid deposited on the sampler surfaces. The ratio between the two drone displacement speeds was determined to be 1.65. Comparing the ratios of total liquid volume deposited on the samplers at lower and higher speeds (as indicated in Table 2, at "Total relatios") showed close to this value proportions. However, when examining the liquid volume ratios deposited at each sampler level, deviations from the expected relationship between velocities were observed. Specifically, deviations occurred at the "B" level of the "branch" samplers with non-rotating propellers, as well as at the "C" level of the "branch" samplers, both with rotating and non-rotating propellers. These deviations made it challenging to establish a consistent rule to describe the relationship between drone speed and liquid volume deposition in these cases. If confirmed by other studies, the aerodynamic phenomena that caused these differences may be worth explaining.

Comparing the air velocity passing through a spruce tree at sampler levels with the air velocity measured at the same places as samplers but without the presence of the tree (Figure 6a,b) revealed noteworthy observations. Firstly, the leaf area index (LAI) at each tree level strongly impacted reducing the air stream's velocity. Secondly, the distance of the tree level from the drone also contributed to the reduction in airspeed, which is particularly evident in Figure 6a,b when measuring without a tree. Therefore, the changes in air velocity observed within the tree crown (as also depicted in Figure 6a,b) were influenced by both the distance of the measurement level from the drone and the presence of the leaf canopy, which dampened the airflow. Whereas significant is a range of changes in airstream velocity. The distance between levels "A" and "C,", when measured without the tree, caused a decrease in speed in the range of 26%. The tree crown decreased air stream velocity about six times, dividing by each other the velocity values at point "C" in measurements without and with the tree. The reason for the reduction in air velocity in the lower part of the tree may have been due not only to the flow-dampening structure of the tree but also to a change in the direction of airflow close to the ground [50,51]. In the study, only the vertical velocity was measured; the horizontal velocity was not measured.

In general, it can be concluded that the results of studies on the quality of liquid deposition in the central part of the crown of a young spruce tree and at the levels of its canopy confirmed the results of earlier studies that the effectiveness of liquid penetration of the tree crown, is influenced by factors such as the speed of the drone above the tree crown and the shape of the tree crown [26–29,40]. After studies, it can also be sure that tree species and foliage surface and density influence liquid deposition.

Studies have shown that the low altitude above the plants of the displacement of the spraying drone makes sense. If, in the future, unmanned aerial spraying systems at low altitudes become a permanent, legislated component of precision agriculture, and in orchard and ornamental tree crops, the airflow generated by the drone rotors will support spray quality and reduce liquid drift [9,35].

5. Conclusions

Comparing the quality of spray deposition in the tree when the drone's propellers were in operation to when they were not rotating revealed an improvement in spray deposition quality when the propellers were rotating, so the drone was flying and liquid spraying. This suggests that using a drone instead of a field sprayer may yield better results for pest control treatments on cultivations, not only conifer trees.

Low altitude above the plants of moving multi-rotor drones when spraying can not only improve the deposition of the spray on the plants but can also prevent droplets from drifting off with the wind.

Described studies indicate that using the strategy of low altitude over the plants of multirotor drones can successfully replace classic methods of plant spraying with field sprayers and increase the quality and precision of the treatments performed.

The process of emptying the drone's tank while spraying, which leads to changes in the rotations of propellers value due to the weight difference between a full and empty tank, can potentially impact the deterioration of droplet stream deposition in the tree crown.

No significant effect of the rotational speed of the drone's propellers and the resulting air stream on the quality of liquid deposition was observed on the samplers placed at the tree trunk. However, a noticeable effect of the rotations of drone propellers was observed on the samplers located on the branches of the tree at a considerable distance from the trunk.

Author Contributions: Conceptualization, A.P., B.B., J.C. and G.M.; methodology, J.C., G.M. and J.D.; software, J.N., J.K. and T.N.; validation, A.P., B.B., J.C. and M.M.; formal analysis, J.D. and A.K.; investigation, J.C., A.K. and J.K.; resources, J.C., G.M., J.N., T.N. and M.M.; data curation, A.P., A.K., J.K. and T.N.; writing—original draft preparation, A.P., B.B. and J.C.; writing—review and editing, J.C., J.D. and A.K.; visualization, A.P., B.B., T.N. and M.M.; supervision, J.C., J.D., A.K. and J.N.; project administration, J.C., G.M., J.K. and M.M.; funding acquisition, J.C. and J.N. All authors have read and agreed to the published version of the manuscript.

Funding: This project is co-financed by the POLISH NATIONAL AGENCY FOR ACADEMIC EXCHANGE, grant number PPN/BIL/2018/1/00072 and grant number PPN/BIL/2018/1/00074. Research was also supported by project LTI20004 "Environmental Research and Development Information Centre" funded by Ministry of Education, Youth and Sports of the Czech Republic, program INTER-EXCELLENCE, subprogram INTER-INFORM.

Institutional Review Board Statement: Not applicable.

Data Availability Statement: Data available on request.

Conflicts of Interest: The authors declare no conflict of interest. The funders had no role in the design of the study; in the collection, analyses, or interpretation of data; in the writing of the manuscript, or in the decision to publish the results.

References

1. Zhu, H.; Altland, J.; Derksen, R.C.; Krause, C.R. Optimal Spray Application Rates for Ornamental Nursery Liner Production. *HortTechnology* **2011**, *21*, 367–375. [CrossRef]
2. Fuentealba, A.; Dupont, A.; Hébert, C.; Berthiaume, R.; Quezada-García, R.; Bauce, É. Comparing the efficacy of various aerial spraying scenarios using *Bacillus thuringiensis* to protect trees from spruce budworm defoliation. *For. Ecol. Manag.* **2019**, *432*, 1013–1021. [CrossRef]
3. Liu, Z.; Peng, C.; De Grandpr, L.; Candau, J.N.; Work, T.; Zhou, X.; Kneeshaw, D. Aerial spraying of bacterial insecticides to control spruce budworm defoliation leads to reduced carbon losses. *Ecosphere* **2020**, *11*, e02988. [CrossRef]
4. Wallace, D.J.; Picot, J.J.C.; Chapman, T.J. A numerical model for forestry aerial spraying. *Agric. For. Meteorol.* **1995**, *76*, 19–40. [CrossRef]
5. Wodecka, C.; Rowiński, R. Przenikanie oprysku lotniczego przez korony drzew. *Sylwan* **1997**, *7*, 107–120.
6. Richardson, B.; Strand, T.; Thistle, H.W.; Hiscox, A.; Kimberley, M.O.; Schou, W.C. Influence of a young *Pinus radiata* canopy on aerial spray drift. *Trans. ASABE* **2017**, *60*, 1851–1861. [CrossRef]
7. Liao, J.; Zang, Y.; Luo, X.W.; Zhou, Z.; Lan, Y.; Zang, Y.; Gu, X.; Xu, W.; Hewitt, A.J. Optimization of variables for maximizing efficacy and efficiency in aerial spray application to cotton using unmanned aerial systems. *Int. J. Agric. Biol. Eng.* **2019**, *12*, 10–17. [CrossRef]
8. Wang, S.L.; Song, J.L.; He, X.K.; Song, L.; Wang, X.N.; Wang, C.L.; Wang, Z.H.; Ling, Y. Performances evaluation of four typical unmanned aerial vehicles used for pesticide application in China. *Int. J. Agric. Biol. Eng.* **2017**, *10*, 22–31. [CrossRef]
9. Chen, S.; Lan, Y.; Zhou, Z.; Ouyang, F.; Wang, G.; Huang, X.; Deng, X.; Cheng, S. Effect of Droplet Size Parameters on Droplet Deposition and Drift of Aerial Spraying by Using Plant Protection UAV. *Agronomy* **2020**, *10*, 195. [CrossRef]
10. Berner, B.; Pachuta, A.; Chojnacki, J. Estimation of liquid deposition on corn plants sprayed from a drone. In Proceedings of the 25th International PhD Students Conference (MendelNet 2018), Brno, Czech Republic, 7–8 November 2018; pp. 403–407.
11. Huo, L.; Lindberg, E.; Bohlin, J.; Persson, H.J. Assessing the detectability of European spruce bark beetle green attack in multispectral drone images with high spatial- and temporal resolutions. *Remote Sens. Environ.* **2023**, *287*, 113484. [CrossRef]
12. Mazur, P.; Moitzi, G.; Wagentristl, H.; Zdanowicz, A. Winter oil seed rape monitoring with unmanned aerial vehicles. In Proceedings of the 10th International Scientific Symposium on Farm Machinery and Process Management in Sustainable Agriculture (FMPMSA 2019), Lublin, Poland, 19–23 November 2019; pp. 145–150.
13. Wavrek, M.T.; Carr, E.; Jean-Philippe, S.; McKinney, M.L. Drone remote sensing in urban forest management: A case study. *Urban For. Urban Green.* **2023**, *86*, 127978. [CrossRef]
14. Ni, J.; Yao, L.; Zhang, J.; Cao, W.; Zhu, Y.; Tai, X. Development of an Unmanned Aerial Vehicle-Borne Crop-Growth Monitoring System. *Sensors* **2017**, *17*, 502. [CrossRef] [PubMed]
15. Balaji, B.; Chennupati, S.K.; Chilakalapudi, S.R.D.; Katuri, R.; Mareedu, K. Design of UAV, Drone for Crops, Weather Monitoring and For Spraying Fertilizers and Pesticides. *Int. J. Res. Trends Innov.* **2018**, *3*, 42–47.
16. Kharim, M.N.A.; Wayayok, A.; Mohamed Shariff, A.R.; Abdullah, A.F.; Husin, E.M. Droplet deposition density of organic liquid fertilizer at low altitude UAV aerial spraying in rice cultivation. *Comput. Electron. Agric.* **2019**, *167*, 105045. [CrossRef]
17. Liu, W.; Zhou, Z.; Xu, X.; Gu, Q.; Zou, S.; He, W.; Luo, X.; Huang, J.; Lin, J.; Jiang, R. Evaluation method of rowing performance and its optimization for UAV-based shot seeding device on rice sowing. *Comput. Electron. Agric.* **2023**, *207*, 107718. [CrossRef]
18. Garcia-Ruiz, F.; Campos, J.; Llop-Casamada, J.; Gil, E. Assessment of map based variable rate strategies for copper reduction in hedge vineyards. *Comput. Electron. Agric.* **2023**, *207*, 107753. [CrossRef]
19. Hafeez, A.; Husain, M.A.; Singh, S.P.; Chauhan, A.; Khan, M.T.; Kumar, N.; Chauhan, A.; Soni, S.K. Implementation of drone technology for farm monitoring & pesticide spraying: A review. *Inf. Process. Agric.* **2023**, *10*, 192–203. [CrossRef]
20. Raptis, E.K.; Englezos, K.; Kypris, O.; Krestenitis, M.; Kapoutsis, A.C.; Ioannidis, K.; Vrochidis, S.; Kosmatopoulos, E.B. CoFly: An automated, AI-based open-source platform for UAV precision agriculture applications. *SoftwareX* **2023**, *23*, 101414. [CrossRef]
21. Su, J.; Zhu, X.; Li, S.; Chen, W.H. AI meets UAVs: A survey on AI empowered UAV perception systems for precision agriculture. *Neurocomputing* **2023**, *518*, 242–270. [CrossRef]
22. Abdullah, M.N.; Dagher, K.E.; Wali, S.S. Improved airborne computer system strategy for swarm drones flying based on skybrush suite and inspired technique. *Meas. Sens.* **2023**, *27*, 100766. [CrossRef]
23. Xiao, Q.; Du, R.; Yang, L.; Han, X.; Zhao, S.; Zhang, G.; Fu, W.; Wang, G.; Lan, Y. Comparison of Droplet Deposition Control Efficacy on Phytophthora capsica and Aphids in the Processing Pepper Field of the Unmanned Aerial Vehicle and Knapsack Sprayer. *Agronomy* **2020**, *10*, 215. [CrossRef]
24. Lou, Z.; Xin, F.; Han, X.; Lan, Y.; Duan, T.; Fu, W. Effect of Unmanned Aerial Vehicle Flight Height on Droplet Distribution, Drift and Control of Cotton Aphids and Spider Mites. *Agronomy* **2018**, *8*, 187. [CrossRef]
25. Zheng, Y.J.; Yang, S.H.; Zhao, C.J.; Chen, L.P.; Lan, Y.B.; Tan, Y. Modelling operation parameters of UAV on spray effects at different growth stages of corns. *Int. J. Agric. Biol. Eng.* **2017**, *10*, 57–66. [CrossRef]
26. Zhang, P.; Wang, K.; Lyu, Q.; He, S.; Yi, S.; Xie, R.; Zheng, Y.; Ma, Y.; Deng, L. Droplet distribution and control against citrus Leafminer with UAV spraying. *Int. J. Robot. Autom.* **2017**, *32*, 299–307. [CrossRef]
27. Meng, Y.; Su, J.; Song, J.; Chen, W.H.; Lan, Y. Experimental evaluation of UAV spraying for peach trees of different shapes: Effects of operational parameters on droplet distribution. *Comput. Electron. Agric.* **2020**, *170*, 105282. [CrossRef]

28. Özyurt, H.B.; Duran, H.; Çelen, I.H. Determination of the Application Parameters of Spraying Drones for Crop Protection in Hazelnut Orchards. *J. Tekirdag Agric. Fac.* **2022**, *19*, 819–828. [CrossRef]
29. Wang, C.; Liu, Y.; Zhang, Z.; Han, L.; Li, Y.; Zhang, H.; Wongsuk, S.; Li, Y.; Wu, X.; He, X. Spray performance evaluation of a six-rotor unmanned aerial vehicle sprayer for pesticide application using an orchard operation mode in apple orchards. *Pest Manag. Sci.* **2022**, *78*, 2449–2466. [CrossRef]
30. Leroy, B.M.L.; Gossner, M.M.; Lauer, F.P.M.; Petercord, R.; Seibold, S.; Jaworek, J.; Weisser, W.W. Assessing Insecticide Effects in Forests: A Tree-Level Approach Using Unmanned Aerial Vehicles. *J. Econ. Entomol.* **2019**, *112*, 2686–2694. [CrossRef]
31. Ogilvie, S.; McCarthy, A.; Allen, W.; Grant, A.; Mark-Shadbolt, M.; Pawson, S.; Richardson, B.; Strand, T.; Langer, E.R.; Marzano, M. Unmanned Aerial Vehicles and Biosecurity: Enabling Participatory-Design to Help Address Social Licence to Operate Issues. *Forests* **2019**, *10*, 695. [CrossRef]
32. Krasylenko, Y.; Rydlo, K.; Atamas, N.; Sosnovsky, Y.; Horielov, O.; Maceček, I.; Šamajová, O.; Ovečka, M.; Takáč, T.; Šamaj, J. Druid Drone—A portable unmanned aerial vehicle with a multifunctional manipulator for forest canopy and mistletoe research and management. *Methods Ecol. Evol.* **2023**, *14*, 1416–1423. [CrossRef]
33. Yang, F.; Xue, X.; Cai, C.; Sun, Z.; Zhou, Q. Numerical Simulation and Analysis on Spray Drift Movement of Multirotor Plant Protection Unmanned Aerial Vehicle. *Energies* **2018**, *11*, 2399. [CrossRef]
34. Wen, S.; Han, J.; Ning, Z.; Lan, Y.; Yin, X.; Zhang, J.; Ge, Y. Numerical analysis and validation of spray distributions disturbed by quad-rotor drone wake at different flight speeds. *Comput. Electron. Agric.* **2019**, *166*, 105036. [CrossRef]
35. Chen, P.; Douzals, J.P.; Lan, Y.; Cotteux, E.; Delpuech, X.; Pouxviel, G.; Zhan, Y. Characteristics of unmanned aerial spraying systems and related spray drift: A review. *Front. Plant Sci.* **2022**, *13*, 870956. [CrossRef]
36. Directive 2009/128/EC of the European Parliament and of the Council of 21 October 2009 Establishing a Framework for Community Action to Achieve the Sustainable Use of Pesticides (Text with EEA Relevance). Available online: https://eur-lex.europa.eu/eli/dir/2009/128/oj (accessed on 2 May 2023).
37. Sánchez-Fernández, L.; Barrera, M.; Martínez-Guanter, J.; Pérez-Ruiz, M. Drift reduction in orchards through the use of an autonomous UAV system. *Comput. Electron. Agric.* **2023**, *211*, 107981. [CrossRef]
38. Vu, N.A.; Dang, D.K.; Le Dinh, T. Electric propulsion system sizing methodology for an agriculture multicopter. *Aerosp. Sci. Technol.* **2019**, *90*, 314–326. [CrossRef]
39. Tang, Y.; Houa, C.J.; Luoa, S.M.; Lina, J.T.; Yangb, Z.; Huanga, W.F. Effects of operation height and tree shape on droplet deposition in citrus trees using an unmanned aerial vehicle. *Comput. Electron. Agric.* **2018**, *148*, 1–7. [CrossRef]
40. Tang, Q.; Zhang, R.R.; Chen, L.P.; Xu, M.; Yi, T.C.; Zhang, B. Droplets movement and deposition of an eight-rotor agricultural UAV in downwash flow field. *Int. J. Agric. Biol. Eng.* **2017**, *10*, 47–56.
41. Zhang, Y.; Li, Y.; He, Y.; Liu, F.; Cen, H.; Fang, H. Near ground platform development to simulate uav aerial spraying and its spraying test under different conditions. *Comput. Electron. Agric.* **2018**, *148*, 8–18. [CrossRef]
42. Černý, K.; Pešková, V.; Soukup, F.; Havrdová, L.; Strnadová, V.; Zahradník, D.; Hrabětová, M. Gemmamyces bud blight of *Picea pungens*: A sudden disease outbreak in Central Europe. *Plant Pathol.* **2016**, *65*, 1267–1278. [CrossRef]
43. Sakalidis, M.L.; Medina-Mora, C.M.; Shin, K.; Fulbright, D.W. Characterization of Diaporthe spp. associated with spruce decline on Colorado blue spruce in Michigan. *Phytopathology* **2021**, *111*, 509–520. [CrossRef]
44. Samek, M.; Modlinger, R.; Bat'a, D.; Lorenc, F.; Vachová, J.; Tomášková, I.; Pešková, V. Gemmamyces piceae Bud Blight Damage in Norway Spruce (*Picea abies*) and Colorado Blue Spruce (*Picea pungens*) Forest Stands. *Forests* **2022**, *13*, 164. [CrossRef]
45. Tang, Q.; Zhang, R.; Chen, L.; Deng, W.; Xu, M.; Xu, G.; Li, L. Influence of the atmospheric boundary layer stability on aerial spraying studied by computational fluid dynamics. *Biosyst. Eng.* **2022**, *215*, 170–187. [CrossRef]
46. Available online: https://www.lechler.com/fileadmin/media/datenblaetter/agrar/EN/lechler_agrar_datenblatt_idk-idkn_en.pdf (accessed on 30 June 2023).
47. Lipiński, A.J.; Lipiński, S. Binarizing water sensitive papers—How to assess the coverage area properly. *Crop Prot.* **2020**, *127*, 104949. [CrossRef]
48. M4E TTAviation. Available online: https://www.ttaviation.org/procat/agriculture-uav (accessed on 15 May 2023).
49. Berner, B.; Chojnacki, J. Evaluation of the thrust force of a multirotor agricultural drone. *J. Res. Appl. Agric. Eng.* **2018**, *63*, 6–8. Available online: https://tech-rol.eu/images/Archiwum_X/2019/05/2018_3_JCH.pdf (accessed on 2 May 2023).
50. Zhang, H.; Qi, L.; Wu, Y.; Musiu, E.M.; Cheng, Z.; Wang, P. Numerical simulation of airflow field from a six-rotor plant protection drone using lattice Boltzmann method. *Biosyst. Eng.* **2020**, *197*, 336–351. [CrossRef]
51. Zhan, Y.; Chen, P.; Xu, W.; Chen, S.; Han, Y.; Lan, Y.; Wang, G. Influence of the downwash airflow distribution characteristics of a plant protection UAV on spray deposit distribution. *Biosyst. Eng.* **2022**, *216*, 32–45. [CrossRef]
52. Law, B.E.; Van Tuyl, S.; Cescatti, A.; Baldocchi, D.D. Estimation of leaf area index in open-canopy ponderosa pine forests at different successional stages and management regimes in Oregon. *Agric. For. Meteorol.* **2001**, *108*, 1–14. [CrossRef]
53. Chojnacki, J.; Pachuta, A. Impact of the Parameters of Spraying with a Small Unmanned Aerial Vehicle on the Distribution of Liquid on Young Cherry Trees. *Agriculture* **2021**, *11*, 1094. [CrossRef]

agriculture

MDPI

Article

Experimental Field Tests of the Suitability of a New Seeder for the Soils of Northern Kazakhstan

Mubarak Aduov [1], Saule Nukusheva [1], Talgat Tulegenov [1], Kadirbek Volodya [1], Kanat Uteulov [1], Bolesław Karwat [2] and Michał Bembenek [2,*]

1 Department of "Agricultural Engineering and Technology", Technical Faculty, NJSC "Saken Seifullin Kazakh Agrotechnical Research University", Zhenis Ave. 62, Astana 010011, Kazakhstan
2 Faculty of Mechanical Engineering and Robotics, AGH University of Science and Technology, A. Mickiewicza 30, 30-059 Krakow, Poland
* Correspondence: bembenek@agh.edu.pl

Abstract: Kazakhstan is historically a livestock country, and the production of feed requires no less attention than the production of grain. To improve the forage base, one solution is the sowing of high-yielding fodder seeds. An experimental seeder was developed with new design solutions for the sowing machine, with three blades installed at an angle of 120° relative to the lower part of the blower shaft, deviated vertically by 8–10°, along with components with a radius vector of 10–15° and the blower shaft attached to the top of the sowing cylinder. The closing part of the disc coulter contained the press rollers with a disc diameter measuring 350 mm. The field tests were conducted with the parameters between the discs set to $\alpha = 10°$, a disc vanishing point of $\beta = 40°$, a coulter angle of 32°, and an individual 320-mm press roller with a cylindrical 60-mm rim, a leash, and a section for setting the seed placement depth. The wheatgrass varieties "Burabay" and awnless brome "Akmola emerald" were sown. The research showed the higher efficiency of the experimental seeder with seeding units and sowing parts compared to a serial seeder in terms of agricultural performance. The increase in seed germination was 3.56%. The experimental seeder surpassed the regular seeder by 4.95% in terms of the depth uniformity of the seed placement, in terms of yield increase by 5.361 cwt/ha, with reductions in traction resistance of 12.3%, and in fuel consumption by 10%. The economic efficiency from the fuel reduction and yield increase was estimated at around 7700 USD/ha per year.

Keywords: mounted seeder; seeding unit; ridge breaker; non-flowing seeds; traction resistance; sowing depth

Citation: Aduov, M.; Nukusheva, S.; Tulegenov, T.; Volodya, K.; Uteulov, K.; Karwat, B.; Bembenek, M. Experimental Field Tests of the Suitability of a New Seeder for the Soils of Northern Kazakhstan. *Agriculture* **2023**, *13*, 1687. https://doi.org/10.3390/agriculture13091687

Academic Editors: Mustafa Ucgul and Chung-Liang Chang

Received: 23 June 2023
Revised: 19 August 2023
Accepted: 21 August 2023
Published: 26 August 2023

1. Introduction

Kazakhstan is a country with an area of 190 million ha. According to the Kazakhstan Land Resources Agency, it contains 188 million ha of natural forage land [1]. Problems in rangeland farming have been studied by scientists from various countries in different ways [2–4]. To improve the forage base, one solution is to sow high-yielding seeds of fodder crops [5–7]. This problem is so urgent that scientists, breeders [6–8], agronomists [9–11], and agricultural equipment engineers [12–14] are working to solve it [9,15,16]. The development of seeders with new working tools, e.g., coulters, was conducted by scientists from different countries [17–19]. In addition, the application of new technologies requires the development of new technical solutions capable of high-quality seeding [15,20,21] and increasing yields [22]. Studies on the development and justification of technological parameters and the testing of seeders for the cultivation of grain crops have also been conducted by scientists from Kazakhstan [23].

A research and production experience analysis shows that the actual state of the fodder base can be found on the pages of "KazakhZerno.kz" [24] and in recommendations by employees of LLP, "NPZH, named after A.I. Barayev." According to A.I. Barayev, the fodder base does not meet the needs of animals. Therefore, fodder production requires the same attention as grain production. Although significant research has been conducted on the development of cultivation machines [25–27], unfortunately, there is a large gap in terms of the research into the development of versatile seed drills with automatically variable designs and the adjustment of their technological parameters. These can be used in different climatic zones with different soil characteristics [28–30]. An analysis of the scientific literature shows an urgent need for the development of new devices designed for seed sowing [31–33]. One challenge in the development of new seed drills is the investigation of the influence of sowing principles on the germination of seeds in mixed grassland [34–36]. The effects of improved seed drills, which implement different seeding principles, on seeding quality, seed germination, and yields are also under study [37–39]. Another task in the development of new seed drills is the improvement of seed germination by applying the best possible seed-drop height and tractor speed [40,41]. In addition, attention is focused on increasing the seeding rate in fields through improved seed drills [42,43].

Only cooperation between science, production, the development of new technologies, and recommendations can solve the above-mentioned problems. However, obtaining high yields of fodder crops directly depends on both sowing methods and the implementation and adoption of these technologies.

According to Zhulmukhametova's speech at the Kazakhstan Ministry of Agriculture in 2017, the decisive factor in the sustainable development of livestock is the provision of fodder. However, the country is facing of a field-unit deficit of 8.7 million tons, which is the reason for the 13% shortfall in gross livestock production. To reduce this deficit by 2025, specific targets have been set using global experience and rapidly implementing it in agriculture [44]. Kazakh scientists drafted a program for developing fodder production, which considers the peculiarities of regions and makes calculations on the content of fodder units, depending on the nutritional value of the fodder crops produced in the area [45]. Furthermore, recommendations are prepared based on the structures of sown areas and by considering the natural and climatic characteristics of regions. In the northern part of Kazakhstan, between 60% and 70% of grain sowing and forage crops are obtained through SZS-2.0, SZTS-6, and SZTS-12 seeders–cultivators [46]. The rest of the acreage is sown by seeders like SKP-2,1 as well as foreign seeders, like those manufactured by John-Deere, Horsch, Argentinian Crucianelli, Pionera Flexi-Coil, Concord, and Amazonia, etc., or readjusted seeders. Furthermore, the domestic seed-drill park is not being renewed, and Kazakhstan has become completely dependent on foreign agricultural equipment suppliers. It is not possible to purchase expensive equipment for small companies, at least for medium-sized farmers. Moreover, the application of new technologies is not possible without the appropriate technical support. Currently, in Kazakhstan, it is necessary to modernize the existing and create new equipment, in light of domestic and foreign experience. Considering the zonal conditions of northern Kazakhstan, imported seed drills do not always correspond to the catalogue indicators [47].

The problems with forage shortages discussed in this article can be solved by creating new high-tech seeding machines. The advantage of the newly developed seeder is the completely new design of the sowing unit, which is able to evenly distribute seeds over areas with small depths of placement of non-flowing grass seeds without injuring them. The literature review established that only disc coulters provide the necessary uniformity of non-flowing grass seeds at a shallow placement depth. Additionally, they provide low traction resistance.

Drawing upon an analysis of the coulters found in current grass seeders and recognizing the limitations within the coulter's technological process, along with considerations for the distinctive physical and mechanical properties of the soil in Northern Kazakhstan, the authors formulated and put forth the design and technological parameters for enhancing the closing component of the disc coulter seeder. However, the analysis indicates that the primary drawback observed after seed sowing is the inadequate performance in rolling the seeds. This results in uneven and insufficient soil compaction both on the sides and above the seeds. An investigation into the operation of the rollers within the sowing sections of seeders with various designs has led to the conclusion that soil compaction quality is influenced by its physical and mechanical properties. This dependence extends to the specific pressure, speed, and design parameters of the rollers. To address this issue, an individual press roller was developed. This roller features a cylindrical rim, a leash, and a sector for adjusting the depth of seed placement. The seeder has a grip width of 3.60 m, row spacing of 0.30 m, and a seeding rate ranging from 10.0 to 30.0 kg/ha, with a seed placement depth of 2.0–8.0 cm. Adhering to international standards, a 350 mm diameter disk was proposed with recommended parameters: $\alpha = 10°$, $\beta = 40°$ for the vanishing point of the disk, and $\alpha = 32°$ for the coulter angle. Additionally, a press wheel with a diameter of 320.0 mm and a rim width of 60.0 mm was introduced. The proposed configuration, with specified parameters between the discs and the coulter angle, enhances the uniform distribution of seeds across the row's width. It also optimally positions the seeds at the bottom of the furrow with a consistent seeding depth, facilitated by the corresponding wheel diameter for effective seed rolling. The disc travel distance is ±10 cm. Employing the suggested equipment enables more efficient seed rolling, establishing firm contact between seeds and soil. This practice, in turn, enhances plant development, growth, and ultimately, increases the yield of grain and forage crops.

The newly developed seeder underwent testing using the wheatgrass varieties "Burabay" and the awnless brome "Akmola emerald." These outcomes were subsequently juxtaposed with tests conducted using a series seed drill. Finally, an economic evaluation was performed, considering the results obtained during the sowing process in relation to both yield augmentation and reductions in fuel and lubricant costs.

2. Materials and Methods

To increase the indicators of quality for the seeding of non-flowing seeds, the authors substantiated and developed a grass seeder equipped in a technological process controller. The novelty of the developed seeder is confirmed by patents and applications for intellectual property (Aduov et al., 2020, 2015, 2016 (patents)) [22,43]. The detailed description of the seeder, the improved seeding unit, the seeding unit with an intelligent control unit, and the entire technological process of seeding are presented in articles [21,23].

The improved seeder for non-flowing grass seeds (Figure 1) consists of the main frame (1) and auxiliary bars (2). Two pneumatic wheels (7) are installed on the main bar of the frame (1) with an automatic hitch (3). Twelve sowing sections are pivotally attached to the rear part of the frame (4). The sowing section is folded from a double-disk coulter (5) and a roller (6). Six seed hoppers (8) are rigidly connected to the sowing machines (9), and each one has two distribution sleeves (10), connected to the sowing section via a seed tube (11).

Figure 1. The view of the experimental seeder for sowing non-flowing grass seeds: 1—frame; 2—seed box with sowing machines; 3—hinged device; 4—sowing section; 5—seeding machine drive; 6—running gear.

The prototype seeder with an intelligent control unit of the technological process for sowing non-flowing grass seeds is characterized by: a working width of 3.6 m, a width between the rows of 0.30 m, a seeding rate range from 10.0 to 30.0 kg/ha, and a seeding depth of 2.0 to 8.0 cm.

- Based upon theoretical studies of the technological process of the proposed seeding unit, the analytical relationships of seed movement at various stages in the seeding unit were obtained. This analysis facilitated the identification of optimal values of the technological and design parameters of the device. Those values ensured high-quality non-flowing grass seed sowing based on the following: the height of the tedder should be from 6 to 8 mm;
- the angle of the vertical blades from 8 to 10°;
- the angle of the vector radius from 10 to 15°; and
- the helix angle α must not exceed 17° and the radius must be within 0.02 m.

The structural dimensions of the distributor's head were considered based on the screw dimensions and the coefficient of the friction of non-flowing seed on steel.

When sowing, the main task is to distribute the seed evenly over the area.

The international standard discs of diameter at 350 +/− 5 mm were used. Recommended parameters are as follows: disc angle α = 10°, disc vanishing point position β = 40°, and coulter arm angle γ = 32°.

The diameter of the press wheel is determined with Equation (1) as follows:

$$D \leq \frac{2 \cdot h}{1 - \cos\alpha} \tag{1}$$

where h is the depth of the track, mm, and α is the angle of the roller rim around the soil (α = 15 to 20°).

The following recommendations were used when selecting and dimensioning the coulter bar and the coulter bar with packer roller:

- the lifting height of the coulters was assumed to be equal to the maximum seeding depth plus 6 to 7 cm for adaptation to the field topography; and
- the force on the lever for lifting and moving coulters to the working position must not exceed 196.2 N.

Considering the technological and physical-mechanical properties of soil and seeds of cultivated crops and the analysis of kinematics and forces of the sowing mechanism, the following design dimensions of the coulter and packer roller are determined [33]:

- the disc diameter at 350 mm;
- the angle between the discs at α = 10°;
- the disc vanishing point position at β = 40°;
- the coulter angle at γ = 32°; and
- the packer roller at a diameter of 320 mm and a rim width of 60 mm.

The intrinsic flow rate of hard-to-loose seeds of forage grasses without vibration ranges from 0.116 to 0.176 m/s, which is not enough for them to flow out of the hopper with no external force. In this connection, a new sowing apparatus with three blades was created to diminish the resulting arches of the sown material. The blower shaft (Figures 2 and 3) is attached to the top of the sowing cylinder via a threaded connection. See also Figure 4.

The control unit includes an electronic control unit, a path-detecting sensor, a seeding control sensor, and the hopper level sensor. The data from all the sensors can be presented on the tractor display (Figure 5).

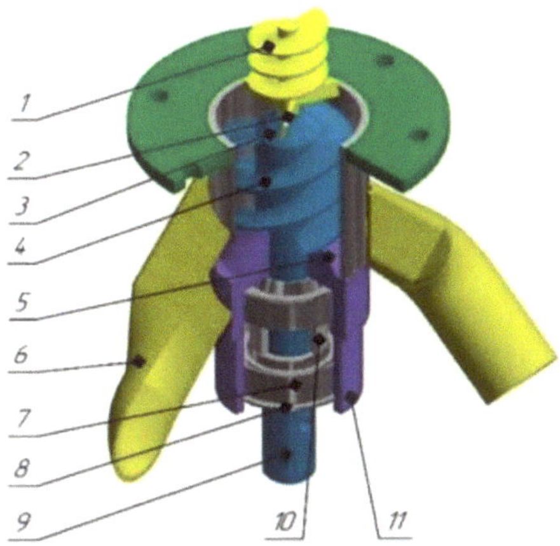

Figure 2. The scheme of a sowing unit for non-flowing grass seeds: 1—rotator; 2—blade supercharger; 3—cone; 4—helical spiral; 5—lower cone; 6—sleeve; 7—bearings; 8—retaining rings; 9—shaft; 10—intermediate ring; 11—bunker.

Figure 3. The image of a sowing unit for non-flowing grass seeds: 1—rotator; 2—blade supercharger.

The bench tests were conducted to objectively assess the technical level and quality of manufacturing of the experimental seeder. They showed the following:

- the experimental seeder sowing capacity was 8 to 30 kg/ha;
- the uneven seeding rate was 4.6% for wheatgrass and 4.8% for awnless bromegrass;
- the total seeding instability was 2.9% for wheatgrass and 2.7% for awnless bromegrass; and
- seed crushing was 0.1%.

Thus, the bench test results of the seeder experimental prototype met the technical specification requirements and Kazakhstan state standards for seeding machines.

For laboratory and field seeder tests, the following was determined: a driving speed of 5, 7, and 9 km/h and a seed sowing depth of 2 to 5 cm. The spacing between rows was 30 cm, and the rate of seeding was within the minimal and maximal values for the chosen crop's range.

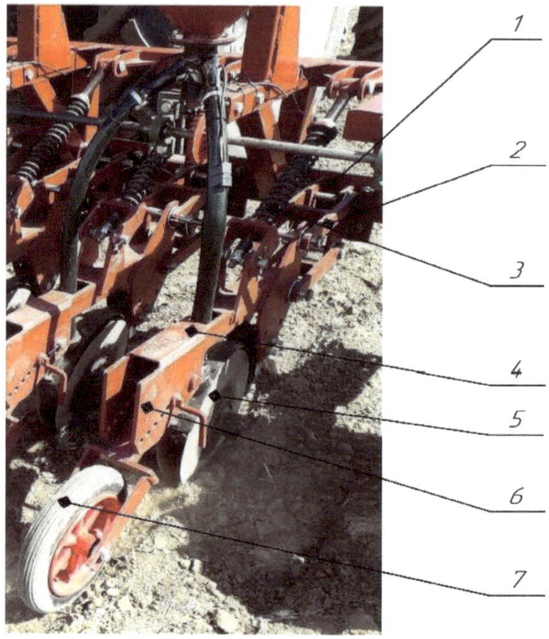

Figure 4. Closing part of the seeder: 1—bracket; 2—parallelogram mechanism; 3—rod with a spring; 4—longitudinal beam; 5—closing disc; 6—sector; 7—wheel.

Figure 5. The placing of the sensors: 1—sensors for monitoring the level of the bunkers; 2—sensors for monitoring the sowing of the seed material.

The traction resistance determination of the experimental seeder was performed under the GOST20915-2011 and GOST31345-2007requirements.

The data collection and processing traction resistance and fuel consumption was performed with a IP 264 device manufactured by KubNIITiM (Passport UV 404176.029 PSOOO "Vector-PM, Russia Uralves K-R-20G-10t) manufactured on the individual order of the scientists from S. Seifullin Kazakh Agrotechnical University.

For data processing, Statistica (StatSoft) was used. For modelling and design, the Inventor (Autodesk) was performed.

Economic tests of the seed drill were carried out on a field with an area of 20 hectares in the farm "Guldana" Settlement: Yalta. District: Gabita Musrepova. Region: North Kazakhstan (52°27′21.6″ N 67°11′05.4″ E).

The experimental plot, measuring 150.0 m in length and 14.4 m in width, was situated within the premises of S. Seifullin Kazakh Agrotechnical University's research and production campus. Based on the results of the selection of perennial cereal grasses recommended by the Research and Production Center for Grain Farming A.I. Baraev, for cultivation in the conditions of Northern Kazakhstan, according to the growing season length, the yield of fodder mass and seeds, the quality of the fodder, and the resistance to adverse environmental factors, wheatgrass varieties "Burabay" (Figure 6) and awnless brome "Akmola emerald" (Figure 7) were sown. They are characterized by poor flowability, preventing high-quality sowing. In the pictures, it can be seen that the seeds of these crops are loaded into a pile or lump; since the seeds are not smooth, they have poor flowability.

Figure 6. Images of the wheatgrass variety "Burabay".

Figure 7. Images of the awnless brome "Akmola emerald".

The expected friction coefficient for loose seeds should be around 0.57, while for the wheatgrass variety "Burabay" it is 0.38 (which is 0.19 lower) and awnless brome "Akmola emerald" is 0.46 (which is 0.11 lower). The proper coefficient of internal friction for loose seeds is 0.97, while for wheatgrass variety "Burabay" is 2.18, which is 1.21 higher; for awnless brome "Akmola emerald" is 1.35, which is 0.38 higher. The value of the seed compaction coefficient ranges from 1.06 to 1.17, while for the wheatgrass variety "Burabay" it is 1.12 and for awnless brome "Akmola emerald" it is 1.17.

In addition, the zoned varieties of perennial grasses are highly productive, resistant to abiotic (frost, low temperatures in winter, and high temperatures in summer, etc.) and biotic (diseases and pests) environmental factors. Wheatgrass varieties "Burabay" and awnless brome "Akmola emerald" are able to provide hay yields of up to 30–45 cwt/ha, and with regular irrigation more than 60 cwt/ha. The seeds of these crops practically do not crumble, and therefore the second advantage is longevity. The term of economic use of the steppe regions is from 4 to 6 years. However, these two crops exhibit poor flowability. This characteristic prompted the authors to select them for research purposes, conducting both laboratory and field tests. The seeding rate and the depth of seed placement were established based on the recommendations of the agronomist of the farm. In this case, the size of the experimental plot was 3.6 × 37 m. The collected phenological observations of the experimental plots were entered in the observation log.

For the agrotechnical and energy evaluation of the seeder under development, its traction resistance was previously determined in the analytical form presented by Aduov et al. [21,23]. The research program for the experimental seeder includes both the bench and laboratory field tests. The sown material prepared for testing was determined and evaluated based on its purity, damage, germination, moisture, and weight of 1000 seeds according to Gosudarstvennyy tandard (GOST). 20290-74 "Seeds of agricultural crops. Determination of sowing qualities of seeds. Terms and Definitions."

The performance quality of the seeder's technological process was evaluated at a speed of 7 km/h, and subsequently, it was enhanced by increasing the speed of the sowing machine by 25 to 30%. The tests were conducted at various speeds while maintaining the same settings of the sowing units.

The parameters of work quality during the sowing process of grass seeds with the new seeder for sowing non-loose grass seeds were compared with the standard SZT-3.6 (Astra, Ukraine) seeder, which until now sowed more than 50% of forage crops with loose and non-loose grass seeds from the main bunkers through disc coulters.

The unevenness and instability of seed sowing were determined with the economic sowing rate. The number of seeds that passed through sowing units (seed tubes) was determined by comparing the source material to the fragmented seeds. Tests were conducted using the Labor Code and GOST 26711-89 at the same working speed. Samples were collected at least three times.

3. Results and Discussion

An experimental model of a mounted seeder was developed and its main indicators were substantiated. Alongside the seeding depth, the design, and technological parameters of the seeding unit within the seeder, the authors also established the height of the ridge breaker for seeds within the hopper, the inclination angle of the blade concerning the vertical, and the angle of rise and the radius of the screw spiral. The test proved the corrected design and technological parameters of the seeding part of the seeder including the disc diameter, disc angle, position of a disc's convergence point, coulter leash angle, and diameter of the packer roller and its rim width. The experimental model of economic tests qualitatively seeded hardly friable and non-flowing seeds of fodder crops and provides the constant and unobstructed movement of seeds without piling up (and consequently clogging) the sowing part, and uniformly and favorably distributed seeds on the area in the soil for germination (Figure 8).

The results of the agronomic performance of the plots are shown in Tables 1–5.

Table 1. The results containing the density of planting.

Type of Seeder	The Type of the Crop	Number of Plants, pcs/m^2	Field Germination, %
Tractor HS1204 + seeder for sowing non-flowing grass seeds	wheatgrass variety "Burabay"	295	90
	awnless brome "Akmola emerald"	331	89
Tractor HS1204 + SZ-3.6 (Astra)	wheatgrass variety "Burabay"	283.3	86.44
	awnless brome "Akmola emerald"	320.4	86.15

Figure 8. Planting of Burabay wheat grass: (**a**) Crops of wheatgrass "Burabay" on the experimental plot. (**b**) Crops of wheatgrass "Burabay" in the control plot.

Table 2. The comparative performance indicators of the experimental and serial SZT-3.6 (Astra) seeders when sowing grass seed.

Name of Indicators	Experimental Seeder Sample	SZ-3.6 (Astra)	Experimental Seeder Sample	SZ-3.6 (Astra)
Culture	Wheatgrass Burabay	Wheatgrass Burabay	Awnless bromegrass Akmola emerald	Awnless bromegrass Akmola emerald
Speed, km/h	7.0	7.0	7.0	7.0
Seeding rate, kg/ha actual	8.61	8.81	13.72	14
Sowing depth, cm	4	4	4	4
Sowing depth:				
(a) medium, cm	4.11	4.07	4.08	4.04
(b) standard deviation, ±cm	0.2	0.39	0.25	0.44
(c) variation coefficient, %	4.81	9.76	6.06	10.95
(d) seeds sown in the average actual layer depth and two adjacent layers, %	91	86	90	84.3

Table 2. *Cont.*

Name of Indicators	Experimental Seeder Sample	SZ-3.6 (Astra)	Experimental Seeder Sample	SZ-3.6 (Astra)
Seeds not sown into the soil, pieces/m^2	No.	No.	No.	No.
Distributed plants: (a) quantity of plants in a 5 cm section, pieces (b) standard deviation, ±pcs (c) variation coefficient, %	 4.9 3.04 62.0	 4.8 3.07 68.6	 5.5 3.52 64	 5.6 3.56 69.2

Table 3. The comparative indicators of the yield structure of wheatgrass sown on the experimental plot with the experimental seeder and sown on the control plot with the SZ 3.6 seeder (Astra).

	Sprouted Plants, units/m^2	The Height of the Plant, cm	Green Matter Yield, kg/ha
Seeder for non-flowing grass seed sowing	293	85	52.989
	295	81	50.840
	315	83	55.628
	319	86	58.370
	320	79	53.787
$\overline{X}$	308.4	82.8	54.323
Seed drill SZ-3.6 (Astra)	280	82	48.851
	317	83	55.981
	316	77	51.770
	295	79	49.585
	308	76	49.804
$\overline{X}$	303.2	79.4	51.198

Table 4. Yield structure of awnless bromegrass on the experimental plot with the experimental model of a non-driven grass seed drill and on the control plot with seed drill SZ-3.6 (Astra).

	Sprouted Plants, pcs/m^2	The Height of the Plant, cm	Green Matter Yield, kg/ha
Seeder for non-flowing grass seed sowing	310	105	70.761
	326	102	72.287
	290	103	64.935
	315	110	75.326
	324	108	76.070
$\overline{X}$	313.0	105.6	71.876
Seed drill SZ-3.6 (Astra)	270	95	55.761
	305	102	67.630
	279	108	65.504
	290	102	64.304
	326	112	79.374
$\overline{X}$	294.0	103.8	66.515

The Table 1 analysis shows that the germinating capacity of Burabay grass seeds was 90% when sown in the experimental plot and 86.44% in the control plot. The difference in the germinating capacity was 3.56%. The germination of awnless bromegrass seeds sown in the experimental plot was 89.0% and 86.15% in the control plot, and the increase in germination was 2.85%. An increase in the germinating ability of grass seeds is reached at the expense of the high quality of work of an experimental sowing unit and the seeding part of a prototype seeder. The experimental seeding unit does not damage the grass seed and creates a uniform flow of seeds.

Table 5. Experimental results for the energy assessment of the experimental non-flowing grass seed drill and the serial SZ-3.6 (Astra).

Composition of the Unit	Sowing Depth, cm	Machine Speed, km/h	Theoretical Traction Resistance, kN	Traction Resistance, kN	Average Fuel Consumption, kg/hour	Average Slipping Ratio, %
HS1204 tractor with the experimental seeder	2	5	3.11	3.1	13.88	19.37
		7	3.13	3.15	13.75	22.12
		9	3.26	3.4	13.67	24.78
	3	5	3.39	3.5	13.96	19.96
		7	3.4	3.6	13.94	22.56
		9	3.69	4.2	13.69	25.45
	4	5	3.66	3.7	14.05	20.54
		7	3.68	4	14.04	23.13
		9	4.15	4.6	14.35	25.31
	5	5	4.2	4.52	14.25	21.67
		7	4.3	4.84	14.27	22.22
		9	5.1	5.89	14.58	26.40
HS1204 tractor with the series seeder	2	5	3.48	3.53	15.4	21.40
		7	3.49	3.59	15.26	24.47
		9	3.5	3.87	15.20	27.45
	3	5	3.75	3.99	15.53	22.00
		7	3.77	4.1	15.53	24.90
		9	3.79	4.78	15.25	28.16
	4	5	4.02	4.22	15.66	22.61
		7	4.04	4.56	15.68	25.48
		9	4.07	5.24	16.09	27.83
	5	5	4.57	5.15	15.96	23.76
		7	4.59	5.52	16.01	26.60
		9	4.67	6.71	16.43	28.94

The analysis presented in Table 2 indicates that during the sowing of Burabay wheatgrass seeds, the seeding depth uniformity achieved by the experimental plot seeder is 4.81%, whereas the uniformity of the seeding depth on the control plot is 9.76%. Thus, the seeder from the experimental model for seeding non-flowing seeds surpasses the series seeder for seeding uniformity by 4.95%. On the experimental plot, the number of seeds sown to actual depth was 91.0% and was 86.0% on the control plot. Therefore, the experimental seeder outperforms the standard one by 5%. When sowing awnless bromegrass seeds "Akmolinsky emerald", the sowing depth uniformity of the seed was 6.06% on the experimental plot sown with an experimental seeder. On the other hand, on the control plot, the seeding depth uniformity was 10.95%. Consequently, the experimental seeder demonstrates an improvement of 4.89% in seeding depth compared to the standard seeder.

The analysis of the yield structure of wheatgrass (Table 3) showed that the quantity of germinated seeds in the experimental field was at 308.4 pcs/m^2, while in the control field it was at 303.2 pcs/m^2, resulting in 5.2 pcs/m^2. The average plant height on the experimental plot was 3.4 cm higher than the control plot counterparts (82.8 and 79.4 cm). Consequently, the Burabay yield rapeseed oil on the experimental plot sown with the experimental seeder was 54.323 cwt/ha and 51.198 cwt/ha on the control plot, resulting in a 3.125 cwt/ha increase in yield. Similar data were also obtained for awnless bromegrass "Akmola emerald" (Table 3).

The analysis revealed that the quantity of sprouted plants on the experimental plot exceeded the quantity of sprouted plants on the control plot (313 and 294 pcs/m^2). In addition, the height of the plant on the experimental plot was higher than on the control plot (105.6 and 103.8 cm). The yield of awnless bromegrass "Akmolinsky emerald" on the experimental plot sown with an experimental seed drill was 71.876 cwt/ha, while the control was 66.515 cwt/ha. Consequently, the increase in yield was 5.361 cwt/ha. Similar data were also obtained for awnless bromegrass "Akmola emerald" (Table 4).

The laboratory and field tests were conducted to assess the energy consumption of the experimental seed drill for seeding non-flowing grass and the serial SZ-3.6 (Astra). The results are shown in Table 5.

The results of analytical and laboratory and field studies of the experimental seeder for non-flowing grass seed and the serial seeder SZT-3.6 (Astra) [21] are presented in Figures 9–15. An analysis of Figure 9 shows that the resistance of the traction of the experimental seed drill increases with increasing seed placement depth. At the 2 cm placement depth, the traction resistance equals 3.4 kN, at 3 cm–4.2 kN, at 4 cm–4.6 kN, and at 6 cm–5.89 kN. Moreover, the experimental (Re) dependence between the traction resistance of the experimental seed drill and the depth of seeding at 9 km/h working speed exceeds the values of the theoretical (Rt) dependence of the traction resistance. This difference ranges from 4.2% at a seeding depth of 2 cm, escalating to 13.41% at a seeding depth of 5 cm.

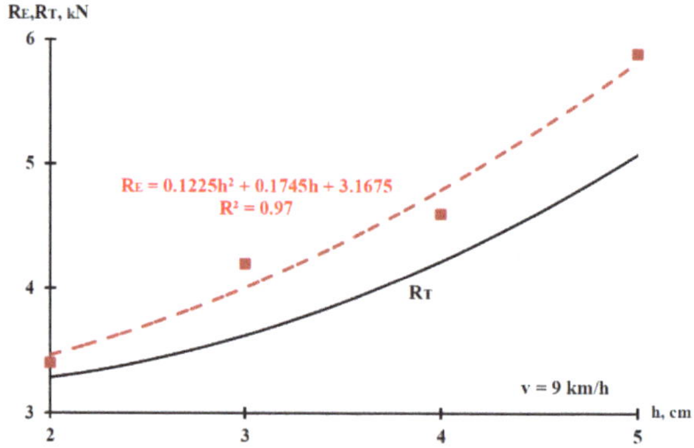

Figure 9. Experimental (Re) and theoretical (Rt) dependencies of draft resistance of the developed seeder on sowing depth at a working speed of 9 km/h.

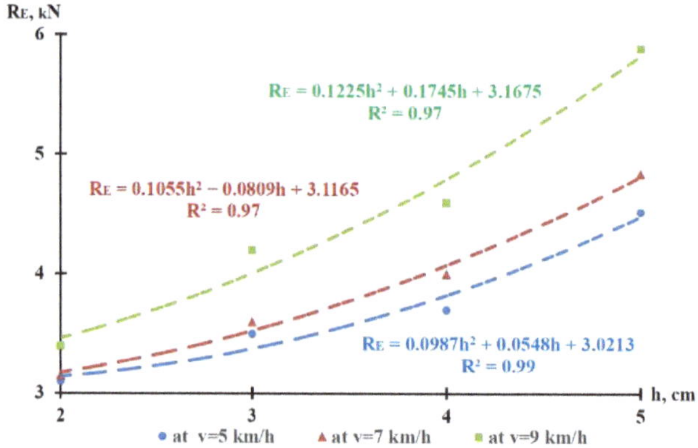

Figure 10. The experimental (Re) dependencies of draft resistance of the developed seeder on sowing depth at different aggregate speeds.

The experimental (Re) dependencies of the traction resistance of the experimental seeder on the depth of seed placement at different unit speeds are shown in Figures 10 and 11. The increase in traction resistance on the depth of seed placement is observed at various operating speeds of the sowing unit. The increase in the sowing depth causes the increase in traction resistance of the developed seed drill.

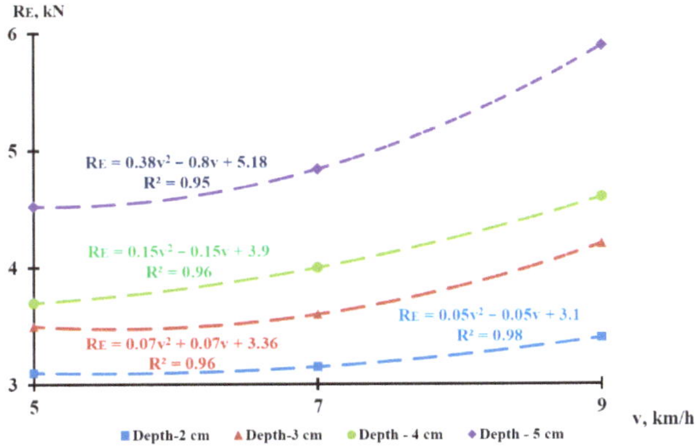

Figure 11. The experimental (Re) dependencies of draft resistance of the developed seeder on the aggregate speed at different sowing depths.

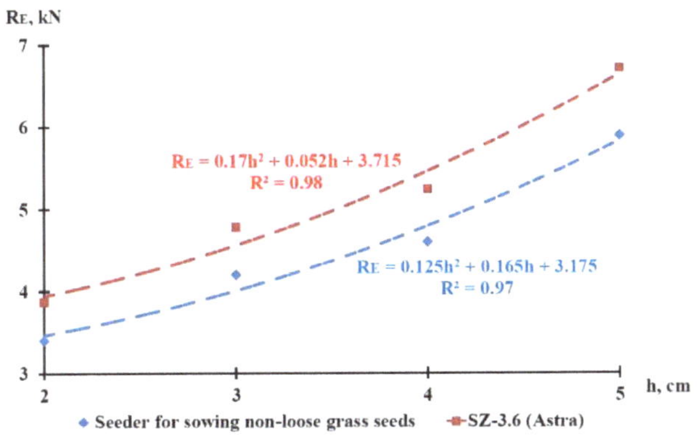

Figure 12. The experimental (Re) dependencies of draft resistance of the developed and standard seeder on sowing depth at a speed of 9 km/h.

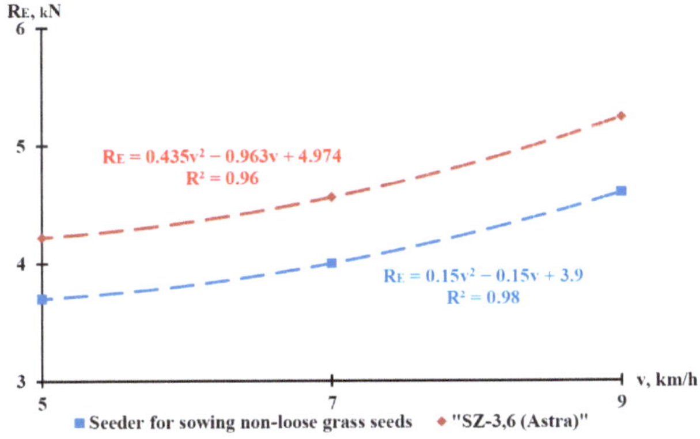

Figure 13. The experimental (Re) dependencies of traction resistance of the developed and serial seeder on the unit speed at a sowing depth of 4 cm.

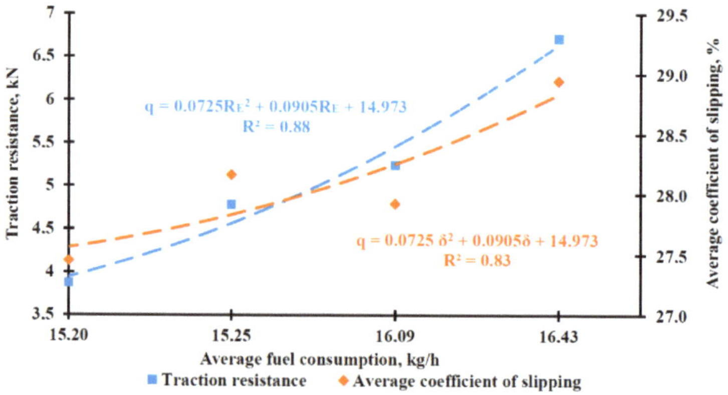

Figure 14. The average fuel consumption during the sowing of non-flowing grass seed on the traction resistance of the developed seeder and average tractor slip coefficient at a working speed of 9 km/h.

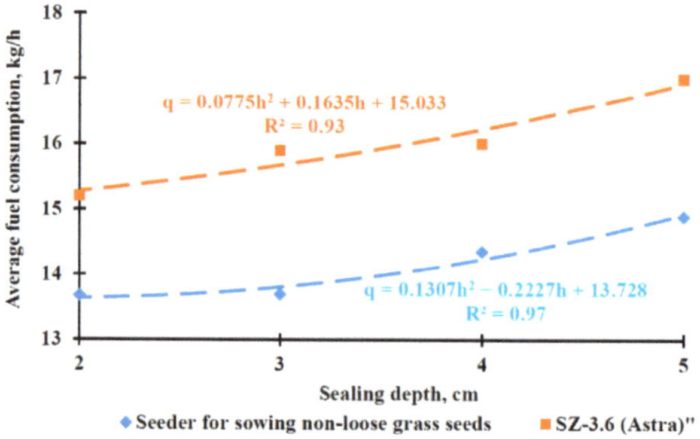

Figure 15. The dependencies of average fuel consumption on the sowing depth of developed and serial seeders at a working speed of 9 km/h.

An analysis shows that the experimental seeder traction resistance is lower than the traction resistance of the serial one by around 12.2% (Figure 12).

The analysis of Figure 13 shows that the traction resistance of the experimental seed drill is lower than the serial one by around 12.3%.

The increase in traction resistance of the developed seed drill results in heightened fuel consumption. A similar relationship is observed between fuel consumption and the coefficient of tractor slippage (Figure 14).

Figure 15 illustrates that the average fuel consumption rises along with the increase in the sowing depth. The difference of the average fuel consumption between the developed and the serial seeder increases from 8 to 10.2%. The same correlation exists between fuel consumption and the coefficient of tractor slippage.

For each type of seed drills for sowing seeds with low flowability and increased connectivity, a large number of both mechanical and pneumatic sowing systems with various technological techniques have been developed to ensure strict norms. For sowing such valuable grasses as rump, granary, wheatgrass, and fescue (sowing with diluents), seeders SZS-2.1, SZT-3.6, and SZP-3.6B have been developed. The North Caucasus Research Institute of Mechanization and Electrification of Agriculture has developed seeders using replaceable coulter blocks and vibro-discrete seeding devices that allow for the sowing of the entire range of agricultural crops [48]. Currently, among all the approaches aimed at improving (creating) hayfields and pastures, the method involving root cultivation has gained the widest acceptance. This technique involves sowing highly valuable grasses into the existing vegetation using specialized combined machinery suitable for both compact and loose types of grasses. [49]. However, all developed machines are adapted to their soil and climatic conditions, which makes it impossible to compare the results with the obtained data.

4. Conclusions

The results of the economic tests established that:

- The germinating capacity of Burabay wheat grass seeds on the plot sown by the experimental seeder was 3.56% higher than the germinating capacity of seeds on the control plot. The improvement of grass seed germination was caused by the high quality of the experimental seeding unit and the seeding part of the prototype seeder.
- The experimental seeder outperformed the standard one by 4.95% in sowing wheat grass in terms of uniform seed placement depth.
- The yield increased by 5.361 kg/ha on the test plot sown with a non-drifted grass seed drill compared to the control plot.
- The traction resistance of the experimental seeder was 12.3% lower than that of the series seeder.

According to the technical characteristics of the prototype seeder, the operating efficiency was calculated for sowing non-flowing grass seeds. The cost-effectiveness of the experimental seeder sample was assessed using costs incurred during sowing compared to the increase in yields and the reduction in fuel and lubricant costs. The annual financial effect of the developed seeder due to an increase in grass yields and a reduction in fuel costs was 7714 USD/ha per year.

Author Contributions: Conceptualization, M.A. and S.N.; methodology, M.A.; software, T.T.; formal analysis, B.K. and M.B.; investigation, T.T., K.V. and K.U.; data curation, M.B. and T.T.; writing—original draft preparation, M.A., M.B. and S.N.; writing—review and editing, M.A., M.B., B.K. and S.N.; visualization, B.K. and M.B., supervision, M.A., M.B. and S.N.; project administration, M.A. All authors have read and agreed to the published version of the manuscript.

Funding: This research was funded by the Ministry of Science and Higher Education of the Republic of Kazakhstan, grant number AR19676894.

Institutional Review Board Statement: Not applicable.

Data Availability Statement: The data presented in this study are available upon request from the corresponding author.

Acknowledgments: The authors thank Jan Pawlik for proofreading the article.

Conflicts of Interest: The authors declare no conflict of interest.

References

1. Smailov, A. In 2023, the State Plans to Withdraw 5 Million Hectares of Unused Agricultural Land. Available online: https://primeminister.kz/ru/news/v-2023-godu-gosudarstvo-planiruet-izyat-5-mln-ga-neispolzuemyh-selhozzemel-1702730 (accessed on 15 June 2023).
2. Alexandratos, N.; Bruinsma, J.; Boedeker, G.; Schmidhuber, J.; Broca, S.; Shetty, P.; Ottaviani, M.G. *World Agriculture: Towards 2030/2050*; FAO: Rome, Italy, 2006; p. 147.
3. Alhassan, W.S.; Barnes, P. Problems and Prospects for Forage Production and Utilisation in Ghana. In Proceedings of the XVII International Grassland Congress, Palmerston North, New Zealand, Rockhampton, Australia, 8–21 February 1993.
4. Armstrong, K.; de Ruiter, J.; Bezar, H. Fodder Oats in New Zealand and Australia—History, Production and Potential. *Fodd. Oats World Overv.* **2004**, 33, 153–177.
5. Barnes, R.F.; Nelson, C.J.; Moore, K.J.; Collins, M. *Forages, The Science of Grassland Agriculture*; Blackwell Publishing: Ames, IA, USA; Oxford, UK, 2007; p. 808. [CrossRef]
6. Franco, W.; Barbera, F.; Bartolucci, L.; Felizia, T.; Focanti, F. Erratum to Developing Intermediate Machines for High-Land Agriculture. *Dev. Eng.* **2021**, *6*, 100084. [CrossRef]
7. Franco, W.; Anastasio, D.; Ferraresi, C.; Gondino, F.; Quaglia, G.; Soprana, L. A New Human Powered Press for Producing Straw Bales for Load Bearing Constructions (Anpilpay 2.0). *Agric. Eng. Int. CIGR e-J.* **2017**, *19*, 98–107.

8. HE, J.; LI, H.; McHugh, A.; WANG, Q.; LI, H.; Rasaily, R.; Sarker, K. Seed Zone Properties and Crop Performance as Affected by Three No-Till Seeders for Permanent Raised Beds in Arid Northwest China. *J. Integr. Agric.* **2012**, *11*, 1654–1664. [CrossRef]
9. Murray, J.R.; Tullberg, J.N.; Basnet, B.B. *Planters and Their Components: Types, Attributes, Functional Requirements, Classification and Description*; ACIAR Monograph: Canberra, Australia, 2006; p. 178.
10. Serrano, J.; Peça, J.; Pinheiro, A.; Carvalho, M.; Nunes, M.; Ribeiro, L.; Santos, F. The Effect of Gang Angle of Offset Disc Harrows on Soil Tilth, Work Rate and Fuel Consumption. *Biosyst. Eng.* **2003**, *84*, 171–176. [CrossRef]
11. Vamerali, T.; Bertocco, M.; Sartori, L. Effects of a New Wide-Sweep Opener for No-till Planter on Seed Zone Properties and Root Establishment in Maize (*Zea mays*, L.): A Comparison with Double-Disk Opener. *Soil Tillage Res.* **2006**, *89*, 196–209. [CrossRef]
12. Zhu, G.; Li, W.; He, J. Design and Experiment on 2BFML5 No-till Planter for Permanent Raised Bed. *Trans. Chin. Soc. Agric. Mach.* **2008**, *39*, 51–54.
13. Sysuev, V.; Demshin, S.; Cheremisinov, D.; Doronin, M. Theoretical Justification of the Main Parameters of the Coulter Group of the Sod Seeder for Strip Sowing. *Agric. Sci. Euro-North-East* **2020**, *21*, 321–331. [CrossRef]
14. Sysuev, V.; Demshin, S.; Doronin, M. Method of Seeding Grass Seeds into Sod and Seeding Machine for Its Implementation. Patent of the Russian Federation 2641073, 15 January 2018.
15. Zaitsev, A.; Solodun, V.; Gorbunova, M. Comparative Evaluation of Seeding Spring Wheat Methods When Using Different Types of Coulters. *IOP Conf. Ser. Earth Environ. Sci.* **2020**, *421*, 062017. [CrossRef]
16. Tessier, S.; Saxton, K.E.; Papendick, R.I.; Hyde, G.M. Zero-Tillage Furrow Opener Effects on Seed Environment and Wheat Emergence. *Soil Tillage Res.* **1991**, *21*, 347–360. [CrossRef]
17. Arifa, W.; Oleh, H. Production Tests of a Seed Drill CPH 2000 for Direct Sowing. *Inmateh-Agric. Eng.* **2018**, *56*, 31–38.
18. Besharati, B.; Navid, H.; Karimi, H.; Behfar, H.; Eskandari, I. Development of an Infrared Seed-Sensing System to Estimate Flow Rates Based on Physical Properties of Seeds. *Comput. Electron. Agric.* **2019**, *162*, 874–881. [CrossRef]
19. Dhir, D.K.; Rajan, P.; Verma, S. Seed Drill Discharge Rate Variation Due to Varietal Differences Using an Automated Calibration Test Rig. *AMA Agric. Mech. Asia Afr. Lat. Am.* **2019**, *50*, 43–47.
20. Hayes, R.C.; Newell, M.; Pembleton, K.; Peoples, M.; Li, G. Sowing Configuration Affects Competition and Persistence of Lucerne (Medicago Sativa) in Mixed Pasture Swards. *Crop Pasture Sci.* **2021**, *72*, 707–722. [CrossRef]
21. Aduov, M.; Nukusheva, S.; Esenalikaspakov, E.; Kazbekisenov, K.; Volodya, K. Analysing the Results Field Tests of an Experimental Seeder with Separate Introduction of Seeds and Fertilizers. *Int. J. Mech. Prod. Eng. Res. Dev.* **2019**, *9*, 589–598. [CrossRef]
22. Aduov, M.A.; Kapov, S.N.; Matyushkov, M.I.; Nukusheva, S.A.; Kaspakov, E.Z.; Kusainov, R.K.; Volodya, K.; Rakhimzhanov, M.R.; Soroka, A.A. A Sowing Machine. Patent of the Republic of Kazakhstan 30503, 16 November 2015.
23. Aduov, M.; Nukusheva, S.; Kaspakov, E.; Isenov, K.; Volodya, K.; Tulegenov, T. Seed Drills with Combined Coulters in No-till Technology in Soil and Climate Zone Conditions of Kazakhstan. *Acta Agric. Sect. B-Soil Plant Sci.* **2020**, *70*, 525–531. [CrossRef]
24. Yurchenko, V.A. The Way to Create a Solid Food Base in Kazakhstan. Available online: https://kazakh-zerno.net/112038-put-sozdaniya-prochnoj-kormovoj-bazy-v-kazakhstane/ (accessed on 15 June 2023).
25. Hayes, R.C.; Li, G.; Sandral, G.; Swan, T.; Price, A.; Hildebrand, S.; Goward, L.; Fuller, C.; Peoples, M. Enhancing Composition and Persistence of Mixed Pasture Swards in Southern New South Wales through Alternative Spatial Configurations and Improved Legume Performance. *Crop Pasture Sci.* **2017**, *68*, 1112–1130. [CrossRef]
26. Verma, A.K.; Pandey, M.K. Automatic Seed Cum Fertilizer Drill: Modification and Performance Evaluation for Intercropping. *AMA Agric. Mech. Asia Afr. Lat. Am.* **2019**, *50*, 44–48.
27. Cheema, M.J.; Nauman, M.; Ghafoor, A.; Farooque, A.; Haydar, Z.; Ashraf, M.U.; Awais, M. Direct Seeding of Basmati Rice through Improved Drills: Potential and Constraints in Pakistani Farm Settings. *Trans. ASABE (Am. Soc. Agric. Biol. Eng.)* **2021**, *37*, 53–63. [CrossRef]
28. Derevjanko, D.; Holovach, I.; Bulgakov, V.; Ihnatiev, Y.; Nozdrovický, L. Mathematical Model of Uniform Cereal Crops Seeding Using a Double-Disk Coulter. *Acta Technol.* **2020**, *23*, 195–200. [CrossRef]
29. Gumarov, G.; Konovalov, V.; Sarsenov, A.; Kubasheva, Z.; Rakhimov, A. Mathematical Modelling of Traction Resistance of the Improved Opener of Grain Seeder. *BIO Web Conf.* **2020**, *17*, 00044. [CrossRef]
30. Jamil, M.; Hussain, S.S.; Qureshi, M.A.; Mehdi, S.M.; Nawaz, M.; Javed, Q. Rice Yield Improvement through Various Direct Seeding Techniques on Moderately Salt Affected Soil. *J. Anim. Plant Sci.* **2017**, *27*, 848–854.
31. Liu, W.; Zhao, X.; Pan, H.; Lakhiar, I.; Wang, W.; Jianping, H. Development and Experimental Analysis of an Intelligent Sensor for Monitoring Seed Flow Rate Based on a Seed Flow Reconstruction Technique. *Comput. Electron. Agric.* **2019**, *164*, 104899. [CrossRef]
32. Rafiq, M.; Ahmad, R.; Jabbar, A.; Munir, H.; Hussain, M. Influence of Different No-till Techniques at Varying Heights of Standing Rice Stubbles on the Wheat Performance. *Int. J. Agric. Biol.* **2017**, *19*, 410–416. [CrossRef]
33. Singh, S.; Singh, M.; Ekka, U.; Singh, M.K. E-Powered Multi-Purpose Two-Row Seeder for Smallholders. *Indian J. Agric. Sci.* **2019**, *89*, 2091–2096. [CrossRef]
34. Yatskul, A.; Lemiere, J.-P.; Cointault, F. Influence of the Divider Head Functioning Conditions and Geometry on the Seed's Distribution Accuracy of the Air-Seeder. *Biosyst. Eng.* **2017**, *161*, 120–134. [CrossRef]
35. Kushwaha, H.; Dass, A.; Khura, T.; Sahoo, P.K.; Singh, K.; Mani, I. Design, Development and Performance Evaluation of Manual Planter for System of Wheat Intensification. *Indian J. Agric. Sci.* **2019**, *89*, 678–687. [CrossRef]

36. Shambhu, V. Design and Development of Low Cost Multi-Row Manual Jute Seed Drill. *Ama-Agric. Mech. Asia Afr. Lat. Am.* **2020**, *51*, 46–51.
37. Shambhu, V.; Thakur, A. Laboratory and Field Performance of Manual Seed Drill for Sowing Jute and Tiny Seeds. *Indian J. Agric. Sci.* **2019**, *89*, 129–132. [CrossRef]
38. Jamil, M.; Hussain, S.S.; Qureshi, M.A.; Mehdi, S.M.; Nawaz, M. Impact of Sowing Techniques and Nitrogen Fertilization on Castor Bean Yield in Salt Affected Soils. *J. Anim. Plant Sci.* **2017**, *27*, 451–456.
39. Kuş, E.; Yildirim, Y. Effects of Seed Drop Height and Tillage System on the Emergence Time and Rate in the Single Seed Planters. *Alinteri J. Agric.* **2020**, *35*, 69–76. [CrossRef]
40. Manea, D.; Marin, E.; Mateescu, M.; Dumitraşcu, A. Theoretical and Experimental Research on Sowing Rate Optimization of Grassland Drills. *E3S Web Conf.* **2019**, *112*, 03003. [CrossRef]
41. Riegler-Nurscher, P.; Karner, J.; Huber, J.; Moitzi, G.; Wagentristl, H.; Hofinger, M.; Prankl, H. A System for Online Control of a Rotary Harrow Using Soil Nurscher, Roughness Detection Based on Stereo Vision. *AgEng* **2017**, *1*, 559–566.
42. Wang, B.; Luo, X.; Wang, Z.; Zheng, L.; Zhang, M.; Dai, Y.; Xing, H. Design and Field Evaluation of Hill-Drop Pneumatic Central Cylinder Direct-Seeding Machine for Hybrid Rice. *Int. J. Agric. Biol. Eng.* **2018**, *11*, 33–40. [CrossRef]
43. Aduov, M.A.; Nukusheva, S.A.; Kaspakov, E.Z.; Volodya, K.; Tulegenov, T.K.; Isenov, K.G. Grass Seeder. Patent of the Republic of Kazakhstan 35326, 22 October 2021.
44. The Development of Animal Husbandry in Kazakhstan Is Hindered by Insufficient Feed Supply. Available online: https://dairynews.today/kz/news/razvitie-zhivotnovodstva-v-kazakhstane-tormozit-ne.html (accessed on 15 June 2023).
45. On Approval of the State Program for the Development of the Agro-Industrial Complex of the Republic of Kazakhstan for 2017–2021. Available online: https://adilet.zan.kz/rus/docs/U1700000420 (accessed on 15 June 2023).
46. Kurach, A.A.; Amantayev, M.A. To Set up a Seeder—To Get a Harvest. Available online: https://agroinfo.kz/naladit-seyalku-poluchit-urozhaj/ (accessed on 15 June 2023).
47. Latyshev, N. Soybeans in Northern Kazakhstan. Available online: https://agrosektor.kz/assets/files/archive/40_06_19.pdf (accessed on 15 June 2023).
48. Semenikhina, Y.A. Improving the Quality of Sowing Seeds of Fodder Grasses with a Vibro Discrete Seeding System. 2013. Available online: https://www.dissercat.com/content/povyshenie-kachestva-vyseva-semyan-kormovykh-trav-vibrodiskretnoi-vysevayushchei-sistemoi (accessed on 18 August 2023).
49. Arslanov, M.A. Design Parameters of the Sowing Part of the Seeder for Sowing Non-Loose and Slightly Loose Grass Seeds in a Wide-Row Way. 2007. Available online: https://tekhnosfera.com/konstruktivnye-parametry-vysevayuschey-chasti-seyalki-dlya-poseva-nesypuchih-i-slabosypuchih-semyan-trav-shirokoryadnym-s (accessed on 18 August 2023).

agriculture

MDPI

Article

Design and Experiment of Row Cleaner with Staggered Disc Teeth for No-Till Planter

Xin Li, Yinping Zhang *, Haojie He, Bin Wang, Hua Zhou, Duanyang Geng and Yuzi Zhang

School of Agriculture Engineering and Food Science, Shandong University of Technology, Zibo 255049, China; 20503030302@stumail.sdut.edu.cn (X.L.); 21503030306@stumail.sdut.edu.cn (H.H.); 20503030298@stumail.sdut.edu.cn (B.W.); zhouhua850@sdut.edu.cn (H.Z.); dygxt@sdut.edu.cn (D.G.); zhangyuzi@sdut.edu.cn (Y.Z.)

* Correspondence: zhangyinping@sdut.edu.cn

Abstract: In view of the large amount of wheat straw coverage and the difficulty with the high-speed working process used in traditional rotary tillage row cleaner. A row cleaner with staggered disc teeth for no-till planting was designed. It was determined that installation with a forward inclination α, horizontal declination β, and forward speed v_0 of the machine were the main factors affecting the straw cleaning rate Y_1 and working resistance Y_2, and the range of values for structural parameters and motion parameters of the row cleaner were determined. Taking α, β, and v_0 as the factors and Y_1 and Y_2 as the response indexes, using EDEM 2018 software to simulate the straw cleaning process under different parameters and determine the influence of each parameter on the straw cleaning performance. After performing a soil bin test, the results showed that there was no straw entanglement and blockage, and the passability was better than that of the traditional flat disc separated row cleaner. When α was 70°, β was 30°, v_0 was 8 km/h, and the embedded depth h of the soil (straw) was 55 mm, the average straw cleaning rate was the highest, which was 90.59%. This study provides a new idea for the design of high-speed corn no-till planters in the Huang-Huai-Hai area of China.

Keywords: no-till sowing; row cleaners; straw mulching; straw cleaning discs; seedbed cleaning; discrete element

Citation: Li, X.; Zhang, Y.; He, H.; Wang, B.; Zhou, H.; Geng, D.; Zhang, Y. Design and Experiment of Row Cleaner with Staggered Disc Teeth for No-Till Planter. *Agriculture* **2023**, *13*, 1373. https://doi.org/10.3390/agriculture13071373

Academic Editors: Mustafa Ucgul and Chung-Liang Chang

Received: 28 May 2023
Revised: 6 July 2023
Accepted: 9 July 2023
Published: 10 July 2023

1. Introduction

No-tillage seeding is an effective measure to retain straw mulching, reduce soil tillage, and protect soil structure [1,2]. It is the key technology and basic requirement for the implementation of conservation tillage [3]. The quality of the no-tillage planter's straw-cleaning and anti-blocking performance are key to high-quality no-tillage planting operations, especially in the Huang-Huai-Hai double-cropping area of China, where the corn sowing period is short, and the wheat straw covers the surface without decomposing, with high toughness and with a tendency entangle or block the planter [4,5]. At present, no-tillage in the double-cropping area is mainly based on crushing and rotary tillage row cleaners [6,7]. The high-speed rotating rotary tiller smashes the straw and mixes it with the soil, and then sows. This straw-clearing method has large soil disturbance and high energy consumption, and the soil moisture is lost quickly; the rotary tiller is severely squeezed on the bottom and side of the soil, forming a plow bottom; the crushed straw is mixed with the soil, because of the compaction is not enough, the seeds cannot be in close contact with the soil, which affects the emergence of seedlings [8,9]. It is the focus of no-tillage sowing research in double-cropping areas to change the traditional method of crushing and rotary tillage to clear straw and prevent blockage, and provide a clean seedbed for no-tillage sowing under the conditions of small soil disturbance and low power consumption.

Mahmoud et al. [10] proposed a new promising process by applying magnetic treatment to water, which can alleviate salinity stress and improve crop productivity. The demanding use of agricultural machines would lead to grave problems concerning the quality of agricultural soils. Jalel et al. [11] proposed a permanent bed technology to improve soil conditions, and Hassen et al. [12], based on a Bayesian network, established a decision support system for risk assessment of soil compaction, which can help users to conduct causal analysis and diagnostic analysis of soil state. However, in the Huang-Huai-Hai double cropping area of China, with scattered plots and a large amount of straw coverage, an efficient straw cleaning device is more suitable for the current fieldwork environment. The straw-scraping wheel-type row cleaner adopts two oppositely installed straw cleaning discs to allow for straw cleaning of the seedbed during the no-tillage operation. The soil disturbance is small, and the operation resistance is low; it is widely used in no-tillage planters at domestic and foreign [13]. Domestic research on straw-scraping wheel-type row cleaners mainly focuses on the optimization of the structure of the straw-cleaning discs and claw teeth. Liu et al. [14] added grass allocation grids after the straw scrapers, but the straw cleaning rate was only 57.5%. Jia et al. [13] designed a stubble-cleaning mechanism with a concave claw structure based on the improvement of the plane straw cleaning discs, and the straw cleaning rate of the seedbed was 83.61%. However, there is a leakage area when using the oppositely installed straw cleaning discs, which limits the improvement of the straw cleaning rate. In view of the problems of low straw cleaning rate, unstable operation performance, and reduced working quality and working efficiency of the planter, Wang et al. [15,16] designed a star-toothed concave disk row cleaners, it adopted the front and rear staggered arrangement to reduce the leakage area, and the straw cleaning rate of the seed bet reached 92.2%. However, the processing of the star-toothed concave structure is complicated, and the front and rear staggered arrangement increases the longitudinal size of the machine.

In view of the above problems, this study shows the design of a disc tooth staggered row cleaner under the conditions of small soil disturbance and low power consumption. Upon optimizing the structure and layout of the straw cleaning discs, the leakage area can be reduced, and the straw cleaning effect can be improved. Through theoretical analysis, simulation experiments, and field performance experiments, the optimal combination of structural parameters and structural parameters was determined. Finally, through soil tank comparative experiments, the operating effectiveness of the optimal parameter combination of the device was verified, providing a theoretical reference for the design of double-season corn zero tillage seeders.

2. Materials and Methods

2.1. Overall Structure and Working Principle

The overall structure of the teeth of the disc staggered row cleaner is shown in Figure 1, which mainly includes a frame, a mounting frame of straw cleaning discs, a connecting frame, and two straw cleaning discs. The claws are evenly distributed around the circumference of the straw cleaning disc, and both the straw cleaning discs are symmetrically installed on both sides of the mounting frame at a certain angle to the ground. The claws of the two straw cleaning discs are staggered at the front and bottom and above the ground contact point, forming a "scissors" overlapping distribution pattern to reduce the leakage area. The straw cleaning discs mounting frame can be adjusted up and down in the connecting frame to set a suitable soil depth and reduce soil disturbance.

When the straw cleaning disc is installed, there is a certain forward inclination and horizontal declination so that the row cleaner has a certain outward expansion angle from bottom to top and from front to back. It generates thrust to both sides of the straw so that the straw moves to both sides of the opener to form a seedbed with consistent straw cleaning. The claw teeth of the straw cleaning disc contact the ground and rotate backward under the action of the friction force between the soil and the straw to achieve the picking and throwing of straw and turning of topsoil. During the working process, if one side of the straw cleaning disc is entangled with straw, the other side of the straw cleaning disc staggered with respect to the claw teeth can apply a force to it so that it can continue to rotate, effectively reducing the occurrence of straw entanglement, blockage, and stoppage.

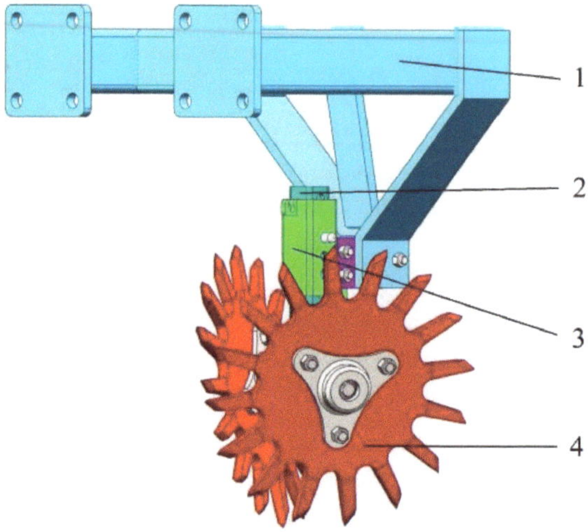

Figure 1. Structure diagram of row cleaner with staggered disc teeth. 1. frame; 2. mounting frame of straw cleaning discs; 3. connecting frame; 4. straw cleaning discs.

In the double-cropping area of wheat and corn, crop stubble is the main factor that affects the quality of no-till sowing, increases the power consumption of seedbed preparation, and limits the speed of operation. This study adopts a method of avoiding stubble sowing, as shown in Figure 2. Disc teeth staggered row cleaners are installed in front of the opener of the corn no-tillage planter, as shown in Figure 3. During operation, row cleaners clean the stalks between the rows of wheat stubble and provide a clean seedbed for corn no-tillage sowing, improve the quality of sowing, reduce the power consumption required for seedbed preparation, and improve the speed of sowing operations [17,18].

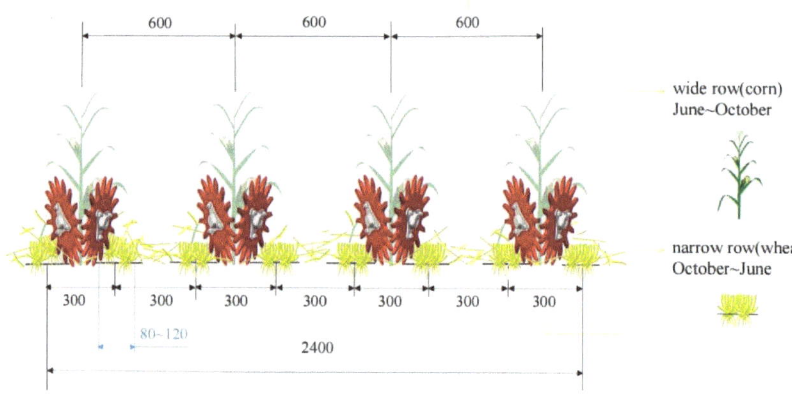

Figure 2. Schematic diagram of stubble avoidance sowing mode.

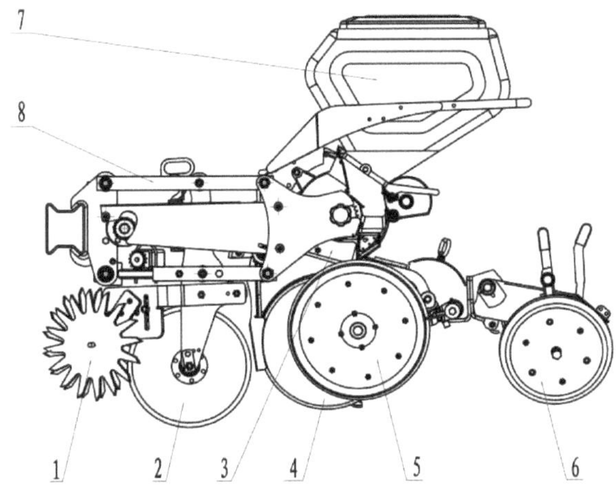

Figure 3. Schematic diagram of row cleaner and opener installation. 1. straw cleaning discs; 2. stubble-cutting disc with ripples; 3. seeding apparatus; 4. opener; 5. depth limiting wheels; 6. soil compacting wheels; 7. seeds box; 8. four-bar profiling mechanism.

2.2. Design and Analysis of Straw Cleaner

2.2.1. Analysis of Kinematic Parameter

In the process of no-till sowing operations, the effect of row cleaners on straw is related to factors such as the forward speed of the machine, the immersion depth of the straw cleaning discs, and the installation angle [19–21]. The kinematics of the straw cleaning tray is analyzed, the center of the straw cleaning disc is taken as the origin, the movement direction of the machine is the x-axis direction, and the $Oxyz$ space cartesian coordinate system is established, as shown in Figure 4. The forward inclination α is the angle between the horizontal plane (xz) and the line connecting the rotation center and the meshing point of the straw cleaning disc. Horizontal deflection β is the angle between the vertical plane (xy) and the line connecting the rotation center of the straw cleaning disc and the meshing point of the straw cleaning disc. The immersion depth is h, and the two intersection points of the straw cleaning discs and the ground are m and n, respectively, then:

$$L_{mn} = 2r\sin\gamma = 2\sqrt{r^2 - (r-h)^2} \tag{1}$$

In the formula: L_{mn}—the length of the connection line between the front and rear claw teeth and the ground intersection, mm;

r—radius of straw cleaning disc, mm;

h—immersion depth of the straw cleaning discs, mm;

γ—the angle between the line connecting the center of the straw cleaning disc and the intersection of the ground and the vertical direction, (°).

Assuming that the working width of a single straw cleaning disc is b, it can be seen from Figure 4 that the calculation formula of the working width b is [22,23]:

$$b = L_{mn}\sin\beta = 2\sin\beta\sqrt{r^2 - (r-h)^2} \tag{2}$$

In the formula: b—working width of one straw cleaning disc, mm;

β—straw cleaning disc horizontal installation, (°).

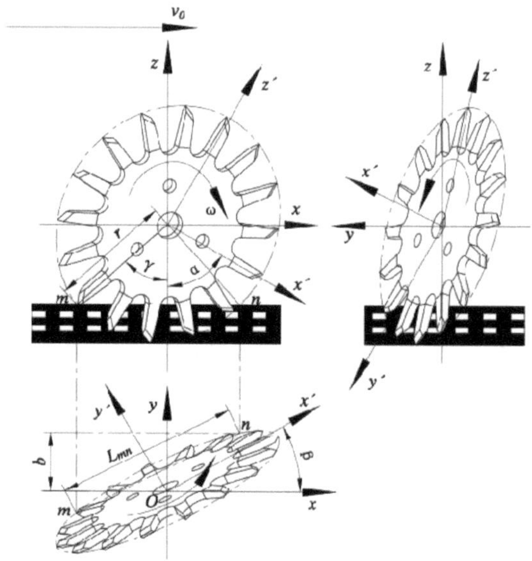

Figure 4. Installation diagram of straw cleaning disc.

The straw cleaning discs are installed in a staggered and opposite manner with respect to the claw teeth, and the straw cleaning area is shown in Figure 5:

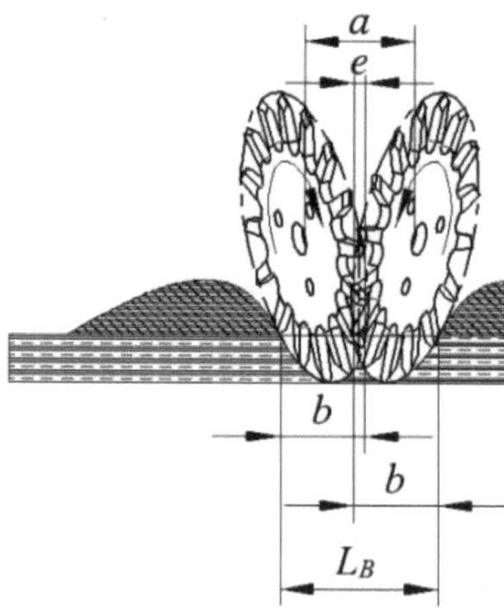

Figure 5. Schematic diagram of the width of the row cleaner.

The theoretical straw cleaning width on both sides of the straw cleaning discs is:

$$L_B = 2b - e = 4\sin\beta\sqrt{r^2 - (r-h)^2} - e \quad (3)$$

In the formula: L_B—the theoretical straw cleaning width of the double-sided straw cleaning discs, mm;

e—the distance between the double intersections between the ground and the front of the double-sided straw cleaning discs, that is, the width of the overlapping area of the two straw clearing discs, mm.

It can be seen from Formula (3) that when the depth h of the straw cleaning discs is determined, the working width of the double-sided straw cleaning discs L_B depends on the horizontal declination β and the width of the overlapping area of the two straw cleaning discs e.

Assume that the contact start point of the two straw cleaning disc teeth is P, the contact end point is Q, point C is the center position of the staggered contact area, and the width of the contact area is L_{PQ}, as shown in Figure 6. L_{PQ} is related to the center distance a, and its relationship can be represented by Formula (4).

$$L_{PQ} = 2\sqrt{L_{OP}^2 - L_{OC}^2} = 2\sqrt{r^2 - \frac{a^2}{4\sin^2\beta}} = 2\sqrt{r^2 - \frac{(2r\sin\alpha\sin\beta - e)^2}{4\sin^2\beta}} \quad (4)$$

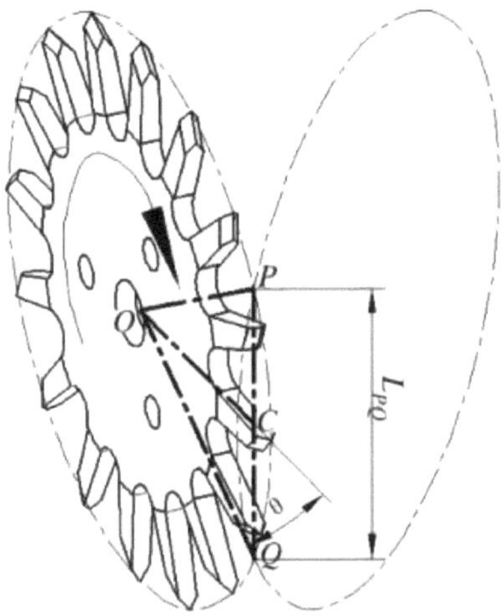

Figure 6. Schematic diagram of straw cleaning discs.

In the formula: a—the center distance of the two straw cleaning discs, mm.

There is a certain overlapping area of straw cleaning between the two straw cleaning discs, which can reduce leakage area. It can be seen from Formula (4) that the smaller the center distance a, the larger the overlapping area width e and the staggered contact area L_{PQ}, the straw cleaning discs contacts at the root of the tooth, and the disc teeth are prone to interference during rotation, resulting in them becoming stuck. The larger the center distance a, The smaller the overlapping area width e and the staggered contact area L_{PQ}, the straw cleaning disc contacts at the tip of the tooth, which increases the moment at the root of the tooth and causes the claw teeth to break. In order to ensure that the two straw cleaning discs are staggered and contacted effectively, the contact point of the disc teeth should be located in the range above the center of the claw tooth and below the tooth top, as shown in Figure 7. Taking into account the tooth thickness and other factors, the contact point of the two claw teeth is designed to be located at $d_c/6$ above the center point of the claw tooth, and we attain:

$$a = 2\sin\beta\left(r - \frac{1}{3}d_c\right) \quad (5)$$

In the formula: d_c—claw tooth height, mm.

From Formulas (2)–(5), it can be known that when the diameter r, claw tooth height d_c, and soil (straw) depth h of the straw cleaning discs are determined, the working width L_B and the contact area width L_{PQ} of the two straw cleaning discs are related to the forward inclination α and the horizontal declination β of installation. The average thickness of stalks covered on the ground during corn planting in the Huang-Huai-Hai double-cropping area of China is about 30~40 mm [24,25]; therefore, the thickness of the stalks is taken as 35 mm. In order to make the straw cleaning discs achieve the best straw cleaning operation effect when the soil disturbance is as small as possible, the straw cleaning disc's teeth can be slightly immersed in the soil. The maximum depth of the design is 20 mm, and h is 55 mm. According to reference [26], the design radius r of the straw cleaning disc is 180 mm, and the claw tooth height d_c is 70 mm.

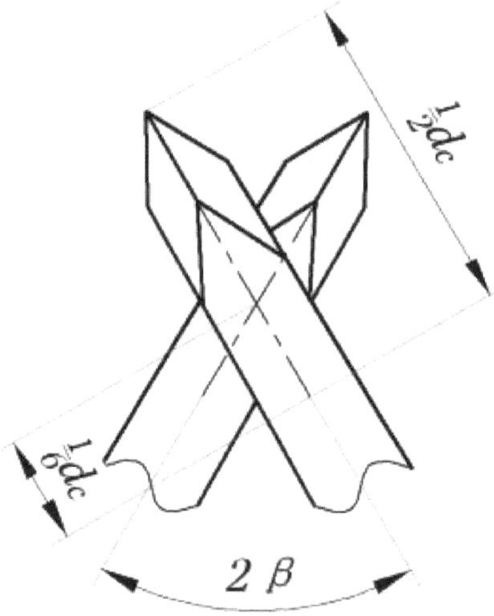

Figure 7. Schematic diagram of teeth.

In the actual operation process, when the forward inclination α installation is too large, a leakage area may be formed in the front, and the leakage area increases with the increase in α; if α is too small, the resistance of the straw cleaning discs will be too large. At the same time, the contact between the claw teeth and the straw will also be reduced. When the horizontal declination β is too large, the thrust of the straw cleaning discs in the forward direction of the machine will be greater than the side thrust on both sides, resulting in the resistance of the machine being too large and the stubble-picking ability reduced; when β is too small, the side thrust performance will be reduced, the straw cleaning discs cannot move the straw to the sides in a timely and effective manner, and it is easy to throw the straw to the rear, resulting in blockage of the machine. Therefore, the forward inclination α and the horizontal declination β installations should not be too large or too small. Referring to the "*Agricultural Machinery Design Manual*" [23], the value range is determined to be $50^\circ \leq \alpha \leq 70^\circ, 30^\circ \leq \beta \leq 45^\circ$.

2.2.2. Number of Claw Teeth

Too many claw teeth will easily lead to too many teeth in the soil, which will lead to excessive pressure on the straw cleaning discs and increased power consumption. At the same time, the distance between the claw teeth will be reduced, and the staggered process will easily interfere. When the number of claw teeth is too small, a larger leakage area will be generated, resulting in a reduced straw-cleaning effect. Therefore, the appropriate number of claw teeth is a key factor in reducing the power consumption of the straw cleaning operation and allowing for the interlacing of the claw teeth. Refer to the "*Agricultural Machinery Design Manual*" [25] to calculate the number of claw teeth:

$$i = \frac{2\pi(r - d_c)}{s_c} \tag{6}$$

In the formula: r—turning radius of straw cleaning disc, mm;
s_c—root chord length of adjacent claw teeth, mm.

The number of claw teeth required for the straw cleaning disc can be calculated from Formula (6). In order to ensure the working strength of the row cleaner and the claw teeth can have a good effect on the straw on the ground, the number of claw teeth $i = 16$.

2.2.3. Shape of Claw Teeth

The distribution of the claw teeth of the straw cleaning disc is generally divided into forward inclination, backward inclination, and radial direction [13,27]. The forward-inclined claw teeth can easily cause the straw to fall between the seedbeds or cause high-speed straw entanglement. However, the backward-inclined claws are easy to stubble when rotating at a low speed, and the straw is difficult to fall off when rotating at a high speed, resulting in a blockage. Radial claw teeth are good for straw removal, picking, and other operations, especially in the form of staggered distribution of disc teeth; no matter whether a low speed or high speed is used, it is not easily blocked. It is difficult to allow for the interleaving of the claw teeth with the forward or backward claw teeth, which is easy to cause the straw cleaning discs to be stuck and stopped. Therefore, the radial claw tooth distribution is selected in this paper to achieve the preset working effect.

The radial claw teeth are designed in a trapezoidal pattern, with the left and right edges symmetrical along the center line of the teeth. As shown in Figure 8, the claw tooth gradient angle τ should not be too large, which will lead to too large friction f_1, so that it is easy to accumulate straw here and cause blockage.

$$F_s \sin(90^\circ - \sigma - \tau) \geq f_2 + f_1 \cos(90^\circ - \sigma - \tau) \tag{7}$$

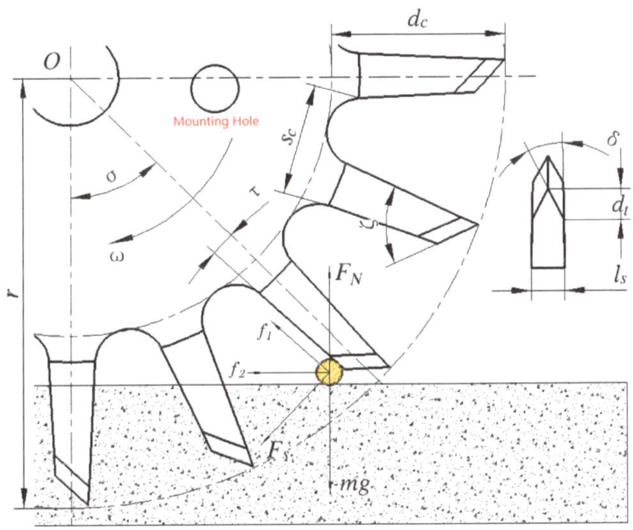

Figure 8. Diagram of claw teeth.

In the formula: F_s—The pressure of the claw tooth on the straw, N;

τ—claw tooth gradient angle, (°);

σ—at this moment, the angle between the claw tooth and the normal line is shown in Figure 6, (°);

f_1—the friction of the claw tooth on the straw, N;

f_2—the friction of the ground on the straw, N.

It can be seen from Formula (7) that if the claw tooth gradient angle τ is too large, the pressure F_s on the straw will decrease, thereby reducing the throwing effect on the straw. Therefore, combined with the existing studies [15,28,29], set the claw tooth gradient angle $\tau = 5^\circ$.

2.2.4. Blade Shape of Claw Teeth

The top of the claw teeth is designed as a blade, which is helpful for breaking the solid soil layer and cutting off the straw. As shown in Figure 8, the blade height d_t is:

$$d_t = \frac{l_s}{2\tan\delta} \tag{8}$$

In the formula: d_t—claw tooth blade height, mm;
l_s—claw tooth thickness, mm;
δ—blade bevel, (°).

The larger the tooth thickness, the larger the contact area of soil breaking, and the greater the resistance to soil entry; the smaller the tooth thickness, the smaller the strength of the claw teeth, and it is easy to press the straw into the soil. Therefore, according to the references [16,30] and the actual soil conditions in the Huang-Huai-Hai area of China, set claw tooth thickness l_s = 13 mm. In order to achieve the best ground-breaking effect and reduce power consumption, set the blade bevel δ = 30° and the blade inclination angle ζ = 60°.

2.3. *Experiment Method*

In order to analyze and obtain the optimal design parameter combination, explore the characteristics of the working quality and working resistance of the row cleaner, and design a virtual simulation orthogonal test for the row cleaner. According to the value range of the installation angle of the straw cleaning disc, the forward inclinations of the straw cleaning disc are designed to be 50°, 60°and 70°, and the horizontal declinations are 30°, 37.5°and 45°. The machine's forward speed is designed to be 6 km/h, 8 km/h, and 10 km/h, using the Design-Expert 10, using the three-factor and three-level Box–Behnken test; the test factors and levels are shown in Table 1.

Table 1. Factors level coding table.

Code Value	The Test Factors		
	α/(°)	**β/(°)**	**v_0/(km/h)**
−1	50	30	6
0	60	37.5	8
1	70	45	10

2.4. *EDEM Simulation Test*

The 3D model of the disc teeth staggered row cleaner was established, and the multi-body coupling dynamics model of the interaction of "row cleaner-soil-straw" was established by discrete element simulation software EDEM 2018. In the actual operation process, the effect of straw cleaning is not only related to the forward inclination and horizontal declination of the installation of the straw cleaning discs, but it is also affected by the forward speed of the machine v_0 [31–33]. Therefore, taking the forward inclination, horizontal declination, and forward speed as the test factors and taking the straw cleaning rate and working resistance as the main response indexes, the operation performance of the straw cleaning device was simulated. Moreover, the rationality of the design of the structure parameters and motion parameters of the row cleaner was verified by the three-factor and three-level rotation orthogonal test in pursuit of the optimal combination of structural parameters and making preliminary preparations for field experiments.

2.4.1. Setting of Simulation Test Conditions

Use the software SolidWorks 2016 to carry out 1:1 solid modeling of the key components of the row cleaner, delete the parts that are not important to the operation process to simplify the row cleaner, and save it in *.STL* format, and import it into the Geometry of

EDEM, the simulation model shown in Figure 9 is obtained. The material is 45 steel, the Poisson's ratio is 0.31, the shear modulus is 7.8×10^{10} Pa, and the density is 7800 kg/m^3.

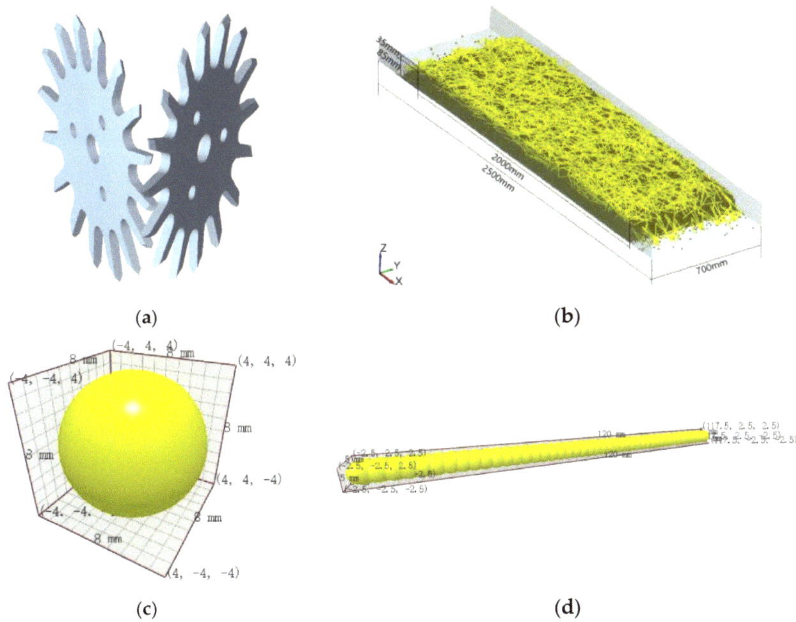

Figure 9. Geometric model of row cleaner and virtual model of soil bin. (**a**) Simulation model of straw cleaner; (**b**) model of soil bin; (**c**) soil particle model; (**d**) straw particle model.

2.4.2. Construction of the Simulation Model

In order to simulate the field test conditions and optimize the simulation process, the soil particles, straw particles, and soil bin models were reasonably simplified. A small ball with a diameter of 8 mm was used as the soil particle model with a Poisson's ratio of 0.38, a shear modulus of 1×10^6 Pa, and a density of 1850 kg/m^3. According to the actual straw coverage in the field, the straw was sampled and observed. Finally, 47 small balls with a diameter of 5 mm and a distance between the centers of 2.5 mm were selected to form particles with a length of 120 mm as the particle simulation model of straw. Its Poisson's ratio is 0.4, the shear modulus is 1×10^6 Pa, and the density is 241 kg/m^3. The soil of the Huang-Huai-Hai double-cropping area of China is loess type, and the soil is loose. Based on a comprehensive literature review [26,34], the mechanical relationship model between particles is set as the Hertz–Mindlin non-sliding contact model. The contact parameters of the simulated materials are shown in Table 2.

Table 2. Material contact parameters for simulation.

Parameter	Rolling Friction Coefficient	Static Friction Coefficient	Coefficient of Restitution
part-soil	0.05	0.6	0.6
part-straw	0.01	0.3	0.6
soil-straw	0.05	0.3	0.5
soil-soil	0.4	0.6	0.6
straw-straw	0.3	0.3	0.5

The EDEM 2018 is used to establish a virtual simulation soil bin. According to the measurement of the field straw coverage before corn sowing in the Huang-Huai-Hai area of China, the size of the particle factory of the soil layer and the straw layer is set to

2000 mm × 700 mm, the thickness of the soil layer is set to 95 mm, and the thickness of the straw layer is set to 35 mm. The size of the soil bin is set to 2500 mm × 700 mm × 140 mm to ensure that sufficient particles can be generated in the soil bin for simulation, as shown in Figure 9. Removal of the upper plane of the soil trough and the two planes in the forward direction of the machine was conducted to ensure a more accurate particle-cleaning effect. The soil particles are only free to settle and stack under their own weight, and the required load is calibrated above the soil particles so that it can achieve the same effect as the actual soil, ensuring the accuracy of the simulation process [35,36].

2.4.3. Simulation Test Process

According to the design requirements, set the straw cleaning discs into the soil (straw) depth of 55 mm; set the grid size to automatic identification and setting; set the fixed time step of the virtual simulation process to the automatic time step, and the total time is 11 s.

As shown in Figure 10, at the beginning of the virtual simulation process, the row cleaner is located on the side of the soil bin, soil particles are generated in 0~3 s, straw particles are generated in 4~8 s, and the straw cleaning process is performed in 9~11 s. Among them, 3~4 s and 8~9 s are set as intermittent times to ensure that the performance of soil particles and straw particles can reach the preset effect.

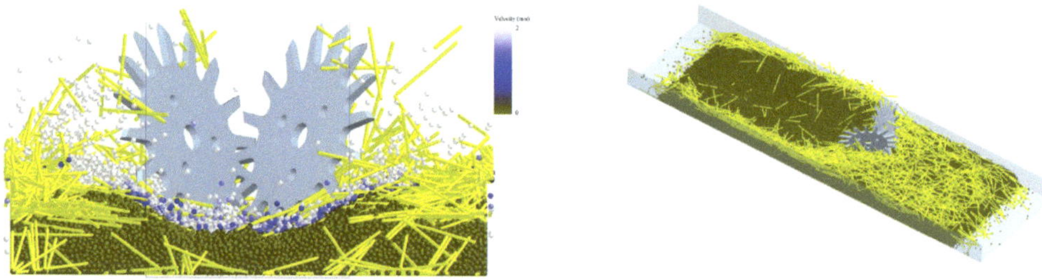

Figure 10. EDEM simulation of the working process of the straw cleaner.

2.4.4. Simulation Testing Methods

Referring to "*Technical Specifications of Quality Evaluation for No-tillage Drilling Machinery*" (NY/T1768-2009) [37], combined with the actual corn sowing operation requirements, the straw cleaning rate and the working resistance of the straw cleaning discs were selected as the evaluation level of the test.

1. Working resistance

The working resistance of the straw cleaning discs can be monitored by using the graph module in the analyst option of the EDEM 2018, and the statistical change graph of its "Force-Time" can be drawn, and the average value of its force can be calculated, as shown in Figure 11.

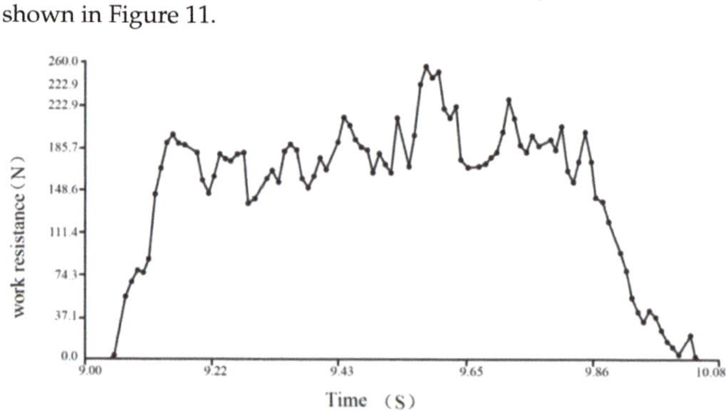

Figure 11. Force-Time chart of row cleaner.

2. Straw cleaning rate

The quantitative change of straw particles before and after the virtual simulation operation can be obtained in the solve report module of the EDEM 2018. As shown in Figure 12, a suitable cleaning area is selected for quantitative calibration, and the straw cleaning rate in the area is calculated.

$$\eta = \left(1 - \frac{N_1}{N}\right) \times 100\% \quad (9)$$

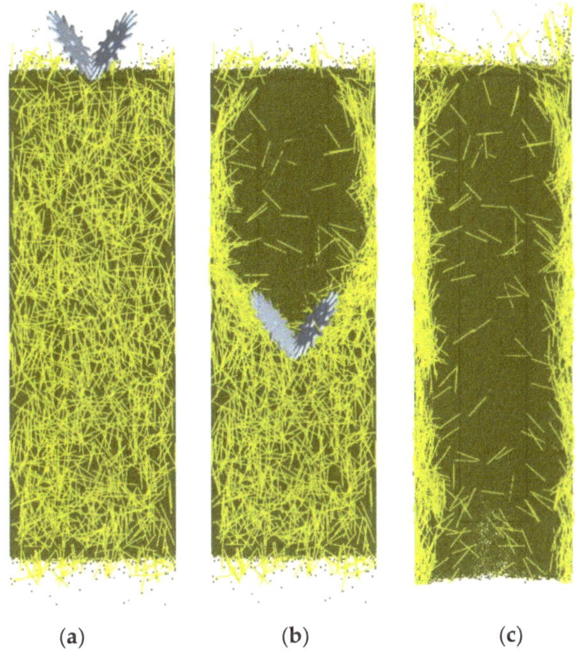

Figure 12. Force-Time chart of row cleaner. (**a**) 8.99 s; (**b**) 9.54 s; (**c**) 10.34 s.

In the formula: η—straw cleaning rate, %;
N—quantity of straw particles before virtual simulation operation;
N_1—quantity of straw particles after virtual simulation operation.

3. Results and Analysis

3.1. Discrete Element Simulation Test Results and Optimization

The simulation test scheme and results are shown in Table 3, and Design-Expert was used for data processing and statistical analysis.

Table 3. Test plan and experimental.

Test Serial Number	Test Factors			Test Index	
	Forward Inclination α/°	Horizontal Declination β/°	Forward Speed v_0/m·s^{-1}	Straw Cleaning Rate Y_1/%	Working Resistance Y_2/N
1	0	0	0	90.50	184.0
2	1	0	1	91.94	259.8
3	−1	0	−1	89.49	220.7
4	−1	−1	0	88.91	152.7
5	−1	1	0	89.34	167.5
6	0	0	0	91.10	186.2
7	1	−1	0	90.79	138.4
8	0	−1	−1	85.87	112.2
9	0	0	0	90.70	180.5
10	0	−1	1	92.52	166.7
11	0	1	1	91.80	193.2
12	−1	0	1	91.22	231.6
13	1	1	0	92.52	226.9
14	0	1	−1	90.64	198.8
15	1	0	−1	90.50	181.1
16	0	0	0	90.90	179.7
17	0	0	0	90.20	181.8

Quadratic regression analysis and multiple regression fitting were carried out on test data, significant influencing factors were screened out, the regression equation of the test index straw cleaning rate Y_1 and working resistance Y_2 was obtained, and their significance was tested. The constant in the equation is the result of software fitting.

The variance analysis of the straw cleaning rate is shown in Table 4. The insignificant factors in the quadratic term and the square term are excluded ($p = 0.05$), and the overall model of the test is extremely significant ($p < 0.01$). The factors α, v_0, and βv_0 were extremely significant, and the factor β was significant at $p = 0.05$. The significance of each factor from large to small is the forward speed, the forward inclination, and the horizontal declination of the straw cleaning disc. The regression equation of straw cleaning rate Y_1 is:

$$Y_1 = 90.53 + 0.85\alpha + 0.78\beta + 1.37v_0 - 1.37\beta v_0 \tag{10}$$

Table 4. Variance analysis of cleaning rate.

Source of Variance	Sum of Square	Degrees of Freedom	Mean Square	*F*	*p*
model	33.19	4	8.30	15.74	0.0001
α	5.76	1	5.76	10.93	0.0063
β	4.82	1	4.82	9.14	0.0106
v_0	15.07	1	15.07	28.58	0.0002
βv_0	7.54	1	7.54	14.29	0.0026
residual	6.33	12	0.53		
lack of fit	5.84	8	0.73	5.98	0.0508
sum	39.52	16			

Upon analyzing the straw cleaning rate Y_1, it can be seen from Formula (10) that the horizontal declination and the forward speed of the machine have an interactive effect on the straw cleaning rate. The fixed forward inclination is 0 level, it can be obtained that within the preset value range, the straw cleaning rate is positively correlated with the forward speed of the machine and negatively correlated with the horizontal declination; that is, the smaller the horizontal declination, the faster the forward speed of the machine, the higher the straw cleaning rate, and the more significant the effect of the forward speed of the machine on it. By selecting different levels of forward declination angle, it can be obtained that the straw cleaning rate is positively correlated with horizontal declination.

The variance analysis of working resistance Y_2 is shown in Table 5, and the overall model of the test is extremely significant ($p < 0.01$). The factors β, v_0, $\alpha\beta$, αv_0, α^2, β^2, and ${v_0}^2$ were extremely significant, and the factor βv_0 was significant at $p = 0.05$. The significance of each factor from large to small is the horizontal declination β, the forward speed of the machine v_0, and the forward α. The regression equation is:

$$Y_2 = 182.44 + 4.21\alpha + 27.05\beta + 17.31v_0 + 18.43\alpha\beta + 16.95\alpha v_0 - 15.03\beta v_0 + 22.26\alpha^2 - 33.32\beta^2 + 18.60v_0^2 \quad (11)$$

Table 5. Variance analysis of operation.

Source of Variance	Sum of Square	Degrees of Freedom	Mean Square	F	p
Model	19,623.49	9	2180.39	23.61	0.0002
α	141.96	1	141.96	1.54	0.2549
β	5853.62	1	5853.62	63.40	<0.0001
v_0	2397.78	1	2397.78	25.97	0.0014
$\alpha\beta$	1357.92	1	1357.92	14.71	0.0064
αv_0	1149.21	1	1149.21	12.45	0.0096
βv_0	903.00	1	903.00	9.78	0.0167
α^2	2085.41	1	2085.41	22.59	0.0021
β^2	4674.62	1	4674.62	50.63	0.0002
${v_0}^2$	1457.46	1	1457.46	15.78	0.0054
residual	646.35	7	92.34		
lack of fit	618.10	3	206.03	29.17	0.0035
sum	20,269.84	16			

Upon analyzing the working resistance Y_2, it can be seen from Formula (10) that the interaction factors between the factors have a significant impact on the working resistance. The forward speed of the machine is 0 level, and it can be obtained that within the preset value range, the larger the forward inclination and the smaller the horizontal declination, the smaller the working resistance will be. When the forward speed is smaller, the contact time between the straw cleaning discs and the soil is longer, and it is easier for the soil to accumulate in the front, resulting in an overall increase in the working resistance. However, when the forward speed is higher, the throwing effect of the straw cleaning discs on the soil is reduced, and the forward dozing effect is enhanced, which leads to an increase in the working resistance.

In order to obtain the optimal combination of structure and motion parameters of the disc teeth staggered row cleaner, the optimal design of the experimental factors was carried out, and the principle of improving the straw cleaning rate and reducing the working resistance was followed, according to the boundary conditions of the experimental factors and the actual operation experience, the multi-objective variable optimization method is used to establish the parameter-constrained objective function model.

$$\begin{cases} \max y_1(\alpha, \beta, v_0) \\ \min y_2(\alpha, \beta, v_0) \\ s.t. \begin{cases} 50^\circ \leq \alpha \leq 70^\circ \\ 30^\circ \leq \beta \leq 45^\circ \\ 6\ \mathrm{km/h} \leq v_0 \leq 10\ \mathrm{km/h} \end{cases} \end{cases} \quad (12)$$

Based on the above analysis, it can be seen that the larger the forward inclination, the smaller the horizontal declination, the faster the forward speed of the machine, and the higher the straw cleaning rate; but at the same time, the larger the forward inclination, the smaller the horizontal declination, and the speed of the machine is 0 level, working resistance is minimal. Based on Design-Expert, the parameters in the objective function are optimized and solved, and a set of reasonable optimization parameter combinations are selected from the results according to the actual operation requirements. The optimal combination obtained is $\alpha3\beta1v_02$, that is, the forward inclination is 70°, the horizontal declination is 30°, and the forward speed of the machine is 8 km/h. At this time, the straw cleaning rate is 90.60%, and the working resistance is 130 N. According to the optimization results, the virtual simulation verification test was carried out. The straw cleaning rate was 90.87%, and the working resistance was 136.4 N, which was basically consistent with the optimization results.

3.2. Field Test Verification

In order to verify the actual operating performance of the disc teeth staggered row cleaner and its advantages compared with the traditional flat disc separated row cleaner, a soil bin test was conducted in the soil bin laboratory of Shandong University of Technology in October 2021. The soil type was loam, and the temperature in the laboratory was 20~26 °C during the experiment.

Before the test, the straw was pre-laid on the ground, and the laying range was a long strip area of 20 m × 1 m, and then the parameters of straw and soil were measured. Table 6 shows the main parameters obtained in the soil bin. Among them, soil moisture content and soil temperature were measured with a TZS-IIW soil moisture and temperature measuring instrument; soil compaction was measured with a TJS-450G soil compaction instrument.

Table 6. Soil tank test.

	Parameter	Numerical Value
0~100 mm soil layer	compaction/MPa	0.658
	moisture content/%	50.03
	test weight/(g·cm^{-3})	1.52
	temperature/°C	27.78
surface straw	length/mm	50~200
	Cover thickness/mm	30~80
	moisture content/%	26.73
	unit coverage/(kg·m^{-2})	0.9375

3.2.1. Test Methods

As shown in Figure 13, the disc teeth staggered row cleaner and the flat disc separated row cleaner were installed on the frame of the soil bin testing vehicle, respectively, so that the test could be carried out under the same working conditions. The test was performed by aligning the row cleaner with the preset ditching area, adjusting the immersion depth of 55 mm, and setting the forward speed of the soil bin testing vehicle to 8 km/h. The following indicators were measured and calculated accordingly during and after the test.

1. Straw cleaning rate of seedbed

Before the start of the test, the five-point sampling method was used to select points in the stable operation area randomly, and the electronic scales were used to weigh all the straws in the same area of each sampling point before and after the operation to obtain the initial mass G_0 and the post-operation mass G_1, the straw cleaning rate in the measurement area is further calculated. Each group of row cleaner test was repeated five times, and the average cleaning rate of each group was obtained [38].

$$\lambda = \left(1 - \frac{G_1}{G_0}\right) \times 100\% \tag{13}$$

In the formula: λ—straw cleaning rate of the seedbed, %;
G_0—straw quality in the frame before operation, kg;
G_1—straw quality in the frame after operation, kg.

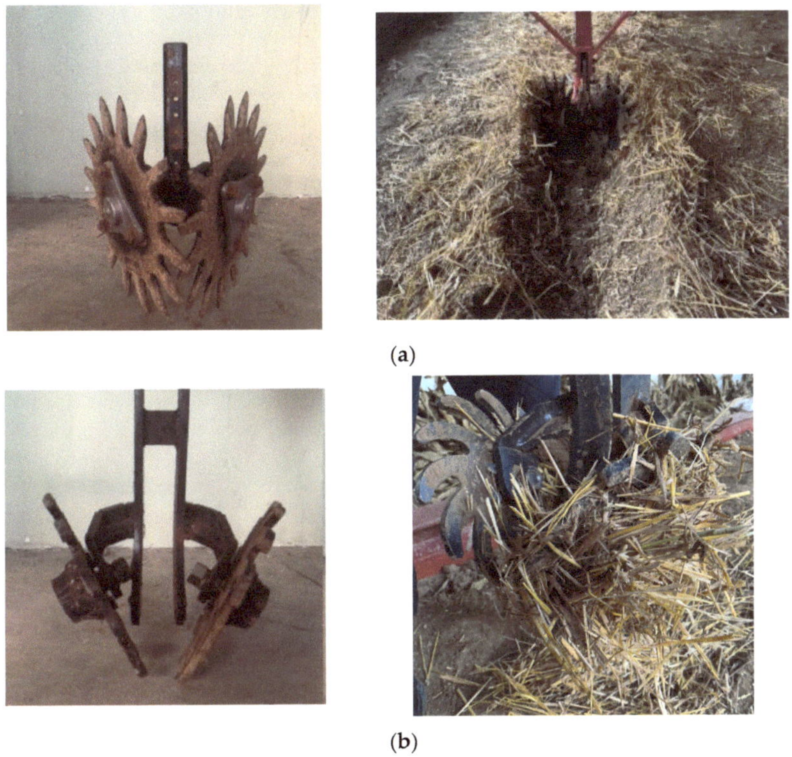

Figure 13. Passability comparison. (a) Disc teeth staggered row cleaner and straw cleaning operation; (b) Traditional flat disc separated row cleaner and straw cleaning operation.

2. Working width

After the test, use the five-point sampling method to select five observation points, measure the working width of the points with a tape measure, and take the average value, as shown in Figure 14.

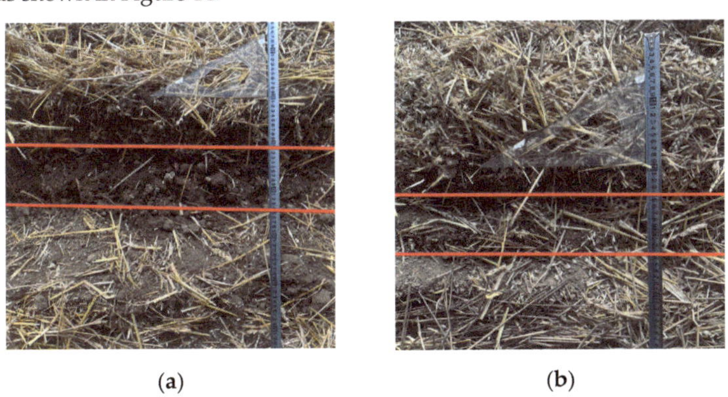

Figure 14. Row cleaning effect comparison. (a) Coil-tooth staggered; (b) Flat wheel split type. (The red line in the figure indicates the width of the operation).

3. Passability

During the process, the row cleaner is at the preset speed, and the operation process of the teeth of the disc staggered row cleaner and the flat disc separated row cleaner are monitored. According to the agricultural industry standard "*Technical Specifications of Quality Evaluation for No-tillage Drilling Machinery*" (NY/T1768-2009) [37] and the performance testing requirements of no-till planters at the agricultural machinery appraisal station of the Ministry of Agriculture, observe whether the two sets of row cleaners can complete the operation normally, and the number of times of blockage or stoppage. The number of rotations was carried out five times for each group of row cleaners, and their performance was compared.

3.2.2. Test Results and Analysis

The test results are shown in Table 7. During the soil bin test, the disc teeth staggered row cleaner did not become blocked and stopped, the average cleaning rate was 90.59%, and the working width was 169.6 mm; There were two minor blockages and one moderate blockage during the operation of the flat disc separated row cleaner, the average cleaning rate was 65.65%, and the working width was 90.8 mm. It can be seen that the disc teeth staggered row cleaner forms a meshing structure similar to "scissors" in the middle of the seedling belt, forming a staggered and overlapping state, which is more conducive to tearing the straw in the middle and throwing it to the two sides. Therefore, the leakage area easily formed by the traditional flat disc separated row cleaner is eliminated, and the straw within the working width is more fully cleaned so as to further improve the straw cleaning rate of the seedbed. At the same time, the cutting edge set on the top of the claw teeth of the straw cleaning discs of the disc teeth staggered row cleaner is more helpful for cutting the straw and cutting into the soil, and the operation performance is better.

Table 7. Test results.

Device Type	Straw Cleaning Rate/%		Working Width/mm		Passability
	Measurements	**Average Value**	**Measurements**	**Average Value**	
disc teeth staggered row cleaner	91.33	90.59	156	169.6	No blockage
	89.57		188		No blockage
	92.31		169		No blockage
	88.94		162		1 minor blockage
	90.80		173		No blockage
flat disc separated row cleaner	64.53	65.65	93	90.8	1 minor blockage
	73.06		90		No blockage
	70.27		85		1 moderate blockage
	60.90		88		1 minor blockage
	61.49		98		No blockage

In addition, since the straw cleaning discs are set so that the claws are staggered, if one side of the straw cleaning disc is blocked and stopped during the operation, the other sides discs will be able to provide torque to it through the meshing point, causing it to break the blockage and resume the rotation again. Therefore, compared with the multiple blockages found in the control group during the test, the experimental group did not have such a situation.

Upon comparing data obtained after the soil bin test with the results obtained from the previous virtual simulation experiment, it can be seen that the results of the two are basically the same, but the straw cleaning rate of the soil bin test is slightly lower than that of the simulation experiment. The reason may be that the soil moisture content is too large during the soil bin test, while the straw moisture content is slightly smaller, but the error mean is acceptable. The soil bin test results show that the disc teeth staggered row cleaner can achieve a good operation effect, and the operation quality can meet the agronomic requirements of no-tillage seeding operation.

4. Discussion

This article studies methods to improve the effectiveness of straw cleaning and blockage prevention during the zero-tillage sowing process of corn in the double cropping area. As far as we know, there is currently relatively little research on interlocking disc tooth type-opposed straw-cleaning devices, which have a certain degree of innovation and scientificity at present. Jia et al. [13] designed and developed a concave disc-type straw cleaning and anti-blocking device. This device uses a front and rear misaligned concave disc, which can increase the overlapping operation area, reduce the missed cleaning area, and throw the straw away along the concave curve, effectively carrying out straw cleaning operations. The similarity with this design lies in the use of opposed discs as key components and the addition of overlapping work areas in the middle through different methods to reduce missed areas. The difference lies in the structures of the straw cleaning disc, and the principle of reducing the leakage area is also different. The specific performance is as follows: This design innovation has designed the installation method of the straw cleaning disc, with the disc teeth alternating with each other, reducing the leakage area and improving the occurrence of congestion. Due to the differences in straw coverage and soil conditions, the experimental results are not comparable. From the soil tank comparison experiment, it can be seen that the average straw cleaning rate of the traditionally opposed disc non-staggered straw cleaning device is 65.65%, and there has been congestion during the experiment process; The average straw cleaning rate of the disc tooth staggered straw cleaning device is 90.80%, and there was no significant congestion during the experimental process, resulting in better operational performance.

This study has conducted some innovative device structural designs, but there are also some limitations. The specific content is as follows.

The mainstream discrete element model of wheat straw is currently a rigid body. In the future, detailed parameters should be determined through physical property tests, and a flexible test model should be set to further improve simulation accuracy.

The passive straw cleaning device designed in this article still carries the risk of straw congestion. It is necessary to further design a low-energy consumption and high-performance active straw cleaning device to further reduce congestion and improve sowing accuracy.

5. Conclusions

1. Complete the overall structural design of the straw cleaning and anti-blocking device of the disc teeth staggered seeder, which is mainly composed of two circular straw cleaning disks installed alternately by disc teeth through appropriate installation angle adjustment. Through kinematics analysis, establish its collaborative parameter model, verify the feasibility of the device through discrete element model establishment and simulation tests, and find out the best operating parameter combination of the combinations available. It can greatly improve the congestion situation during the corn no-tillage sowing process;
2. The key collaborative operation parameters of the device were optimized through regression analysis. When $\alpha = 70°$, $\beta = 30°$, and $v_0 = 8$ km/h, the device has the highest straw cleaning rate (95.87%);
3. In order to verify the reliability, scientificity, and feasibility of the straw cleaning and blocking prevention device of the disc tooth staggered seeder, a soil groove test was conducted. Compared with the traditional double disc opposed non-staggered straw cleaning device, when the straw coverage is 0.9375 kg/m^2, the disc tooth staggered device does not block, and the average cleaning rate is 90.59%. The operating effect is significantly better than the traditional double disc opposed non-staggered straw cleaning device, verifying the feasibility of the disc tooth staggered device. It is basically consistent with the results of the discrete element simulation test, which verifies the scientificity, feasibility, and accuracy of the quadratic polynomial regression model.

Author Contributions: Conceptualization, Y.Z. (Yinping Zhang); methodology, Y.Z. (Yinping Zhang) and X.L.; software, X.L., H.H. and B.W.; validation, Y.Z. (Yinping Zhang), H.Z. and D.G.; formal analysis, Y.Z. (Yinping Zhang) and Y.Z. (Yuzi Zhang); writing—original draft preparation, X.L.; writing—review and editing, Y.Z. (Yinping Zhang). All authors have read and agreed to the published version of the manuscript.

Funding: The work was supported by the National Key R&D Program of China (2021YFD2000401-2), the Innovation Capability Enhancement Project for Technological Small and Medium-sized Enterprises (2023TSGC0339), the Modern Agricultural Industrial System of Shandong Province (SDAIT-02-12).

Institutional Review Board Statement: Not applicable.

Data Availability Statement: The data presented in this study are available on request from the corresponding author.

Conflicts of Interest: The authors declare no conflict of interest.

References

1. Zhang, Y.P.; Wang, Z.W.; Diao, P.S.; Du, R.C.; Li, H.W.; Yao, W.Y. Short-term effects of mechanical and ecological tillage pattern on physical and chemical characteristics of soil in arid area. *Trans. Chin. Trans. Chin. Soc. Agric. Mach.* **2018**, *49*, 45–55.
2. Malasli, M.C.; Ahmet, C. Disc angle and tilt angle effects on forces acting on a single-disc type no-till seeder opener. *Soil Tillage Res.* **2019**, *194*, 104304. [CrossRef]
3. He, D.; Li, H.W.; Chen, H.T.; Lu, C.Y.; Wang, Q.J. Research progress of conservation tillage technology and machine. *Trans. Chin. Soc. Agric. Mach.* **2018**, *49*, 1–19.
4. Hu, H.N.; Li, H.W.; Li, Q.J.; Wang, J.; He, C.Y.; Lu, Y.B.; Liu, W.P. Anti-blocking performance of ultrahigh-pressure waterjet assisted furrow opener for no-till seeder. *Int. J. Agric. Biol. Eng.* **2020**, *13*, 64–70. [CrossRef]
5. Zhang, Y.P.; Du, R.C.; Diao, P.S.; Yang, S.D. Experiment of no-tillage and drought direct sowing rice and feasibility analysis in Shandong Province. *Trans. Chin. Soc. Agric. Eng.* **2016**, *32*, 24–30.
6. Shinoto, Y.; Otani, R.; Matsunami, T.; Maruyama, S. Analysis of the shallow root system of maize grown by plowing upland fields converted from paddy fields: Effects of soil hardness and fertilization. *Plant Prod. Sci.* **2021**, *24*, 297–305. [CrossRef]
7. Mairghany, M.; Yahya, A.; Adam, N.M.; Su, A.S.; Aimrun, W.; Elsoragaby, S. Rotary tillage effects on some selected physical properties of fine textured soil in wetland rice cultivation in Malaysia. *Soil Tillage Res.* **2019**, *194*, 104318. [CrossRef]
8. Zhang, X.R.; Guo, L.; Li, H.W.; He, J.; Zhang, L.R. Experiment on the anti-blocking technology in no-tillage wheat seeding. *Adv. Mater. Res.* **2012**, *1671*, 472–475. [CrossRef]
9. Latifmanesh, H.; Deng, A.; Nawaz, M.M.; Li, L.; Chen, Z.; Zheng, Y.; Wang, P.; Song, Z.; Zhang, J.; Zheng, C.; et al. Integrative impacts of rotational tillage on wheat yield and dry matter accumulation under corn-wheat cropping system. *Soil Tillage Res.* **2018**, *184*, 100–108. [CrossRef]
10. Hozayn, M.; Elaoud, A.; Attia, A.A.; Ben, S.N. Effect of magnetic field on growth and yield of barley treated with different salinity levels. *Arab. J. Geosci.* **2021**, *14*, 701. [CrossRef]
11. Jalel, R.; Elaoud, A.; Ben, S.N.; Chehaibi, S.; Ben, H.H. Modeling of soil tillage techniques using Fruchterman–Reingold Algorithm. *Int. J. Environ. Sci. Technol.* **2021**, *18*, 2987–2996. [CrossRef]
12. Ben, H.H.; Elaoud, A.; Masmoudi, K. Modeling of agricultural soil compaction using discrete Bayesian networks. *Int. J. Environ. Sci. Technol.* **2020**, *17*, 2571–2582.
13. Jia, H.L.; Liu, X.; Yu, H.B.; Lu, Y.; Guo, C.J.; Qi, J.T. Simulation and experiment on stubble clearance mechanism with concave claw-type for no-tillage planter. *Trans. Chin. Soc. Agric. Mach.* **2018**, *49*, 68–77.
14. Liu, Z.P.; Tian, M.; Yang, S.X.; He, R.Y. Design of a grass separation equipment with driven rotating tooth disk. *J. South China Agric. Univ.* **2018**, *39*, 120–124.
15. Wang, Q.; Jia, H.L.; Zhu, L.T.; Li, M.W.; Zhao, J.L. Design and experiment of star-toothed concave disk row cleaners for no-till planter. *Trans. Chin. Soc. Agric. Mach.* **2019**, *50*, 68–77.
16. Wang, Q.; Tang, H.; Zhou, W.Q.; Wang, J.W. Design and experiment of automatic width control row cleaners. *Trans. Chin. Soc. Agric. Mach.* **2021**, *52*, 25–35.
17. Zhang, Y.P.; Du, R.C.; Diao, P.S.; Yang, S.D.; Wang, Z.W. Design and experiment of wide band seeding rice seeder with reversed stubble cleaning and anti-blocking. *Trans. Chin. Soc. Agric. Eng.* **2017**, *33*, 7–13.
18. Wang, Q.J.; Li, H.W.; He, J.; Li, W.Y.; Liu, A.D. Effects of wide-ridge and narrow-row no-till cultivation on soil water and maize yield. *Trans. Chin. Soc. Agric. Eng.* **2010**, *26*, 39–43.
19. Galibjon, M.; Dimitris, S.; Alim, S.; Hans, W. Dynamic performance of a no-till seeding assembly. *Biosyst. Eng.* **2017**, *158*, 64–75.
20. Zhang, B.P. *Sowing Machinery Design Principle*; China Machine Press: Beijing, China, 1982; pp. 389–403.
21. Matin, M.A.; Fielke, J.M.; Desbiolles, J.M. Torque and energy characteristics for strip-tillage cultivation when cutting furrows using three designs of rotary blade. *Biosyst. Eng.* **2015**, *129*, 329–340. [CrossRef]
22. Lin, J.; Liu, A.D.; Li, B.F.; Li, B.; Zhao, D.F.; Lu, C.Y. 2BG-2 type corn ridge planting no-till planter. *Trans. Chin. Soc. Agric. Mach.* **2011**, *42*, 43–46.

23. Bao, W.Y. Study on Key Parts and Holistic Device of the No-till Planter of the Ridge Cropping System in Northeast Area of China. Ph.D. Thesis, Shenyang Agricultural University, Shenyang, China, 2009.
24. Wang, W.W.; Zhu, C.X.; Chen, L.Q.; Li, Z.D.; Huang, X.; Li, J.C. Design and experiment of active straw-removing anti-blocking device for maize no-tillage planter. *Trans. Chin. Soc. Agric. Eng.* **2017**, *33*, 10–17.
25. Chinese Academy of Agricultural Mechanization Sciences. *Agricultural Machinery Design Manual;* China Agricultural Science and Technology Press: Beijing, China, 2007; pp. 288–299.
26. Cao, X.P.; Wang, Q.J.; Li, H.W.; He, J.; Lu, C.Y. Combined row cleaners research with side cutter and stubble clean disk of corn no-till seeder. *Trans. Chin. Soc. Agric. Mach.* **2021**, *52*, 36–44.
27. Tagar, A.A.; Ji, C.Y.; Adamowski, J.; Malard, J.; Qi, C.S.; Ding, Q.S.; Abbasi, N.A. Finite element simulation of soil failure patterns under soil bin and field testing conditions. *Soil Tillage Res.* **2015**, *145*, 157–170. [CrossRef]
28. Wang, Q.; He, J.; Li, H.; Lu, C.; Rabi, G.R.; Su, Y. Design and experiment on furrowing and anti-blocking unit for no-till planter. *Trans. Chin. Soc. Agric. Eng.* **2012**, *28*, 27–31.
29. Jiang, J.L.; Gong, L.N.; Wang, D.W.; Wang, G.P. Design and experiment for driving double coulters anti-blockage device of no-till planter. *Trans. Chin. Soc. Agric. Eng.* **2012**, *28*, 17–22.
30. Fan, X.H.; Jia, H.L.; Zhang, W.H.; Yang, H.T.; Gu, Y.Q.; Li, H.G. Parametric analysis of finger-type anti-blocking residue-cleaner for no-till planting. *Trans. Chin. Soc. Agric. Mach.* **2011**, *42*, 56–60.
31. Hou, S.Y.; Chen, H.T.; Zou, Z.; Wei, Z.P.; Zhang, Y.L. Design and test of lateral stubble cleaning blade for corn stubble field. *Trans. Chin. Soc. Agric. Eng.* **2020**, *36*, 59–69.
32. Cao, X.P.; Wang, Q.J.; Li, H.W.; He, J.; Lu, C.Y.; Yu, X.X. Design and experiment of active rotating collective straw-cleaner. *Trans. Chin. Soc. Agric. Eng.* **2021**, *37*, 26–34.
33. Jia, H.L.; Zhao, J.L.; Jiang, X.M.; Jiang, T.J.; Wang, Y.; Guo, H. Design and experiment of anti-blocking mechanism for inter-row no-tillage seeder. *Trans. Chin. Soc. Agric. Eng.* **2013**, *29*, 16–25.
34. Zeng, Z.W.; Chen, Y. Simulation of straw movement by discrete element modelling of straw-sweep-soil interaction. *Biosyst. Eng.* **2019**, *180*, 25–35. [CrossRef]
35. Niu, M.M.; Fang, H.M.; Chandio, F.A.; Shi, S.; Xue, Y.F.; Liu, H. Design and experiment of separating-guiding anti-blocking mechanism for no-tillage maize planter. *Trans. Chin. Soc. Agric. Mach.* **2019**, *50*, 52–58.
36. Yao, W.Y.; Zhao, D.B.; Xu, G.F.; Chen, M.Z.; Miao, H.Q.; Diao, P.S. Design and experiment of anti-blocking device for strip to row active corn no-tillage seeding. *Trans. Chin. Soc. Agric. Mach.* **2020**, *51*, 55–62.
37. Ministry of Agriculture and Rural Affairs of the People's Republic of China. *Technical Specifications of Quality Evaluation for No-tillage Drilling Machinery;* Standards Press of China: Beijing, China, 2009; Available online: https://www.chinesestandard.net/PDF/English.aspx/NYT1768-2009 (accessed on 12 June 2022).
38. Chen, H.T.; Wei, Z.P.; Su, W.H.; Hou, S.Y.; Ji, W.Y.; Shi, N.Y. Design and experiment of cleaning and anti-blocking of front-mounted seed bed preparation device for grand ridge with raw stubble. *Trans. Chin. Soc. Agric. Mach.* **2021**, *52*, 51–60.

agriculture

MDPI

Article

Tribological Study of Chisel Knives in Sandy Soil †

Laurențiu Constantin Vlăduțoiu [1], Georgiana Chişiu [2,*], Andrei Tudor [2], Nicolae-Valentin Vlăduț [1], Lucian Fechete Tutunaru [3], Eugen Marin [1] and Iulia-Andrea Grigore [1]

1 National Institute for Research, Development of Machines and Installations Designed for Agriculture and Food Industry, INMA Bucharest, 077190 Bucharest, Romania; laurentiuvladutoiu82@gmail.com (L.C.V.); vladut@inma.ro (N.-V.V.)

2 Department of Machine Elements and Tribology, University Politehnica of Bucharest, Splaiul Independenței 313, 060042 Bucharest, Romania; andreitudor1947@gmail.com

3 Department of Automotive Engineering and Transports, Technical University of Cluj-Napoca, 400641 Cluj-Napoca, Romania

* Correspondence: georgiana.chisiu@upb.ro

† This paper is a part of the PhD Thesis of Laurențiu Constantin Vlăduțoiu, presented at University Politehnica of Bucharest (Ro).

Abstract: This paper presents the interaction system within the mechanical soil processing process, consisting of two large elements, the metal of the tool and the soil. Due to the two main forces acting on the chisel knives—friction and impact with the sandy soil—the wear of these chisel knives was determined. To determine the wear, a stand was used which allowed testing chisel-type knives in laboratory conditions by changing their functional parameters: working depth, angle of the knives to work the soil, working speed, humidity and granulation of the test environment. The present paper presents an application of the Archard-type wear law to the contact between a chisel-type knife and sandy soil (wet and dry sand). The theoretical model regarding the Archard wear coefficient considered three forms of surface damage (shake down, ratcheting and micro-cutting). The sand was considered spherical and rigid and the surface of the knife was flat. The experimental model considered real steel knives with different surface hardness and operation under controlled conditions of sand granulation, humidity, attack angle, depth of penetration and speed of sliding. The theoretical and experimental results highlight the wear behavior of chisel knives (Archard coefficient) in wet and dry sand.

Keywords: soil; abrasion wear; active parts; shakedown; ratcheting; micro-cutting; Archard wear coefficient; sandy soil

Citation: Vlăduțoiu, L.C.; Chişiu, G.; Tudor, A.; Vlăduț, N.-V.; Fechete Tutunaru, L.; Marin, E.; Grigore, I.-A. Tribological Study of Chisel Knives in Sandy Soil. *Agriculture* **2023**, *13*, 1235. https://doi.org/10.3390/agriculture13061235

Academic Editors: Chung-Liang Chang and Mustafa Ucgul

Received: 15 May 2023
Revised: 5 June 2023
Accepted: 7 June 2023
Published: 12 June 2023

1. Introduction

The problem of globalization with the development of agriculture is more than necessary to solve. It is supposed that one will have good agricultural machines with high technology, but sometimes the performance depends on the good working of active elements by making them from materials with long-term lifespans which are resistant to wear. Sandy soils are good for some plants, but with a good percent of water content. This is mainly found in arid and semi-arid regions of the planet, but it is also found in humid areas.

The quality of these soils is mostly relatively easy to work with because it has aerated particles in the composites [1,2].

During soil works, the active elements of agricultural machines such as blades, chisels and discs are exposed to abrasive wear with the soil they come into contact with [3,4]. For this reason, the active elements need to be verified for wear resistance in different situations of working to estimate an average life expectancy of the wear resistance, to ensure the timely replacement of parts [5]. The harder particles a soil contains, the more abrasive it is [6]. Furthermore, the abrasivity of the soil could be increased if it consists of particles with high hardness. In most cases, this hardness is greater than that of the working tool [7].

This fact determines the premature wear of the tiller knives via suffering modification to the geometry of their cutting part. This causes the necessity for replacement and reconditioning works, which can be led to low productivity and an increase in the costs of agricultural works [8–10].

Wear appears when two forces, the friction force and the impact force, act on the surface of the chisel knives during the working process [11].

The mechanical tillage of the soil is a complicated working process due to the resistance to soil breakdown and the intense abrasive wear of chisel-type knives [12,13]. Premature tool wear leads to low productivity and high reconditioning and replacement costs [14].

The current concerns of researchers are directed towards finding and applying modern methods of abrasion wear protection [15] and prolonging the life of parts through reconditioning [16], determining the mechanical and wear characteristics of materials [11], characterization of new materials for hardening, etc.

The aim of this study is to make a theoretical and experimental analysis regarding the wear of chisel-type knives operating in sandy soils. Real knives made of steel with surfaces of different hardness and functioning in wet and dry sand are taken into consideration.

2. Theoretical Aspects

2.1. The Contact Area

If the contact pressure is greater than the "shakedown" limit of the blade material and there is relative movement, then the blade material will plastically deform and a mark will form on the surface [17].

For particles with a spherical body—Figure 1a, of radius R_1 and F_n, the normal force taken up by a particle—the definition of Striebeck-type pressure, p_s, can be as the ratio between force and diametrical area:

$$p_s = \frac{F_n}{\pi R_1^2} \tag{1}$$

The maximum pressure in the contact center, p, is:

$$p = \frac{F_n}{\pi a_p^2} = p_S \frac{R_1^2}{a_p^2} = \frac{p_s}{a_{pa}^2} = \frac{H_s}{2} \tag{2}$$

where the radius of the plastic contact circle, a_p, is:

$$a_p = \sqrt{\frac{2F_n}{\pi H_s}} = R_1 \sqrt{\frac{2p_s}{H_s}} \tag{3}$$

$a_{pa} = \frac{a_p}{R_1}$ is the dimensionless radius of contact between the particle and the knife surface.

H_s is the hardness of the knife surface.

For particles with cylindrical segments ("lying" particles):

$$p_s = \frac{F_n}{2R_1 L_p} \tag{4}$$

where L_p is the length of the particle.

In this case, the maximum pressure in the contact center, p, is:

$$p = \frac{F_n}{a_p L_p} = p_s \frac{2}{a_{pa}} = \frac{H_s}{2} \tag{5}$$

where the radius of the plastic contact circle, a_p, is:

$$a_p = \frac{2F_n}{H_s L_p} = \frac{p_s}{H_s} \cdot \frac{2R_1}{L_p} \quad (6)$$

In the case of plastic deformations, the material of the knife is perfectly plastic, and the abrasive particle is perfectly rigid [18]. The scratch resistance of abrasive particles (quartz) is significantly higher (7 on the Mohs scale) than the scratch resistance of steel (4–5 on the Mohs scale). The mechanical behavior of the knife material is assessed by the surface hardness H_s after the material has been worn and the bulk hardness H_b, which remains constant during wear [17,18]. Figure 1b shows the angle of attack of the abrasive particle in contact with the surface of the knife and the schematical Hertzian contact sphere on a flat plane. For the analysis of the knife wear process in the three states of deformation ("shakedown", "plowing", cutting"), the attack angles of the particles are defined [16].

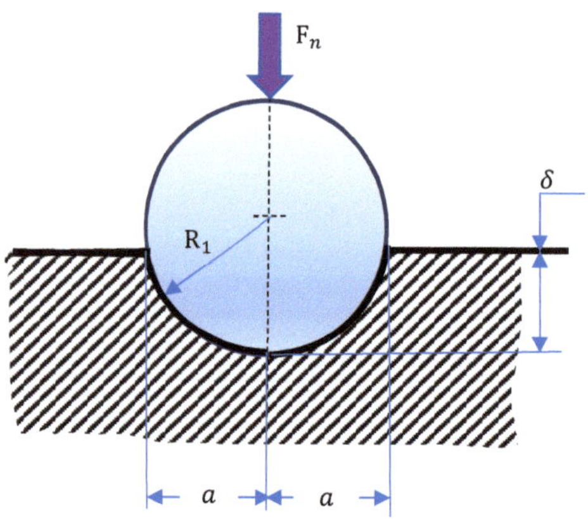

(a)

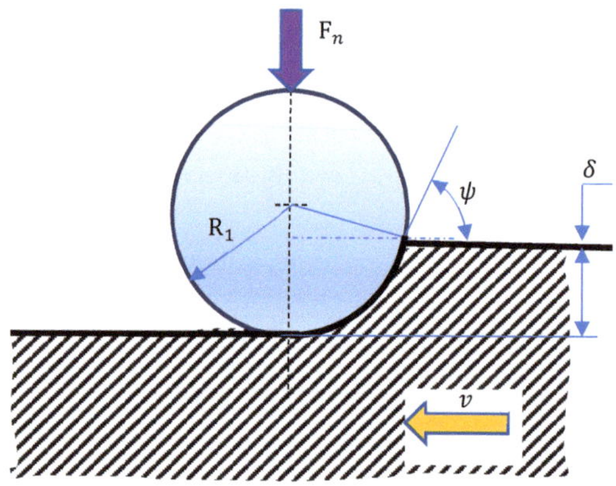

(b)

Figure 1. The contact of a sphere against a surface of chisel knife. (**a**) Hertzian contact sphere on flat plane, when there is an elastic sphere and the flat surface is a rigid plan of a chisel knife. a is contact area radius, R_1 is the radius of a sphere particle and δ is the penetration depth; (**b**) attack angle of the particle with the surface of a chisel knife, ψ v – speed of chisel knife.

For defining the attack angle of a particle with the surface of a chisel knife, there are three types of angles:

ψ_e—the critical angle of attack, up to which the deformation is elastic;

ψ_b—the minimum angle of attack at which furrowing (bordering) occurs;

ψ_a—the minimum angle of attack at which microchipping occurs;

L_c¯ the dimensionless transition distance between two neighboring traces that makes the transition from plowing to micro cutting. Can be written as:

$$L_c = \frac{L_p}{2a} \quad (7)$$

For steels, the solution given by Xie [17] can be applied:

$$\Psi_e(p_{as}) = acos\left[1 - \left(\frac{3\pi}{4}p_{as}\right)^{\frac{2}{3}}\right] \quad (8)$$

where $p_{as} = \frac{p}{\tau_c}$ is dimensionless pressure and τ_c is shear yield strength.

f is the adhesion component of the friction coefficient and depends on the state of lubrication (soil moisture).

$H_r = \frac{H_b}{H_s}$ is the relative hardness of the knife material and indicates the roughening capacity.

The minimum angle of attack at which furrowing (bordering) occurs is:

$$\Psi_b = \frac{1}{10f} \cdot \left[18.6 - (H_r)^4 \cdot 10\right] \cdot \frac{\pi}{180} \tag{9}$$

for $\psi_b \leq \psi \leq \psi_a$

The dimensionless distance between two adjacent traces corresponding to the transition from ploughing to cutting is:

$$L_c = \left(\frac{\psi - \psi_b(f \cdot H_r)}{\psi_a(f \cdot H_r) - \psi_b(f \cdot H_r)}\right)^{H_r} \tag{10}$$

Figure 2 illustrates the dependence of the critical distance between two adjacent traces on the angle of attack, for different ratios of volume and surface hardness of steel.

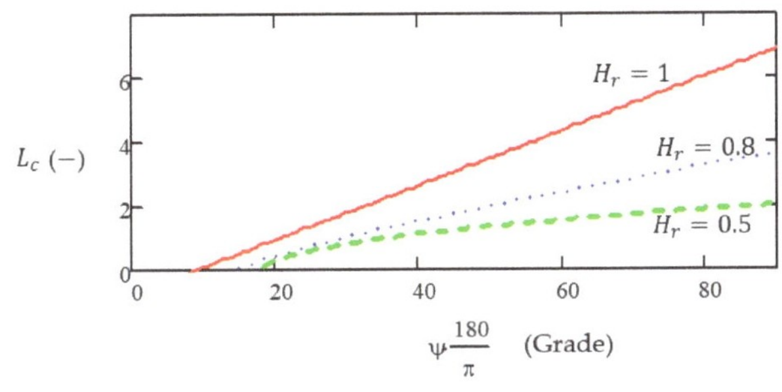

Figure 2. The critical distance between two adjacent tracks of the hard particle as a function of the angle of attack.

The ratio $\frac{H_s}{H_b}$ represents the ability of the hardening material in operation.

For metals [17,18], $H_s = 6 \cdot \tau_s$ τ_s is the shear strength of the material in the surface area; $H_b = 6 \cdot \tau_b$ τ_b is the shear strength of the bulk material.

2.2. The Archard Wear Coefficient

In the wear process, the basic equation is the Archard equation [19]:

$$V = k \cdot \frac{F_n \cdot s}{H_s} \tag{11}$$

where V is the volume of material removed and/or displaced from the contact area, s is the total sliding length and k is the Archard dimensionless wear coefficient (wear intensity).

The Archard wear coefficient [19–21] can be determined analytically when the geometrical characteristics of the abrasive are known, in the case of abrasive-type wear, or the characteristics of the microgeometry of the surfaces, the state of lubrication and the mechanical properties of the material being worn are known, for other forms of wear. The complexity of the phenomena in the friction and wear process requires, in most cases, the experimental determination of the Archard wear coefficient.

With some assumptions regarding the idealization of the shapes of the abrasive particles fixed in the soil, the analytical determination of the deformation component of the friction coefficient and the Archard wear coefficient is proposed. Three forms of deformation of the knife material are defined: folding surface—elastic "shakedown", edging—"ratcheting" and micro-cutting—"micro-cutting". These shapes depend on the geometry of the particles and the mechanical characteristics of the knife material (elastic limit, shear strength and hardness).

Thus, for the three forms of deformation of the knife material (folding surface—elastic "shakedown", edging—"ratcheting", micro-cutting—"micro-cutting"), the coefficient of friction and the coefficient of wear were adapted according to the works [17,18,21–24].

2.2.1. Elastic "Shakedown"

The sliding friction coefficient between the hard particle fixed in the soil and the knife surface [25,26] has only the adhesion component f and is only dependent on the state of lubrication and hysteresis phenomena: $k_{ue} = 0$.

2.2.2. Micro-Cutting

The coefficient of friction depends on the attack angle, the adhesion and the relative position of the adjacent tracks.

Thus,

$$\mu_1 = \left(\frac{2}{\pi}\right)^{0.5} \cdot \frac{tan(\psi)}{{L_a}^{0.25}} \cdot \left[1 - f \cdot \left(1 + \frac{\pi}{4tan(\psi)^2}\right)^{0.5}\right] if\ \psi < \frac{\pi}{3} \tag{12a}$$

$$\mu_2 = \frac{1.382}{{L_a}^{0.25}} \cdot (1 - 1.23f) if\ \psi \geq \frac{\pi}{3} \tag{12b}$$

Or

$$\mu_2 = \begin{vmatrix} \mu_1(L_a, f, \psi)\ if\ \psi < \frac{\pi}{3} \\ \mu_2(L_a, f)\ if\ \psi \geq \frac{\pi}{3} \end{vmatrix} \tag{12c}$$

where μ_1 is the coefficient of friction for attack angle values under $\frac{\pi}{3}$ and μ_2 is the coefficient of friction for attack angle values over $\frac{\pi}{3}$.

For steel, the critical distance between two adjacent traces is $L_a = 0.2$.

Thus, the global friction coefficient at microchips has the expression:

$$\mu_a = \begin{vmatrix} \mu_3(L_a, f, \psi)\ if\ L_a \leq 0.2 \\ \mu_3(L_a, f, \psi)\ if\ L_a > 0.2 \end{vmatrix} \tag{13}$$

where μ_a is the global friction coefficient, which depends on the attack angle ψ and the adhesion component f.

μ_3 is coefficient of friction for values of two adjacent traces.

The variation in the microchip friction coefficient depending on the position of the particles in the soil (angle of attackș) for different relative distances between the traces is exemplified in Figure 3.

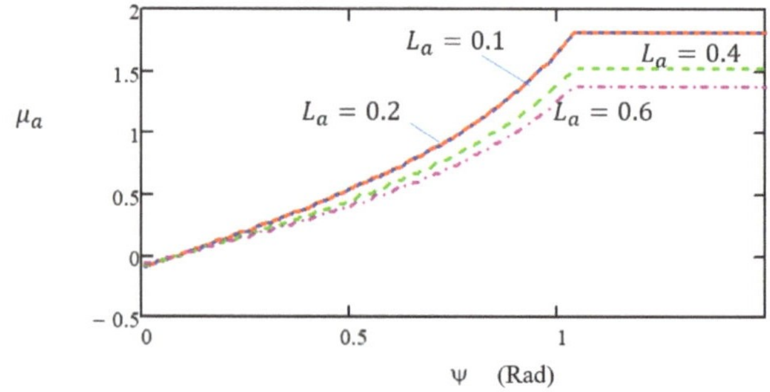

Figure 3. Global abrasive coefficient of friction in microchip wear of the knife by the particle.

The wear coefficient (wear intensity) has the following expressions for different values of the attack angle ψ (Figure 4):

$$k_1 = 0.018\frac{tan(\psi)^3}{f \cdot {L_a}^{0.5}} \cdot (H_r)^{0.5}\ \psi \le \frac{\pi}{4} \tag{14a}$$

$$k_1 = 0.018\frac{1}{f \cdot {L_a}^{0.5}} \cdot (H_r)^{0.5}\ \psi > \frac{\pi}{4} \tag{14b}$$

$$k_3 = \begin{vmatrix} k_1(0.2, f, H_r, \psi)\ if \psi < \frac{\pi}{4} \\ k_2(L_a, f, H_r, \psi)\ if\ \psi \ge \frac{\pi}{4} \end{vmatrix} \tag{14c}$$

$$k_a = \begin{vmatrix} k_3(0.2, f, H_r, \psi)\ if\ L_a \le 0.2 \\ k_3(L_a, f, H_r, \psi)\ if\ L_a > 0.2 \end{vmatrix} \tag{14d}$$

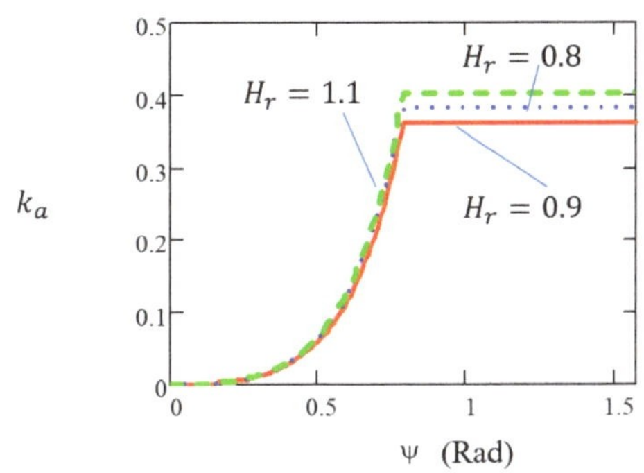

Figure 4. The coefficient of wear by microchipping, depending on the angle of attack.

The critical angle of attack, at which furrow wear separates from microchip wear and with the distance between traces $L_a = 0.2$, is:

$$\psi_{cr} = \psi_b + \psi_a - \psi_b \cdot 0.2^{\frac{1}{H_r}} \tag{15}$$

ψ_{cr} is noted with a critical angle of attack and it can be obtained by mixing the minimum angle of attack at which furrowing (bordering) occurs (ψ_b) with the minimum angle of attack at which microchipping occurs $((\psi_a))$.

2.2.3. Wear by Plowing (Edging, Ratchetting)

In this case, the material is arranged on the edges without detaching the chips. If the hard particle repeatedly penetrates the soft material in the same direction, the plastic deformation for each cycle is accumulated. In this way, the wear particles are very fine, and it is a ductile breaking mechanism [17,27].

The coefficient of friction has the expression:

$$\mu_b = \left(\frac{2}{\pi}\right)^{0.5} \cdot tan(\psi) \cdot \left[1 + f \cdot \left(1 + \frac{\pi}{4tan(\psi)^2}\right)^{0.5}\right] \tag{16}$$

where μ_b is the coefficient of friction in the case of wear by plowing (edging, ratchetting).

The evolution of the global friction coefficient in the plowing wear process is highlighted in Figure 5 as a function of attack angle and different values of the frictional adhesion component.

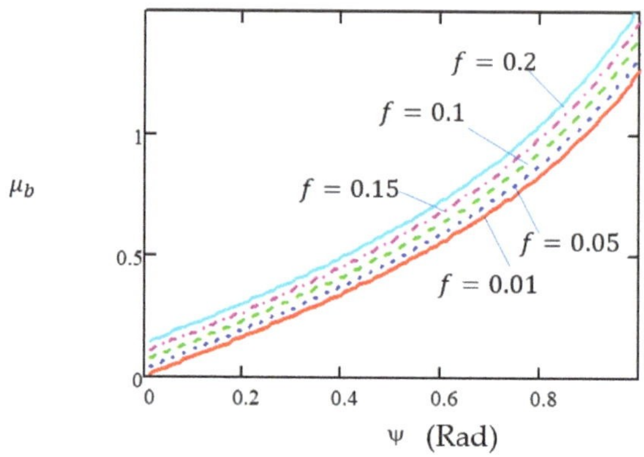

Figure 5. The global coefficient of plowing friction, depending on the attack angle.

In this case, the Archard wear coefficient–approximate empirical relationship can be written as:

$$k_b = 0.225 \cdot 6 \cdot f \frac{tan(\psi)^3 L_a^{0.5}}{\varepsilon_f} \quad (17)$$

where ε_f is the adhesion between soft and hard asperities, which come into contact between soil particles and the surface of the knife.

The variation in the Archard wear coefficient with attack angle by plowing is shown in Figure 6.

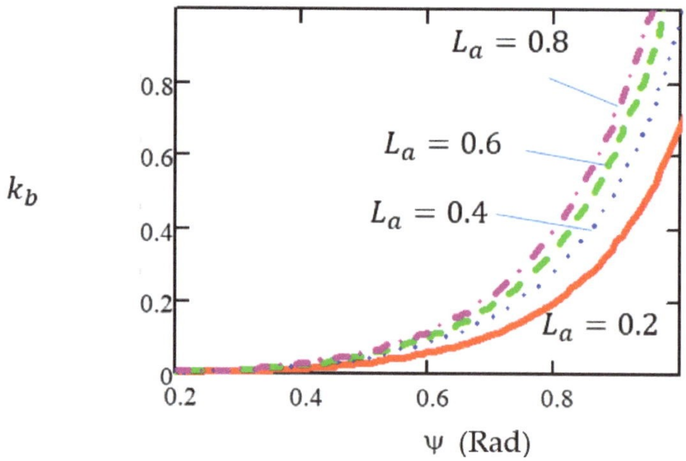

Figure 6. Coefficient of wear by plowing, depending on the attack angle ψ.

3. Materials and Methods

3.1. Test Materials

In this research, two materials were analyzed and compared with a control of C45 steel. These were made of C45 steel heated by hardening and steel grade E295 (Figure 7).

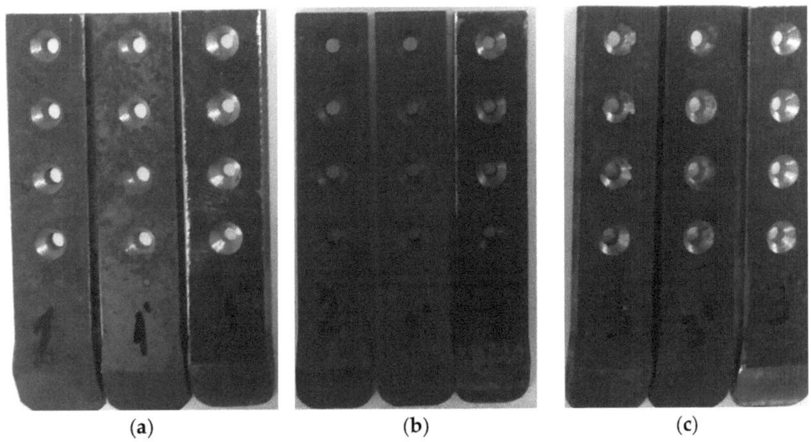

Figure 7. Three types of steel materials were tested: (a)—untreated, (b)—heat-treated C45 and (c)—E295.

The geometrical form of samples tested was maintained with the same chisel knives from the real equipment, but in the smaller dimensions and with a similar angle of attack of 27°. The samples were cut to a length of 157.5 mm and a width of 30 mm with a thickness of 8 mm.

The processing of sandy soils (high quartz content) led to the wear of the knives predominantly through abrasion and superficial fatigue due to the interaction between the abrasive particles and the surface of the knife. The sand used for the tests was sand from the sorting of a soil in the south of Oltenia, close to the Danube, also called "Oltenia Sahara" with a continental, slightly Mediterranean climate; in the past, this was an intensively exploited agricultural area. Additionally, it can be added that sandy soil can induce the greatest linear wear on the furrow of agricultural machines compared with other types of soil, as found in the detailed studies of Braharu D [28–30].

Accordingly, the physical composition of soils and the rheological modeling of soils are extremely complex.

Because of the very high abrasive conditions of agricultural soils, they can amplify the wear of the chisel-type knives; therefore, the experimental tests in this work were conducted in working environments close to those to observe up close the interactions that are created between the knife and the soil [31]. For this reason, we chose a sandy soil with a composition of fine quartz with diameters of particles of around 0.3 mm. The tests were carried out in two laboratory conditions: in sand moistened with water and in dry sand.

Soil can be soft or hard because of the water content. In view of this, the specific humidity of the wet sand was determined using the ratio:

$$W = \frac{q_a}{q_{su}} 100\% \quad (18)$$

where W is the relative humidity; q_a is the water content; and q_{su} is the unit of dry matter.

3.2. Test Equipment

Figure 8 shows an experimental equipment used for testing chisel-type knives [31,32]. The equipment can record all the parameters in the work process, such as the depth, speed and angle of the knives, even when it changes.

Figure 8. Experimental stand for testing chisel-type knives. Close-up view of the chisel knife on the experimental stand.

The technical details of this experimental stand have been described in detail in the previous study of the author, Vlăduțoiu et al., 2020 [31,33,34]. The power of the electric motor was 7.5 kW at a rotation speed of 1460 rpm. Gauge dimensions: the outer diameter of the pool—2000 mm and pool height—1000 mm.

This equipment permits testing the samples of knives at a maximum depth of 300 mm. The circular trajectory traveled by the sample knives was 1600 mm.

The mounting of the knives on the support of the experimental stand was performed one by one after each test; this can be seen in Figure 9. During the testing, the knives worked in the sand at a depth of 220 mm and at an angle of attack of 27° (Figure 10). In this way, we were able to quantify the wear.

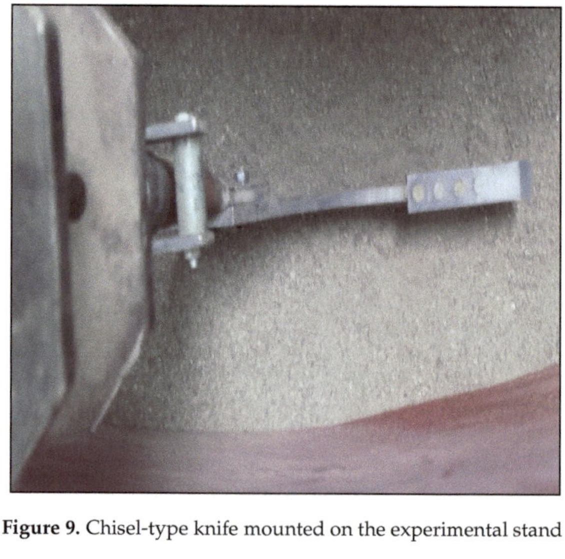

Figure 9. Chisel-type knife mounted on the experimental stand [31].

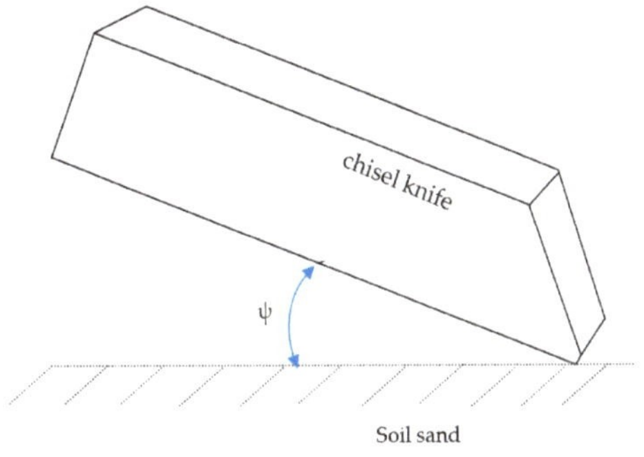

Figure 10. Attack angle of the chisel knife with the soil sand.

For each test, care was taken to keep the same working conditions, the same depth and the same speed of work, with the duration of the working test being one hour.

3.3. Wear Experiments

Weighing of the samples began before the experimental tests were conducted and then continued to measure the weight after one hour. The process was repeated 8 times for each sample knife, so that we could quantify material losses through abrasive wear.

In Figure 11 it is the sensor Theta Probe type ML2x with which the humidity of the soil sand was measured. He is a very important parameter in wear tests. The precise estimation of volumetric soil moisture was ±1%. For storing information from the sensor, we used an apparatus Data logger HH2.

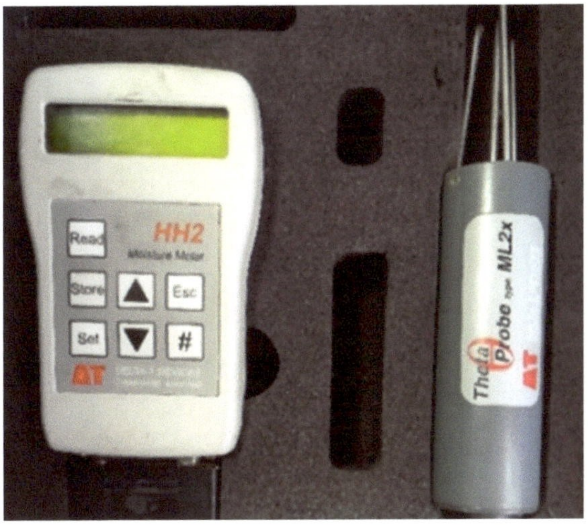

Figure 11. Theta Probe soil moisture sensor.

To determine the weight of the samples tested in this work, a precision scale was used. To determine global wear, the gravimetric method was used. This method incorporates a difference of the initial mass of the sample and the mass obtained after wear test.

4. Results

Tables 1–4 show the effective wear for each hour of operation.

Table 1. Evolution of the weight loss of three types of steel in dry sand.

Knife Type	Weight of the Chisel Knife, after Weighing at a Test Time Interval on the Experimental Stand (g)									
	Before	After 1 h	After 2 h	After 3 h	After 4 h	After 5 h	After 6 h	After 7 h	After 8 h	Total Wear
C45	259.19	258.91	258.66	258.23	257.91	257.68	257.25	256.96	256.76	2.43
C45 heat-treated by quenching	259.66	259.49	259.25	259.03	258.89	258.79	258.64	258.51	258.4	1.26
E295	240.33	240.11	239.95	239.81	239.71	239.57	239.46	239.37	239.28	1.05

Table 2. Mass differences (effective wear) after each hour of testing in dry sand.

Knife Type	Wear over Time in Dry Sand (g):							
	After 1 h	After 2 h	After 3 h	After 4 h	After 5 h	After 6 h	After 7 h	After 8 h
C45	0.28	0.53	0.96	1.28	1.51	1.94	2.23	2.43
C45 heat-treated by quenching	0.17	0.41	0.63	0.77	0.87	1.02	1.15	1.26
E295	0.22	0.38	0.52	0.62	0.76	0.87	0.96	1.05

Table 3. Evolution of the weight loss of three types of steel in wet sand.

Knife Type	The Weight of the Chisel-Type Knife, after Weighing at a Test Time Interval on the Experimental Stand (g)									
	Before	After 1 h	After 2 h	After 3 h	After 4 h	After 5 h	After 6 h	After 7 h	After 8 h	Total Wear
C45	256.76	254.91	252.92	251.19	249.58	247.8	246.25	244.51	242.46	14.3
C45 heat-treated by quenching	258.4	256.67	254.99	253.38	251.86	250.37	248.71	247.14	245.85	12.55
E295	239.28	238.14	236.74	235.72	234.54	233.24	232.15	230.85	229.74	9.54

Table 4. Mass differences (effective wear) after each hour of testing in wet sand.

Knife Type	Wear over Time in Wet Sand (g)							
	After 1 h	After 2 h	After 3 h	After 4 h	After 5 h	After 6 h	After 7 h	After 8 h
C45	1.85	3.84	5.57	7.18	8.96	10.51	12.25	14.3
C45 heat-treated by quenching	1.73	3.41	5.02	6.54	8.03	9.64	11.21	12.55
E295	1.14	2.54	3.56	4.74	6.04	7.13	8.43	9.54

During experiments, which were 8 h per sample in working conditions such as dry sand, the results obtained were 1.05 g of wear for the E295 sample, the quenched C45 saw 1.26 g and the chisel-type knife from C45 suffered a wear of 2.43 g.

High values were noted at two samples of steel, treated and untreated, while the smallest values could be observed at the sample of E295 steel.

Figure 13 represents the experimental data for dry working conditions; the sample of C45 steel suffered the higher loosening of the weight compared with the other two types of steel. The E295 sample was the most resistant to wear in dry working conditions for a period of 8 h. However, in addition, the C45 sample heat-treated by quenching had a wear close to that of E295 steel.

Figure 12 shows the variation in the weight of the samples with time, in dry sand, after 8 h of testing.

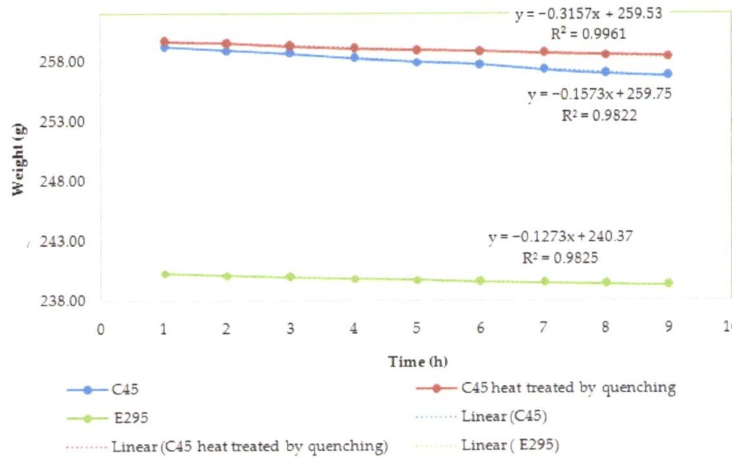

Figure 12. Variation in the weight with time for 3 types of steel in dry sand.

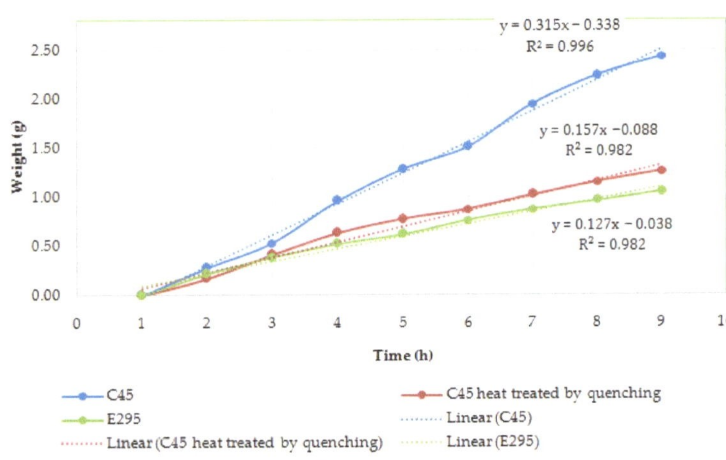

Figure 13. Evolution of the wear in time for 3 types of steel in dry sand.

The results obtained in Table 3 show that the sample of E295 had 9.54 g of wear after 8 h working in wet sand, and it was the best. The biggest loss was untreated C45 steel with 14.3 g of wear and with an appropriate value, followed by C45 heat-treated by quenching with 12.55 g of wear.

In Table 4 are the experimental data of wear over time in wet sand through 8 h of testing for three types of steel, treated (C45 heat-treated by quenching) and untreated (C45), in comparison with E295 steel.

The abrasion resistance of E295 steel was much better than that of the C45 heat-treated by quenching, but was still inferior to that of the untreated C45 steel (Figure 14).

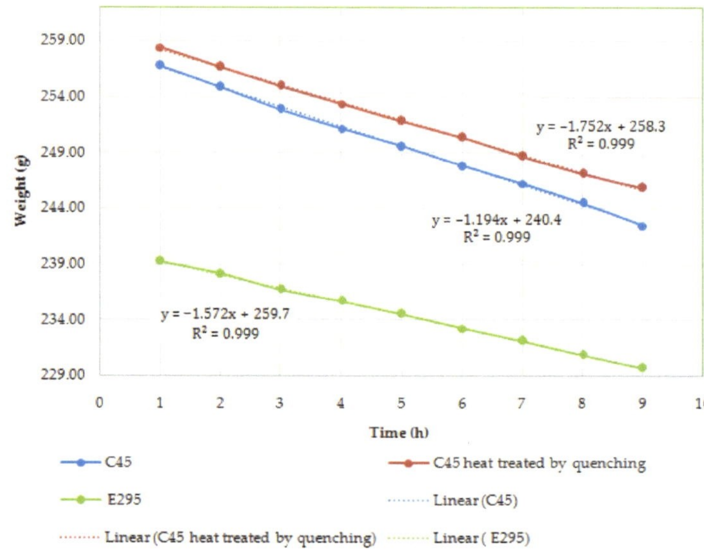

Figure 14. Variation in the weight with time for 3 types of steel in wet sand.

The weight loss of the samples after wear exposure for 8 h is given in the lower corner of Figure 15. The experimental data in this figure are for wet working conditions.

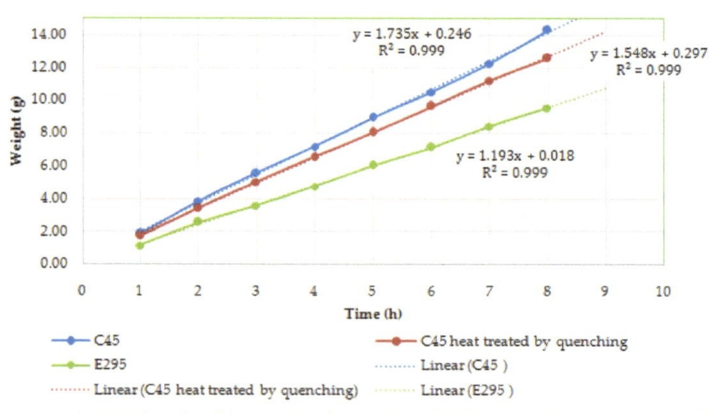

Figure 15. Evolution of wear in time for 3 types of steel in wet sand.

It can be observed, even under these conditions, that the E295 steel sample had the lowest weight loss, while the other two test steels had higher weight losses with close values.

The sample of E295 steel had good tribological behavior in comparison with two samples of treated and untreated steel in any conditions of working. It suffered a weight loss of 9.54 g in wet sand and 1.05 g in dry sand.

For experimental data processing and to determine the experimental Archard coefficient of wear, it was necessary to appeal to the theoretical equations mentioned in the previous chapter.

In Archard's Equation (11), the volume of material removed can be replaced by, $V = \frac{m}{\rho}$:

$$\frac{m}{\rho} = k \cdot \frac{F_n \cdot s}{H_s} \tag{19}$$

where V is the volume of material removed; m is the mass of the chisel knives measured after the test, from the experimental data; ρ is the density of material of knives, $\rho = 7800 \cdot 10^{-9} \frac{\mathrm{Kg}}{\mathrm{mm}^3}$; s is the total sliding length, $s = 733,876$ mm, at the speed of the working machine, of the knife holder at $n = 146$ rpm and the 1600 mm circular path traveled by the knife; F_n is the normal load force (Table 5); and H_s is the hardness of the material of the chisel knife (Table 6).

Table 5. Normal load force.

	Depth, mm					
	1 50	2 100	3 150	4 200	5 250	6 300
Resistance force to soil penetration [N]	20	40.8	70.4	100	119.4	139.8
F_n [N]	32	61	94	252	456	747

Table 6. Hardness for three types of knives.

Knife Type	Hardness, HRC
C45	25
C45 heat-treated by quenching	40
E295	50

By determining the Archard wear coefficient from the experimental data using the values from Tables 1 and 3, the following values from Table 7 can be obtained.

Table 7. Results of the experimental Archard coefficient.

	C45		C45 Heat-Treated by Quenching		E295	
The Experimental Archard Coefficient	**Dry Sand**	**Wet Sand**	**Dry Sand**	**Wet Sand**	**Dry Sand**	**Wet Sand**
k_1	2.2732×10^{-5}	0.00013377	1.47334×10^{-5}	0.00014674	6.13891×10^{-6}	0.0002980
k_2	3.7238×10^{-5}	0.00021913	2.41356×10^{-5}	0.00024039	1.00565×10^{-5}	0.0003489
k_3	6.7383×10^{-5}	0.00039653	4.3674×10^{-5}	0.00043500	1.81975×10^{-5}	0.0004166
k_4	0.00018064	0.00106304	0.00011708	0.00116618	4.87848×10^{-5}	0.0005918
k_5	0.00027837	0.00163813	0.00018042	0.00179707	7.51765×10^{-5}	0.0010212
k_6	0.00053064	0.00312268	0.00034393	0.00342567	0.000143305	0.0020832

The experimental results from Figure 16 can provide valuable information about the variation in Archard wear coefficient of the three chisel-type knives in wet sand during the 8 h of operation; thus, it can be seen that the chisel-type knife from E295 suffered less wear during the entire period of operation, followed by the heat-treated C45 chisel knife, and the C45 chisel knife experienced the most wear.

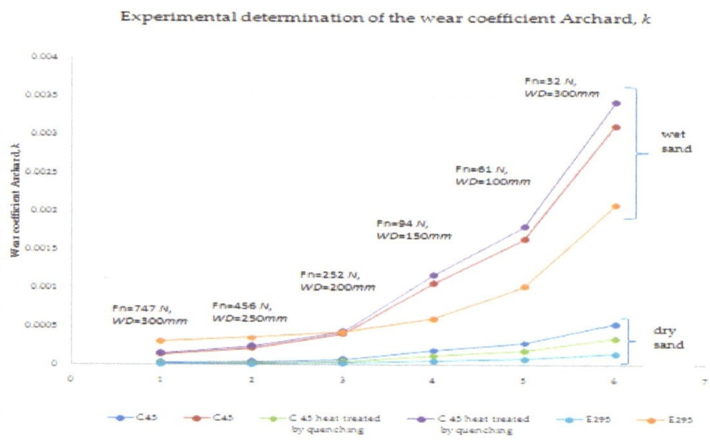

Figure 16. Experimental determination of the Archard wear coefficient, *k*, for different loads (F_n) at different working depths (*W.Ds*).

Additionally, in dry conditions the results obtained were smaller than in wet conditions, and the values for the Archard wear coefficient were greater for the untreated steel, C45 chisel knife. Notably, the E295 steel had a smaller value for the Archard wear coefficient in dry conditions.

5. Discussion

The experimental results regarding the wear of the knives with time (implicitly, the friction length between the knife and the sand particles) confirmed the linear dependence, with different slopes (angles) for the three steels with different harnesses operating in dry sand and wet sand.

Thus, the C45 steel with the lowest hardness had the highest slope (global wear coefficient 1.735 mm/h for wet sand and 0.315 mm/h for dry sand).

The chisel knife in E295 had the minimum wear rate for both wet sand and dry sand. All experimental results regarding the variation in knife wear over time were approximated using linear regressions with a statistical confidence coefficient (R2) greater than 0.98. The slopes of the lines indicated the rate of wear.

The effect of sand humidity on the wear coefficient was essential. This effect can be explained by the fixing ("locking") of the particles between them because of the adhesion generated by the water (internal sand friction, natural slope angle) and the predominant existence of the sliding movement between the particle and the knife. In the case of dry sand, the particles predominantly rolled and with forced sliding.

The analysis of the theoretical wear coefficient of the Archard type, for the simple case-rigid spherical particles–steel contact, allowed highlighting the dependence on the surface hardness, on the friction coefficient (predominantly sliding in wet sand and rolling with sliding in dry sand) and the state of deformation (the contact angle between the particle and the knife).

The Archard wear coefficient of the knife can be explained theoretically by considering the state of stress and deformation in the contact area between the work material (soil in the present case) and the knife.

To highlight the parameters with significant influence on the wear coefficient, the case of sandy soil was accepted. Sand particles were considered perfectly spherical and rigid and were fixed in the soil matrix.

The material of the knife was elastic with ecruisation. The friction between the abrasive particle and the knife was assessed with two components, the adhesion component and the deformation component. The weight of one or another of the components depended on the state of deformation (shakedown, ratcheting or micro-cutting).

Furthermore, some of the important parameters were established, especially the critical distance between two adjacent traces on the attack angle of the particle with the surface of the chisel knife, which linked up the topography of the harder surface to the material properties of the softer one. For higher values of the attack angle of the particle, the values for the distance between two adjacent traces increased, and the transition from plowing to cutting was closer.

6. Conclusions

The abrasive wear mechanism between soil and chisel-type knives from mechanical tillage was analyzed based on experiments and theoretical approaches.

Tribological parameters of the knives for three types of steels with different hardness were determined.

The abrasive behavior of these materials was studied in two different working environments: in dry and wet conditions. Evaluating the energies of the penetration process in dry sand and wet sand for different points allowed us to highlight the increase in force in the presence of water adsorbed by the sand.

The theoretical model of abrasion wear of the knife was modeled when hard particles (considered rigid) were fixed in the matrix and acted as "micro-knives".

From the experimental results and from the analytical modeling, it appeared that the wear of a chisel knife, when processing sandy soils, is of the Archard type.

The forms of damage through wear of the knife were shakedown ("elastic" fatigue), ratcheting and micro-cutting.

The rate of wear of the knife, evaluated by the slope of the wear-time curves, was significantly higher for wet sand than for dry sand.

The hardness of the knife surface was the essential material characteristic for the durability of the knife.

For the working conditions, in the process of soil sliding with abrasive particles on the knife surface, the Kapoor sliding plasticity index—"shakedown pressure"—was proposed.

The dimensionless Archard wear coefficient was determined for spherical abrasive particles with random variable radii, variables characterized by mean and root mean square deviation. The existence of a minimum of the wear coefficient with the mean square deviation of the particle radius was highlighted.

The durability of the furrow knives was very important in the correct and long-lasting operation of the machines, and thus the appropriate maintenance intervals can be established.

The theoretical and experimental modeling of the wear process of chisel knives operating in sandy environments can be extended for different soils considering the components of friction through oxidation and elastic fatigue. The generalization is performed with caution, through the detailed analysis of the contact pressure and the relative speed between the knife and the soil.

Author Contributions: Conceptualization, L.C.V. and G.C.; methodology, L.C.V., G.C. and A.T.; software, L.F.T. and E.M.; validation, N.-V.V., G.C. and L.C.V.; formal analysis, I.-A.G.; investigation, L.F.T.; resources, N.-V.V. and L.F.T.; data curation, G.C.; writing—original draft preparation, L.C.V. and G.C.; writing—review and editing, G.C.; visualization, L.C.V.; supervision, N.-V.V. and A.T.; project administration, L.C.V.; funding acquisition, G.C. All authors have read and agreed to the published version of the manuscript.

Funding: This work was funded by the Romanian Ministry of Education and Politehnica University of Bucharest through the PubArt program.

Institutional Review Board Statement: Not applicable.

Data Availability Statement: Not applicable.

Conflicts of Interest: The authors declare no conflict of interest.

References

1. Samuel, A.; Dines, L. 3—Soil health and management. In *Lockhart and Wiseman' s Crop Husbandry Including Grassland*, 10th ed; Woodhead Publishing Series in Food Science, Technology and Nutrition; Woodhead Publishing: Cambridge, UK, 2023; pp. 49–79, ISBN 9780323857024.
2. Keller, L.; Idowu, O.J.; Ulery, A.; Omer, M.; Brewer, C.E. Short-Term Biochar Impacts on Crop Performance and Soil Quality in Arid Sandy Loam Soil. *Agriculture* **2023**, *13*, 782. [CrossRef]
3. Guo, Z.; Qiu, Y.; Yan, X.; Wang, J.; Zhang, Y.; Zhang, P.; Zhang, F. Self-Excited-Resonance of Soil-Engaging Surface Spectrum: A New Method of Soil Cutting Resistance Reduction. *Agriculture* **2023**, *13*, 1154. [CrossRef]
4. Aramide, B.; Pityana, S.; Sadiku, R.; Jamiru, T.; Popoola, P. Improving the durability of tillage tools through surface modification—A review. *Int. J. Adv. Manuf. Technol.* **2021**, *116*, 83–98. [CrossRef]
5. Chenarbon, H.A.; Minaei, S.; Arabhosseini, A. Replacement age of agricultural tractor (MF285) in Varamin region (case study). *J. Am. Sci.* **2011**, *7*, 674–679.
6. Dexter, A.R. *Soil Mechanical Notes, Course in Agricultural Soil Mechanics*; Swedish University of Soil Sciences: Uppsala, Sweden, 2002.
7. Rostami, J.; Ehsan, A.G.; Palomino, A.M.; Mosleh, M. Development of soil abrasivity testing for soft ground tunneling using shield machines. *Tunn. Undergr. Space Technol.* **2012**, *28*, 245–256. [CrossRef]
8. Canarach, A. *Fizica Solurilor Agricole*; Editura Ceres: Bucureşti, Romania, 1990. (In Romanian)
9. Jafari, R.; Tavakoli, H.T.; Minanee, S.; Raoufat, M.H. Large deformation modeling in soil-tillage tool interaction using advanced 3D nonlinear finite element approach. In Proceedings of the 6th WSEAS International Conference on Simulation, Modelling and Optimization, Lisbon, Portugal, 22–24 September 2006.
10. Ucgul, M.; Fielke, J.M.; Saunders, C. Defining the effect of sweep tillage tool cutting edge geometry on tillage forces using 3D discrete element modelling. *Inf. Process. Agric.* **2015**, *2*, 130–141. [CrossRef]
11. Matache, M.; Ganga, M.; Mihai, M.; Postelnicu, E.; Bajenaru, S. Researchs regarding determination of mechanical and wear characteristics for fricţion materials. *INMATEH-Agric. Eng.* **2008**, *28*, 120–123.
12. Moitzi, G.; Neugschwandtner, R.W.; Kaul, H.-P.; Wagentristl, H. Effect of tillage systems on energy input and energy efficiency for sugar beet and soybean under Pannonian climate conditions. *Plant Soil Environ.* **2021**, *67*, 137–146. [CrossRef]
13. Kushawaha, R.L.; Chi, L.; Roy, C. Investigation of agricultural tools with plasma-sprayed coatings. *Tribol. Int.* **1990**, *23*, 297–300. [CrossRef]
14. Gill, W.R.; Vandenberg, G.E. *Soil Dynamics in Tillage and Traction, U.S.A*; Handbook 316; Department of Agriculture: Washington, DC, USA, 1968.
15. Iovanas, R.; Iovanas, D.M. *Reconditionarea și Remanierea Produselor Sudate*; Transilvania University Publishing House: Brașov, Romania, 2006; ISBN 973-635-416-2.
16. Lacey, P.; Torrance, A.A. The calculation of wear coefficients for plastic contact, Elsevier Sequia-Nederlands. *Wear* **1991**, *145*, 367–383. [CrossRef]
17. Xie, Y.; Wilfams, J.A. The prediction of friction and wear when a soft surface slides against a harder rough surface. *Wear* **1996**, *196*, 21–34. [CrossRef]
18. Challen, J.M.; Oxley, P.L.B.; Doyle, E.D. The Effect of Strain Hardening on the critical angle for abrasive (Chip Formation), Elsevier Sequoias -Netherland. *Wear* **1983**, *88*, 1–12. [CrossRef]
19. Archard, J.F. Contact and rubbing of flat surfaces. *J. Appl. Phys.* **1953**, *24*, 981–988. [CrossRef]
20. Varenberg, M. Adjusting for running-in: Extension of the Archard wear equation. *Tribol. Lett.* **2022**, *70*, 59. [CrossRef]
21. Kapoor, A.; Franklin, F.J.; Wong, S.K.; Ishida, M. Surface roughness and plastic flow in rail wheel contact. *Wear* **2002**, *253*, 257–264. [CrossRef]
22. Williams, J.A.; Xie, Y. The generation of wear surfaces. *J. Phys. D Appl. Phys.* **1992**, *25*, A158–A164. [CrossRef]
23. Shmulevich, I.; Asaf, Z.; Rubinstein, D. Interaction between soil and a wide cutting blade using the discrete element method. *Soil Tillage Res.* **2007**, *97*, 37–50. [CrossRef]
24. Wolffsohn, J.S.; Drew, T.; Dhallu, S.; Sheppard, A.; Hofmann, G.J.; Prince, M. Impact of soft contact lens edge design and midperipheral lens shape on the epithelium and its indentation with lens mobility. *Investig. Ophthalmol. Vis. Sci.* **2013**, *54*, 6190–6196. [CrossRef]
25. Zmitrowicz, A. Wear Patterns and Laws of wear—A review. *J. Theor. Appl. Mech.* **2006**, *44*, 219–253.
26. Wang, B.; Mengji, C.; Jinfeng, W.; Guanghua, L.; Kunyong, L. Bionic Optimization Design of Rotary Tiller Based on Fuzzy Algorithm. *Mob. Inf. Syst.* **2022**, *2022*, 1203973. [CrossRef]
27. Kapoor, A.; Williams, J.A.; Johnson, K.L. The steady state sliding of rough surface. *Wear* **1994**, *175*, 81–92. [CrossRef]
28. Zelenak, M.; Valicek, J.; Brumek, J.; Hlavacek, P.; Haluzikova, B.; Vylezik, M.; Kubena, V. Measurement and analysis of the hardness of aluminium surface layers by the nanoindentation and scratch tests. *Chem. Listy* **2011**, *105*, 680–683.
29. Braharu, D.; Gângu, V.; Vlăduţ, V.; Postelnicu, E.; Băjenaru, S. *Fundamental Research Regarding the Methods of Anti-Wear Deposit, Adaptable to Components for Agricultural Equipments*; Annals of the University of Craiova—Agriculture, Montanology, Cadastre series: Craiova, Romania, 2007; Volume XXXVII, pp. 36–47, ISSN 1841-8317.

30. Schroth, M.H.; Istok, J.D.; Ahearn, S.J.; Selker, J.S. Characterization of Miller-Similar Silica Sands for Laboratory Hydrologic Studies. *Soil Sci. Soc. Am. J.* **1996**, *60*, 1331–1339. [CrossRef]
31. Vladutoiu, L.C.; Fechete, T.L.; Grigore, I.A.; Sorica, E.; Petre, A.; Cristea, O.D. Experimental research on determination of wear of working part of chisel type made of three types of materials. *INMATEH-Agric. Eng.* **2020**, *62*, 269–276. [CrossRef]
32. Grigore, I.A.; Petrea, A.A. Experimental research on determining the wear of chisel knife made of three types of materials. *INMATEH-Agric. Eng.* **2020**, *62*, 3.
33. Vladutoiu, L.; Persu, C.; Grigore, A.I.; Sorică, E.; Petre, A.A.; Cristea, M.; Cristea, O.D.; Grigore, I. Experimental research on determination of deformations in the cutting process in wet sandy soils. *Ann. Univ. Craiova-Agric. Mont. Cadastre Ser.* **2020**, *50*, 603–608.
34. Vladutoiu, L.; Tudor, A.; Fechete, L.; Marin, E.; Vladut, V.; Matache, M.; Dumitru, I. Determining the wear of active organs for processing soil depending on the working depth. *Ann. Univ. Craiova-Agric. Mont. Cadastre Ser.* **2015**, *45*, 253–258.

MDPI AG
Grosspeteranlage 5
4052 Basel
Switzerland
Tel.: +41 61 683 77 34

Agriculture Editorial Office
E-mail: agriculture@mdpi.com
www.mdpi.com/journal/agriculture

www.ingramcontent.com/pod-product-compliance
Lightning Source LLC
Chambersburg PA
CBHW041423210926
51850CB00013B/3338
* 9 7 8 3 7 2 5 8 3 7 9 7 7 *